COMPUTER INTEGRATED MANUFACTURING AND ENGINEERING

COMPUTER INTEGRATED MANUFACTURING AND ENGINEERING

U. Rembold
University of Karlsruhe

B.O. Nnaji
University of Massachusetts at Amherst

A. Storr
University of Stuttgart

Prentice Hall

An imprint of **Pearson Education**

Harlow, England · London · New York · Reading, Massachusetts · San Francisco
Toronto · Don Mills, Ontario · Sydney · Tokyo · Singapore · Hong Kong · Seoul
Taipei · Cape Town · Madrid · Mexico City · Amsterdam · Munich · Paris · Milan

Pearson Education Limited
Edinburgh Gate
Harlow
Essex CM20 2JE
England

and Associated Companies throughout the world

Visit us on the World Wide Web at:
http://www.pearsoneduc.com

Cover designed by Chris Eley
and printed by The Riverside Printing Co. (Reading) Ltd.
Typeset by P & R Typesetters Ltd (Salisbury, UK).

Printed and bound by Antony Rowe Ltd, Eastbourne

First printed in 1993

ISBN 0-201-56541-2

British Library Cataloguing-in-Publication Data
A catalogue record for this book is available from the British Library.

Library of Congress Cataloging-in-Publication Data is available

10   9   8   7
10   05   04

Contents

Preface   xv

Introduction   xxi

Chapter 1	**The Manufacturing System**	**1**
1.1	A short historical excursion	1
	1.1.1 The role of the computer in manufacturing	3
1.2	An introduction to a basic manufacturing system	4
	1.2.1 A model factory	4
	1.2.2 The organizational and engineering components of a manufacturing system	8
	1.2.3 Definition of the computer terminology in manufacturing	9
1.3	The importance of flexibility in manufacturing	11
	1.3.1 Types of manufacturing operations	11
	1.3.2 The requirements of a flexible manufacturing system	16
	1.3.3 Automation potential in manufacturing	17
1.4	The functions of a manufacturing system	19
	1.4.1 Market research	21
	1.4.2 Long-range forecasting	21
	1.4.3 Capital equipment and facility planning	22
	1.4.4 Customer order servicing	22
	1.4.5 Engineering and design	23
	1.4.6 Manufacturing process planning	24
	1.4.7 Marketing	26
	1.4.8 Production order scheduling and manufacturing monitoring and control	26
	1.4.9 Purchasing and receiving	28
	1.4.10 Inventory management	30
	1.4.11 Quality control	31
	1.4.12 Maintenance	31
	1.4.13 Accounting	32
1.5	The hierarchical planning and control concept	32
	1.5.1 Types of hierarchical control structures	33
	1.5.2 Building a control hierarchy for a manufacturing system	35
	1.5.3 Data exchange and information flow between control tiers	39

1.6   Future developments   40
      1.6.1   Which CIM strategy should be followed?   42
      1.6.2   Future outlook   43
1.7   Problems   46

Chapter 2   CIM Models and Concepts   47
2.1   Introduction   47
2.2   Existing CIM models   48
      2.2.1   The IBM concept of CIM   49
      2.2.2   The NIST–AMRF hierarchical model   50
      2.2.3   The Siemens concept of CIM   53
      2.2.4   The Digital Equipment Corporation CIM concept   55
      2.2.5   ESPRIT–CIM–OSA Model   58
2.3   The Amherst–Karlsruhe CIM model   63
      2.3.1   Interconnecting data processing activities   63
      2.3.2   The manufacturing system model   66
      2.3.3   The product model   68
      2.3.4   Process planning model   71
      2.3.5   Manufacturing scheduling   75
      2.3.6   Manufacturing control and monitoring   78
      2.3.7   CIM network of the model factory   84
      2.3.8   CIM database   89
2.4   Problems   102

Chapter 3   Analysis Tools for Manufacturing   104
3.1   Introduction   104
3.2   An integrated approach to manufacturing system planning   105
      3.2.1   The planning cycle   106
      3.2.2   The requirements analysis   108
      3.2.3   Data preparation and analysis   113
      3.2.4   System configuration and technical evaluation   114
      3.2.5   Return on investment analysis   115
      3.2.6   Final system design and documentation   118
      3.2.7   System realization and testing   118
3.3   Simulation   119
      3.3.1   Simulation tools   119
      3.3.2   Methods of modeling   119
      3.3.3   Simulation methods   121
      3.3.4   The simulation cycle   123
      3.3.5   A comprehensive simulation tool   125
3.4   Petri nets   127
      3.4.1   Introduction   127
      3.4.2   Fundamentals of Petri nets   127
      3.4.3   A planning example using a Petri net   132
3.5   Artificial intelligence methods for manufacturing   135
      3.5.1   Introduction   135

	3.5.2	Basic components of an expert system	136
	3.5.3	Building an expert system	144
	3.5.4	Exert systems for manufacturing	151
	3.5.5	Knowledge-based configuration of a manufacturing system	167
	3.5.6	Operation of a knowledge-based planning and control system	172
3.6	Problems		174

Chapter 4   Flexible Manufacturing and Assembly Equipment   176

4.1	Introduction		176
4.2	Numerical control and design features of NC machines		177
4.3	Flexible manufacturing equipment		181
	4.3.1	Machining centers	181
	4.3.2	Flexible manufacturing cells	186
	4.3.3	Flexible manufacturing systems	191
4.4	Planning and introduction of interconnected manufacturing systems		198
4.5	Flexible assembly systems		206
4.6	Economic considerations		210
4.7	Problems		212

Chapter 5   Control Structures for Manufacturing Systems in the CAM area   213

5.1	Introduction		213
5.2	Function-oriented structure		214
5.3	Hardware structures		217
5.4	Software-oriented structures (program building blocks and files)		224
5.5	Computer-aided organization of manufacturing accessories		231
5.6	Programming NC equipment		242
	5.6.1	NC machines	242
	5.6.2	Industrial robots	251
	5.6.3	CAD/NC integration	253
	5.6.4	Cells	255
5.7	Problems		257

Chapter 6   Communication Nets and Protocol Standards   258

6.1	Introduction		258
6.2	Communication topologies		259
	6.1.1	Star topology	260
	6.1.2	Ring topology	260
	6.1.3	Bus topology	260
	6.1.4	Tree topology	260

viii   Contents

6.3   Access procedures   261
      6.3.1   Polling procedure   261
      6.3.2   Time-division multiplexing procedure   261
      6.3.3   Carrier sense multiple access   262
      6.3.4   Token passing   262
6.4   The ISO/OSI reference model   263
      6.4.1   Introduction   263
      6.4.2   Fundamentals   263
      6.4.3   Standards for the physical layer and the data
              link layer   267
      6.4.4   Application-layer standards   268
6.5   Communications profiles   280
      6.5.1   MAP/TOP   280
      6.5.2   Mini MAP   282
      6.5.3   CNMA   283
6.6   Field bus   283
      6.6.1   Application areas   284
      6.6.2   Requirements   285
      6.6.3   Incorporation in the ISO/OSI reference model   286
6.7   CNMR pilot installation as an example of the use of
      open communications   287
      6.7.1   Description of the pilot installation   287
6.8   Problems   292

Chapter 7   CAD: Its Role in Manufacturing   294
7.1   Introduction   295
7.2   Historical perspective   295
7.3   The design process   297
7.4   Design hierarchy   298
7.5   The role of the computer in the design process   299
      7.5.1   Problem definition   299
      7.5.2   Geometric modeling   299
      7.5.3   Engineering analysis   300
      7.5.4   Design evaluation   301
      7.5.5   Automated drafting   301
7.6   Methods of constructing geometric elements in CAD   301
7.7   Transformation in 2-dimensions   304
      7.7.1   Translation   304
      7.7.2   Rotation   304
7.8   Transformations in 3-dimensions   306
7.9   Computer graphic aids   307
      7.9.1   Software   307
      7.9.2   Vector graphics   308
      7.9.3   Raster graphics   308
      7.9.4   Graphics display terminals   308

7.10 CAD modeling and database   310
    7.10.1 Organization of database   311
7.11 Solid representation schemes   313
    7.11.1 The desired CAD information and its extraction   313
    7.11.2 Pure primitive instancing scheme   314
    7.11.3 Cell decomposition   314
    7.11.4 Constructive solid geometry (CSG)   315
    7.11.5 Boundary representation (B-Rep)   316
7.12 Representation schemes   318
    7.12.1 Adjacent topology for reasoning about objects   318
    7.12.2 Uniqueness of representation scheme   321
7.13 Bill of materials   321
7,14 Interfaces for CAD/CAM   324
7.15 Types of interfaces   326
7.16 Description of various interfaces   329
    7.16.1 IGES (Initial Graphics Exchange Specification)   329
    7.16.2 PDDI (Product Definition Interface)   332
    7.16.3 PDES (Product Data Exchange Specification)   333
    7.16.4 SET (Standard d'Echange et de Transfert)   334
    7.16.5 VDAFS (Verband der Automobilindustrie
           Flächenschnittstelle)   335
    7.16.6 CAD*I Interface   336
    7.16.7 STEP (Standard for External Representation of
           Product Data)   337
    7.16.8 EXPRESS language   341
7.17 Requirements of a product model   342
7.18 Design features   343
    7.18.1 Genus and the Euler formula   343
    7.18.2 Feature classification   344
    7.18.3 Generic feature classification   346
    7.18.4 Generic classification methodology   346
    7.18.5 Feature patterns   348
7.19 Feature classification by application   349
    7.19.1 Feature classification for assembly   350
    7.19.2 Sheet metal application   353
    7.19.3 Sheet metal feature classification   354
7.20 Product level classification   356
7.21 Concurrent engineering   356
7.22 The product modeler   361
    7.22.1 Design for automated assembly   362
    7.22.2 Product assembly modeler   363
7.23 Quality methods in design   366
    7.23.1 Value engineering   366
    7.23.2 Taguchi method   366
7.24 Life cycle costs in design   367

	7.25 Conclusions	369
	7.26 Problems	369
Chapter 8	**Process Planning and Manufacturing Scheduling**	**371**
	8.1 Introduction	371
	8.2 Process planning	372
	8.3 Automated process planning	373
	8.3.1 Generative technique	374
	8.4 The operational sheet	375
	8.5 Group technology	377
	8.6 Coding structure	379
	8.7 Available coding systems	383
	8.7.1 The Opitz coding system	383
	8.7.2 The KK-3 system	384
	8.7.3 The MICLASS system	385
	8.8 Design data and automated process planning	386
	8.8.1 Feature extraction research	387
	8.8.2 Process selection	388
	8.8.3 Fixturing selection	391
	8.8.4 Tool selection	393
	8.8.5 Cutting parameters	394
	8.8.6 Cutting path	398
	8.8.7 Examples of existing automated process planning systems	399
	8.9 Manufacturing resource planning	401
	8.9.1 Resource planning	401
	8.10 Material inventory systems	404
	8.10.1 Introduction	404
	8.10.2 Vilfredo Pareto's law (1897) and the ABC inventory	405
	8.10.3 Types of manufacturing inventory	406
	8.11 Material requirements planning	407
	8.11.1 Introduction	407
	8.11.2 The product structure	408
	8.12 Lot sizing techniques	411
	8.12.1 Economic order quantity (EOQ)	412
	8.12.2 Wagner–Whitin algorithm	415
	8.13 Sequencing of operations	418
	8.13.1 Background	420
	8.13.2 Problem definition	420
	8.13.3 Assumptions	421
	8.13.4 Operational restrictions	423
	8.13.5 Mathematical model objective	424
	8.13.6 Cost function	424
	8.14 Scheduling systems	430
	8.14.1 The $n/1$ problem	433
	8.14.2 The $n/2$ flow shop	433

8.14.3 Calculation of idle time in $n/2$ flow shop 434
8.14.4 $n/2$ job shop 435
8.14.5 $n/3$ job shop 436
8.14.6 Two jobs, m machines (2/m job shop) 436
8.14.7 n/m job shop 437
8.15 Project scheduling 439
8.16 Assembly line balancing 443
8.17 Summary 447
8.18 Problems 447

Chapter 9 Robotics **449**
9.1 Introduction 449
9.2 Industrial robots 450
9.3 Classification of robots 450
9.3.1 Geometric classification 450
9.3.2 Control classification 455
9.4 Major components and functions of robots 455
9.4.1 Drive system 456
9.4.2 Control system 456
9.4.3 Sensors 457
9.5 End-effector 461
9.5.1 Different types of gripper systems 462
9.5.2 Gripper actuators 462
9.5.3 Gripper designs 463
9.6 Robot coordinate systems 464
9.6.1 Position and orientation 464
9.6.2 Derivation of rotation and translation matrices 466
9.6.3 Relating the robot to its world 475
9.7 Manipulator kinematics 476
9.7.1 Parameters of links and joints 476
9.7.2 Denavit–Hartenberg representation 478
9.7.3 Kinematic chains 480
9.8 Dynamics of kinematic chains 482
9.8.1 Attributes of Lagrange–Euler formulation 482
9.8.2 Attributes of Newton–Euler formulation 482
9.8.3 Attributes of generalized d'Alembert approach 483
9.8.4 Lagrange–Euler formulation 483
9.8.5 Trajectory planning and control 484
9.9 Robot programming 492
9.9.1 Explicit robot programming 493
9.10 Task-level programming 499
9.10.1 World modeling 501
9.10.2 Task specification 503
9.10.3 Robot program synthesis 506
9.10.4 Collision-free motion planning 510

9.11 Applications 515
  9.11.1 Material transfer and machine loading 515
  9.11.2 Processing operations 516
  9.11.3 Assembly application 518
9.12 Problems 520

Chapter 10 Material Handling   **522**
10.1 Preview and summary 522
10.2 Modern material handling concepts 523
  10.2.1 A simple material handling problem 523
  10.2.2 Logistics of a material handling system 525
  10.2.3 Material handling strategies 531
10.3 Controlling the material flow 535
  10.3.1 Computer architecture for material handling 535
  10.3.2 Input/Output devices 537
  10.3.3 Data media for automatic material flow control 539
  10.3.4 Tracking of objects 543
  10.3.5 Control of material transportation vehicles 547
10.4 Hardware for material handling 558
  10.4.1 Ground conveyors 558
  10.4.2 Overhead material transportation systems 559
  10.4.3 Material transportation vehicles 561
10.5 Warehousing 565
10.6 Problems 570

Chapter 11 Quality Assurance   **572**
11.1 Introduction 572
11.2 An integrated quality assurance concept 573
  11.2.1 A comprehensive quality assurance system 573
  11.2.2 Quality assurance in engineering 576
  11.2.3 Quality assurance in manufacturing 576
  11.2.4 The cost of securing quality 578
11.3 Tasks of quality assurance 580
  11.3.1 Quality planning 581
  11.3.2 Testing the product 582
  11.3.3 Controlling the manufacturing process 583
  11.3.4 Controlling the quality control operations 584
11.4 Performing the quality control operations 585
  11.4.1 Quality control planning 585
  11.4.2 Testing 587
  11.4.3 Evaluation of the test results 589
  11.4.4 Quality reporting 589
11.5 Components of a computer-controlled test system 590
  11.5.1 Introduction 590
  11.5.2 The test set up 591
  11.5.3 Coordinate measuring machines 595

11.5.4 Self-learning test systems   599
11.6 Hierarchical computer system for quality assurance   601
    11.6.1 Quality assurance supervising tier   601
    11.6.2 Quality assurance planning tier   602
    11.6.3 Quality assurance control tier   603
    11.6.4 Interfaces of a quality assurance system   603
11.7 Programming of test systems   605
    11.7.1 The programming languages   605
    11.7.2 Low-level programming   606
    11.7.3 High-level programming   607
    11.7.4 Application oriented programming   608
11.8 Problems   612

References and further reading   **614**

Index   **629**

Preface

For the industrialized nations, the manufacturing industries have become the most important contributors to prosperity. However, it becomes increasingly difficult to meet customers' demands and compete on the international market. Thus, manufacturing industries must be able to react quickly to prevailing market conditions and to maximize the utilization of resources. Conventional means of 'hard' automation are now no longer able to meet these challenges, since they are very poor in information processing. Within recent years, computer technology, in conjunction with software technology, has made available to the manufacturer tools which can greatly improve their reaction to a new market situation, speeding up the design of products, improving process planning, maximizing resource scheduling, and streamlining production flow through factories. When the computer has become a major component of a manufacturing system and helps to plan and operate it, we call it Computer Integrated Manufacturing (CIM).

The manufacturing systems of the future have to be flexible, and for this reason they must be reprogrammable. But any increase in flexibility will entail higher installation costs. Thus, it is necessary to provide a streamlined and uninterrupted production process which is highly efficient and reliable. Information processing plays a major role in obtaining these goals. Information is considered an important resource whose true value is often difficult and impossible to estimate. Information gathering, processing and evaluation is a key task in many new business concepts, such as rapid prototyping, simultaneous engineering, design for manufacturing and assembly, modularity, programmability and standardization.

There are several definitions of CIM. In this book we define CIM as it commonly used in industry. CIM conveys the concept of a semi- or totally automated factory in which all processes leading to the manufacture of a product are integrated and controlled by computer. It includes computer-aided design (CAD), computer-aided process planning (CAP), production planning and control (PP&C), computer-aided quality control (CAQ) and computer-aided manufacturing (CAM). CIM is concerned with common data models which can be used for the entire design and manufacturing cycle. Thus, CIM is centered around the decisions regarding the planning and controlling of the data flow, data processing and data dissemination in a plant. The tools are models, algorithms, artificial intelligence methods, software engineering aids, computers, data communication systems, interfaces to man and machines as well as interfaces between machines.

Since CIM starts with CAD, basic knowledge of this technology is required.

This book only covers new CAD technologies such as standardization of design data needed for manufacturing. Conventional CAD methods are already covered adequately in the existing literature.

This book, which introduces the reader to CIM, is intended to be an aid to the senior and graduate student who wishes to understand CIM systems, and to help engineers design and implement such systems. Of course, CIM is a concept which can be applied to many manufacturing operations such as the making of aircrafts, cars, machine tools, plastic toys, garments, footwear and so on. For this reason, the book emphasizes concepts, methods, structures, standards, interfaces and protocols. Examples of flexible manufacturing cells and systems are given. In particular, the chapters on *Flexible Manufacturing and Assembly Equipment* and *Control Structures of Manufacturing Systems* in the CIM area discuss typical parts manufacturing problems.

The reader interested in applying CIM technologies in a particular plant must be familiar with the product and the processes already in place. This book will aid in the conception and design of CIM components and planning and control systems. As the design and manufacture of different products varies considerably – for example, the manufacture of cars compared with shoes – students should broaden their knowledge with specific case study material. This will enable them to supplement their knowledge of the general principles of CIM gained from reading this book with a specific knowledge of how an individual product is manufactured. For this reason, the reader should become familiar with manufacturing process of particular interest, learning in detail how its products are made and shaped.

In Chapters 1 and 2, we introduce the reader to an overall concept of manufacturing. This is because a factory can only be understood as a complex system consisting of many activities and components. It makes no sense to consider an activity or a component by itself. In the real world, there is a strong interdependence between all manufacturing functions, and information is passed back and forth between them. Every function and subfunction must be operating as an integral part of the whole system. The reader who is not fully familiar with manufacturing is encouraged to revisit these chapters once they are more comfortable with all activities described later in the book in order to gain a deeper understanding of a system concept of manufacturing. The multiplicity of components entering into the design and operation of a plant makes necessary a good knowledge of the global behavior of a total manufacturing system; this is what we are trying to convey to the reader in the two introductory chapters.

The chapters of the book have the following contents. Chapter 1 introduces the reader to the operational principles of a complete manufacturing system. The role of the computer and its contribution to flexibility are discussed. The manufacturing system is divided into functional modules and, for each module, the input data, data processing activities and output data are explained. An effort is made to show the information flow through an entire manufacturing operation. In the conclusion, the chapter emphasizes the importance of hierarchical planning and control of a manufacturing system.

Chapter 2 discusses the role of manufacturing models for information technology in computer-operated factories. With the emergence of modern manufacturing technology, attempts are being made to build manufacturing systems from generic hardware and software modules. These modules are configured to a specific manufacturing system with the help of a generic manufacturing model. Several models are discussed, and their particular features are pointed out. One of the most important existing developments is the ESPRIT-CIM-OSA model, which is supported by the European Commission. This work will greatly influence the conception of future models for manufacturing. In the second part of this chapter, the Karlsruhe–Amherst model is discussed. An attempt is made to show how the most important manufacturing activities can be presented by one model. Details of many of these activities are described in later chapters.

Chapter 3 is intended to give the reader an overview of modern aids for planning and setting up complex manufacturing systems. Planning tools are quite numerous and employ operations research and artificial intelligence methods, simulation tools, Petri nets and other examples. Emphasis is placed on the application of planning tools, and it is assumed that the reader will obtain a basic knowledge of them from the existing literature. Planning methods like simulation, Petri nets and expert systems are the structuring tools for the design of complex systems.

Chapter 4 is an excursion into the design and operation of modern flexible manufacturing and assembly systems. After a short introduction to NC machines, flexible manufacturing cells and flexible manufacturing systems, numerous real machining systems are introduced. The presentation starts with simple manufacturing machines and shows how they can be interconnected to complex production facilities with automatic tool selection and material feeding, and transportation devices. The discussion includes the selection of workpiece spectra for flexible manufacturing.

Chapter 5 is about control structures for computer-aided manufacturing systems. The chapter starts with a concept of structuring control into hierarchical layers. For this layer concept, the control hardware and software is explained. There is a thorough discussion of the computer-aided organization of manufacturing resources, such as tools and manufacturing accessories. The chapter concludes with an overview on NC machine and robot programming. This chapter builds on experience gained from existing flexible manufacturing systems, and numerous examples are given.

Chapter 6 is concerned with in-plant communication. A CIM system can only work well if planning and control information is correctly distributed in a timely manner. There are several topologies from which a communication system can be constructed, and communication must follow well-defined access protocols. Modern communication systems use the ISO/OSI reference model to specify the topology, access methods and formats of the information to be exchanged. The various layers and most important standards of the OSI reference model are discussed in relation to manufacturing applications. Of specific interest are the FTAM (File Transfer, Access and Management) and MMS (Manufacturing Message Specification) standards. Examples show how these standard are applied. Other subjects of interest are MAP (Manufacturing Automation Protocol), TOP (Technical and Office Protocol), CNMA (Communication Network for

Manufacturing Applications) and the fieldbus. Examples are given to show their application.

Chapter 7 gives an overview on new developments in computer-aided design and their impact on manufacturing. It starts with an explanation of the product model as the basic source of engineering and manufacturing documents. Second, the most important features of the various CAM standards and interfaces of interest to engineering and manufacturing are discussed. Third, thought is given to design for manufacturing, assembly, inspection and testability. This chapter does not go into the details of CAD. The reader interested in this topic should become familiar with books on basic CAD practices.

Chapter 8 discusses planning and scheduling of manufacturing operations. When a production lot is prepared for manufacturing, parts are grouped to facilitate setup for machining and assembly; thereafter, process planning determines the production processes and their sequence; as the next step, the tools and fixtures are selected; and finally, machining data and programs are generated. For ongoing production, the machines and assembly lines are selected and a load balance is done. Material requirements planning initiates the ordering, purchasing and distribution of parts and raw materials. When everything is ready for production, the orders are released to manufacturing.

Chapter 9 gives an overview on the basics of robotics. The different types of robot design are explained, together with the mathematics needed to describe robot motions. The topic of robot programming is divided into explicit and implicit programming. Whereas most presentday robots are programmed by explicit languages, in the future the implicit method will gain importance. However, the final language for programming robots will be based on a hybrid approach using implicit and interactive graphic features.

Chapter 10 gives an introduction to material handling. The distribution of parts, raw material, tools and manufacturing accessories is vital to ascertain an uninterrupted production. The principles of logistics and supporting computer structures for material handling are explained. In material handling, the identification of parts and their tracking through production is essential to assure that the right parts are at the right time at the right station. An overview is given on material transportation and storage systems and the role of the computer in their control. Topics like push-and-pull production principles are also part of this chapter.

Chapter 11 concludes the book with a presentation of computer-aided quality control (CAQ) methods. Quality assurance comprises the quality that is engineered into a product, the quality of parts manufacture, and the post-auditing of quality at the customer's site. All quality problems must be observed, analyzed and corrected by an integrated quality control system. The chapter discusses computer-aided quality planning, testing and evaluation. The operation of computer-controlled tests in ongoing production is discussed and it is shown how computers can be programmed to learn their own test limits. Particular attention is paid to the operation of coordinate measuring machines. The chapter concludes with a discussion on programming languages for test applications.

This work was conceived, structured and written by three authors who have factory and university experience. An effort was made to write the book in a coherent way so that the reader will be obtaining an overall concept of CIM.

We believe that we reached this goal in most aspects. For the structuring of the book we had many meetings to coordinate its contents. The material was carefully discussed and edited several times. Various manufacturers helped to structure the outline of the book. We are indebted to the former Dean of Engineering of the University of Massachusetts at Amherst, Dr James E.A. John, and his successor Dr Keith R. Carver as well as Ministiralrat Karl-Heinz Kammerlohr of the Ministry of Science and Art of the state of Baden-Württemberg, for the encouragement and help they gave us to realize this book. There are many people who contributed to the success of this work. We are thankful to Manfred Gärtner, Ulrich Hänerle, Wolfgang Sperling, Dr Manfred Härdtner (Universität Stuttgart), Georg Näger, Bärbel Seufert (Universität Karlsruhe), Dr Klaus Linke (Deutsches Institut für Fernstudien, Tübingen), Mehren Kamran, Jyh-Haw Kang, T.S. Lang, Hsu-Shang Liu, Jennifer Savickis and Pam Stephan (University of Massachusetts at Amherst). Tim Pitts of Addison-Wesley also deserves our thanks for his help and patience.

The authors hope that the work will help the reader to get an insight into the general philosophy and concepts of CIM and the operation of the basic building blocks of the business function of the factory. We wish our readers much luck with their studies.

Prof. Dr B.O. Nnaji, *University of Massachusetts at Amherst, Mass., USA.*
Prof. Dr U. Rembold, *Universität Karlsruhe, Germany*
Prof. Dr A. Storr, *Universität Stuttgart, Germany*

Introduction

This book is concerned with the application of the computer to the support of manufacturing. The computer has only recently been introduced to the performance of planning and control tasks in manufacturing systems. Before the advent of the computer, many other types of control aids and methods were used. Originally, the artifacts that mankind used in daily life and work were handmade by skillful craftsmen. These craftsmen often had a great dexterity and ingenuity, and were capable of making extremely complex devices such as the movement of a watch or the mechanisms of astronomical instruments. Many of our present-day manufacturing methods date back to the time when man started to use tools: for example, turning and drilling are basic machining operations that have been simplifying the work of craftsmen for over 6 millennia. The first known turning machine for wood was drawn on an inside wall of an Egyptian pyramid. Typically, with this type of machine, man was responsible for controlling and powering it. The skills of the operator were essential in controlling the machine; he was responsible for generating the desired workpiece contour and also for controlling the feed and speed of the cutting tool. As objects became increasingly complex to manufacture, the control of the machining operation became increasingly difficult. When man started to use metals, then other means of power had to be found. New sources of power were water and, later on, steam. To make machine tools more adaptable to different metals and workpiece sizes, they were equipped with reduction gears to control speed and torque. A further improvement was the invention of the lead screw to drive the machining tables at different speeds. All these improvements entailed the incorporation of manufacturing know-how into machine control, and it relieved the craftsman from performing some of the control tasks. With the introduction of tracing templates, master cams and linkages, it was even possible to produce automatically predefined workpiece contours and shapes. In other words, the control devices were components with memory, and they were capable of reproducing a predetermined operation consistently. However, they had a severe problem; for new types of contours and shapes new gears, cams, linkages and templates had to be produced, which usually was very expensive. Thus, this type of automation was not very flexible.

Until almost the middle of the Nineteenth century, all artifacts were custom-made. The increasing wealth of the population stimulated the demand for conventional and new products, thus it became more and more difficult to satisfy the market. It needed figures like Frederick Taylor and Henry Ford to invent the division of labor and to define the principles of mass production before manufacturing could be revolutionized at the end of the last and the beginning of this century. With their ideas, many machine tools were tied together to form a flow-line manufacturing concept. The power was supplied by central power sources, and often it was very difficult to arrange the machines in an optimal way. The early machining and assembly lines used for mass production were rigid and expensive, and they only could serve the market as long as there was a need for uniform products. As soon as a market was saturated, customers changed their desire for products where there was a choice of designs and features. Conventional mass production methods could not handle this variation in product type, since it was virtually impossible to alter the rigid machine setup.

The repetitive machining and assembly lines had process knowledge incorporated in them in a rather permanent way. First, cams, linkages and templates had insufficient memory capacity. Second, there was a limit to the control mechanisms that could be built on mechanical principles. Third, the control of the power supply imposed severe restrictions on the configurability of the production system.

The invention of the electric motor and electric switching and control devices started a new era of machine control. These devices led to the conception of individual powered and controlled machine tools and production machines. Thus, manufacturing facilities could be configured toward the requirements of making a given product.

With the greater number of product variants, new problems arose like time scheduling and full utilization of manufacturing resources. To solve these organizational tasks, graphical, mathematical and heuristic planning and control methods were introduced. It turned out that, for a large workpiece spectrum which needed the use of many different machine tools, good planning and control became extremely difficult – the manufacturing engineer was quite happy if the production run never had to be changed. The required organizational flexibility could only be improved with the introduction of the computer in the manufacturing system. However, the computer underwent its own evolution. The first computers using vacuum tube technology had little impact on improving the organizational control of manufacturing systems. The invention of the transistor led to the development of complex computers and large-capacity memories, which are the basis for computer-integrated manufacturing (CIM) systems. The first computer-operated controllers were numerical controllers (NC) and programmable logical controllers (PLC). Both were very successful in controlling the making of the product on the manufacturing level. An efficient organization control, however, was only possible with the introduction of distributed computer systems and communication buses suitable for operating in a noisy manufacturing environment. One of the great challenges in conceiving and building CIM systems became the construction of software.

When we look at the emergence of the first computer-controlled factory and observe the developments which will lead into the next millennium, it is surprising how fast future CIM concepts have been evolving. With the computer-controlled manufacturing concept of the Seventies, standalone NC machines or conventional machine tools typically were employed, and material movement was done with simple transportation means. The organizational planning and control was performed with the help of a centralized computer system. For communication with the factory floor and the feedback from the process, manufacturing documents in the form of written reports, tapes and punched cards were used. This type of operation was very slow, and reactions to market or production changes often imposed severe bottlenecks. The incorporation of the computer into the manufacturing equipment controllers simplified standalone control of machine tools and provided an improved utilization of resources.

However, a real CIM system can only be engineered when all activities which contribute to the making and shaping of the product are integrated within the factory concept of the Nineties. Engineering, process planning, manufacturing scheduling, order release and control, material movement and quality control must all be interconnected and the total system must operate as one unit.

The new CIM systems are being constructed with the help of standardized machine interfaces, communication buses, computers and software. The greatest challenge of the control system will be the conception and production of modular software which can be used for various applications. Today, the software often amounts to 50–60% of the total installation cost of a CIM component. This cost increases when whole CIM systems are implemented. The future of successful CIM applications will greatly depend on the availability of standardized hardware and software components and on configuration tools with which these components can be assembled into manufacturing systems.

CIM can be considered as a philosophy in which the computer plays a central role for planning and controlling the manufacturing process. However, the manufacturing systems and processes are in general of great complexity and have evolved using experience gathered over many years. For this reason, it is virtually impossible to configure, build and run a completely automatic factory in which the computer contains all the manufacturing knowledge needed to guide the whole plant. The computer can take on routine work in a plant under normal working conditions. Strategic planning and control, as well as handling of disturbances, however, have to be done by well trained and experienced managers who know how to use the computer as a supporting tool for achieving the goals of the manufacturing system. The success of a company in its marketing environment depends on how well a CIM system is used and operated. CIM is not a cure-all for every manufacturing problem; it must be applicable and justifiable. thus, in many factory situations CIM will only solve problems in parts of the overall function; an attempt must be made to integrate these parts into an operable manufacturing system.

1

The manufacturing system

CHAPTER CONTENTS

1.1	A Short Historical Excursion	1
1.2	An Introduction to a Basic Manufacturing System	4
1.3	The Importance of Flexibility in Manufacturing	11
1.4	The Functions of a Manufacturing System	19
1.5	The Hierarchical Planning and Control Concept	32
1.6	Future Developments	40
1.7	Problems	46

1.1 A Short Historical Excursion

Manufacturing is a vital source of wealth in every industrialized nation. Although manufacturing has played an important role during the evolution of man, the most rapid development of manufacturing skills occurred during the last two centuries. The first complex products, like the movement of a watch, were made by highly skilled craftsmen who had little or no theoretical knowledge about the static and dynamic behavior of the mechanism they created. In addition, they used crude tools to shape a product, and had to rely on materials whose composition and properties could not be controlled or were unknown. For the making and shaping of the product, mainly human energy was used; in some cases wind or water power was available. Many simultaneous developments had to take place to give the craftsman the basic building blocks upon which a modern manufacturing technology could be founded. Adequate power to operate a manufacturing process was provided with the invention of the steam engine. The basic understanding of the static and dynamic behavior of rigid bodies and of the properties of materials were other prerequisites for the conception of the first automated manufacturing equipment. These original machines were controlled by cams and linkages, and were driven by belts via common drive shafts. They became the backbone of automated factories to produce a variety of complex, mostly custom made, products. A further improvement of the production equipment could be achieved with the invention of the electromotor and electrical switching components. Now it became practical to provide each machine tool with its own power drive and to configure a factory in various ways, depending on the required layout of the manufacturing system. Thus, the factory could be tailored towards an application.

The concept of part interchangeability and division of labor triggered the invention of the rigid flowline principle. With this manufacturing method, high-quality products were made available to many people who hitherto could not afford the expensive custom made products. Because of the enormous investment, the rigid flowline production could not be altered easily; thus the customer had very few options to choose from. When the market became saturated with standard products, the manufacturers introduced numerous product variants with special features. This implied that the manufacturing system had to be made more flexible. With the invention of the vacuum tube and transistor, it became possible to build machine controllers which could be reprogrammed to a limited degree; thus the manufacturing machines were able to produce various product variants.

A major breakthrough in programmable automation was achieved in 1947 with the invention of the numerical controller (NC) at the Massachusetts Institute of Technology. It was the first time that hardware and software technologies were combined successfully in one control unit. The NC allowed machining of complex low volume parts, thus providing a very high flexibility to a manufacturing system.

With the integration of the transistor and other electronic components into very compact circuits (very large-scale integration (VLSI)), families of micro, mini and mainframe computers became available for controlling manufacturing processes. These computers led to the development of programmable controllers, process computers and business computers, all basic elements of programmable automation. A special feature of these computers was that they can be easily adapted to a specific manufacturing task by a dedicated control algorithm. As automation proceeded it became obvious that conventional automation technologies had been exhausted and that computer hardware and software had the greatest potential of further rationalizing manufacturing operations and obtaining truly programmable factories for a product family.

It was soon realized that special skills were required for the design of the computer systems and the control software and that it was very difficult and time consuming to implement them in a factory, and to combine them with conventional manufacturing technology.

For the control of a manufacturing process by computer, the production machines had to be provided with special interfaces to make them compatible with the computer. In addition, man had to be given tools to be capable of communicating with the computer and the process. For this purpose, various basic technologies were developed for the man–machine communication, graphic data processing, picture processing, construction of databases, knowledge engineering, factory communication, microelectronics, modeling, simulation, programming, and sensor systems.

These basic technologies have short life cycles, and for this reason it is often difficult to assess their importance and to implement them. In many cases, the evolution of such a new technology was so rapid that applications which had been started several years ago had become obsolete and had to be done over again.

1.1.1 The role of the computer in manufacturing

The computer has had a substantial impact on almost all activities of a factory. Often, the introduction of the computer changed the organizational structure of a department and made necessary the adoption of completely new management structures. Since the computer is capable of doing repetitive work very efficiently, the task of many management functions has also changed drastically.

The future development of computer technology in manufacturing cannot be exactly predicted. It will depend on various aspects, including the development of hardware and software, the skill of merging computer technology with conventional manufacturing know-how, the possibility of simplifying and standardizing manufacturing processes and procedures and so on. One important aspect is the efficient cooperation of the various engineering skills necessary to configure a computer-integrated manufacturing (CIM) system.

The operation of a CIM system may give the user substantial benefits compared with conventional systems. Experience has shown the following typical benefits:

- Reduction of design costs by 15–30%
- Reduction of the in-shop time of a part by 30–60%
- Increase of productivity by 40–70%
- Better product quality; reduction of scrap by 20–50%
- Improved product design; for example, the use of the finite elements method in conjunction with a computer allows for calculating 3–30 times as many design variants, compared with conventional methods

Depending on the market strategies pursued, available labor, wages, market size and so on, CIM is being used in different ways by various countries (Wolf, 1988).

In Europe, the main emphasis is on the achievement of flexible automation whereby the entire manufacturing system is included. The market in the various European countries is relatively small, therefore there is a demand for smaller lot sizes. Europe has had considerable experience with flexible manufacturing systems and has developed various integration concepts. Also, an attempt is being made to utilize the well-trained engineers, managers, and craftsmen and to integrate them into the CIM system.

Japan follows a different strategy. The Japanese companies think in terms of mass production and for this reason automation is mainly concerned with the operative on the plant floor where programmable machines, flexible machining centers and adaptable assembly systems are employed. With this strategy, it is not possible to provide a great number of product variants.

The American philosophy hinges upon the quarterly performance reports which have to be given to the stockholder of a company. There is a need to even out production cycles and to show uniform utilization of the equipment. In addition, American industry cannot resort to a pool of well trained craftsmen because in their educational system there is no real provision for apprenticeships. For this reason the American philosophy is to implement as much human manufacturing know-how as possible into a CIM system, thereby reducing the risk of human errors and minimizing the need for intervention with the manufacturing process.

1.2 An Introduction to a Basic Manufacturing System

1.2.1 A model factory

From the previous section, we have learned that a modern manufacturing operation is very complex and has to be evolved from various engineering disciplines. The computer is being used as a planner and scheduler to organize and supervise the production process, as a controller to steer the manufacturing equipment, and as a verifier to assure a specified quality.

We now introduce a small manufacturing plant to discuss its principle functions and to explain the most important terminology which is being used in connection with a computer-operated manufacturing facility. Unfortunately, the terminology is not uniformly used. However, an attempt will be made to be concise in this book and to explain all functions, if they are not obvious from their names. It is our task in the following chapters to discuss the individual activities in detail.

The model factory produces custom made and standard valves. Figure 1.1 shows the components of the factory, including engineering, planning, scheduling, control and the manufacturing facility. The latter has three machine tools for processing of parts; one robot for the assembly of the product; one measuring station and a material management unit consisting of a parts storage, an automatically guided vehicle (AGV) for material transportation and a finished goods storage. The flow of the product is from left to right, whereby the raw material is entered in the parts storage and from there brought successively by the transport vehicle from one machine to another for processing. The completed parts are deposited in the finished goods storage for shipping to the customer. The planning and control of this factory is done with the aid of a computer network which interconnects all manufacturing functions of the plant, including the manufacturing equipment. We assume that this factory is highly automated and that it has a database containing master files on the product, engineering methods, manufacturing resources and methods, factory resources, quality objectives and productivity goals. In order to process the data, there are algorithms for engineering activities, planning and scheduling methods, make and buy decisions, and equipment and quality

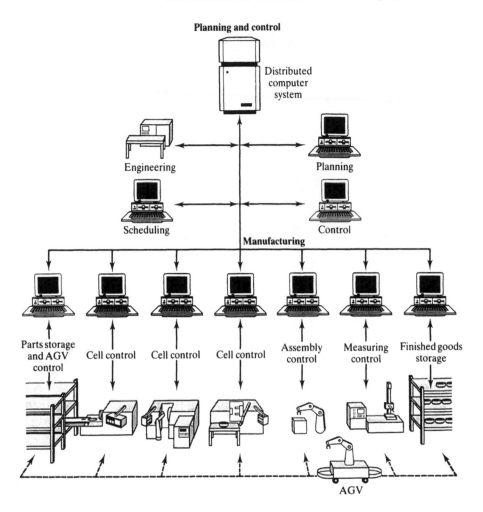

Figure 1.1
A simple computer-controlled manufacturing system.

control operations. The principal supervisory activities of the system are engineering, planning, scheduling, and control, as shown in Figure 1.2 and in Rembold *et al.* (1985). The activities order release and verification are designated functions needed in the processing of an order and could be incorporated in other functions. Each activity is represented by a box. Permanent data is entered from the top, whereas order related data and information on new processes and equipment are entered from the left. The activities are triggered by one or several events and activated by control algorithms. The output of the control algorithms is sent to the next lower activity. In a real system, information will also be fed back to the higher level activities, however, for simplicity this is only shown for the feedback data on manufacturing resource utilization and product quality. A box may contain one or several subactivities which are executed in parallel, in sequence, or in a hybrid fashion. An activity can be either processing of information or processing of material. Figure 1.3 shows the functions of an activity box.

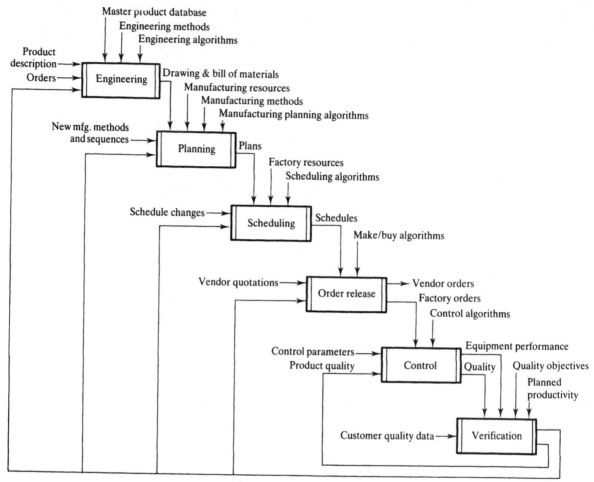

Figure 1.2

Planning and control activities of a manufacturing system. (Courtesy of Marcel Dekker, Inc.)

Let us first explain an information process where parts are scheduled through a factory. The input to the activity is the part spectrum to be manufactured and the available machine tools. This information is located in the buffer (computing memory). The activity is triggered by sending a start signal to the decision point 1 to signal the availability of the information. When the results have been obtained a ready signal will be issued at decision point 2 which may trigger another activity. The results of the calculations are stored in the buffer. A material processing activity is done in a similar fashion. Let us assume that the process is a turning operation and that the raw material is located in the entrance buffer. Factory control sends a start signal to decision point 1. There is also a feedback signal available at this point from the machine tool that it is ready for turning. The part is now sent from the buffer to the machine tool for turning. Upon completion of the part, a signal is generated at decision point 2 to indicate to the factory control system that the part has been manufactured. This information is used to bring the

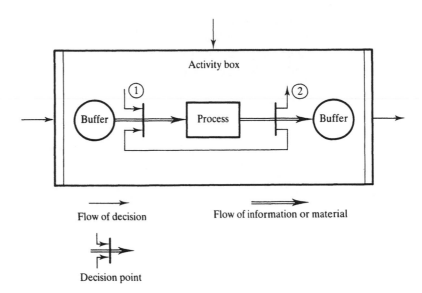

Figure 1.3
Flow of information
or material through
an activity box.

part to the output buffer. After this short introduction to the activity block, we will explain the function of the planning and control system in Figure 1.2.

An order entered by a customer may consist either of standard or custom made valves. In the case of a custom made valve, a product description is given to engineering. Engineering will be designing the product with the help of the computer using known methods and algorithms. The output of this activity includes the engineering documents, consisting of the drawing, parts program and the bill of materials. The next activity is planning the manufacturing process. The manufacturing database for resources and methods is searched to draft the process plan containing the description of the machining processes and sequences. With this plan, the scheduling function will be activated. At this point, the customer order will be competing with other orders. Thus, a schedule has to be found which tries to satisfy a strategy to meet all product delivery dates. The schedule is entered into the order release activity. First, a make or buy decision is made, which may be the result of a favorable vendor quotation for special parts or which may be necessary due to tight delivery schedules.

Second, the parts to be manufactured inhouse are released for production. The raw materials are brought to the machine tools and sequentially machined to their defined form and properties; and finally, the robot assembles them. The machining of the part is simulated on the display and the parts programs are entered into the machine tools to produce the required contours. The control algorithms for the machine tools will be supplied with the control parameters to produce the required quality. The verification activity compares the quality obtained with the quality objectives and customer quality data and, if necessary, proposes corrections to control. In addition, verification checks the performance of the manufacturing equipment to pinpoint problems and suggest corrections.

In the case of the customer ordering standard valves, the engineering and planning activities can be bypassed because all engineering and planning documents are already available.

Feedback information on the manufacturing resources utilization and product quality is routed back to all activities. This information can be utilized as follows. First, the reason for a product defect could be due to a poor design, thus the design would have to be improved. Second, poor utilization of manufacturing equipment may be caused by selection of improper manufacturing processes. This may necessitate new process planning. Third, a schedule cannot be adhered to because of a machine tool failure. This may require rescheduling by sending parts to outside vendors. Fourth, the manufacturing process may have slowly degraded and it becomes difficult to obtain specified dimensions or other quality parameters. In this case, the machine tool may need an adjustment or new control parameters.

1.2.2 The organizational and engineering components of a manufacturing system

In this section we will be looking at our factory system from another aspect; namely, the organizational engineering and manufacturing environment, as shown in Figure 1.4 and in CADCAM Labor (1987).

On the left side of the figure are the organizational activities which are necessary to channel an order from its entry to its completion. The functions are order planning, order release and order control as well as manufacturing control. In the center of the figure is the database containing order processing, engineering and manufacturing know-how. In the third chapter, we will be looking in more detail at the database. The order-specific activities have a connection with engineering and manufacturing to coordinate these functions. The strategic planning activity is concerned with the long-range operation. It will be explained in a later section of this chapter.

On the right side of the figure are the functions which engineer and manufacture the product. There must be close cooperation, however, between engineering and manufacturing to ensure that the product can be made economically. In other words, engineering must have a thorough understanding of the manufacturing process. In theory, it should be possible for engineering to generate automatically all documents needed for manufacture, including the drawing, bill of materials, process and quality control plan.

The lightly shaded area bears the name logistics. This is an activity concerned with the data and material flow through the plant. It tries to tie the data and material flow together to obtain an unobstructed and optimized product stream though the plant, from the design to the product delivery. This is a very important activity, it must assure in a flexible way that the right part is at the right station at the right time. It provides the information needed for processing for all parts.

The dark area indicates the quality control activities which must be performed during the making of the product. The ultimate judge to set the quality standards

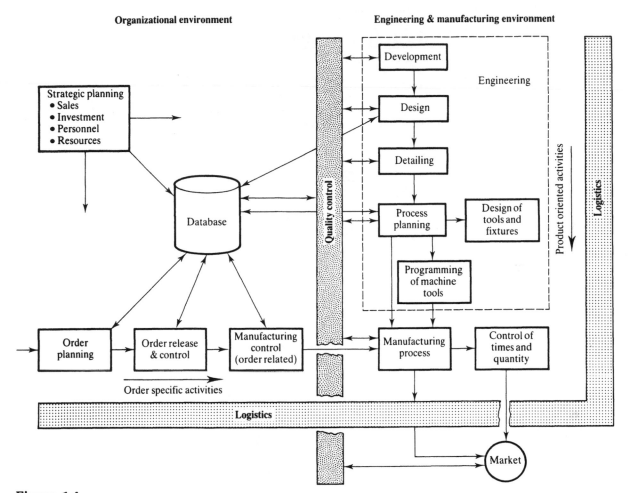

Figure 1.4
Organizational, engineering and manufacturing environment of an enterprise.

is the customer. Engineering tries to determine the threshold separating a good product from a bad one. Problems with the product are reported by the customer (market) back to quality assurance. With this information, an attempt is made to engineer good quality into the product. The effort of building high quality products runs like a red line through the entire manufacturing process, where at each stage the desired quality is verified.

1.2.3 Definition of the computer terminology in manufacturing

With the advent of the computer for the planning and control of manufacturing operations, various new terms have been created. The most important ones are shown in Figure 1.5 and discussed in AWF (1985). They will be explained below:

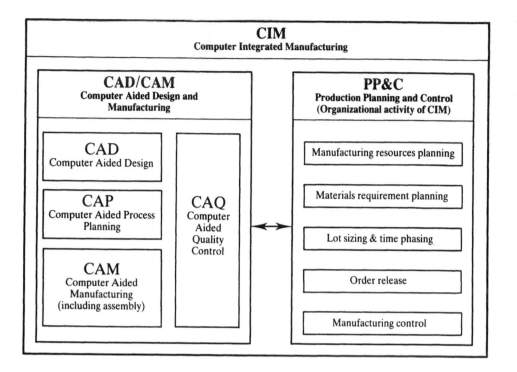

Figure 1.5
Computer integrated manufacturing activities.

- CAD (computer-aided design). This activity comprises computer supported design, drafting, and engineering calculations. Since engineering is also involved in product testing, NC program generation, and other computer supported functions, the term CAE (computer-aided engineering) is often used.
- CAP (computer-aided planning). This activity is concerned with the computer-aided generation of a technological plan to make the product. The process plan describes the manufacturing processes and sequences to make a part.
- CAM (computer-aided manufacturing). This activity defines the functions of a computer to control the activities on the manufacturing floor, including direct control of production equipment and management of material, cutting tools, fixtures and maintenance.
- CAQ (computer-aided quality control). This activity combines all ongoing quality control work of a manufacturing system. In some cases it is termed CAT (computer-aided testing), which is somewhat restrictive in its meaning.
- CAD/CAM designates the sum of the activities CAD, CAP, CAM, and CAQ.
- PP&C (production, planning and control). This function is the organizational activity of CIM. It is concerned with manufacturing resources planning, materials requirement planning, gross requirements planning, time phasing, order release and manufacturing control.
- CIM (computer-integrated manufacturing). This combines the activities CAD, CAP, CAM, CAQ and PP&C in one system.

1.3 The Importance of Flexibility in Manufacturing

1.3.1 Types of manufacturing operations

The manufacturing industries make a wide variety of products, which on the one side of the spectrum may be custom made and on the other mass produced. The tools used for making a product have to be selected for the specific application and tailored to production requirements. Seldom are there two manufacturing systems which are alike; even when they make the same product. However, in many cases it is possible to employ standard machine tools and auxiliary equipment that are configured to a production system. For a special product it may be necessary to custom build manufacturing machines, which may be unique to that industry.

Within the last few years, competitive pressure has forced many companies to adapt a plant quickly to changing market conditions. This problem led to the conception of flexible manufacturing systems (FMS). Of course, flexibility means that the equipment can only be reprogrammed for a variant or a product similar to that for which it was originally designed. It is unlikely that the plant can be quickly adapted to an unrelated product without making major changes to the production equipment and the layout of the system.

Modern manufacturing systems, basically, use five manufacturing principles, as shown in Figure 1.6. In this figure, the yearly production rate is plotted over the

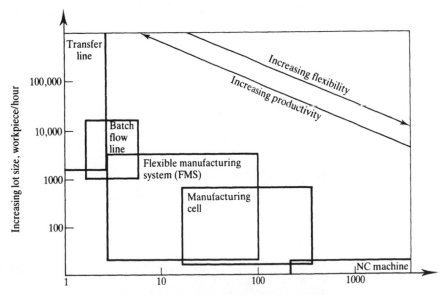

Figure 1.6
Modern manufacturing concepts.

number of different workpieces which can be manufactured by the various methods. The numbers on the x and y axes are approximate values. They depend on the production rate, lot size, workpiece complexity, similarity of the workpiece variants manufactured, and many other parameters. The flexibility of a system increases from left to right and the productivity from right to left. These five principles use either the batch type or flowline production method.

The numerically controlled (NC) machines perform typical job shop operations, as shown in Figure 1.7. When a workpiece is to be machined, the raw part and the tools are set up and the NC program is entered. The latter selects and directs the tool to machine the desired surface contour. The flexibility of this machine tool is very high; tools can easily be changed and when a new type of part is to be machined, a new part program is entered into the controller. Due to its high set-up cost, the production method is only suitable for low price rates.

The next manufacturing method is performed by the manufacturing cell: it also represents a batch type operation. This cell may have several NC machines or it may contain a multifunction machine, as shown in Figure 1.8. In such a cell, various machining operatives can be performed on the same workpiece. Handling of the workpiece can be done by a special set-up device or a robot. Programming of the system is similar to that of NC machines, only the part handling has to be included. With such a system, higher economical production rates can be achieved at the cost of reduced flexibility, since automatic set-up devices can only be designed for a limited workpiece spectrum.

The flexible manufacturing system consists of a number of NC machine tools which are interconnected by a transportation system, as illustrated in Figure 1.9.

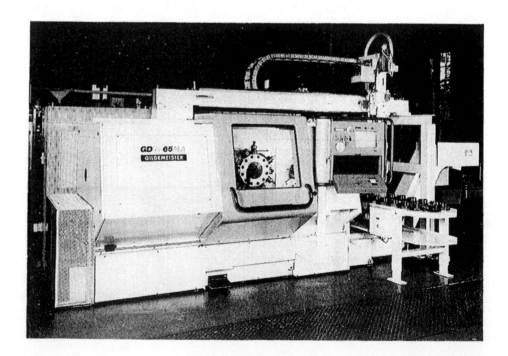

Figure 1.7
NC lathe (courtesy of
Gildermeister AG,
Bielefeld, Germany).

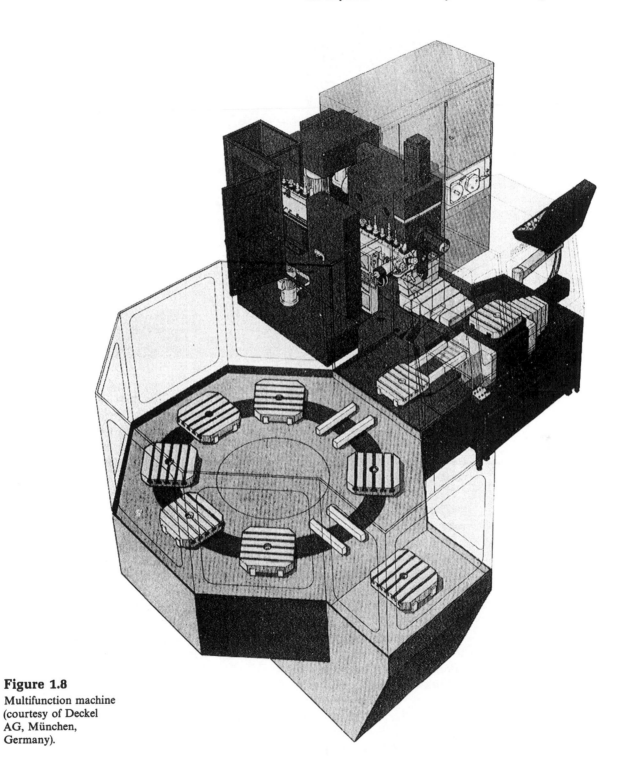

Figure 1.8
Multifunction machine
(courtesy of Deckel
AG, München,
Germany).

Figure 1.9
Flexible
manufacturing system.
(Courtesy of MBB
Flugzeuge, Augsburg,
Germany)

These machines use the flowline principle. A machine tool can have single or double multifunctions. The individual workpiece is routed to selected machines for processing. Programming is done as with other machine tools, however, the designated route and handling of the part has to be included. Since this operation tries to approximate a flowline principle, high production rates can be obtained economically.

The batch flowline principle is used in connection with the flexible manufacturing system. Here, the parts are grouped together in batches to minimize set-up and tool change times.

The transfer line uses the oldest mass production principle, it is also a flowline operation, as shown in Figure 1.10. The workstations are usually custom made, fixed type automation machines, which, once they have been set up, are seldom reconfigured. A change of machining operations is only done when design changes have been made to the product. Of course, with this principle, high production rates can be obtained at minimum cost. Whereas in all previous manufacturing methods the computer plays an important role, for programming and controlling the machines, it is not needed for the operation of a transfer line. However, the computer has found its place for monitoring such lines in order to identify machine disturbances quickly and to reduce repair time.

When we plot the relative manufacturing cost/workpiece over the production rate for the various automation principles, we obtain the curves shown in Figure 1.11. This figure shows that each technology has its place in manufacturing. There is also a distinct point where the cost of each method is at its minimum.

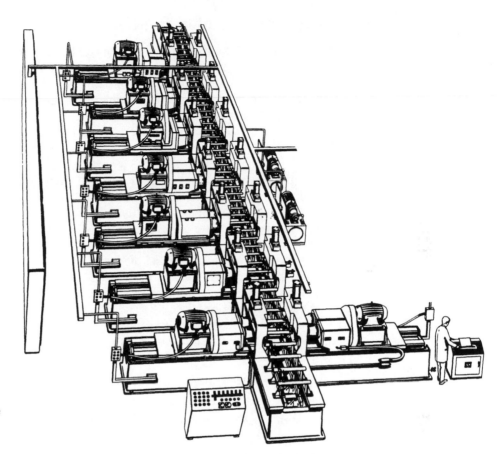

Figure 1.10

A typical transfer line monitored by a computer.

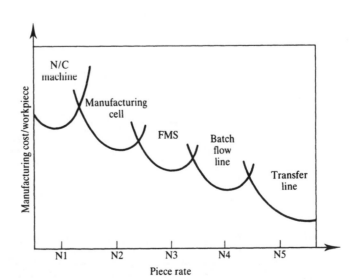

Figure 1.11

Relative manufacturing cost of workpieces over piece-rate for various manufacturing technologies (the value of N changes with increasing complexity of the part).

1.3.2 The requirements of a flexible manufacturing system

In the highly industrialized countries, a trend can be observed where the customers have a preference to purchase more and more products which have specific options; in other words, the number of product variants which have to be built increases steadily. Thus, the old transfer line principle becomes too rigid. For this reason, flexible manufacturing systems are becoming economically attractive. The requirements for a flexible manufacturing system are shown in Figure 1.12 and discussed in Hirschberg (1989). The key subjects of this figure are productivity, quality, and flexibility.

A high productivity can be achieved with relatively low investment costs and high production rates. Investment costs, however, are usually very high for FMS systems. This implies that the availability and utilization of the machines must be very high. In addition, there is the necessity to obtain low in-process inventories and short production cycles. We have seen from the previous discussions that the flowline principle is more economical than the job shop principle. For this reason, the flowline production is used by most FMS installations. Particular attention in this system must be paid to the material flow and the division of work elements between machining stations.

The quality is manufactured into the product by the automation system. To obtain a high quality, the following requirements are mandatory, see also Horn (1987).

- The selection of technologies which can be handled securely
- Unique identification of parts and tools
- Automation of the process and assembly
- Automation of material transportation and handling
- Limitation of number of manufacturing tools, for example standardization of pallets, fixtures and tools
- Automatic testing after each manufacturing process.

In essence, these requirements indicate that man as the least reliable link should be taken out of the manufacturing process.

Figure 1.12

Requirements of flexible production automation.

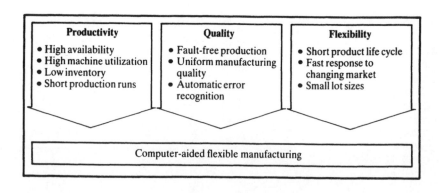

The requirement for a high flexibility has to be seen from the following aspects, see also Putz and Luck (1986). They are:

- Flexibility with regard to orders. This implies that the FMS must be capable of producing efficiently orders of various lot sizes without retooling.

- Flexibility with regard to the product. This requirement states that the FMS must be designed to handle various product variants and to react quickly to changing customer demands. Retooling and reprogramming of the system must be minimized.

- Flexibility with regard to the system. This flexibility is needed when the system is to be redesigned for higher productive rates or when either the addition or the change of existing manufacturing processes is necessary.

1.3.3 Automation potential in manufacturing

Historically, most automation efforts were directly concerned with the improvement of the physical manufacturing process; thereby, many good results had been obtained, for example, the introduction of the NC technology greatly improved the manufacturability of complex 3D parts and their quality. However, by looking at the introduction of the NC principle over the last 30 years it becomes apparent that this technology has not achieved the manufacturing potential it had promised. What is the reason for this problem? We will be discussing two typical areas where substantial inroads can still be made, as explained in Lutz (1987). They are order processing and manufacturing control.

In recent years, conventional manufacturing methods have changed considerably. The most important goal of the old-fashioned mass production method was the maximum utilization of the installed equipment, see Figure 1.13 and Neunheuser (1984). In the future, manufacturing organizations will be trying to operate the equipment in a market-oriented way with high emphasis on flowline production and no inventory. This goal can only be obtained with a high flexibility and an instant response to the market.

Let us now look at problems with present order processing methods. Figure 1.14 shows the flow of an order from sales to manufacturing; see also Milburg and Bürstner (1986). Typically, an order goes through the factory in a batch type mode because the various order processing activities are departmentalized, and there is a lack of good synchronization. This problem leads to the replication of basic data in the various departments and introduces many unnecessary errors into order processing. In addition, sequential processing of data prevents parallel processing which can be done efficiently by a computer system. Another problem is the assessment of the order processing costs, which is needed to verify the calculated costs. A flexible manufacturing concept requires the extensive use of the computer for order processing, thereby improving the following aspects:

- Improved synchronization of the individual order processing activities,

- Continuous updating of all orders; also geometry, technology, and operating data,

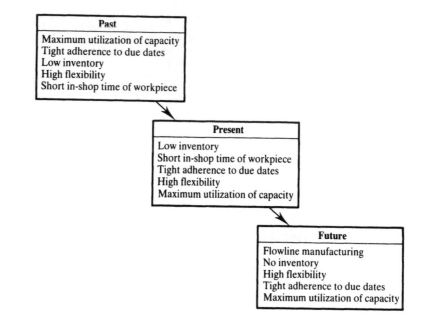

Figure 1.13

Change of manufacturing goals with time.

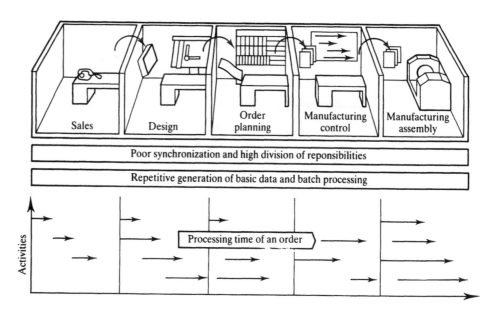

Figure1.14

Conventional flow of an order.

- Provision of an instant access method to all order processing data by all departments,

- Abolition of the departmentalized structure of the organization and introduction of a flowline processing structure,

- Improvement of data transparency to facilitate the determination of the actually incurred costs and the adherence to due dates.

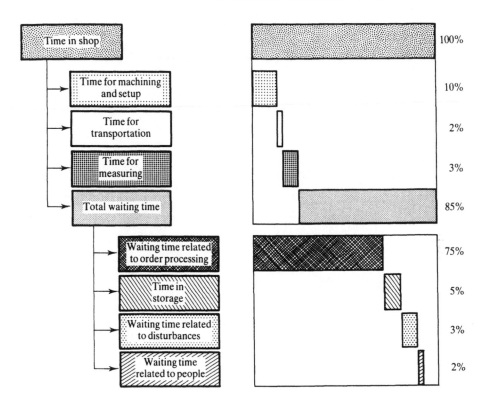

Figure 1.15
Life of an average
workpiece in the
factory.

Related problems occur when a part is being manufactured. The situation is depicted in Figure 1.15 and discussed in Wiendahl (1986). It shows the (well publicized) life cycle which an average part has in a factory. This illustration, however, is more detailed than is normal. It shows how a part spends 10% of its factory residence time in the machine tool, 5% of the time is used up for transportation and measuring, and 85% is waiting time. Waiting time is very expensive for the manufacturer because valuable operating capital sits idle in inventory. This situation can also be improved by organizational methods with fast and more accurate data processing means.

1.4 The Functions of a Manufacturing System

A manufacturing system consists of a multitude of functions which are interconnected by a complex computer-controlled communication system. It supports strategic and technologic planning, organizational planning and scheduling, manufacturing control and monitoring and accounting functions. In this system, the flow of information, of funds and of material has to be controlled in a precise manner to service the customer market with a high quality product and to assure the financial soundness of the enterprise. There are numerous models used to represent the interactions of the various functions of a manufacturing system. We use the model shown in Figure 1.16 which gives a good overview of a complete manufacturing system, see also VDI-Bericht (1971). The interactions between the various functions are not as simple as shown. It is virtually impossible to describe all microscopic communication channels and control loops of a manufacturing organization. Later in this book we will get acquainted with other models of a manufacturing system.

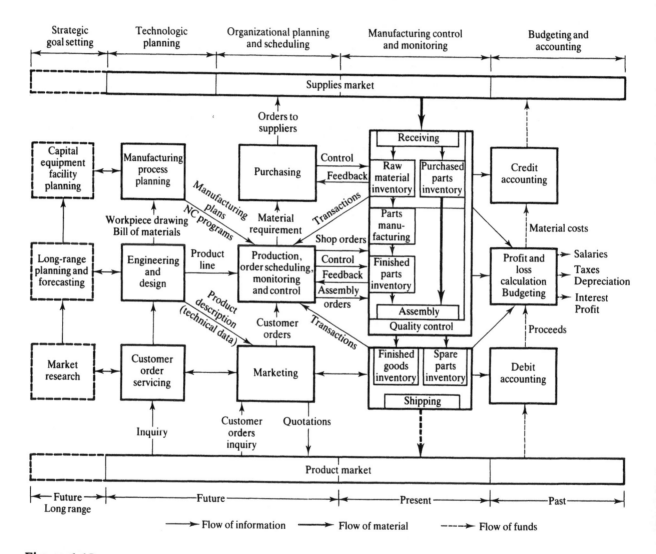

Figure 1.16

Basic functions of a manufacturing facility for small and medium-sized production lots.

Although some of them may show the interaction of the various functions in a more logical way, in general, they are not as detailed.

The model in Figure 1.16 has two horizontal scales. The scale on the top indicates how the activities can be grouped together by planning and control topics. The scale on the bottom shows the time frames for each activity. The arrows used in the figure indicate the flow of the information, materials and funds.

In practice, models vary and are not necessarily the same as the one shown in Figure 1.16. Every organization will be structured according to a specific model, which can be developed in house, set up by a consulting firm, or copied from another manufacturing organization. The structure of a model depends on the manufacturing method used, the type of product and product mix made, and the environment in which the company builds its products and markets them. In manufacturing industries, there is no standard terminology being used. As far as possible we selected common terms for our model which the reader should be able to understand. The functions of the individual manufacturing activities will be described in the next section.

1.4.1 Market research

A manufacturing organization must define its place in the market and the goals it wants to pursue. For this reason, it must know the market potential, demographic developments, technical and commercial trends, product innovations, future supplies of resources, availability of labor and finances, and its competitive status. Market research is a function where basic data, which is needed to secure a strong future market position, is systematically collected from publications and experts. It is then evaluated with mathematical and scientific forecasting methods, such as statistics or techniques and decision theory. The results are used by long-range planning and forecasting to make strategic decisions. Often this forecasting predicts 10 to 20 years into the future. Of course, the accuracy of this data will lessen with the increase in the number of years for which the market situation is being evaluated. This activity must have a strong tie to customer order servicing and marketing to evaluate new sales ideas and customer wishes and complaints.

1.4.2 Long-range forecasting

In long-range forecasting, the product or product line to be marketed will be determined. The company needs to know the yearly production rate, product features, location and layout of the manufacturing facility and the type of manufacturing equipment needed. The production rate will determine the manufacturing method, processes, type of raw material being used, and quality control procedures to be applied. Usually, an existing plant will have to be modernized and equipped with the most up-to-date processes. Ways have to be found that will not interrupt ongoing manufacturing operations to keep up the supply of the present product. Where a new plant is being planned, strategic considerations have

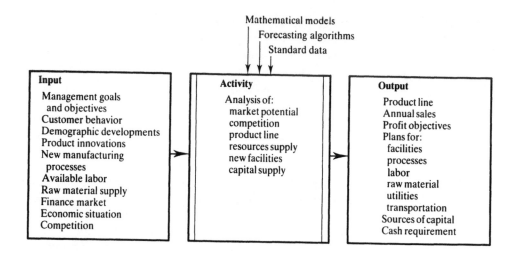

Figure 1.17
Long-range
forecasting.

to take account of its location, to optimize transportation routes, and to assure a good labor market, raw material supply and low cost utilities. The effect of forecasting will be felt in every function of the company. For this reason, the implementation of the forecast strategy is a corporate-wide endeavor and must involve all departments of the company. Forecasting will draw heavily on information from customer servicing, marketing, engineering, manufacturing, and quality control. Figure 1.17 shows the information flow through the forecasting activity.

1.4.3 Capital equipment and facility planning

The selection of hardware and software is the task of this activity. All existing production machines, warehouse facilities and material movement equipment are investigated to determine their usefulness for the production. Plans are made for the layout of the manufacturing floor; old and new equipment is incorporated in the plant. The computer equipment and the software to plan and control the production is selected. This activity is supported by expert systems, computer programs for the layout of facilities, and the simulation of manufacturing operations. Numerous plant alternatives are tried out, and return on investment calculations are made. An attempt will be made to find an optimal plant configuration which can be extended easily for future products or changes. Typical input and output parameters of capital equipment and facilities planning are shown in Figure 1.18.

1.4.4 Customer order servicing

This is an activity which helps the customer to follow and expedite his order through the factory. It represents a plant-wide control system to follow an order through a plant, and from which a customer can find out the status of an order.

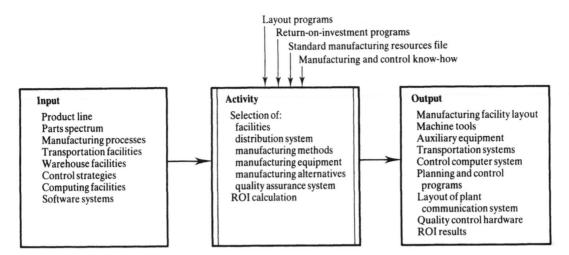

Layout programs
Return-on-investment programs
Standard manufacturing resources file
Manufacturing and control know-how

Input	**Activity**	**Output**
Product line	Selection of:	Manufacturing facility layout
Parts spectrum	facilities	Machine tools
Manufacturing processes	distribution system	Auxiliary equipment
Transportation facilities	manufacturing methods	Transportation systems
Warehouse facilities	manufacturing equipment	Control computer system
Control strategies	manufacturing alternatives	Planning and control
Computing facilities	quality assurance system	programs
Software systems	ROI calculation	Layout of plant
		communication system
		Quality control hardware
		ROI results

Figure 1.18
Capital equipment
and facility planning.

The manufacturer can enter through this service any engineering or delivery date changes he deems to be necessary. Of particular interest to the customer are the progress of his order, and delays due to manufacturing problems or parts shortage. This activity will locate the problems and will try to expedite the order if necessary. Figure 1.19 shows how this service interacts with other plant activities, see also Rembold *et al.* (1985). Expediting can be done for any manufacturing activity, however, it usually interacts with order planning, order processing, and production control.

1.4.5 Engineering and design

Engineering is a key activity in a manufacturing system. It determines the function and design of the product and has a major influence on the manufacturing process to be selected. Once the product has been engineered 70% of the cost of making it has been fixed. For this reason, there must be a close cooperation between engineering and manufacturing. Engineering relies extensively upon textbooks, standards, catalogs, product databases, simulation programs and CAD support. An important source of knowledge for engineering is marketing, quality control, and customer order servicing. These activities must be constantly monitored for new product ideas and customer complaints. Engineering conceives, designs and tests the product. For mass produced products, it is also responsible for a pilot production run to determine if the product is marketable. The most important documents coming from engineering are:

● The drawing, describing the product, its dimensions, tolerances, materials, surface treatments and quality control procedures,

● The bill of materials, describing the structure of the product and its components.

With advanced engineering systems it is possible to automatically produce

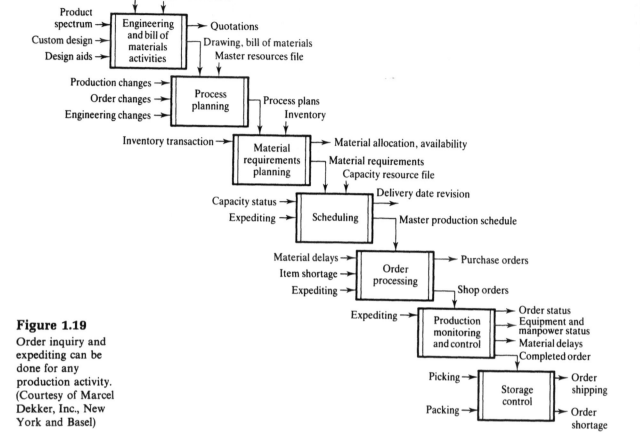

Figure 1.19
Order inquiry and
expediting can be
done for any
production activity.
(Courtesy of Marcel
Dekker, Inc., New
York and Basel)

documents which are normally done by manufacturing process planning, for
example:

- Preparation of the process sheet, describing the manufacturing processes and
  sequences,

- Generation of the NC programs.

In the future, it will also be possible to automate the engineering tasks even further
and to perform some of the production scheduling functions. The main activities
of engineering are depicted in Figure 1.20.

1.4.6 Manufacturing process planning

Manufacturing process planning takes the workpiece description of engineering
design and determines the processes and processing sequences needed to produce
the workpiece, as shown in Figure 1.21. The processes to be selected are highly

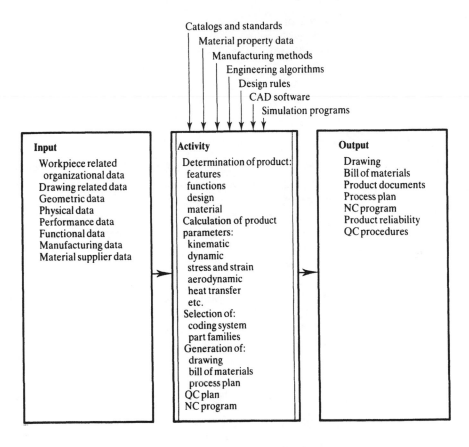

Figure 1.20
Engineering and
design activities.

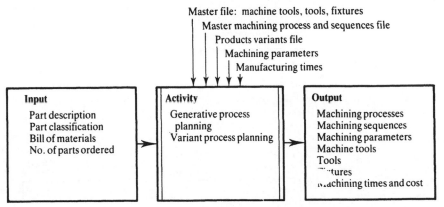

Figure 1.21
Manufacturing process
planning.

dependent on the size of the workpiece, its design, the material to be used, and the piece rate. An important factor is the labor rate. With a high labor rate, an attempt will be made to perform the processing with multipurpose machines, requiring little or no workpiece handling. Many manufacturing organizations have standard production processes and procedures already worked out for part variants and stored in a manufacturing methods file. In this case, the planner will enter a part code, which is determined from the shape of the part and its properties, into the computer. The system automatically suggests a process plan. The output of this activity is a process sheet for the part and the NC program, if these documents have not already been generated by design.

1.4.7 Marketing

Marketing is a concept which is customer oriented, it comprises the entire manufacturing organization and its place in the competitive market. Marketing tries to cater to the customer to satisfy his needs and desires. Marketing is influential in determining the product, product line, and product features. Due to its close contact with the customer, marketing knows the customer preferences, complaints, and quality views. In the context of our manufacturing model (Figure 1.16), we are interested in sales, which is part of marketing. The proceeds coming from sales provide the operating funds to sustain the firm. Thus, sales determines the model mix, volume to be manufactured, pricing strategies, product specification, and its quality. The sales department is the interface to engineering, planning, customer order servicing, manufacturing, and quality control. Figure 1.22 shows the flow of information through sales.

1.4.8 Production order scheduling and manufacturing monitoring and control

When an order is released for manufacturing it competes with numerous other orders for manufacturing resources to meet completion dates; for this reason, a rather complex scheduling procedure must be invoked to uniformly utilize the existing resources and still meet delivery dates (see Figure 1.23). An attempt will be made to perform as many processes as possible on one machine tool to

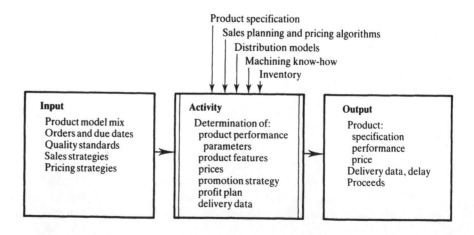

Figure 1.22
Sales activities.

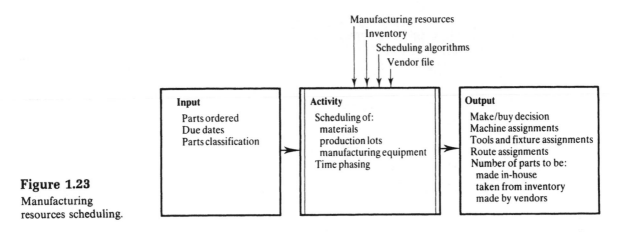

Figure 1.23
Manufacturing
resources scheduling.

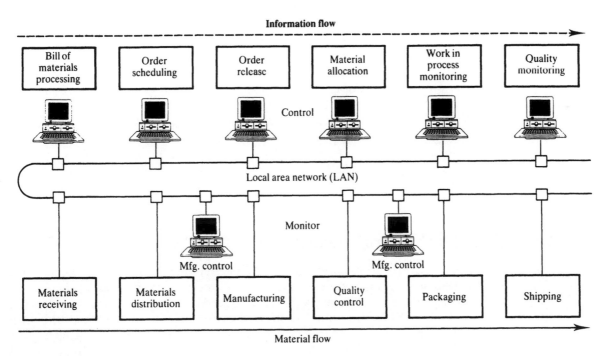

Figure 1.24
Order scheduling and
control requires the
synchronization of
information flow with
the material flow.

reduce set up and handling times. In addition, the scheduling algorithm may try planning a flowline operation to minimize transportation efforts. The part family concept, which will be explained later in this book, plays a major role in streamlining the flow of parts through a plant. Scheduling actually consists of three parts:

(1) Time phasing: to assure timely manufacture of all parts, to meet due dates;

(2) Resource scheduling: to utilize equally the manufacturing machines, and to avoid queues and bottlenecks;

(3) Material distribution: to assure that all parts and materials are at the right station at the right time.

When a part cannot be manufactured on time or if standard parts are required a make/buy decision is made to determine which parts should be sent to a vendor and which should be made in the plant.

To assure a flawless material flow through a plant, a distributed computer system is used as shown in Figure 1.24. The control function ensures that the information flow triggers the material flow in the correct sequence and that all operations are synchronized. Monitoring is the feedback to control to verify that the planned events have actually taken place.

With programmable automation, control is also responsible for distributing the NC programs and operating parameters to the machine tools and the material distribution system.

Monitoring is a very involved function in an ongoing manufacturing process. The principal tasks of monitoring are to follow a job and all parts through the plant, to keep track of the presence of employees and required skills, to monitor the assignment of the manufacturing resources, to observe the correct functions of the manufacturing equipment and to control machine utilization and defects. Monitoring is also needed to keep track of the quality of the parts flowing through the plant. Figure 1.25 shows typical monitoring functions, see also Hoppen (1976).

1.4.9 Purchasing and receiving

This activity performs purchasing of raw material, standard parts and components to be bought from outside vendors. There are various strategies which may be pursued. Parts can be either ordered on demand, or kept in stock. The decision for either of these strategies depends strongly on parts availability and inventory cost. When parts are difficult to get, stocking up the inventory will be a good decision to meet delivery dates. When parts are very expensive, the company will try to minimize inventory costs to free working capital. In most operations, a strategy is being followed where enough inventory is available to provide a smooth production flow. Many mass producers are following a strategy where the suppliers have to deliver the parts at exactly the time (just in time) when they are needed.

When parts and material are requested for production, first, the inventory will

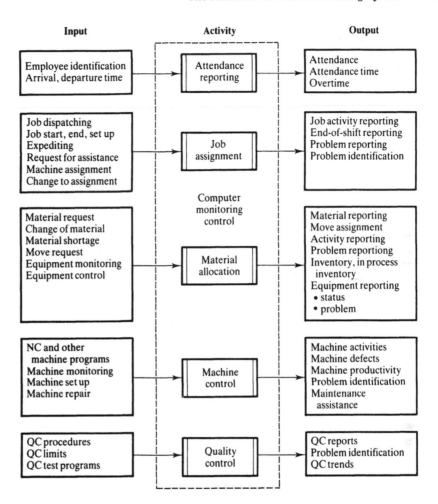

Figure 1.25
Function of computerized plant floor monitoring and control.

be checked; if necessary, requests for quotations are sent to prospective vendors. Upon receipt of the quotations, orders will be sent to the supplier. The order may be issued on price and performance of the supplier.

When the order is received, the number of parts delivered and their quality are checked against the original order. An order will be accepted if the requirements are met. Often, only some parts do not meet the requirements; in this case, a new order for the missing part is issued. Upon inspection, the parts are released to production or inventory.

The inspection procedures are often computer controlled. For this purpose the parts are identified and the corresponding test programs and inspection procedures are sent to the test stations. Testing may be done on a sample basis, or 100% of all parts are tested. The decision to do one or the other depends on the company's philosophy or the vendor performance. If the vendor is completely reliable, tests performed in his plant may suffice. Usually a performance file is kept on each

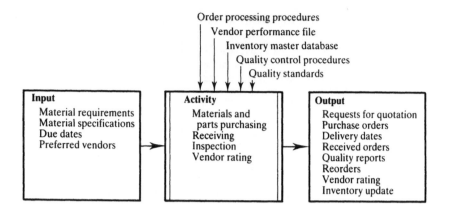

Figure 1.26
Purchasing and receiving activities.

supplier, which is constantly updated. Figure 1.26 shows the purchasing and receiving functions.

1.4.10 Inventory management

A perfectly operating production system should not need any inventory. Money invested in inventory is idle capital and is unavailable for the operation of the plant. In practice, however, inventory is necessary to hedge against part shortage, lost sales opportunities due to unexpected orders, and production difficulties within the plant. Figure 1.26 shows several kinds of inventory. They are: inventory for raw material, purchased parts, finished parts, finished goods and spare parts. In addition, there is an in-process inventory which usually is a buffer between manufacturing operations. All this inventory has to be managed, controlled, and, if possible, avoided. There are numerous algorithms available which can minimize the cost of carrying inventory. To avoid problems with faulty parts, usually more parts are ordered than needed; for this a yield allowance factor is provided.

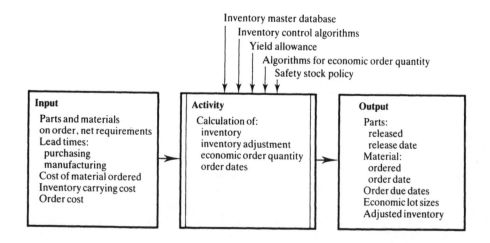

Figure 1.27
Functions of inventory management.

Important inputs to inventory are the purchasing and manufacturing lead times to ensure that the parts are available when required for fabrication. The principal dataflow of inventory management is shown in Figure 1.27.

1.4.11 Quality control

Quality control is a supervisory function which interfaces with most activities of manufacturing. The quality must be designed into a product. Usually, there is a conflict of interest when the quality is considered from various views. The customer wants a versatile and durable product with a long service life. Engineering wants to develop a reliable product having only those functions which make the product useful. The manufacturer wants to employ affordable manufacturing processes and material. The competition wants to increase its market share with a better product. The decision to arrive at a compromise for building a product the customer can afford, which gives him the desired quality and service and which is still competitive is very difficult to make. But despite this fact, quality standards must be set and all manufacturing operations must adhere to strict quality control procedures. There are many quality control stations to be installed along the manufacturing process.

Engineering has to be provided with equipment for short and long term tests. Many quality standards are set by the government to protect the customer from injury or death. Test results often have to be documented and kept on file for a long time. Figure 1.28 shows quality control functions.

1.4.12 Maintenance

Maintenance is responsible for keeping the company's manufacturing equipment in excellent operating condition. For each machine there is a maintenance schedule which has to be followed very thoroughly. Programmable and flexible manufacturing systems are usually extremely complex and must be supervised with computer-

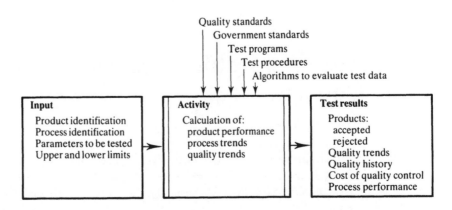

Figure 1.28
Quality control
functions.

controlled monitoring equipment to pinpoint the cause of a breakdown immediately, thereby reducing the search time for locating the fault. With complex equipment, the search time for an average breakdown can be one hour or even more. In recent years, expert systems have become an important tool for fault location. These programs try to draw conclusions from a given failure pattern, similar to a physician trying to find the cause of an illness from its symptoms. Also, preventative maintenance has become a standard procedure for complex production equipment. The computer schedules the maintenance, provides the maintenance program and keeps records of the equipment performance and problems.

1.4.13 Accounting

Accounting in Figure 1.16 is divided into three sections: credit accounting, controlling, and debit accounting. Accounting is responsible for assuring the financial soundness of a company. The operating budgets are made for a year for the entire company and its subfunctions. The operating units are divided into cost centers. Each cost center has its budget and all expenses incurred are constantly checked against the budget. The results are reported and summarized in a quarterly profit and loss statement.

In general, the incoming funds are derived from the sales activities. The outgoing funds are for material, equipment, services purchased, salaries, taxes, and depreciation. The profit is the balance between the incoming and outgoing funds. It usually is very difficult to check the performance of all functional units and cost centers of a company. Some of these units only have a service assignment and it appears that they only use up funds. Another problem in accounting is the distinction between fixed and variable costs. In particular, the variable costs are very difficult to assess.

A financial report must contain the performance of all products. If there is a loser then there must be a good reason why it should be maintained in the product line. Often, some products are only sold because they enable the company to offer a complete product line to the customer.

1.5 The Hierarchical Planning and Control Concept

A manufacturing system must be well planned and controlled to be competitive in its market environment. In the previous section, we have learned that planning is concerned with the future and control with the present of a company. Both planning and control must be well coordinated to ensure that the business goals and objectives are met. Basically, there are two approaches to planning and control.

On the one hand, a centralized approach is taken and on the other the functions are organized according to a hierarchical concept. In general, centralized planning and control only work for a small manufacturing entity. For a large organization, these functions must be divided into well-structured subunits forming self-sustaining planning and control entities which can be easily overseen and can operate independently of one another. In this section, the structure of the hierarchical control concept, and the features and ways of implementing such a control system will be discussed.

It is a common practice to divide the planning and control structure of a manufacturing system in three or four tiers. The tasks performed on each tier are as follows (see also Figure 1.16).

- *Strategic planning.* This is the highest level and is concerned with the long-range planning of the manufacturing system to secure its place in the marketing environment. It usually contains marketing research, long-range planning, capital equipment forecasting, and facility planning.

- *Operations planning.* It comprises the technological and organizational planning activities, customer order servicing, engineering design, manufacturing process planning, marketing, production order scheduling and control, and purchasing.

- *Manufacturing control.* Its functions are: inventory management, material movement, parts manufacturing, assembly, and quality control.

- *Machine control.* It is concerned with the direct control of the manufacturing machines.

We will learn later that the tasks of these control tiers are viewed differently by different manufacturing organizations. Each activity will be directed by personnel with a defined position in the management hierarchy. With a computer-controlled system, there is also a specific type of computer used at each level.

1.5.1 Types of hierarchical control structures

Hierarchical control systems are usually laid out in the fashion of a pyramid and have several tiers in which data of various abstraction levels and time values is being processed. On top of the pyramid, only a few decisions are made, and they are usually made in sequential order. A control decision at this level usually has a major impact on the actions and reactions of the entire system. When a decision is passed down to the bottom, it will be fanned out into a multitude of smaller control decisions which are made in parallel. The division and subdivision proceeds downward until the decision making process has reached the lowest level; there an individual decision only has a small effect on the entire pyramid. For the purpose of feedback control, status information from the lowest level is brought back to the top. On its way up, this information is concentrated in sequential steps at each tier until the top has the factual data needed for making a corrective action. Control and feedback information is available on any tier pertaining to this level and intermediate action and reaction can take place. Thus, control can be taken at any level.

There are certain generic functions a control tier must be able to perform:

- Once it has received an assignment from an upper tier, operate as an independent unit in conjunction with its subordinates.
- Independently decide which tasks should be further divided into subtasks and assign the execution thereof to subordinates.
- Schedule the cooperation of subordinates.
- Monitor the subtasks and schedule new, corrected control parameters, if necessary.
- Report any abnormalities to a supervisor.

When we look at a complex mechanism, like that of a manufacturing operation, we can identify various hierarchical control structures. They are:

(1) The *management hierarchy*, consisting of a president, plant managers, section directors, group leaders, and a foreman, see Figure 1.29. Usually, decision and feedback data is processed in a strict hierarchical order.

(2) The *organizational hierarchy* is used where a multiproduct company operates for every product a unique manufacturing facility. However, similar components for these products are fabricated in common, specialized plants. From the organizational point of view, every manufacturing facility has to ensure that its products are made as though it owns and operates all the manufacturing processes for its product. This hierarchical control requires interleaving with the management structures of the various plants involved in making the product.

(3) The *hierarchical structure of the control system*. This is the hardware and software architecture which is being used for controlling a manufacturing operation. In the ideal situation, there is a one-to-one correspondence between the management and control hierarchy, as shown in Figure 1.30. In most cases, however, this correspondence only partially exists. For example, one of the

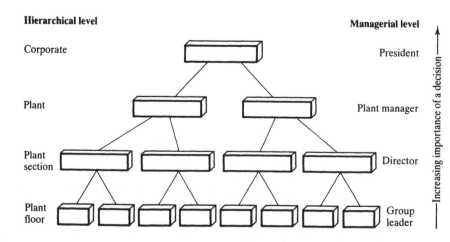

Figure 1.29
Management hierarchy of a company.

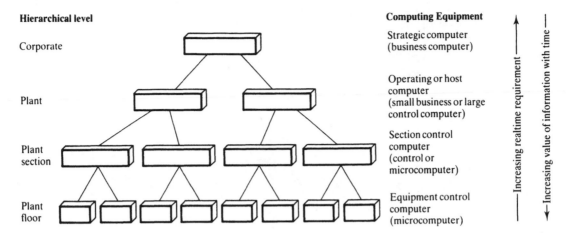

Figure 1.30
Hierarchical computer-based control system.

machine tools on the lowest tier may have a complex controller which uses several control hierarchies.

(4) The *control structure of the database*. A hierarchical data structure is used to store and retrieve data for every application. Often the parts data for a product is structured in such a way that on the top level of the hierarchy the product is identified, on the next level its component, further down the subcomponents and so on. A one-to-one correspondence to levels of other hierarchies will hardly exist. For example, several tiers of a control hierarchy may need access to the same record which has its own unique hierarchical data structure with no relation to any tier.

There are numerous organizational control hierarchies which can be identified in a manufacturing system, for example, in engineering, quality control, communication and so on. During the study of this book, we will get to know several of them.

1.5.2 Building a control hierarchy for a manufacturing system

One of the main problems with building a hierarchical control structure is the definition of the borders between the tiers and the interleaving of the various hierarchies to a functional unit. The structuring of a control hierarchy depends on the application, management philosophy, available computing equipment and software, degree of automation and so on. Figure 1.31 shows the location of the planning and control activities within the control hierarchy of a manufacturing process as a function of the product, order, resources and quality, also discussed in DIN (1987). Quality is separated from the hierarchy to indicate that it is a controlling function for the entire manufacturing system.

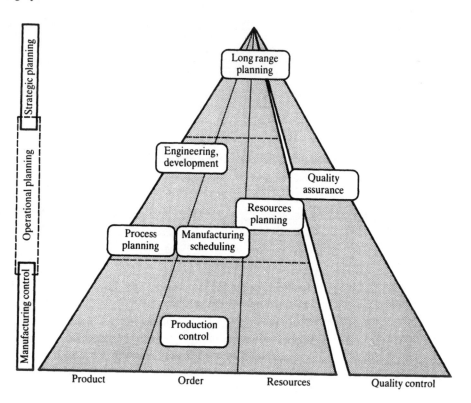

Figure 1.31
Location of the
planning and control
activities within the
control hierarchy of a
manufacturing system.

In Figure 1.32 an attempt is made to show planning and control as a function of the organizational, technical, marketing and quality assurance activities. From this figure it can be seen that there are areas, both on the vertical and horizontal scale, where overlapping occurs of the various activities.

The managerial and computer control structure for a small, hypothetical manufacturing system is shown in Figure 1.33. For the sake of clarity, the strategic planning activities are omitted. Of particular interest in this figure is the assignment of the computers to support these activities. At the corporate level, product planning, manufacturing planning, and administrative functions are located. Design is responsible for developing and testing the product and for generating the engineering documents. Manufacturing planning transfers these documents via a process plan into instructions for the factory. At this level, the concept of a virtual manufacturing system must exist to configure the processes to a functional entity. The orders are given priorities, and, if necessary, other orders are rescheduled to meet due dates. At the corporate level, the business computer is used as the main support to planning. It processes large volumes of data in a time scale of days and weeks. The business computer acts as a host for several plant computers. The communication is via a wide area network.

All steps necessary to manufacture the product are taken at the operational planning level. The required machine tools, NC programs, tools, and fixtures are reserved, and the activities of the material control system are scheduled. As soon

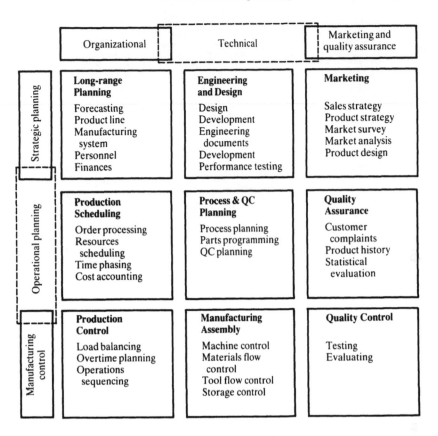

	Organizational	Technical	Marketing and quality assurance

Strategic planning	**Long-range Planning**
	
	Forecasting
	Product line
	Manufacturing
	  system
	Personnel
	Finances
		Engineering and Design
		
		Design
		Development
		Engineering
		  documents
		Development
		Performance testing
			Marketing
			
			Sales strategy
			Product strategy
			Market survey
			Market analysis
			Product design

Operational planning	**Production Scheduling**
	
	Order processing
	Resources
	  scheduling
	Time phasing
	Cost accounting
		Process & QC Planning
		
		Process planning
		Parts programming
		QC planning
			Quality Assurance
			
			Customer
			  complaints
			Product history
			Statistical
			  evaluation

Manufacturing control	**Production Control**
	
	Load balancing
	Overtime planning
	Operations
	  sequencing
		Manufacturing Assembly
		
		Machine control
		Materials flow
		  control
		Tool flow control
		Storage control
			Quality Control
			
			Testing
			Evaluating

Figure 1.32
Overlap of planning and control within the organizational, technical and marketing activities.

as all resources are available, the order can be released. There are mainly small business computers and large control computers used on the plant level. The feedback time for control data at this level ranges from one hour to one day, depending on the factory application. The plant computers in turn act as a host for the shop floor computers. It communicates with its lower level via a local area network.

The manufacturing order is split into several sequential and parallel activities. Each machine is provided with the required programs, materials, and tools. The parallel and sequential flow through the plant floor is controlled and monitored. Most of the data processing on the shop floor level must be done in real time, requiring control or microcomputers with interrupt capabilities. A field bus serves as a communication link between the shop floor and the machine control computers.

At the machine control and monitoring level, the manufacturing equipment is provided with parameters to control the processing of material. The machines are usually controlled by microcomputers which must have a very fast reaction time to respond to any parameter changes or disturbances. Here, the realtime requirements are the most stringent ones in the entire control system.

On the right side of Figure 1.33, the various databases are shown which contain the planning and control algorithms and all manufacturing data.

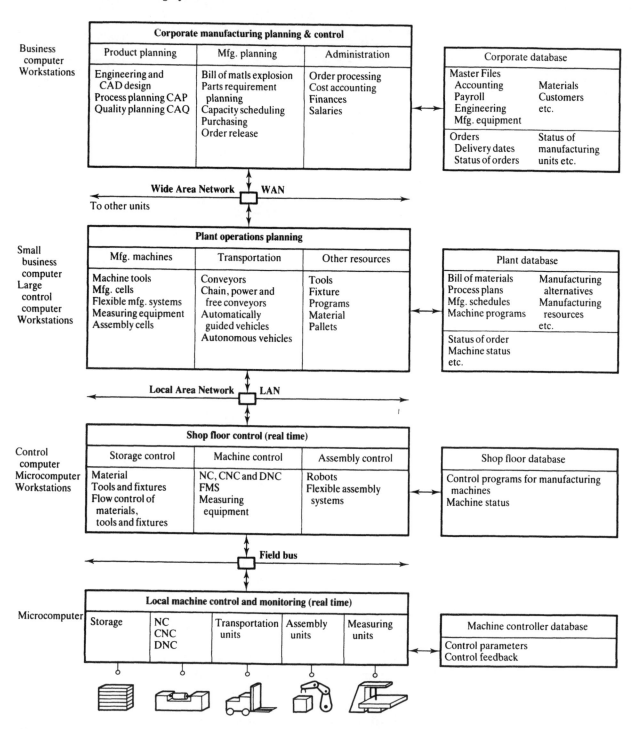

Figure 1.33

Typical assignment of activities to the control tiers of a manufacturing organization.

1.5.3 Data exchange and information flow between control tiers

The control of a manufacturing system requires the processing and exchange of a great amount of data. Usually, the flow of information between the components of a system is of such a magnitude that realtime control is impossible. Such a system can only be controlled effectively if its activities operate independent of one another, and if the data exchange between the activities is kept to a minimum. An attempt should be made to only loosely couple the activities of the manufacturing system with one another. However, this is often contrary to the operational mode of a manufacturing plant where machine tools, material handling systems, robots, and measuring machines have to be synchronized to obtain a linear flow operation. The activities on the shop floor are the most critical ones in regard to synchronization; here material and data are processed simultaneously and the realtime requirements are the highest. When a hierarchy is structured, firstly a level of responsibility is assigned to each tier. Second, the task of each activity is specified. Third, the cooperation with external activities is defined. Fourth, a centralized control strategy is conceived.

Requirements for an activity

Each activity must be given a complete set of information and if necessary the means to perform its function. The function of an activity may start many others in parallel or sequence. Thus, a proper synchronization must be provided. Typical requirements to be met by an activity are as follows:

- Its material and data processing task must be defined.
- The resources for performing a task must reside inside the activity.
- The flow of material and data into and out of the activity must be defined.
- It must be known where material and data comes from and where it goes to.
- The material and data to be used and their availability must be known.
- An activity must be reserved before it can be used.
- The minimum and maximum duration of an activity must be known for control purposes. Thus a time stamp must be provided at the start and at the end of an activity.
- When an activity receives an order it must be verified and when it issues an order it must be acknowledged.
- A buffer for material and data must be available if this is necessary for synchronization.
- An activity must know by which other activity it may be used, or which other activities it is permitted to use.
- A mechanism must be provided for each activity to monitor its utilization and functions.
- A default mechanism must be available if problems arise.

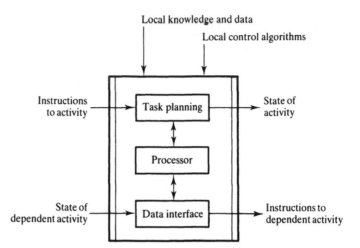

Figure 1.34

Control structure of
an independent
manufacturing activity.

Internal control structure of an activity

Figure 1.34 shows the control structure of an activity, see also Lutz (1987). Its main components are the task planner, processor and data interface.

The task planner receives from an upper level the instructions for the execution of an activity. It examines the instruction and investigates to see if it belongs in the domain of its task. Then, a plan is made for the execution. Resources are checked and reserved, if necessary, processing alternatives are investigated. If the task is executable, it will be given to the processor for further actions. The processor starts the activity and makes the connections with dependent activities. It controls the sequential or parallel execution of dependent activities and supervises the flow of data through the activity. In case of problems, the task planner may look for alternative actions. The data interface ensures that incoming and outgoing data have a common format, protocol and sequence.

The order flow through a manufacturing system was explained in Section 1.1.3 with the help of plant producing valves. It was also shown that processing feedback information is an essential activity of the control system. Figure 1.35 shows how the various planning and control levels can be interconnected with feedback loops. The figure shows an ideal feedback control system. The reader can imagine how important it is to be able to process the feedback information and that every control tier can act as an independent control loop. In addition it is important that protocols, data formats and interfaces are identical for the design of a plant to be able to construct such a control system.

1.6 Future Developments

In the present competitive marketing environment, it is very difficult to predict the future behavior of the market. We can only look into the past and from there predict the future. The most successful companies are those which had considerable

Orders

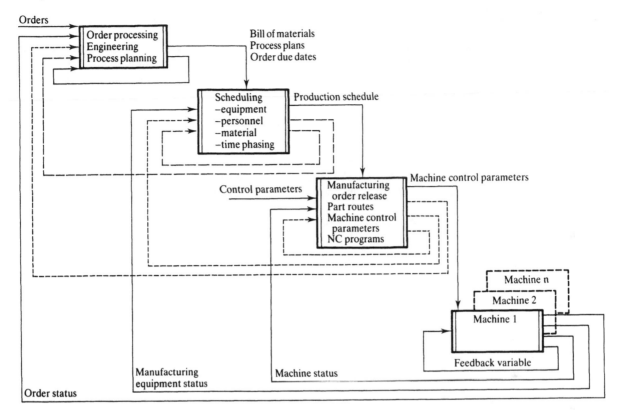

Figure 1.35
Interaction of control
loops in a
hierarchically
organized
manufacturing
operation.

investments in the past to prepare for the future. Very often, more than 50% of the sales of these companies come from products which did not exist 10 years ago. Also, the life cycle of the product becomes shorter and shorter and new tools must be found to speed up the time to develop a new product and to introduce it to the market. We already have enough experience to know that the computer will be one of the principal tools for a factory to be more responsive to the demands of the future market. For this reason, we must find new ways for more efficiently utilizing information and for integrating the computer into the design and manufacturing process. Furthermore, the computer and information technology will require skills which the average engineer does not have. Educational institutions need to prepare both the product designer and manufacturing engineer for their future jobs. We must also become more aware of the fact that a factory cannot be operated without man. As a matter of fact, all modern CIM strategies assume that man remains in key positions in a factory and that the CIM technology is built up around him. Thus, the skills of man and the capability of the automation devices are both needed. We will now look at several development trends in CIM technology.

1.6.1 Which CIM strategy should be followed?

The pursuance of a CIM strategy can be incidental or planned. For example, computer applications may have been developed independently in a factory and later an attempt is made to combine them. This is usually done when the computer is retrofitted to an older plant. Or, a CIM strategy may be followed according to a well-defined plan. The latter approach is the better one and should be pursued when a new plant is designed with the computer in mind. A technology as complex as CIM must be learned and the best way to get acquainted with it is to start with a defined application and to add new applications as time goes on.

The design of a CIM system may be started from the top. This can be done when a new plant is designed. The advantages are as follows:

(1) The information flow through the control modules from the top to the bottom will be defined in a concise and logical manner and the task of the modules can be outlined in a true hierarchical fashion.

(2) All hardware and software modules can be provided with compatible data formats, communication protocols, and interfaces.

(3) Standardization of hardware and software modules is simplified.

(4) The modules can be easily interfaced in a hierarchical fashion and tested with a simulator at the interface to the lower tier.

The disadvantages of this approach are:

(1) Often, modules are purchased from an outside vendor or developed in another plant and they are not compatible with the system being built.

(2) When the system needs to be changed at a future date, a completely different control strategy may be followed, which, when implemented, requires a major revision of the existing architecture.

The bottom-up approach has the following advantages:

(1) Many control modules can be started in parallel and operated individually.

(2) Modules from outside vendors are often easier to implement.

(3) The implementation has fewer overheads than usually associated with most standards.

The disadvantages are:

(1) The interfacing of applications on a control tier or between control tiers may be extremely difficult if no attempt was made to define interfaces, data formats, and communication protocols.

(2) Computing and communication equipment used may not be compatible.

(3) Programming of the computer system, in particular where hierarchical levels are linked together, may become very difficult because no consideration was given to the use of common languages and programming tools.

Usually, in practical applications, neither strategy is followed. The systems grow together from the top and the bottom. The whole matter gets further complicated when a CIM system is linked with either a pure CAM, PP&C, CAD/CAM, or CIM strategy, see also Nuber *et al.* (1989).

- CAM strategy. The computer is introduced by the manufacturing engineer. He tries to automate manufacturing processes, warehouse operations, material movement systems, and quality control standards. The installation of CNC and FMS systems without interface to design is another example.

- PP&C strategy. This is an organizational approach. The main goals are to obtain improved schedules, to shorten set up and lead times, to improve lot-sizing and time phasing, and to reduce the in-shop time of the workpieces.

- CAD/CAM strategy. With this approach, a so-called vertical integration is taken to merge CAD, CAM, and CAQ. This is to reduce the duplication of processing and storing data. Usually, the integration process is via design where all pertinent product data is used for all succeeding manufacturing planning and control operations.

- CIM strategy. This strategy usually builds on one of the strategies mentioned above. It is a consequent integration of an existing computer-controlled operation. There are two possible strategies. First, one starts with a centralized database and an existing application and integrates new applications in a well-defined fashion. Second, one starts with a computer network and tries to coordinate the access to all pertinent data of all manufacturing modules, so the data can be used for common control purposes.

1.6.2 Future outlook

One of the greatest challenges of CIM will be the integration of CAD, CAP, PP&C, and CAM. An open system architecture for computer integration will have to be provided to make common CIM modules, computing equipment and peripherals available to the manufacturing community. There are several ongoing activities to define product, manufacturing and operational models which are the backbone of any standardization effort. These efforts try to tackle a broad spectrum of techniques and components and address a diversity of users. The results are often difficult to understand for an average manufacturing engineer because the models are usually very abstract and use an unfamiliar terminology. The limitation of traditional types of communication causes a dilemma for both users and system suppliers of CIM applications. The spoken and written natural languages are not very useful for concise technical specifications, and ambiguity often leads to ill-defined systems and interfaces. The so-called formal languages used in computer science are difficult to use except by a computer scientist. Graphic methods are also not easy to use; they usually have an excellent presentation on screen, but they suffer severe ambiguity when objects have to be described which are outside their defined format. Efforts are underway to design specification aids for CIM objects which combine the best features of the natural, formal, and graphical representation methods.

One problem with the design of CIM modules is the software. There are many programming and system design languages being used or that are still under development. As yet we do not know which direction they will take. The explicit languages are often too cumbersome to apply. Even simple applications can be difficult to describe. The implicit or task-oriented languages may give the user the solution to answer the problem. The use of expert knowledge, its presentation, and the modeling of the world for which the language is being designed is an extremely difficult task for the system designer. Often such programming systems need expensive computing power, which even larger corporations cannot afford.

Another tool to be mentioned in connection with software aids is simulation. Computer simulation had been a very popular research subject, and many useful products have been created. However, it is still difficult to animate the flow of material and the manufacturing processes as an entity on the screen. Simulation systems which will allow the emulation of a factory from a global view down to the microscopic view should be available to arrange new manufacturing systems, to plan manufacturing runs or to experiment with manufacturing alternatives. Certainly, there are many modular simulation products available. However, they are usually for special purposes and they only represent a limited view of a manufacturing operation. A tool box for simulation aids must include models of all manufacturing processes and configuration means to set up the desired manufacturing environment.

Another promising development that appears on the horizon is artificial intelligence (AI). The tools so far available for manufacturing really do not deserve to be called artificially intelligent. Expert systems are the most useful tools from AI, which may be applied to planning and control of manufacturing systems. However, they use nothing else but knowledge which was extracted from experts. There is no method which, by itself, gives a solution for a given problem. The possibilities of AI in manufacturing have been overestimated, and very few good applications exist. However, the potential of AI is very great and it is necessary to develop methods which can be used for solving manufacturing problems, particularly for operations planning. Most software for operations planning has led to the development of large and complex application packages which are too rigid when manufacturing variables are included. It is believed that with AI methods, it will be possible to take many routine jobs out of the responsibility of the manager so that he can devote his creative ability to solving complex problems. Some consider it possible to merge AI solutions with conventional ones, although this is still not well understood. Occasionally AI techniques will be used in their pure form. They have to be integrated into other technical systems and applications. AI will emerge as a useful tool when more efficient methods for presenting and processing of knowledge have been found and when low cost computer systems are available on which expert systems can be run efficiently.

Undoubtedly, CIM will have quite an impact on the management structure of an organization. Management has to do a lot of rethinking when it tries to apply CIM. When a factory is automated the planner must have an integrated view of

CIM in mind. CIM is not something that can be bought off the shelf, it has to be engineered and configured to an application and the organization has to grow with this new technology. The factory of tomorrow will be using new operational methods and integration techniques; this will require new organizational structures and management philosophies. It will necessitate much experimentation to find the correct control structures for a CIM system. Like any other technology, CIM has to be economically justified. This is a particularly challenging problem for management sciences, since it is usually impossible to tag a value on information.

The implementation of a CIM system is a combined effort by people with many skills. There is knowledge needed from marketing, engineering, manufacturing, computer sciences, industrial engineering and so on. It is obvious that we cannot expect to find all these skills in one person. However, when we talk about integration, people are needed who understand the complex functions of a manufacturing system and who are capable of tying them together. A well-trained manufacturing engineer should know the principles of engineering, manufacturing planning and scheduling, computer technology and programming, factory communication, basic manufacturing technologies, and financial and economic issues. He should have special skills in working with experts from various fields and organizational levels to coordinate their work and to tie it into an integrated functional system.

To summarize the main thoughts of this discussion: we think that the following developments must take place to build well conceived CIM systems.

- The development of universal models of the product, the manufacturing system and the manufacturing operation.
- The development of modeling and animation systems for manufacturing processes.
- The development of standards for communication within a manufacturing system.
- The development of more power computers.
- The investigation of new management strategies for CIM systems.
- The development of new curricula for the manufacturing engineers.
- The exploitation of AI methods for manufacturing.

Only those industrialized countries where CIM can be integrated successfully in existing and new production facilities will be able to maintain an influential position on the international market. They will offer products of high complexity which can be produced in several variants. The production of many conventional mass products, like standard electric motors, will be taken over by emerging industrialized countries which are in the possession of good production technology and for which CIM is too expensive. Because of this, the interdependency between highly developed industrialized countries and less developed ones will increase. It is certain that CIM will become an integrated component of an efficient production facility.

1.7 Problems

(1) How would you define CIM and what is the task of the computer in a CIM concept? How does a CIM concept differ from conventional manufacturing concepts?

(2) Why is information (which is not tangible) considered as one of the most important resources of a manufacturing operation?

(3) What is the basic difference between the CIM philosophies in the USA, Europe and Japan?

(4) Explain the principal functions of the supervisory, control and verification activities of the model factory and the information flow between them.

(5) Explain the flow of information discussed in this chapter between the order-specific activities and the product-specific activities.

(6) If you are the manufacturer of garments (textiles) and you want to computerize your factory, explain the principal functions of the enterprise and show how a computer can be used for planning and control.

(7) Describe the different types of manufacturing principles and where they are used.

(8) Why are flexible manufacturing systems becoming more and more economically attractive? What are their advantages and disadvantages?

(9) In which activities of a manufacturing system can substantial cost reductions be obtained and what are the reasons for that?

(10) Explain the inputs, outputs and data processing tasks of at least four enterprise activities and discuss their role in the manufacturing system.

(11) Show the layout of a flexible manufacturing system for producing special purpose refrigeration units. What kind of machine tools would you use, how would you interconnect them and how would you design the material flow system? How does this plant differ from a conventional mass-production plant for refrigerators?

(12) Why is a well-designed order inquiring system needed for a good manufacturer–customer relationship? Explain the details of such a system for an electronic components manufacturer.

(13) How would you design an order scheduling and control system for a typical automotive parts plant?

(14) Describe the requirements for and the internal control structures of an enterprise activity.

(15) Draw a control structure of a manufacturing system with its levels and columns and locate the various planning and control functions within it (Figure 1.31). Interconnect two adjacent activities and show the dataflow between them. How are these activities grouped together from a global view and in which time frame are these activities done?

(16) What are the different ways of designing a CIM system and according to which strategies can it be introduced into an enterprise?

(17) Design a hierarchical planning and control system, show how the various levels cooperate, and explain what feedback information is needed on each level.

2

CIM models and concepts

CHAPTER CONTENTS

2.1 Introduction   47
2.2 Existing CIM models   48
2.3 The Amherst–Karlsruhe CIM Model   63
2.4 Problems   102

2.1 Introduction

Manufacturing systems of the last decade of this century and the first decades of the next century will be the testbeds of the computer-integrated manufacturing (CIM) concepts of the future. The efficient processing and distribution of information will play a key role in obtaining truly programmable factories for specific product families. One of the major challenges in building these factories is the conception and design of a generic manufacturing model and that of providing a set of standardized hardware and software tools from which a functional factory can be configured. Once a product and its manufacturing methods are defined, the system engineer will be able to present his ideas to the generic model and a specific model can then be built by selecting from a library of tools those needed for making and shaping his product. Of particular interest is the configuration of the data processing and distribution activities. Computers of the programmable factory used for automated planning, scheduling, and control will execute complex algorithms particularly where optimization of conventional computing technology has reached its limits. The computer system of the factory of the future will have a distributed architecture using von Neumann, parallel processor and vector array concepts which can be interconnected by standard buses, protocols and interfaces. Many planning and control algorithms will be cast in firmware to render ultrafast execution times. This chapter discusses the problems of providing the proper information technology needed for the future factories, and several important developments will be shown.

2.2 Existing CIM Models

CIM starts with the conception of a new design, where the functions and features of the product are conceived. The design of the product greatly determines the manufacturing methods and processes. In a CIM concept the computer is considered to be an integrating factor for controlling the interaction of man and machine to create the product from design to manufacture. CIM not only provides information but also controls the distribution of information to assure an unhindered flow of raw material, parts and products in an optimal manner.

The building of a CIM system is a very tedious and lengthy process; it requires expertise and extensive manufacturing experience to tie computers into the various activities of a production system and to interconnect them to a functional entity. It would be of great advantage to the designer and manufacturing engineers if they had a generic CIM model available, providing them with the basic manufacturing system components and configuration tools, to be able to conceive a particular manufacturing system for any product they want to build. Concepts of generic manufacturing models evolved from efforts of various industries and research groups to build models for their particular applications. Since many components in these models have similar functions, it became apparent that a universal generic model would be a great advantage. However, to build a generic model which could be used for an appliance factory, an aircraft factory, a garment factory and so on, is very difficult and may result in a high degree of abstraction. For this reason, it is very important that the generic model is well structured and very comprehensive to provide the user with a tool which allows him to follow up on his own manufacturing strategies.

The activities represented by a CIM model should be triggered by a customer order for a product, whereby various stages of organizational and technical functions are performed in a sequential or parallel manner. The scope of the model must comprise all functions which are needed to reach the goal of an enterprise. For some applications, it may suffice that the model only contains functions internal to the factory whereas in other applications the supplier environment may also have to be included. A very detailed model may even comprise the market, legal requirements, resources supply, sources of finances and so on.

When building a model, many of the aspects discussed next may have to be considered. It will be impossible to include all of them in one model and to obtain a meaningful presentation. Thus there will always be shortcomings in any model. A model may contain:

(1) Presentation of the business functions
     A typical CIM model and its subfunctions are shown in Figure 1.5. The emphasis in this model is on the computer application of the various activities

of a factory. The model shows that CAQ and PP&C are overlapping activities, interfacing with CAD, CAP and CAM.

(2) Integration of the management information database
In this model the master CAD/CAM database performs an integrating function in the CIM enterprise, see Section 2.3.8, CIM Database.

(3) Presentation of the material and product flow
A typical material and product flow through a factory is shown in Figure 1.1.

(4) Presentation of the information flow
With this concept, the flow of information is shown into and out of various manufacturing activities as well as between the activities, as in Figures 1.2 and 1.16. The information flow may be order-related or product-related. Figure 1.14 shows an order-related flow. In Figure 1.4 there is an order-related flow from left to right in a horizontal direction and a product-related flow from top to bottom in a vertical direction.

(5) Description of the interfaces and communication protocols
In this approach the standardization of interfaces and protocols is stressed and the possibility of using modular system components becomes apparent.

(6) The presentation of the hierarchical planning and control functions
Since manufacturing information is processed in a hierarchical manner, it is of advantage if the model shows how the information for the planning and control functions is hierarchically represented and how it is refined from an abstract to a detailed level.

(7) The inclusion of time
In manufacturing the adherence to fixed delivery schedules is of the utmost importance. For this reason, a model showing elapsed and outstanding manufacturing times would be a great advantage.

2.2.1 The IBM concept of CIM

IBM became involved with the manufacturing industries when the first computers were introduced into the factory situation. IBM quickly realized that the manufacturing industry constitutes an enormous market potential for hardware and software. The company started to concentrate on the development of various data processing modules for specific inhouse and customer applications. An attempt was made to conceive products which could be used by various industries with little or no change. Often, this was a very difficult task because it was necessary to change the product and/or the methods and procedures of the plant to be automated.

This effort led in the early 1970s to the development of a general CIM philosophy called Communication-Oriented Production Information and Control System

(COPICS) which comprised all major planning and control activities of manufacturing, as follows and see IBM (1973):

- engineering and production data control
- customer order servicing
- forecasting
- master production schedule planning
- inventory management
- manufacturing activity planning
- order release
- plant monitoring and control
- plant maintenance
- purchasing and receiving
- stores control
- cost planning and control

Figure 2.1 shows the functional flow of information in a COPICS environment. The particular information processing support IBM gives is in administration, decision making and various applications. A strong emphasis is placed on communication, database management and presentation, see also Figure 2.2 and IBM (1989).

An attempt was made to tie together individual data processing solutions into a general concept, and IBM must be credited for pioneering the early CIM activities. Many of the concepts presently being pursued were heavily influenced by this endeavor. Originally, IBM tried to promote, in the framework of COPICS, their own hardware and software solutions. No attempt was made to include other products or to provide interfaces for them. However, with the emergence of numerous rival products, the customer insisted on making them compatible with IBM solutions. Thus, the company broadened their scope and took steps to offer general interfaces to integrate other products into their concept. For example, the work was supported by using communication systems based on LAN and MAP.

Despite early engagement in factory data processing, no general CIM model was developed by IBM. However, the company continuously widened their computer work in CIM and presently offers products for almost any data processing activity in a manufacturing company.

2.2.2 The NIST–AMRF hierarchical model

This model was proposed by the National Institute of Standards and Technology (NIST); the organization was formerly known as the National Bureau of Standards (NBS). At NIST an advanced manufacturing research facility (AMRF) was built and operated to work out hardware and software standards for computer-controlled manufacturing systems, see Albus *et al.* (1981). Experimental results from this facility will also be used to verify the model. The model supports a

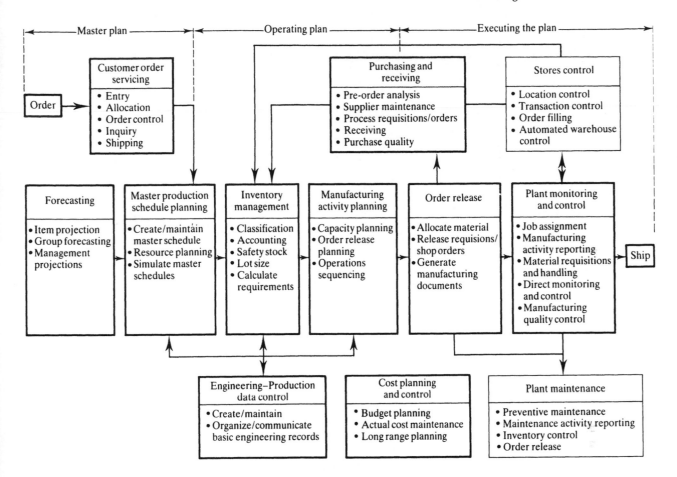

<-- Master plan --> <-- Operating plan --> <-- Executing the plan -->

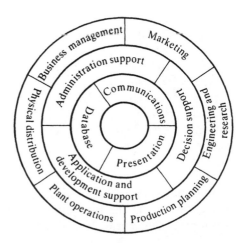

Figure 2.2
CIM architecture elements defined by IBM.

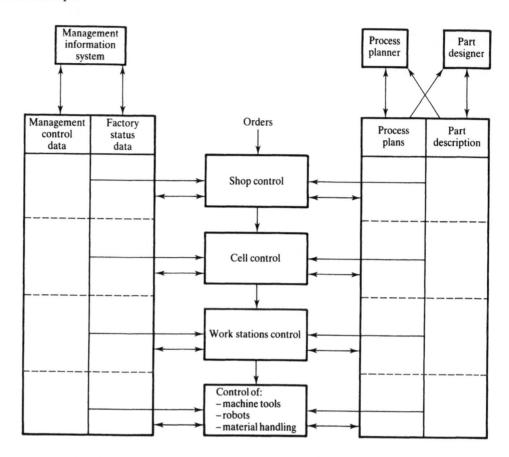

Figure 2.3
The NIST
(NBS)-AMRF (CIM
architecture).

dynamically controlled manufacturing environment, and is a modular design. A hierarchical computer and sensor system is assumed to plan and control the manufacturing operations. A significant amount of this is done in real time; particularly on the lower level. In order to provide transparency to data processing, a communication network interconnects all computing facilities. The individual activities can send and write information from or to virtual memory locations with the help of so called 'mailboxes'; this considerably simplifies the design and operation of the control system. Like the other models described, the NIST model greatly facilitates the configuration of a control system for manufacturing. The model is of particular value to small and medium-sized batch operations.

The model consists of three columns, a management information system, a control system and a design and planning system, as in Figure 2.3. A comprehensive management information system coordinates and controls the entire manufacturing plant. The design and process planning system prepares the manufacturing documents. Information about an order is entered into the highest level of the hierarchy of the system, and global goals and long-range strategies are decided. The order initiates the manufacturing, design and process planning activities. At each lower level the processing of the order is refined successively, until at the bottom of the hierarchy a set of primitive control instructions is generated to directly operate the manufacturing equipment. Each level of control obtains only that information which is pertinent for the fulfillment of its assignment. For example, a process plan has structured information for the cell level, workstation level and machine level. The commands for a manufacturing assignment are passed in sequence through the control levels; they may invoke parallel and sequential operations in lower control levels. Every piece of information needed for making the product is generated by the system, that is, material selection, determination of machining operations and sequences, calculation of machining parameters, part scheduling, part routing, and so on. The management information system knows the order priorities, manufacturing equipment status, material availability, online processing status, quality produced, and so on. The sensors on the factory level collect status information about the manufacturing process and send this information upwards to provide feedback to all levels. The entire system is of modular design including the hardware and software. Components can be added and deleted with a minimum of effort and disturbance to the factory. The central database contains a complete state description of the factory at any time, and thus the system can respond immediately to a production change or disturbance.

The key points in this model are its hierarchical planning and control structure, its transparency, and the partition of the management, planning and control functions.

2.2.3 The Siemens concept of CIM

The computer activities of Siemens in manufacturing are guided by a rather strong and well-structured CIM philosophy. The company both sells data processing products and uses them in their numerous manufacturing activities. To Siemens, CIM is not a ready made product, but a strategy and a concept to reach the marketing goals of an enterprise.

The Siemens model of CIM is shown in Figure 2.4 and discussed in Baumgarter (1989). CIM comprises the main functions of planning, sales, purchasing, PP&C, CAD, CAQ and CAM. These functions are interconnected by an intensive information flow. The task of the information flow is the processing and distribution of data in a concise and timely manner. For the communication of the data, interfaces and protocols have to be provided. To specify the requirements for a factory-wide data processing system, the following questions have to be answered:

- what kind of data is generated?
- what kind of data is needed and where is it used?
- who administers and maintains the data?
- who is responsible for what data?
- what data is kept in a common database?
- where must data be obtained from and where must data be sent to?

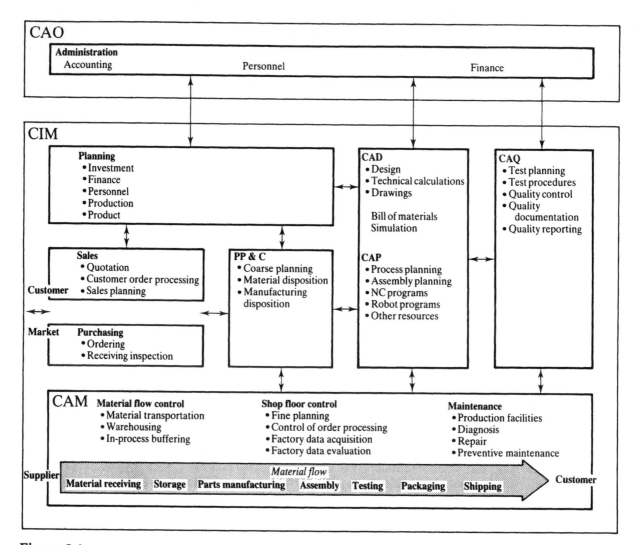

Figure 2.4
CIM concept for
Siemens AG.

A hierarchical model of an enterprise is used to process the data in a comprehensive manufacturing system, similar to the one introduced in Chapter 1. Each hierarchical level has its own data processing requirements and there exists a steady flow of instructions from the upper levels to the lower ones. In order to control and synchronize parallel activities on each level, an intensive horizontal dataflow takes place; this data exchange is fastest on the plant floor level.

A particular feature of the Siemens model is that it incorporates a computer-aided organization (CAO) activity which comprises accounting, personnel and finance. Thus, a strong tie exists between CAO and CIM. The Siemens CIM concept goes into well-structured details of every major module shown in Figure 2.4. For each module its submodules are defined and their interconnections are explained. A description is given on the dataflow, the required interfaces for the data exchange and the contents of the data. Special considerations are given to batch type and mass production; also various layouts of production systems and assembly stations are considered. The Siemens model is well suited for introducing a manufacturer into the general aspects of CIM as well as into its finer intricacies.

2.2.4 The Digital Equipment Corporation CIM concept

Digital Equipment Corporation is also a producer and user of CIM hardware and software. The term CIM means to Digital the improvement of a manufacturing process with the aid of the computer and the integration of the information processing of all enterprise activities, see also Flatau (1988).

A comprehensive CIM model of Digital is shown in Figure 2.5. It actually

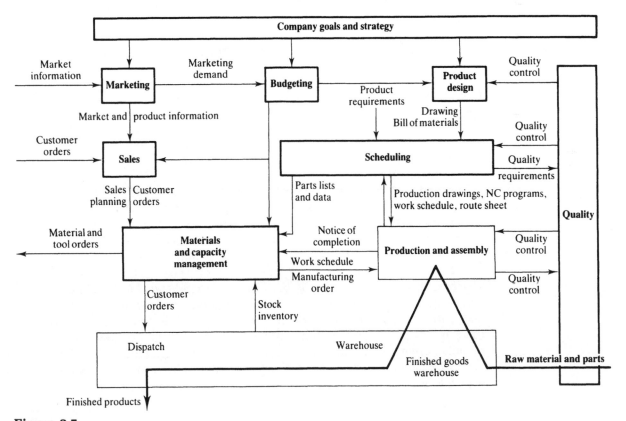

Figure 2.5
The Digital Equipment Corporation CIM model.

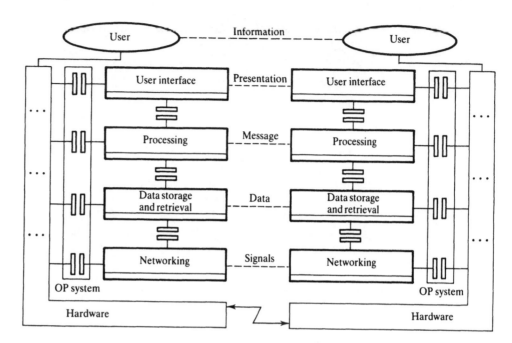

Figure 2.6
Two information
technology models
communicating with
each other.

resembles the Siemens model and has many similar features. The individual activities of this model are supported by a well-structured information technology model. Figure 2.6 shows two information technology models, representing an activity one and activity two; they are interconnected by a communication system. The information technology model is an extension of the OSI (Open Systems Interconnection) model.

The model assumes that CIM is a business-oriented approach to the automation of a computer hierarchical control system and must be tailored to the specific requirements of a factory. The goals and objectives of this model are to define a computer-integrated manufacturing system as consistently as possible and to provide reference documents for planning and implementing distributed, heterogeneous and open CIM systems. With this approach, the entire control system is divided into functional modules which reflect the business and its data. The designer of a CIM system starts with the analysis of all manufacturing activities and identifies their functions and data flow; with this information, the layout of the physical system can be done. A functional model and a physical model of manufacturing system and its subsystems are the result. In addition, the information support technologies are described. Of particular interest are the following, also depicted in Figure 2.6.

The user interface to data sinks and data sources

The user interface connects the user with the processing module. This is possible through device-independent data structures, input and output operations as well as communication functions.

The data processing service

The data processing service provides the tools for the execution of management and application processes. Parallel, heterogeneous processes can automatically exchange data and messages.

The data storage and retrieval service

The data storage and retrieval module is device independent and renders all service functions for data and file management of an open distributed and heterogeneous system. In addition, services of data consistency, security and accuracy are provided.

The networking service for distributed systems

The networking module is the physical and logical connector to an open system. Protocols establish the required conventions between system components, determine the standard communication path and identify standard data elements.

The information technology model strongly supports the possibility of inter-connecting heterogeneous systems. Thus, interfaces and handlers can be proprietary to obtain the required integration and presentation of data messages, whereas the protocols must be of standard design.

The system integration is a strong effort within the framework of this CIM model. The scope of integration comprises the business functions, data structure, hardware/software architecture and information systems. The concept for these functions is as follows:

Business function integration

For this activity, the business functions are defined and all computer programs are assembled to support the functions. The integration is mainly achieved by combining the internal process functions and defining the data exchange.

Data structure integration

Successful system integration is achieved by organizing all data in a defined physical database in which the logical associations are well structured. An attempt is made to set up distinct data structures for the various applications and to define algorithms which establish links between the data structures.

Infrastructure integration

The system architecture is conceived as a distributed environment using hetero-geneous hardware and software components. Features included are transportable software and application packages, device-independent storage and retrieval, unhindered data exchange, unique query languages, unified operating systems, common networks and protocols, high modularity and possibilities to adapt quickly to technology changes.

Information system integration

System integration is considered to be an ongoing incremental process which adopts to the ever present changes of business functions, data structures and infrastructures.

The key points of this model are that it supports all enterprise functions; it gives a good system overview; it modularizes the subfunctions to reusable packages and it defines, very clearly, interfaces for easy system configuration. A vague attempt is made to include the material and product flow.

2.2.5 ESPRIT–CIM–OSA model

ESPRIT is an industrially oriented R&D program with the aim of boosting the level of industrial competitiveness of the European Community. For this purpose, an open system architecture (OSA) is being provided which allows the configuration of manufacturing installations from generic modules.

Within the framework of the CIM–OSA concept, it is possible to construct CIM architectures for various manufacturing industries and applications from basic building blocks according to defined guidelines. The integrating infrastructure will help companies to dynamically organize and schedule their enterprise activities, see also CIM–OSA (1989a and 1989b).

An important part of this standardization effort is the CIM–OSA reference architecture which contains a system building block library of generic functions.

The CIM–OSA concept consists of an enterprise engineering environment and an enterprise operation environment, as shown in Figure 2.7. The engineering environment contains the reference architecture and resources used to structure

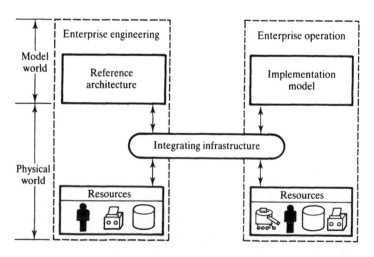

Figure 2.7
CIM–OSA enterprise environments.

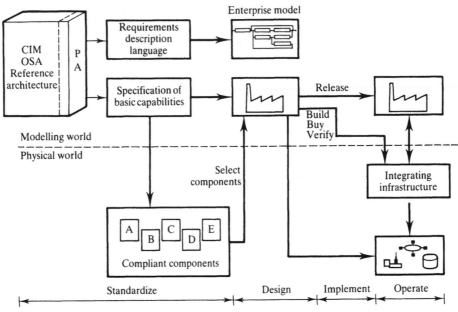

Enterprise model

Modelling world
- - - - - - - - - - - - - - - - - - - -
Physical world

Select
components

Compliant components

Figure 2.8
CIM–OSA
environment.

Standardize   Design   Implement   Operate

PA = Particular architecture

and set up an enterprise model. With the help of the engineering environment the designer is able to consult the CIM–OSA reference architecture and to develop the particular enterprise model. The user can try out various manufacturing strategies and can optimize these with the help of modeling and simulation tools.

A support environment and integrated enterprise operation environment are provided for the enterprise operation and execution of the tasks defined in the operational version of the particular architecture. A very carefully controlled mechanism is defined for passing revisions from the design to the operational environment, so easing the management of change. Both environments use the services of the integrating infrastructure.

Figure 2.8 shows how the user can design a manufacturing operation with the help of the CIM–OSA environment. With the available CIM–OSA reference architecture, the user can describe the basic capabilities which are needed to build the desired manufacturing system.

In order to specify all aspects of an enterprise, the CIM–OSA approach defines four different views, they are as follows:

(1) Function view
     The function view provides a 'language' to describe the static and dynamic functional properties of the enterprise.

     The basic construct of the function view is the enterprise function. It defines a uniform approach to describe enterprise functionality and/or behavior.

     The description consists of three parts. Functionality, behavior and structure are described separately.

(a) Functionality is concerned with the objectives, constraints, function inputs and outputs, control inputs and outputs, resource inputs and outputs.

(b) Behavior is concerned with the flow of control or the way processes and activities are employed in reaction to events to fulfill the business objectives under imposed constraints (financial, administrative or technical contraints). This can be described by means of procedural rules defining control structures such as precedence relationships, parallelism and branching conditions.

(c) Structure is concerned with recording the structure of the functional decomposition of processes into sub-processes and activities.

The functions of the enterprise are defined from an abstract level down to a detailed level.

Domain processes, business processes and enterprise activities are sub-types of the enterprise function and defined as follows, see also Figure 2.9.

(d) Domain processes are at the root of a function decomposition tree. They have a functional part but no control input/output and no resource input/output, a behavior part and a structural part. They belong to a unique domain and must be triggered by at least one event.

(e) Business processes are at intermediate levels of a function decomposition tree. They have a functional part but no control input/output and no resource input/output, a behaviour part and a structural part. They are not directly triggered by events, but their associated set of procedural rules is activated by the calling function.

(f) Enterprise activities are leaves of function decomposition trees. They have a fully defined functional part, a structural part but no behavior part. They cannot be triggered directly by events.

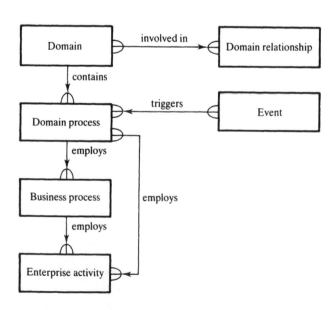

Figure 2.9

Enterprise model building blocks.

(2) Information view
Describes the information needed for each function from the user's point of view.

(3) Resource view
Defines all resources and their relationships to the function, information and organizational view.

(4) Organizational view
It describes the organizational structure of the enterprise by defining the task of individuals responsible for enterprise assets (resources and information) and for its operational entities (business processes, products and so on).
The flow of the information and product through the manufacturing facility can be simulated to determine the practicality of the particular architecture.

Figure 2.10 shows a schema of the CIM–OSA concept. In the upper part of the figure the functional aspects of an enterprise model are shown. In the lower part of the figure the implementation structure is shown. This structure consists of an interface H to the human, two computer systems A and B and a machine controller M.

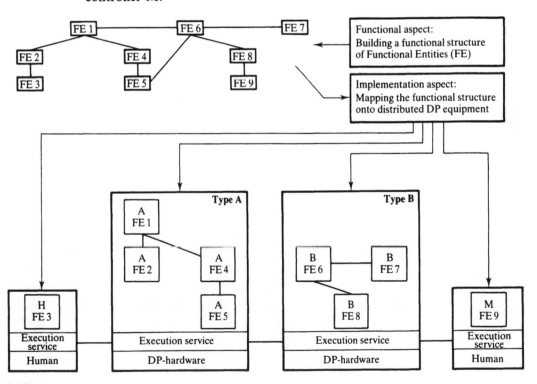

Figure 2.10
Process of functional entities on a distributed execution service system.

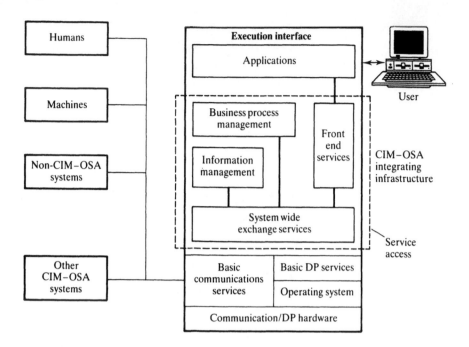

Figure 2.11
CIM–OSA integrating
infrastructure.

The CIM–OSA concept provides an integrating infrastructure to the user. The integrating infrastructure acts as a kind of combined configuration and operating system which ties the various CIM-modules and views together in an operable control system. Figure 2.11 shows the setting of the integrating infrastructure in the manufacturing world.

The CIM–OSA integrating infrastructure has the following components:

- **Application.** The user calls an application via the basic communication services.
- **Front-end service.** This module is the CIM–OSA application interface. It translates the interaction method of any specific external functional entity into the transaction-oriented interaction behavior.
- **Business process management.** This function is responsible for the control structure which is needed for all applications to operate as a unit. The management of the basic business resources, such as humans, manufacturing equipment and processing time, is an additional task of this module.
- **Information management.** This function is responsible for providing an adequate information service, with emphasis on data integrity, consistency, reliability and redundancy. It takes care of the access rights and user priorities, as well as data storage, retrieval and conversion.
- **System-wide exchange service.** This service is responsible for the management of the data communication. During operation, it furnishes the data channels, secures independent data access and synchronizes and manages the data transfer.

The system has various protocols. In the application module, humans and machines communicate with the integrating infrastructure via a specific external

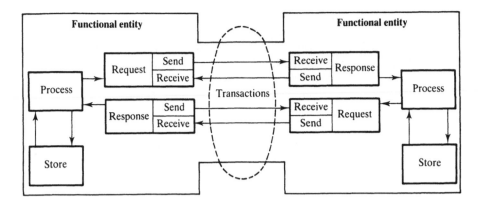

Figure 2.12

The concept of functional entities and their menus of communication.

protocol. The modules within the integrating infrastructure communicate via internal protocols.

Within the CIM–OSA concept a manufacturing system is composed of modules of functional entities, see Figure 2.12. A functional entity is an ideal, implementation-independent object, for example, an algorithm, mathematical expression, control function and so on, describing a conceptual component of a business function. Functional entities may act in sequence or parallel. The data exchange between related functional entities is via a transaction. There are three operations which can be performed during a transaction.

(1) The requester sends a request data unit to a responder.

(2) The responder acts upon a request.

(3) The responder sends a response data unit to the requester.

The functional entities are the basic generic building blocks from which can be configured the functional structure of a planning, control and physical architecture of a manufacturing system.

The key points of the CIM–OSA approach are the generic CIM model, the library of hardware and software modules and the possibility of configuring the particular manufacturing system with the aid of the integrating infrastructures. The tools for the description, configuration and simulation of the manufacturing system are well defined. The CIM–OSA model addresses the hardware and software user as well as the supplier. Thus, the user may have various sources from which he can configure a manufacturing system.

2.3 The Amherst–Karlsruhe CIM Model

2.3.1 Interconnecting data processing activities

When a manufacturing process is configured to a functional entity, the structure of the information process for controlling and monitoring has to be designed. There will be information sources and drains. A source generates information and leads it into the process. Typical sources are the drawing, bill of materials, set points from the output gates of a control computer, sensors and so on. A drain on the other hand, consumes and/or stores information. Typical drains are the inputs to a CRT display, the feedback gates to a controller, the input port to a memory system and so on. Information units may act both as a source and a drain, for example, a data storage device. There are various ways of sending information from a source to a drain. The CIM–OSA model discussed in the previous section uses a channel for sending and receiving information between two functional entities. However, there is nothing said about the type of channel and the media to carry the information. The exchange of information can only be done efficiently if the communication interfaces are compatible, including the following:

- medium, for example fiber optic conductor;
- mechanical connector, for example the dimensions and arrangement of conductors;
- logic levels, for example the length and height of pulses;
- communication protocol, for example the length and arrangement of the various protocol fields;
- data format, for example integer or floating point presentation of numbers.

The most important communication protocols for manufacturing will be discussed in Chapter 5. There are various ways of transferring information between entities of a manufacturing system, as shown in Figure 2.13.

Loose communication link

With this simplest method of communication, data contained on a storage device is carried by hand from one computer to another. The data formats and file structures of the information source and drain should be the same. If this prerequisite cannot be met pre- or postprocessors must be used to adapt the data format and file structures of the source and drain. This, however, may be a very cumbersome and time consuming method because a considerable amount of data redundancy will be created.

Direct communication

A simple solution for improving the loose communication system is the direct connection of the data source and drain. This, however, implies that, in addition to the data formats, the communication interfaces and protocols have to be the same. A further problem is the synchronization of the source and drain. The sender of the information would have to make sure that the receiver is ready for the data transfer at exactly the same time when the data is transmitted. This is a difficult problem when sender and receiver operate at different speeds.

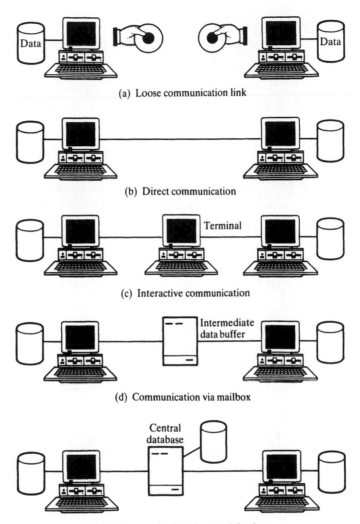

(a) Loose communication link

(b) Direct communication

(c) Interactive communication

(d) Communication via mailbox

(e) Communication via central database

Figure 2.13
Various modes of
factory data exchange.

Interactive communication

In this case, the communication between the two systems is done interactively via
a terminal. The data exchange between source and drain can be facilitated via a
window technique on the same screen where each application gets its own window.
Here, the user must be capable of logging into the databases of both communicating
systems. In order to access the databases a query language must be provided for
both the source and the drain. The databases should have a user friendly format.
Relational databases are suitable for this application.

Communication via mailbox

This method overcomes the synchronization problem of direct communication. The data records to be communicated are deposited into an intermediate data buffer, often called a mailbox; the exchange takes place via this buffer. A prerequisite for the system is the use of the same data formats and structure by both communication units. The data redundancy of this method is very high.

Communication via centralized databases

With this method, the data for both the source and the drain are kept in the same database using the same data format and structure. The user has access to the information via a database manager and a query language where he can manipulate the data from a logical point of view. This method also lends itself to multiuser operation, assuming a high integrity of the manipulated data. The centralized database has a great advantage when changes, expansions or deletions are to be made. An advanced database manager also provides safety measures to overcome data transmission errors, transmission interruptions and system failures.

2.3.2 The manufacturing system model

The Amherst–Karlsruhe model describes the technical activities of manufacturing. It assumes that orders have been entered into the system and that all resources are available for planning and controlling the manufacturing facility. The concept has several layers representing various manufacturing activities, see Figure 2.14. The first layer contains the engineering and design functions where the product is designed and developed. The outputs of this activity are the drawings and the bill of materials. The second layer is process planning. Here the process plans for manufacturing, assembling, and testing are made. The process plans together with the drawing, the bill of materials and the customer orders are the input to scheduling. There is a distributed computer network connecting, via the MAP protocol, the engineering, planning, and scheduling activities. The output of scheduling is the release of the order to the manufacturing floor. Manufacturing is controlled by a hierarchically structured realtime computer system. Its set points are the operating parameters used for starting and controlling the activities on the production floor.

The operations of the machine tools and material movement equipment are monitored by a data acquisition system. The collected data represents the state of the manufacturing system and is the feedback information used for control. The communication on the factory floor is done via a field bus.

The computer system connects all functions of the manufacturing operation; it is laid out as a hierarchically controlled loosely coupled network which can be configured to perform a specific manufacturing function. The configuration actually takes place at the scheduling level. An introduction to hierarchical control and the types of computers assigned to each control level has been discussed in Chapter 1. The subject will also be covered in a later part of this chapter.

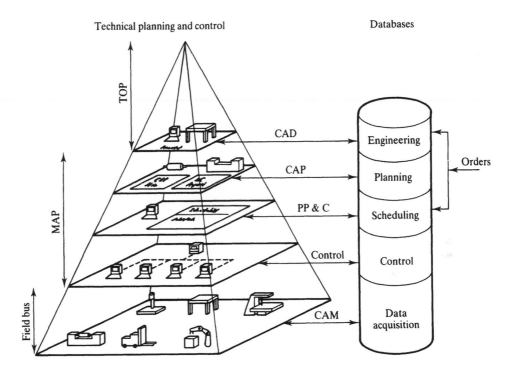

Figure 2.14
The hierarchical
planning and control
concept.

There is a distinction made between the physical control system and the logic control system. The user of the control system will configure logic control modules, each performing a specific control task, with the help of a configuration language, see Figure 2.15. The logic control modules are then interconnected to the physical control system to plan and supervise the actual manufacturing operation, see also Rembold (1986).

A logic control module can be considered as an atomic data processing unit for the planning and control of a specific manufacturing function. There are three ways by which a logic control module can be implemented. First, the module may consist of an algorithm coded and implemented as a program. Second, the algorithm may be cast into a very fast special purpose VLSI circuit. Third, the model may be figured as a hybrid system consisting of hardware and software. Each logic control module is provided with a standard interface and communication protocol to be able to communicate with other modules of the hierarchical control system. On the upper hierarchical level the TOP or MAP protocol may be applied, and on the lower levels is a protocol suitable for field bus communication.

Figure 2.16 shows the software tools needed for programming and configuring a flexible manufacturing system. In this example, three logic control modules are shown which are to be configured to a global control system to operate the manufacturing process. The manufacturing engineer describes his operation with an application-specific language. He may require several languages, for example, one for NC and one for robot programming. The system selects the logic control modules for the various operations to be performed. The modules are configured

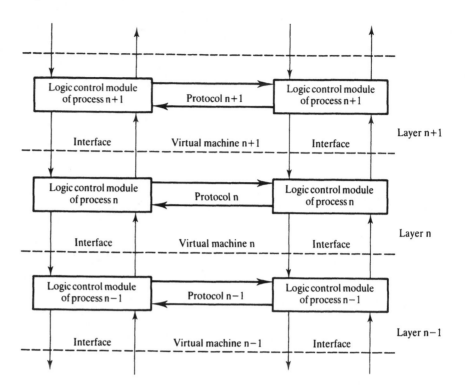

Figure 2.15
The hierarchical concept of a virtual computer control system.

with the help of a manufacturing system configuration language to the global system. The operation of this system can be tried out with the help of the manufacturing system emulator and simulator. There is also a graphic and experimental teaching module to perform graphic programming or to transfer experimental results into the control system. A more advanced system may even have a natural language interface for programming. The computer system configuration language is needed to assemble the physical components required for the logic control modules. Here configuration may only have to be done when new logic modules are created or if changes are made to existing ones.

Expert systems can be implemented and operated with the help of specific AI development tools. The other components shown are conventional software tools and need not be explained any further.

2.3.3 The product model

The aim of the product model is to provide the designer with a modeling tool with which he can describe the product. The model will contain information about the product functions, shape, geometry, surface finish, manufacturing and assembly methods and quality control procedures. Theoretically, the model should have a virtual concept and must provide all information needed to make, shape and test the product. The basic components for building the model must be independent

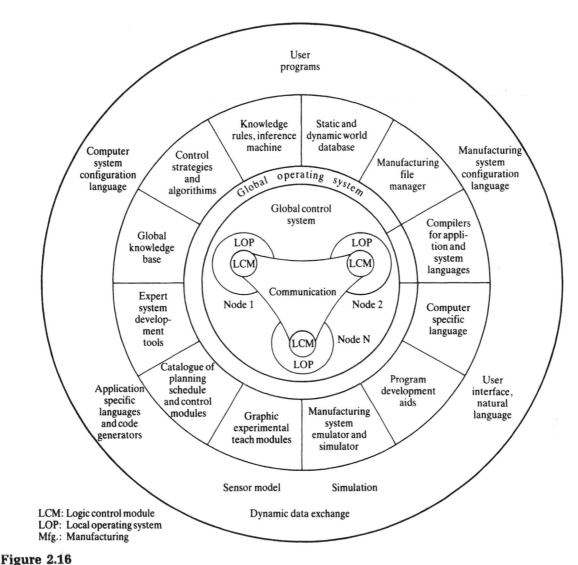

User
programs

Knowledge
rules, inference
machine

Static and
dynamic world
database

Computer
system
configuration
language

Control
strategies
and
algorithims

Manufacturing
file
manager

Manufacturing
system
configuration
language

Global operating system

Global control
system

Global
knowledge
base

LOP

LOP

LCM

LCM

Compilers
for appli-
tion and
system
languages

Expert
system
develop-
ment
tools

Communication

Node 1

Node 2

Computer
specific
language

Node N

LCM

LOP

Catalogue of
planning
schedule
and control
modules

Program
development
aids

Application
specific
languages
and code
generators

Graphic
experimental
teach modules

Manufacturing
system
emulator and
simulator

User
interface,
natural
language

Sensor model

Simulation

LCM: Logic control module
LOP: Local operating system
Mfg.: Manufacturing

Dynamic data exchange

Figure 2.16
Software tools for
building a complete
integrated
manufacturing control
system.

of the product and its manufacturing processes. In addition, the model must be
capable of operating on computing equipment of various manufacturers.

The product model structures the semantic information about the product in
various tiers, for example, a conception, design and manufacturing tier. Figure 2.17
shows a simplified example of a product model.

- The geometric tier has all the information necessary for describing the geometry
  of the product. It contains the various presentation methods used to structure
  a 3D object in the computer.

- The planning tier stores all information on the manufacturing resources and
  the planning and programming algorithms.

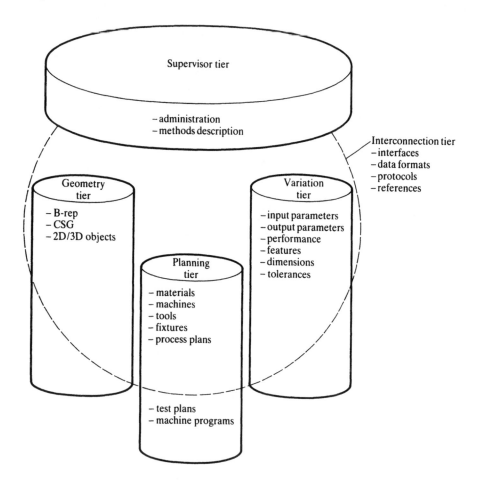

Figure 2.17

Example of a product
model.

- The variation tier is the input/output interface to the user. Through this interface the specific product parameters, dimensions, tolerances, surface finishes and so on, are defined.

- The interconnection tier provides the interfaces between all tiers so that they can exchange information between the various semantic presentations through standardized interfaces and data format translators.

Figure 2.18 shows an example of the type of models that may have to be built for functional, technical, geometric and production modeling, see also Grabowski *et al.* (1988). The problem of interfacing and data format translation becomes apparent. The supervising tier oversees the operation of all tiers and ensures the proper sequential and parallel operations of product engineering.

Most of the standardization efforts in product design are concerned with providing interfaces for connecting various CAD systems and for integrating them with product planning. The most important of these activities are (see also Chapter 7):

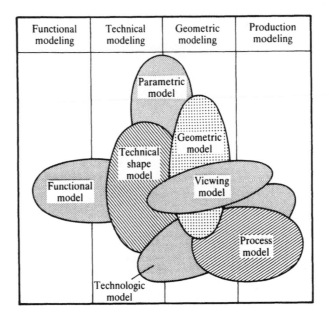

Figure 2.18
Venn diagram of the integrated product model (Grabowski et al., 1988).

- IGES, Initial Graphics Exchange Specification
- SET, Standard d'Éxchange et de Transfert
- VDAFS CAD (VAD-Flächenschnittstelle), Interface of the German Association of Automobile Manufacturers
- CAD* ESPRIT CAD-Interface
- GKS, Graphical Kernel System
- FFIM, Form Features Information Model
- PDES, Product Data Exchange Specification
- STEP, Standard for the Exchange of Product Model Data

Although there are many researchers working on the concept of a product model, so far there are few results which are useful to industry. The difficulties are due to too many variations of product and manufacturing methods. The endeavor of building a product model will be more successful when models are more tailored to specific product classes.

2.3.4 Process planning model

The basic information for planning the manufacture of a product is contained in the CAD database. It consists of graphics and manufacturing data, text and a description of standard components. There are three plans to be made: one for manufacturing, one for assembling, and one for testing, see also Figure 2.19. Many requirements for these plans are similar, but not identical. The planning process

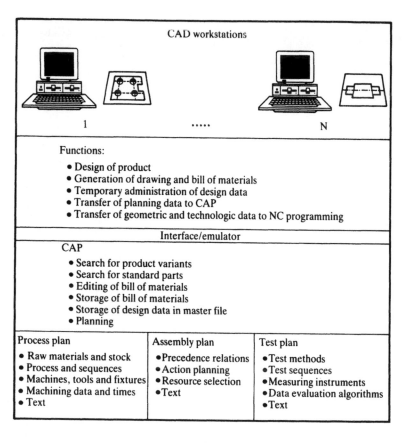

Figure 2.19
CAP activities.

consists of two parts: planning and programming. With conventional planning methods both of these activities are treated independently. However, with implicit or task-oriented programming, the two activities are integrated into one module. In the following discussions we will be treating planning and programming separately, whenever possible.

Planning

In our model, planning will be represented by one or various logical control modules configured to a specific application. The IGES specification can be applied for transferring graphical CAD data to CAP, however, for all other data no standard is presently available. Figure 2.20 suggests various ways of interfacing CAD with CAP.

A part drawing may contain standard, new and variant parts. The process plans for these parts are drafted separately.

Standard components are either purchased, or their process plan can be directly retrieved from a general standard parts database or a factory master file of standard processes. Access to these databases may be provided by an interface containing a pre- or postprocessor.

For new parts, the generative process planning method is applied. Here the

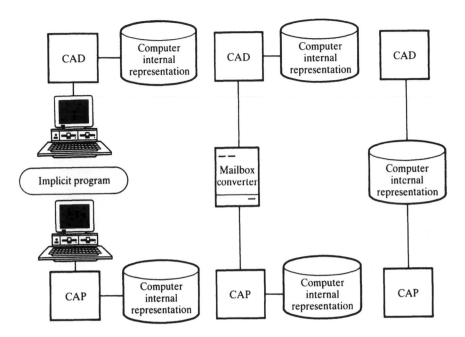

Figure 2.20
Various ways of
interfacing CAD with
CAM.

selection of the raw material, machining processes, machining sequences, tools, fixtures and machining data is inferred with the help of a knowledge-based system from a world model containing all manufacturing knowledge about a specific workpiece spectrum. The knowledge-based system may operate directly with data from the CAD database or infer the process plan from task-oriented instructions describing the global machining operation to be performed.

For variant parts, the machining operations will be inferred from a script or skeleton-based planning method where the workpiece parameters are inserted into empty slots of the script.

Only certain form elements of a part can be planned by the variant method, and other form elements must be planned by the generative method. In this case, a hybrid planning system must be provided, integrating generative and variant planning methods.

The output of planning is the input to scheduling, which in our model is on the next level down. The planning parameters are used by scheduling to initiate actions on the manufacturing floor.

Programming

The data in the CAD database is the source for programming of the manufacturing equipment. Programming has to be done for NC machines, robots, and the test equipment. The programming languages and methods used differ considerably from one another, as shown in Figure 2.21. The machine tools can be programmed with APT or a derivative, or a task-oriented language. There are various ways of configuring a logic control module for programming, as shown in Figure 2.22.

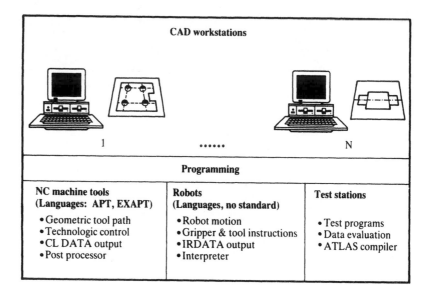

CAD workstations

Programming

NC machine tools (Languages: APT, EXAPT)	Robots (Languages, no standard)	Test stations
• Geometric tool path • Technologic control • CL DATA output • Post processor	• Robot motion • Gripper & tool instructions • IRDATA output • Interpreter	• Test programs • Data evaluation • ATLAS compiler

Figure 2.21
Programming of manufacturing equipment.

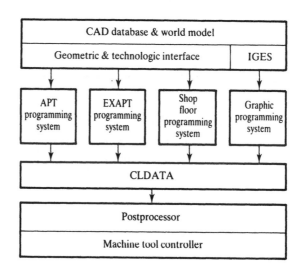

Figure 2.22
Various methods of connecting the CAD database to the NC programming system.

The module will have a geometric and technological interface to the CAD database to obtain the basic workpiece data. The database also contains a world model describing the domain-specific machine tool world. This world model is used in connection with task-oriented programming (not shown in the figure). The output of the programming system is the standardized CLDATA format. Shop-floor programming and graphical programming systems can be interfaced in a similar manner.

A programming environment for robots is shown in Figure 2.23. Since there are no standard programming systems presently available, each programming system used must provide an interface to the CAD database and one to the robot

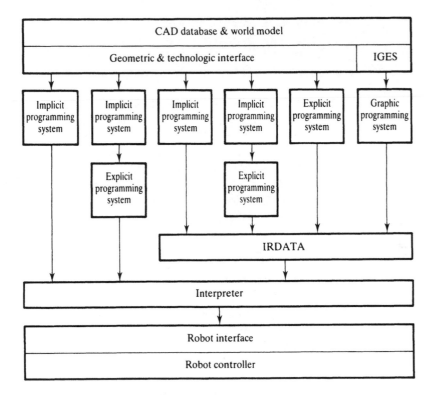

Figure 2.23
Various methods of connecting the CAD database with the robot programming system.

controller. The interfaces to the database must be able to handle geometric as well as technologic data. The interfaces to the robot can be IRDATA code (standardized in Germany). In this case it is assumed that the robot interpreter will understand the code. Thus various types of robots can be handled by this programming environment.

Programming of measuring machines can be done in a similar manner to programming of machine tools. The CLDATA interface is applicable, however, the processes used in the test equipment require a very high intelligence to reduce the flow of measuring data so that this information can be handled efficiently by the interface. If a programming language like ATLAS is being used then it is the task of the system designer to provide the interface for this language.

The generated programs will be sent to the proper machines when the orders are released which use the machines.

2.3.5 Manufacturing scheduling

Customer orders, delivery dates, manufacturing resources, and priority rules are the principal parameters for setting up the manufacturing schedule. The principal documents used for scheduling are the bill of materials and the drawing. The process plan and machine programs provide the basic control parameters for sequencing and controlling the manufacturing processes. With presently available standards, only the geometric interfaces to the CAD database are of relevance to scheduling. Scheduling is one of the most difficult manufacturing activities. There is an enormous repertoire of know-how needed about the various scheduling methods, their advantages and limitations. Scheduling has been a subject of research for many decades. Some of the best known methods are combinatorial switching, branch and bound, dynamic programming, linear and integer programming, heuristic methods, and lately various approaches using knowledge-based techniques. Characteristic of all scheduling techniques is their almost infinite appetite for computational power. The combinatorial problem is of such a magnitude, that even small scheduling tasks require excessive computing capacity.

In our model, the scheduling algorithms will be implemented in logical control models which are stored in a library of logic scheduling algorithms. The algorithms for simple and fast scheduling methods are written in software. Complex algorithms will be implemented in dedicated VLSI circuits (firmware) or in special parallel computer architectures. There will also be a configuration program which can configure hardware and software components to logic control modules to adapt a scheduling method to a specific problem.

Scheduling is actually a very dynamic process, however, most algorithms presently used are only applied to a static environment. In order to make scheduling more responsive to changes of orders, equipment problems and material shortage, it should be performed online. This requires that feedback information from the operating processes must be returned to scheduling to adapt the schedule to the prevailing factory condition.

The scheduling system used in our model will be similar to the one discussed in Yung (1989). A schema of the dynamic scheduling system is shown in Figure 2.24. The heart of the system is a knowledge-based selector for scheduling methods and a library of logic scheduling algorithms. The system knows from a given order and manufacturing status which logic scheduling algorithms have to be used to obtain the desired manufacturing goal and to meet due dates. These algorithms are executed and the corresponding control information is sent to the manufacturing floor. Status feedback from the manufacturing floor is returned to the knowledge-based system to check the validity of the current schedule.

Figure 2.25 shows an integrated scheduling environment consisting of a knowledge-based selector for scheduling methods, a schedule simulator, a learn module, a manufacturing equipment configuration module, and the controllers on the production floor, see also Yung (1989). The knowledge-based selector for scheduling methods sends one or several tentative production schedules to the simulator. If an optimal schedule is found it is sent to the manufacturing equipment configuration module. The scheduling result is also returned to the knowledge-based selector and is sent to the learn module. The learn module tries to infer from this scheduling result improvements for existing scheduling methods, and sends the improved methods to the knowledge-based selector. From here, it is entered into the library of logic scheduling algorithms. The manufacturing equipment configuration module allots the parts to be manufactured to the machines on which they are to be produced; also the part routes are determined. The respective programs for the material flow system and the machine tools are

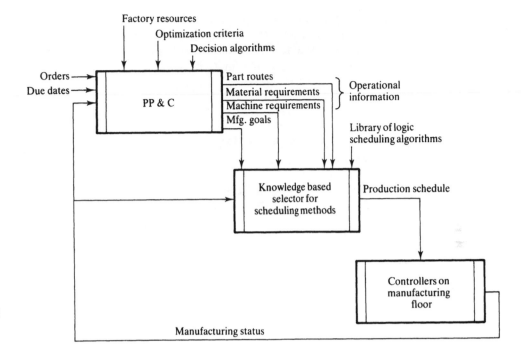

Figure 2.24
Dynamic scheduling
of manufacturing
operations.

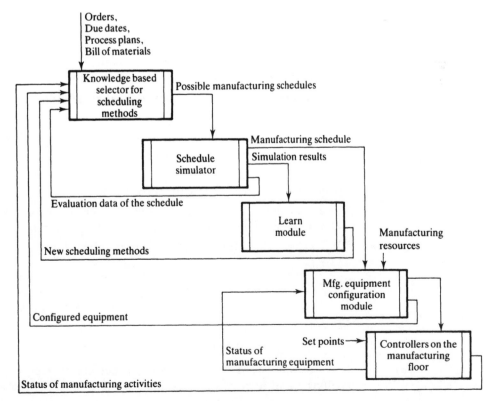

Figure 2.25
The integrated
scheduling
environment of the
generic manufacturing
model.

sent via field bus to the controllers on the manufacturing floor. The status of the manufacturing activities is returned to the knowledge-based selector and the status of the manufacturing equipment is sent back to the configuration module.

In our model, the manufacturing equipment configuration is done by the computers located on the control level of the manufacturing system.

2.3.6 Manufacturing control and monitoring

System control architecture

The manufacturing process will be monitored and controlled by a computer network. The hierarchical structures of this network may have one, two, or three tiers depending on the size and complexity of the manufacturing system. The principle of hierarchical control was discussed in Chapter 1. The communication in this network is done via the MAP protocol on a higher tier and a field bus protocol on the lowest tier for machine control and monitoring. Typical features of our hierarchical computer network are:

(1)   Planning and supervising activities at the upper, and control and monitoring activities at the lower levels.

(2)   The use of process computers at the upper, and microcomputers and programmable controllers at the lower levels.

(3)   TOP and MAP protocols at the upper, and field bus protocols at the lower levels.

(4)   Generically structured components which can be interconnected according to the plug-in principle.

(5)   Open-end systems which can be configured, extended or contracted online automatically or by interactive communication.

(6)   Continuous operability of lower level upon failure of upper level.

(7)   The use of the flying master principle to assure controller redundance.

(8)   The use of black and white or color display screens for operator control panels.

(9)   Simple programmability and configurability of the system via keyboard, light pen or pictorials, using task-oriented programming.

(10)   Pictorial presentation of the process flow (macroscopic and microscopic views).

(11)   User guidance to locate system failures.

(12)   User-controlled access to process data of any level.

(13)   Automatic switch over to standby equipment upon failure.

(14)   Selective presentation of process trends and historical performance data.

(15)   Simulation of the control strategy of every tier.

Figure 2.26 shows a small factory consisting of a parts storage, various NC machine tools, a measuring machine, a finished goods storage and a transport

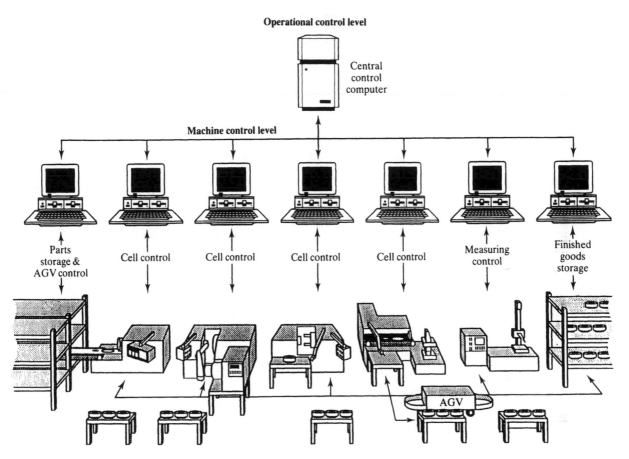

Figure 2.26

Shop floor controlled by computer.

system using an automatically guided vehicle (AGV), see Milburg (1985). Our factory has two control tiers, one for operational planning and control, and one for machine control and monitoring. Upon placement of an order, the system selects the machine tools, tools and fixtures, and the machine programs. When the order is released, the control information is sent to the computers of the individual manufacturing units to prepare for processing. Also, a part route is prepared and sent to the control computer of the AGV. The vehicle retrieves the parts and transports them to the machine tools for processing. Upon completion, the part is measured and brought to the finished goods inventory. The performance of all manufacturing equipment is monitored and reported to the central control computer. Also the quality control data is collected, and evaluated, and the parameters are sent to the central control computer.

The planning and control functions of the central control computer are shown in Figure 2.27. In addition, the contents of the master manufacturing database for this operation are shown.

Typical control and monitoring functions of a control computer for a production machine are shown in Figure 2.28. The assignments to this computer are not as

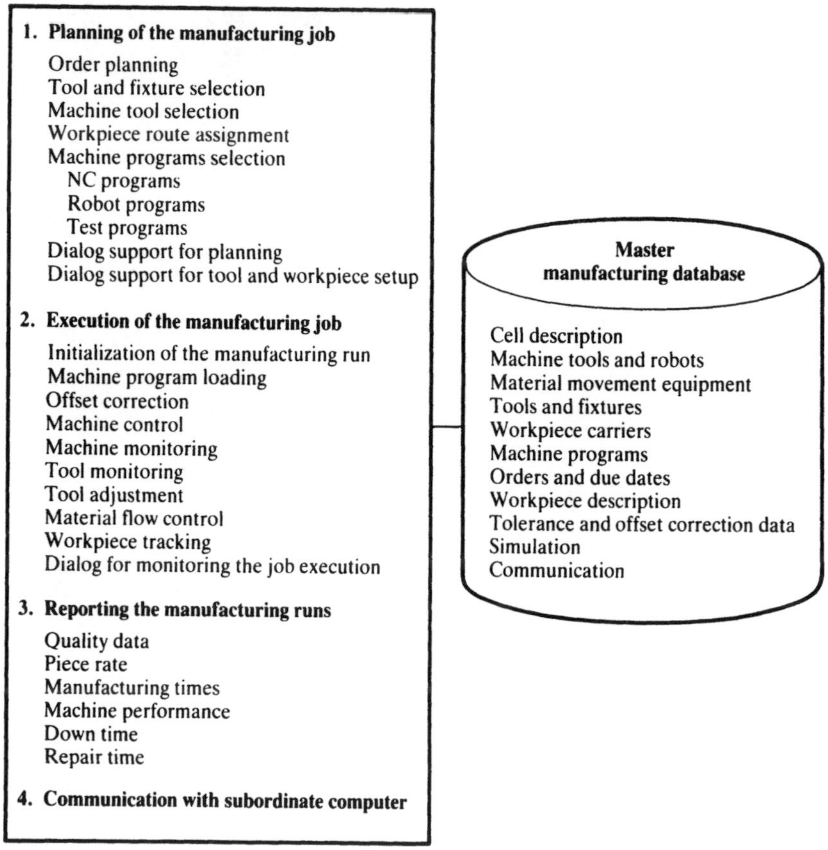

1. **Planning of the manufacturing job**

   Order planning
   Tool and fixture selection
   Machine tool selection
   Workpiece route assignment
   Machine programs selection
      NC programs
      Robot programs
      Test programs
   Dialog support for planning
   Dialog support for tool and workpiece setup

2. **Execution of the manufacturing job**

   Initialization of the manufacturing run
   Machine program loading
   Offset correction
   Machine control
   Machine monitoring
   Tool monitoring
   Tool adjustment
   Material flow control
   Workpiece tracking
   Dialog for monitoring the job execution

3. **Reporting the manufacturing runs**

   Quality data
   Piece rate
   Manufacturing times
   Machine performance
   Down time
   Repair time

4. **Communication with subordinate computer**

Master manufacturing database

Cell description
Machine tools and robots
Material movement equipment
Tools and fixtures
Workpiece carriers
Machine programs
Orders and due dates
Workpiece description
Tolerance and offset correction data
Simulation
Communication

Figure 2.27

Data processing tasks
of the cell control
computer.

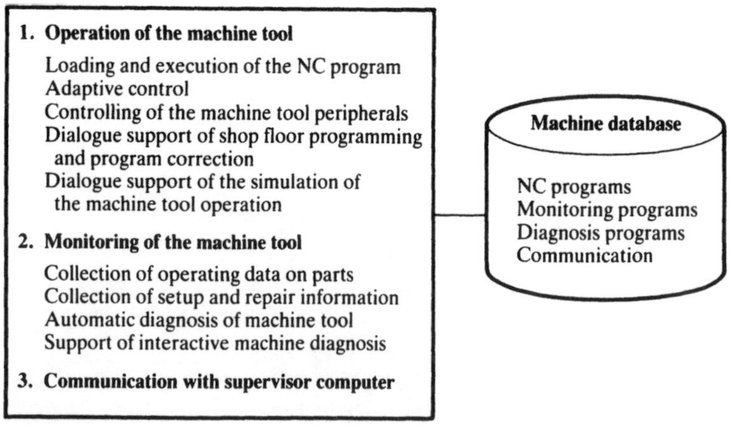

1. **Operation of the machine tool**

   Loading and execution of the NC program
   Adaptive control
   Controlling of the machine tool peripherals
   Dialogue support of shop floor programming
     and program correction
   Dialogue support of the simulation of
     the machine tool operation

2. **Monitoring of the machine tool**

   Collection of operating data on parts
   Collection of setup and repair information
   Automatic diagnosis of machine tool
   Support of interactive machine diagnosis

3. **Communication with supervisor computer**

Machine database

NC programs
Monitoring programs
Diagnosis programs
Communication

Figure 2.28

Data processing tasks
of the control
computer.

numerous as those of its supervisor. Upon scanning the activities and databases it becomes obvious that every manufacturing operation tries to solve the same or similar problems. In other words, the tools for solving a problem on the manufacturing floor can be assembled from logic control modules. For our generic module, this is done automatically by the manufacturing equipment configuration model, as seen in Figure 2.25.

In a more practical factory, however, equipment of various vendors is being installed and manual intervention must be possible. We will be discussing provisions to include these situations with the help of research contributions reported by Weston *et al.* (1989a) and Weston *et al.* (1989b).

Weston *et al.* suggest an automation integration language which allows the user to communicate and interact with any type of process, and to statically or dynamically configure a manufacturing system. For this purpose, discrete manufacturing activity modules are introduced which are akin to our logic control modules. Each module is supplied with the communication capabilities pertinent to its working environment. It is capable of sending and receiving information. The user communicates with the system via very powerful dialogue primitives. A typical manufacturing environment is shown in Figure 2.29. Here, two operator terminals and three processes are connected via a gateway to the computer network. The operators can communicate with the local processes via a cell supervisor or with external processes via the gateway. Figure 2.30 shows how a communication or synchronization between two processes can be established. The connection of various types of manufacturing equipment can be done by two methods. The first method is a backplane solution, whereby a network interface is incorporated into the machine controller, which is very complex, and should be done by the controller manufacturer, as in Figure 2.31. The second solution uses a microcomputer as a communication interface doing the data manipulation and network adaptation, Figure 2.32. This device can be programmed and interconnected by the user.

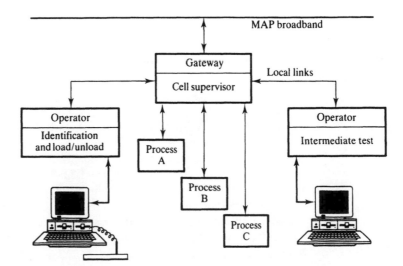

Figure 2.29
Example of a cell controller/gateway function (Weston *et al.*, 1989).

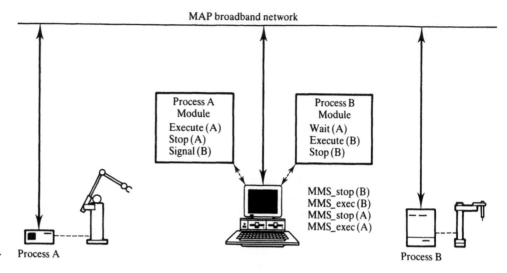

Figure 2.30

Task synchronization
features (Weston, 1989).

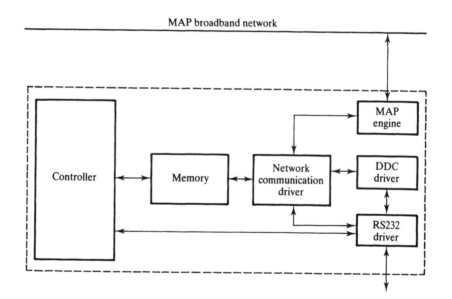

Figure 2.31

Backplane solution for
machine connection
(Weston, 1989).

In order to facilitate the design of the machine controller, the concept of the generic machine controller is introduced by Weston *et al.* (1989b). In this concept, the various control operations of a machine controller are placed in a library of control system modules, as in Figure 2.33. Typical functions are motion control, I/O processing, sequencing, synchronization, programming, integration and so on. With the help of design tools, the user defines his application. The system retrieves the pertinent control modules and assembles them to a logical control architecture. The actual control program is loaded via a mapping procedure into the custom-designed control hardware.

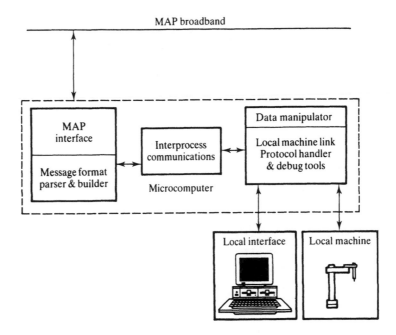

Figure 2.32
Gateway solution for
machine connection
(Weston, 1989).

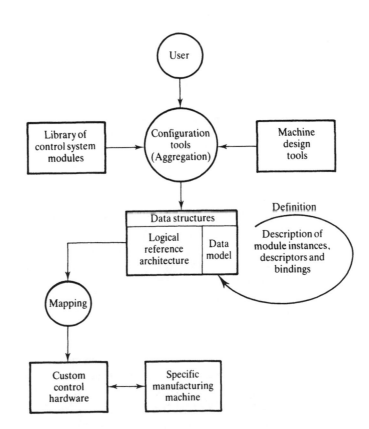

Figure 2.33
Configuration tools
for building a machine
controller (Weston,
1989).

Shop-floor control

The material flow through the manufacturing facility can be controlled by various methods, following a just-in-time strategy or the KANBAN principle (see Chapter 10). An objective of every company will be to hold the in-process inventory at a very low level. For this reason, it is necessary to strictly control the supply and flow of parts. In our generic factory, we provide the operator on the plant floor with simulation aids by which he will be able to try out a manufacturing run by animating it on a graphical terminal. For this purpose, the master database will contain generic models (icons) of all manufacturing operations; they can be configured by the user to a functional manufacturing system on the screen. By means of an interactive dialogue, changes to the process can be made if desired.

The user will be given graphical tools to observe the various fabrication stages of a product along its journey through the plant and to obtain any operating statistics he may need for the completion of a manufacturing run. Figure 2.34 shows such a product flow. The location of the product is indicated by an arrow. The three subpictures represent three manufacturing cycles.

For quality control operations and assembly supervision, model-based vision is necessary; whereby the model of the object to be recognized is retrieved via a standard interface from the CAD database. It will also be possible to observe manufacturing operations, where parts are incrementally added until the entire product is completed.

2.3.7 CIM network of the model factory

The processing and flow of data in the conceptual factory described by the Amherst–Karlsruhe model is controlled by a hierarchical computer system. The computers are interconnected by various types of communication networks, as in Figure 2.35. Each hierarchical level has specific requirements regarding the data rate, communication distance, communication speed, number of participants, configuration of network structure and realtime behavior. We will give a short synopsis on the data transmission system of each hierarchical level. Transmission systems and transmission protocols will be discussed in detail in Chapter 6.

The computers of the corporate level are interconnected by a wide area network (WAN) which allows communication over long distances. This is an open communication system where various types of communication media, such as copper cable, fiber optic line, radio link or satellite transmission may be used. The communication system allows online connection of new participants during the operation of the network. It is possible to communicate with other factories, vendors or even with overseas partners. Standards proposed within the framework of the ISO model may be applied. The WAN must be capable of communicating large amounts of data between the participants.

On the plant control level, a local area network (LAN) interconnects the various computing systems which coordinate and control the factory operations. A LAN net may use a copper cable or optical fibers over a distance of two kilometers. The CSMA/CD and token protocol define the access to the bus.

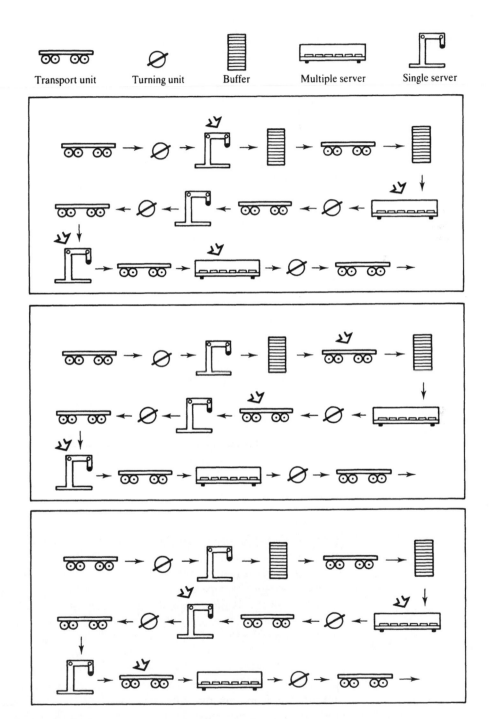

Figure 2.34
Observation of a product flow through a manufacturing facility on a graphic terminal.

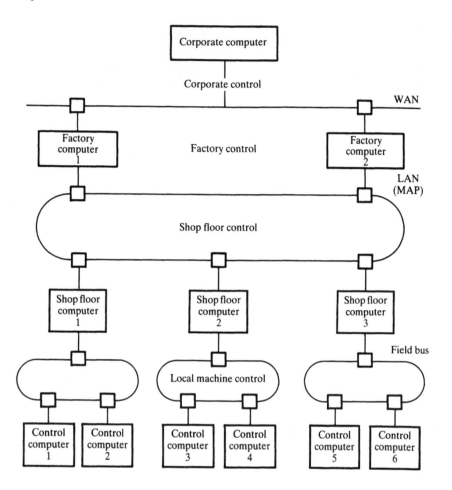

Figure 2.35

A computer network for CIM.

The topology of a LAN may be configured according to a star, bus or ring structure, see Figure 2.36. The wiring costs of ring and bus systems are low. These buses may, however, be subject to noise if electrical data transmission is used. A disturbance may transmit through the entire system. The wiring of a star structure is more expensive because it uses longer connection lines. However, in the event of noise, it transmits through one branch only. A basic disadvantage of the LAN network is that it does not allow realtime operation between participants. In addition, access to the network may be difficult if many users are connected.

The communication on the factory floor is done via intelligent subsystems, for example, microcomputers, programmable controllers and smart sensors. Because of the many controllers, terminals and sensor systems employed on this level, the communication system must be low cost and must be able to support various standards. The participants on this level communicate over a short distance. Typical standardized serial and parallel bus systems are the SCSI bus, IEEE bus and the

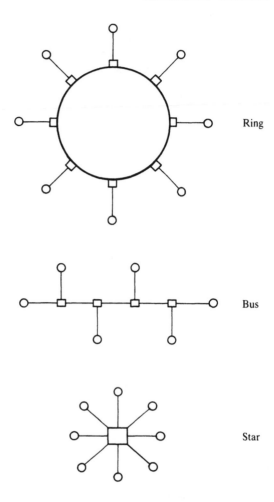

Ring

Bus

Star

Figure 2.36
Basic LAN concepts.

RS 232/422 interface. Of special interest is the field bus which is presently being developed and may use copper cables or optical fibers.

Figure 2.37 shows how networks may be structured in a real factory control system using four tiers. A strict adherence to one communication system at one level may not be practical when systems of various configurations and vendors are interconnected.

MAP/TOP

The variety of problems encountered in network design prompted the development of standards for networks. First, the general network standard called the Open Systems Interconnection (OSI) was developed through the International Standards Organization (ISO). Then other network standards were developed for application purposes to adapt to the OSI reference model (Chapter 6). Prominent among these application standards is the Manufacturing Automation Protocol (MAP) initiated

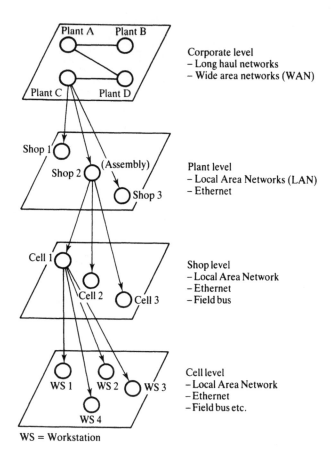

Corporate level
– Long haul networks
– Wide area networks (WAN)

Plant level
– Local Area Networks (LAN)
– Ethernet

Shop level
– Local Area Network
– Ethernet
– Field bus

Cell level
– Local Area Network
– Ethernet
– Field bus etc.

WS = Workstation

Figure 2.37
Network conceptual
levels in CIM.

by the General Motors Corporation in 1980, in cooperation with many other companies. One main feature of a local area network is the ability to connect the various company's computers and control systems which may have been purchased from different vendors to a network so that they can work together.

More recently, an effort has been made to develop a MAP parallel standard for engineering and office areas. The Boeing company has been the initiator and catalyst of this concept with its introduction of the Technical and Office Protocols (TOP) in November 1985.

It is worthwhile to note that MAP and TOP are not standards. Thus, wherever international standards exist, MAP and TOP specify particular elements within the standards which are most appropriate to the particular application areas covered by the specifications, see Kaminski (1986). Where international standards are not yet finished, MAP and TOP have to make the best estimate of the likely structure of those standards and retain the ability to modify those specifications as standards become established. In most respects, the MAP and TOP specifications are, and will remain, identical. They differ only in that they are designed to interface with different kinds of application software and, in some details, the way messages are sent around the network. The two types of network can be connected to each other transparently – so that the user is not aware of the transition – and at low cost.

When fully developed, it is intended that the two systems will be able to satisfy all local area networking needs within our model factory and will have standard procedures for communicating with other local area networks in other plants, whether down the road or on the other side of the world. In order to maintain this universality, some internal subdivisions are becoming necessary while still keeping the ability to communicate between networks. Also, new versions of the MAP specification are being developed which are intended for high-speed communication within manufacturing cells.

2.3.8 CIM database

The management information system

The heart of a computer-integrated manufacturing system is a management information system which processes, handles and controls the shared data needed for administration, design, planning, scheduling, and control. The efficiency of operating a computer-integrated manufacturing plant depends on the quality and integrity of a well-designed management information system. There is a close relationship between data which is being processed and used in the various manufacturing activities, for example, design information is needed for planning, scheduling, machine programming, quality control and so on. For this reason, an information system must be designed as an entity which comprises all manufacturing activities. The components of the information system are the planning and control modules, the common database, the computer network and the communication system. The functionality of the management information system depends on how well these components are integrated and used. The data handled and processed by the management information system comes from external and internal sources of the enterprise. Figure 2.38 gives a global picture of the data which is entered into the system from the world in which the enterprise operates. The outer ring shows activities which project out into the future, to make forecasts which are needed to secure the position of a company in its competitive environment. The inner ring represents the activities of the ongoing manufacturing process.

Manufacturing takes much data from the outer ring, but it also generates its own data. The management information system must process and store data from all levels of a manufacturing operation. Most of the information maintains its value for a longer time period, thus it is necessary to provide ample storage capacity.

Within the context of our generic model of a manufacturing system, the activities of the inner ring are of particular interest. To get a feeling for the magnitude of the problem, the most important planning and control activities needed to channel a product through a plant are shown in Figure 2.39. The reader who wants to study the activities in more detail should start with sales and work his way around

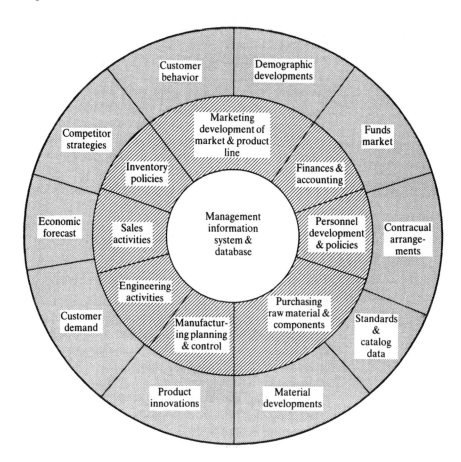

Figure 2.38
Comprehensive
management
information system of
an enterprise.

the figure to shipping. The text in the boxes explains the activities. The database contains the master files and temporary order-related data for all activities.

There are various master files on the customers, suppliers, products, manufacturing processes, personnel and so on. These files should be located in a common database, as seen in Figure 2.39. Access to the master files must be provided for all activities requiring the data. Information located in the master files is of a more permanent nature and need not to be changed often.

The order-related information is composed of data which has a temporary value to manufacturing. It is usually activated with the order entry and can be deleted with the delivery of the product, if it is not needed for future reference. Order-related data is also used by various manufacturing activities and must be accessible to them.

In the remainder of this section, we will be discussing special features which the database must have for our generic manufacturing model. First, attention will be paid to data models. Second, the communication with the database will be explained. Third, requirements for the databases will be discussed. Fourth, an overview will be given on the various database types.

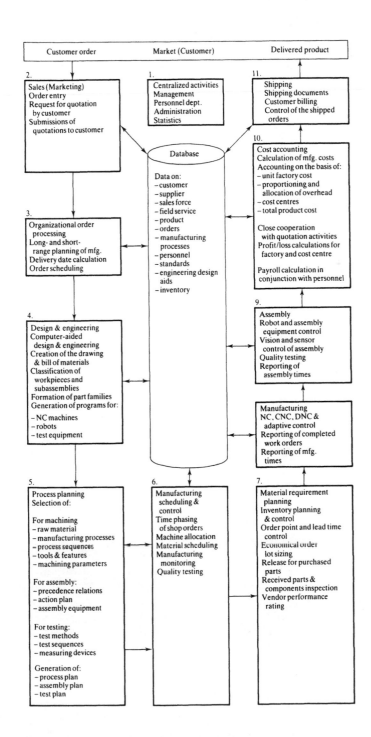

Figure 2.39
A CIM database
must support the
entire flow of an order
through a factory.

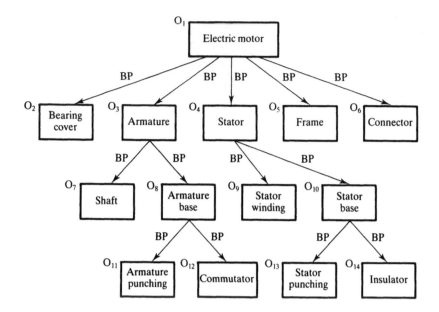

Figure 2.40

Hierarchical
representation of an
electric motor in a
database. Relationship
BP 'a is base of b'.

Data models

Manufacturing information entered into a database can basically be represented
by three classical data models. They are the hierarchical data model, the network
data model and the relational data model. The use of a particular model in our
system depends on the required representation scheme, the available memory, and
the ease of accessing and manipulating data. The relational database is increasingly
getting attention because it offers to the user a very natural way of communicating
with the stored manufacturing information.

The hierarchical data model.

In this data model the information about a product, process or an order is
represented as nodes in a tree with a superimposed class structure. The relations
are represented as edges. The tree has a root and successors. The root may be
called a parent and a successor a child. Each child may be the root of a subtree
having new siblings. The entrance to the information of the tree is through the
root segment and from there one passes down from a parent to a child.

Figure 2.40 shows how an electric motor can be described by a hierarchical
model. The relationship BP indicates that 'part a is the base of part b'.

The features of the hierarchical database model are:

- it represents a 1:n relationship;
- the data structure is simple;
- access to date is fast;
- the model is inflexible with regard to restructuring;
- rather rigid data access;

- the database is highly redundant;
- topological information is difficult to represent.

The network data model.

In the network model objects are also represented as nodes and relations as arcs. However, there can be various relationships between different objects. Figure 2.41 shows the same electric motor represented by a network model. In addition to the hierarchical relationship BP, 'part a is the base of part b' there are the relationships MD, 'part a is mounted before part b' and MO, 'part a is mounted on part b'. Topological information can be presented in this model by a ring structure.

The properties of this database are:

- it allows *m:n* presentation between objects or ring structures;
- access to the database may be from several head nodes of a ring structure;
- complex data structure;
- difficult to implement; depending on complexity;
- difficult to restructure; depending on complexity;
- reduced redundancy.

The relational data model.

This data model does not deal with individual records but with its sets. Figure 2.42 shows a data file containing information on several electric motors. The table presented in this figure is called a relation, a column heading is a relation attribute

Figure 2.41

Network presentation of an electric motor in a database. Relationship BP 'a is base of b'; MB 'a is mounted before b'; MO 'a is mounted on b'.

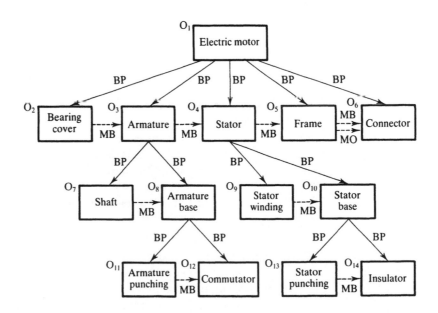

Identifier	KW	RPM	Used in product
Motor 1	2	1800	A
Motor 2	2	3600	B
Motor 3	3	1800	A
Motor 4	5	1800	C
Motor 5	6	900	B

Figure 2.42

A simple relational presentation of various objects.

and a row is presented by a relation tuple. Typical questions which may be asked are:

- find all motors used in product A,
- find all 2 kW motors,
- find all 1800 RPM motors used in product A and C.

The properties of a relational database are:

- it represents $m:n$ relationships,
- it may be entered from any attribute,
- fully dependent on the represented data,
- easy adaptability,
- good flexibility,
- quick implementation,
- response times can be very slow,
- needs lots of memory.

Communication with the database

Personnel with various skills will be communicating with the database. For this reason, it is necessary to provide a communication interface which will greatly facilitate the data access, retrieval, manipulation and storage. The following aspects are of importance to the communication, see Lockemann (1988).

Adequacy of the communication language.
The user should be given a communication language he is familiar with, for example, application-oriented language, process flow charts, drawings and so on. With this requirement, a data model must be available which describes the application in a natural way. An example of an inquiry language unfriendly to the user will be shown with the help of Figure 2.43 and discussed in Dadam *et al.* (1989). In this figure, a robot is depicted and a logical model made of it. If this robot is presented by a relational database, then there will be six tables (relations) scattered in various places of the database. An inquiry to retrieve robot information would be as follows:

SELECT   r.rob_id, r.rob_descr, ar.arm_id, ax.axis_no, mat.r_no, mat.sp1,
          mat.sp2, mat.sp3, mat.sp4, ax.ja_min, ax.ja_max, ax.mass, ax.accel,
          re.eff_id, e.function

FROM   robots r, robot_arms ar, axes ax, matrices mat, robot-endeff re, endeffect e

WHERE   r.rob_id = 'Robl' AND r.rob_id = ar.rob_id AND ar.rob_id = ax.rob_id AND ar.arm_id = ax.arm_id AND ax.rob_id = mat.rob_id AND ax.arm_id = mat.arm_id AND ax.axis_no = mat.axis_no AND r.rob_id = re.rob_id AND re.eff_id = e.eff_id

This example shows that a relational model is not well suited for accessing information of a geometric object; the inquiry has to be very detailed.

Language uniformity.

It is desirable to provide the same communication language to various users who access a database; otherwise they will have problems communicating with one another on the same subject. However, this requirement is difficult to meet since it is almost impossible to represent various subjects by the same language. In addition, we have just learned that a data model imposes limitations on the language. Thus, models for two different domains may result in different requirements for the language.

Openness to communication.

This requirement states that a user should be able to login to and logout from a database at any time without interfering with any other database activities.

Assurance of data integrity.

The communication between the user and the database must ensure that the correct data is actually transferred. This can be obtained by providing a well-structured data model, consistency check, and synchronization and data protection mechanisms. The use of a transaction supervisor is shown in Figure 2.44.

Provision for loss of data.

An important safety aspect is the provision for loss of data. Data may be stored over many years and it is important that no loss of data occurs when data is needed or stored.

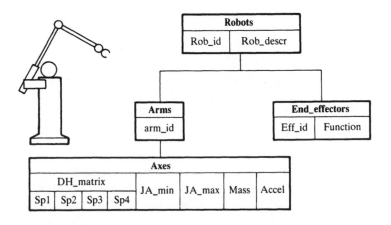

Figure 2.43
Logic model of a simplified robot.

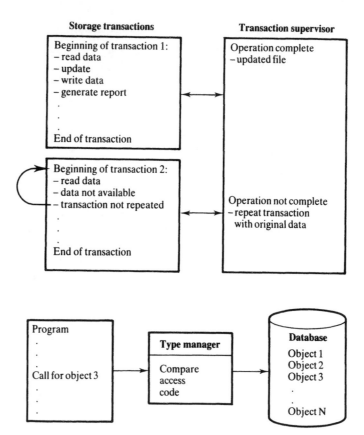

Figure 2.44

Improvement of data consistency with the aid of a transaction supervisor.

Figure 2.45

Protection of data by object-oriented access right.

Protection from interference with parallel communicators.

The communication with a databank must not interfere with parallel communication by multi-access operations.

Independence of physical communication media and computing systems.

The advancement of communication and computer technology brings new equipment to the manufacturing operations. In order to keep up with the progress, the means of communication with the database must be device independent.

Access rights.

There are numerous people participating in the manufacture of the product. Most users need access only to domain specific data. In order to protect the database from misuse, the access rights must be clarified, see Figure 2.45.

Multiuser operation

A database should have multiuser access. The individual user must have the impression that it is his own database. Waiting times must be minimal. The system must be protected by some kind of synchronization mechanism to allow various users to make changes to data at the same time.

Required Features of Databases for CAD/CAM

The previous section has taught us that data sharing between the various manufacturing activities is a basic requirement for obtaining an efficient data processing operation. For this reason, data models have to be provided which have a unique structure, modes of operation and constraints to represent the semantics of the diverse manufacturing data, including engineering, planning, controlling and so on. The semantics defines the architectures, operation rules and integrity of the database. The database must be user-oriented; when it is designed, all data definitions must be analyzed and a semantics must be conceived which allows the orderly implementation and use of the factory data. It is also necessary to define complex data types found in CAD/CAM activities. Most models hitherto applied by manufacturing were built for common EDV operations and are inadequate for manufacturing. Other objects of concern, when designing a database, are the integrity of stored and manipulated data, the user friendliness of the system and a multiuser operation.

Some of these features will be discussed in more detail. The reader who wants to study this subject further is referred to Lockemann (1988) and Su (1985).

Structured object-oriented data presentation.

Most objects in the manufacturing environment have a complex structure and are difficult to model with conventional programming means which only provide basic data types. Figure 2.46 shows an extended relational robot data model for defining a robot, see Dadam *et al.* (1989). In this model the definition of ordered sets and lists are added features. The model uses an extended query language which permits structured access to the robot data or parts of it.

The inquiry:

SELECT X

FROM X IN robots

WHERE X. Rob_Id = 'ROB1'

renders the same information on the robot as the example shown in the previous section.

The inquiry:

SELECT aX

FROM aX in ar.axes, ar in r.arms, r in robots

WHERE position (ax) = 3 and r. Rob_id = 'ROB'

provides access to the information on the third axis of the left arm of robot ROB1.

Modeling of dynamically changing objects.

In a manufacturing environment, objects change dynamically when they proceed through the manufacturing process, for example, a pallet may get successively unloaded while it proceeds from one work center to another, or a workpiece changes its shape during processing at different machine tools. The change of state has to be captured by the data model in an efficient way; recording temporal,

{ROBOTS}											
Rob_ID	Rob Description	{Arms}									{Endeffect}
		Arm_ID	<Axes>								Eff_ID	Function
			<DH_Matrix>				JA_min	JA_max	Mass	Accel		
			Sp1	Sp2	Sp3	Sp4						
Rob1	Robiflex 300 is a robot with two arms, produced by comp. XYZ in XIO. The service unit is located at	left	1 0 0 0	0 0 −1 0	0 1 0 0	0 0 100 0	−180	180	50.0	1.00	E150 E200 E240	Screwdriver Welder Gripper
			1 0 0 0	0 1 0 0	0 0 1 0	70 0 20 1	−250	60	45.5	2.0		
		right										
Rob2	Quickrob 300 is capable of performing several different tasks . . .	left	1 0 0 0	0 1 0 0	0 0 1 1	0 80 0 1	−98	200	40.0	2.0	E210	Gripper
		right										

Figure 2.46

Extended relational data model of a robot. { } sets of objects; ⟨ ⟩ lists or ordered sets.

positional and procedural relationships. A similar situation exists when different versions of a design have to be presented.

Modeling of recursive operations.

In design, process planning and scheduling many recursive problems arise, for example, nesting of hole patterns in a sheet metal part to be stamped.

Modeling of data objects described in various domains.

For processing a part, data is often stored for domain-specific activities. However, it may be necessary for a specific manufacturing operation to combine this data. A model which can capture this data and present it as an entity would be advantageous.

Consistency of the database.

When a manufacturing object is modeled in the database, it must be assumed that the model actually represents the object and that it contains the essential features necessary for data manipulation.

Data independence.

Object models should be constructed as logic entities, so that they can be used for representing a family of physical objects.

Figure 2.47
Collection of
independent databases.

Design of the database

For our generic Amherst–Karlsruhe model, we would like to employ a database having all the features discussed in the previous sections. Some additional features we would need are high access speed, ease of updating and maintaining, and redundancy of storage. Since our model must also accommodate components of realistic manufacturing operations, we will be looking at four different databases. They are:

- a loose collection of independent databases
- a central database
- an interfaced database
- a distributed database

Each of these databases has its advantages and disadvantages.

Collection of independent databases.
For our model, the independent databases only have historical value. They belong to the early pioneer era of computer technology. The concept of independent databases was unintentionally developed out of a basic need for every data processing application which sprung up in a factory to have a database, see Figure 2.47. The need for a central database was not sufficiently recognized at that time and the tools to build and operate one were not available. Usually, little consideration was given to common data, formulas, database structures and protocols. The following advantage and disadvantages are experienced with this approach:

- A fast way of implementing many manufacturing applications.
- The maintenance of individual databases is easy.
- A number of redundant databases will be implemented.
- It is virtually impossible to access other databases.

- The integration of these databases is extremely difficult, and only possible with pre- and postprocessors, thereby reducing the data integrity.
- The maintenance of many independent databases is expensive.
- A change in one database may necessitate changes in several others.

Despite these difficulties, the independent databases are not completely out of date. Typically, an engineer working on a special design will be installing a database just for handling heat transfer, stress, or vibration analyses. This type of data is not relevant to manufacture and will not enter the production process. There are many other examples where independent databases are useful.

Centralized database.

The centralized database has the best structure as a data deposit for our generic model, see Figure 2.48. Ideally, it can be designed as a unified multiuser database, having minimum data redundancy, a high degree of data security, and provisions for data recovery. Presently, there are a multitude of standard databases available; while they handle, for example, PP&C applications very well, they are not suited for CAD applications. For this reason the centralized database, even though it fits our model best, is still far in the future. Such a database will inherently be slow as long as it is used with a von Neumann computer. The administrative effort to operate it in the multiuser mode will be prohibitive. These are some of the features:

- The database fits the generic manufacturing model well.
- The database assures a high degree of data consistency.
- Data manipulation in a multiuser environment will be quite slow, if the database serves many applications.
- Maintenance of such a database is unwieldy.
- Presently there is not enough know-how available to handle the variety of data types used in manufacturing by one system.
- A comprehensive data protection mechanism is needed to prevent users from accessing and changing unauthorized data.

Currently, a centralized database is only of interest as a local database, which serves a specific application, such as market research and long-range forecasting.

Figure 2.48
Central database.

Figure 2.49
Interfaced database.

Interfaced database.

With the interfaced database, the user has access to his own domain-specific database and the central database, as shown in Figure 2.49. Using this concept, the manufacturing activities can be visualized as individual information processing isles, each having their own programs, data, data formats and organizational structure. Communication between the various activities is done with standard protocols, formats and data structures. In this setup local databases may access each other; for this purpose, pre- or postprocessors may be used to make the communication possible. However, if there is a broad spectrum of data types associated with an activity, then it might be difficult to find standardized interfaces. For example, for geometric data the IGES (see Chapter 7) format may be applied, but for other drawing-related data, no standard exists. One of the great advantages of this concept is that it provides access to a centralized database where common data is stored. For example, design may deposit the product data needed for process planning, machine programming, and scheduling in the centralized database. The interfaced database is commonly used for data processing or distributed computer systems and fits well in modern computing concepts. The advantages and disadvantages of this database are:

- The user has his own protected database.
- The user can access common data.
- The database assures good data consistency.
- The central database needs a good protection mechanism for misuse and change of data.
- The system fits well into the distributed computing concept.
- Interfacing of databases can be done with pre- and postprocessors.
- The databases can be hierarchically designed to fit a hierarchical management or control structure.
- Fast access time to local data.

Figure 2.50
Distributed database.

- Local responsibility for databases.
- A high availability during system failure.
- Well suited for maintenance.
- Common formats, communication protocols and interfaces must be provided.
- Difficult to coordinate the interfacing to the central datafile.

Distributed database.
This concept is similar to that of the interfaced database, as shown in Figure 2.50. The difference in this case is that the access to the master database is restricted and local databases cannot directly access each other. Typically, with this type of database, jobs are distributed at fixed time intervals to the local activities, where the data processing takes place. At predetermined times, the central database is updated. The advantages of using distributed databases are similar to those of the interfaced databases. The protection of the centralized database, however, is better since the access to its data can be well controlled. Distributed databases are used in engineering, process control, CNC control and so on.

2.4 Problems

(1) What is a CIM model and which aspects should be coverted by it?
(2) Explain the most important features of the different CIM-models introduced in this chapter.
(3) Why does the CIM-OSA model introduce an enterprise engineering environ-

ment and enterprise operation environment and how can a specific CIM system be constructed with the help of these environments?

(4) Derive from a reference architecture a specific architecture for a typical job shop operation. What would the reference architecture look like and what will your specific architecture look like?

(5) Design the interfaces of a selected module of a reference architecture. Show the flow of control information and the data processing needed for this module.

(6) Explain the structure of the manufacturing system of the Amherst–Karlsruhe model and the features of each layer.

(7) Describe what kind of information must be represented in a product model and show how the product model introduced in this chapter is structured.

(8) How is generative process planning and variant planning done?

(9) Prepare a schema of a novel scheduling concept and explain how it works.

(10) There are various models used for the design and manufacturing cycles of a product. Show, for a toy car, in detail the various activities (business functions) leading from design to manufacturing. Why is it so difficult to interface them?

(11) Explain the data processing tasks of a cell control computer and machine control computer, as shown in Figures 2.27 and 2.28.

(12) Give a short account of important communication systems and standards. What distances can be covered by them and what is the data transmission rate?

(13) What are the features of the data models introduced in this chapter?

(14) Explain the advantages and disadvantages of the different types of databases.

(15) Construct a CIM database for the manufacture of chairs. Take two functions and explain them in detail. Show the data exchange between sales and engineering.

(16) Construct a database for an DNC manufacturing system for rotor blades of an aircraft turbine. Why did you select this particular structure? How can you interface this database with that of engineering?

3

Analysis tools for manufacturing

CHAPTER CONTENTS

3.1 Introduction   104
3.2 An Integrated Approach to Manufacturing System Planning   105
3.3 Simulation   119
3.4 Petri Nets   127
3.5 Artificial Intelligence Methods for Manufacturing   135
3.6 Problems   174

3.1 Introduction

In this chapter we will be discussing various new planning tools and approaches to conceive manufacturing systems and to plan and schedule production runs. Planning is an integral part of any manufacturing operation, it is usually an activity concerned with time planning and in-depth planning. Time planning is concerned with the time frame in which an activity has to take place. In-depth planning is concerned with the provision and optimal utilization of all manufacturing resources. It begins with a rather abstract concept and narrows this down over several hierarchical levels to a very detailed plan. The difficulty of planning is that almost every manufacturing activity requires its own planning methods, and that the methods change dynamically with time. For example, computer-aided planning methods are much different from the manual methods used 30 years ago.

There is no general concept about the contents and structure of each planning tool. For example, a planning tool for software development will look different from one which is used for planning manufacturing processes. The tools which are configured to a planning system are often building blocks which may be applied to various problems. A very complex planning system may use bar charts, critical path methods, decision tables, linear and dynamic programming, simulation, Petri nets and AI methods. Often it is very difficult to determine which tool can be used advantageously for the solution of a given problem. When a large planning system is configured from basic planning tools, the internal interfaces must be carefully designed to provide an efficient communication between the components. A planning task goes through several stages.

Planning starts with the conception of the product and ends with its maintenance. An important part of planning is the availability of a comprehensive database, containing information on the products, processes, manufacturing resources, competition and so on. The planner must have access to all data he needs for his work and he is also responsible for updating the database to ensure that information is always current.

In most cases, planning is a recurring activity and the basic tools will be used many times over. For this reason, a planning system must be designed in a modular fashion, where the modules can be parameterized by an inexperienced user. It may also be of advantage to provide tools for configuring a special-purpose planning system from a library of components.

There are two planning activities concerned with the manufacturing process; facility and operations planning. With the advent of the flexible manufacturing concept the dividing line between these two is not very clear any more. Here, it is of importance to provide a layout of the manufacturing equipment so that the machine tools and transportation equipment can be easily configured to a workable system according to a virtual manufacturing concept.

In this chapter, we will place most emphasis on new planning tools for the planning and configuring of manufacturing systems and components and the integration of these tools into complete planning systems. Several of the more conventional planning tools will be mentioned, but not discussed. The reader who is interested in these tools should look at specific literature, which is abundant. The subjects of process planning, test planning, assembly planning and scheduling will be treated in detail in other chapters. In this chapter, emphasis is placed on the need for good planning tools that must be used for the design and planning of a manufacturing system. The planning methods discussed provide a framework for tools used to analyze the various manufacturing operations presented in subsequent chapters.

3.2 An Integrated Approach to Manufacturing System Planning

The planning of a manufacturing activity may be done for a new or existing plant. In the case of a new plant, the restrictions for configuring the system are given by management philosophy, the market, available resources and the expected return on investment. However, if a plant has to be retrofitted, the restrictions on the available manufacturing space, processes and work-force have to be considered. Often, in such a situation old and new technologies have to be combined. For a

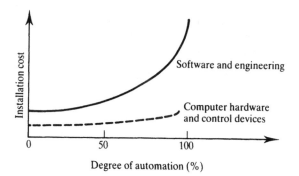

Figure 3.1

Cost of a control computer installation over degree of automation.

CIM system this can impose many problems in terms of equipment–computer communication and interfacing of the various technologies.

The planning cycle will go through various stages and may have to be repeated several times until an acceptable solution has been obtained. It is important that the integrity of the results of every phase is carefully reviewed before the planning enters the next phase. An error introduced in an early phase will have a detrimental effect if it is carried through into the operating manufacturing system. The cost of removing it in the operating manufacturing system will be several times more expensive than removing it in the beginning.

One major problem is to determine the degree of automation which can be economically justified. This depends on factors such as the number of products sold, the cost of available resources, the labor skills available, the complexity of the product, the available automation technology and many others.

Figure 1.11 shows how various manufacturing technologies influence the product cost. Often it is not economical just to use one kind of manufacturing technology; several of them have to be interconnected. For example, a family of products in general, uses common parts. For these parts classical mass-production methods may be used because they are seldom changed. For the final assembly of the product, however, a flexible assembly system may be the most economical solution.

Another problem which must be considered is the cost of software. The cost of computer hardware has been steadily falling over the last 25 years and that of the software has been increasing. Today, it is still very difficult to estimate automation software, in particular for online computer systems. Compared with the hardware cost, the cost of software increases disproportionately with the degree of automation, as shown in Figure 3.1.

3.2.1 The planning cycle

The planning of a manufacturing operation starts with the design of a product or product family and an idea of how many will be sold and how they should be manufactured. The design will be reviewed for its manufacturability. The reviewing is usually an iterative process because various changes have to be made until the experts agree on the production methods to be used. There are various expert skills involved in setting up the manufacturing system. These skills must be properly coordinated and supervised in order to build a functional production facility. For our discussions, we will be defining the phases of the planning cycle as follows:

- requirements analysis,
- data preparation and analysis,
- system configuration and technical evaluation,
- economic analysis,
- final system design and documentation,
- system realization and testing.

The work of the first four phases we usually call the pre-design activities and that of the last two phases the design and implementation activity. There are numerous tools available to support the planning, as shown in Figure 3.2. Some tools are specifically used for the work of one phase, others can be applied to many phases. The early phases are usually the most critical ones because they specify the entire system; unfortunately, for these phases, the development tools are often sparse or difficul to understand and use.

In the later phases there will be extensive use of analysis, modeling and simulation and an attempt should be made to design a common database which can be accessed by all activities; this will enhance the integrity and correctness of the

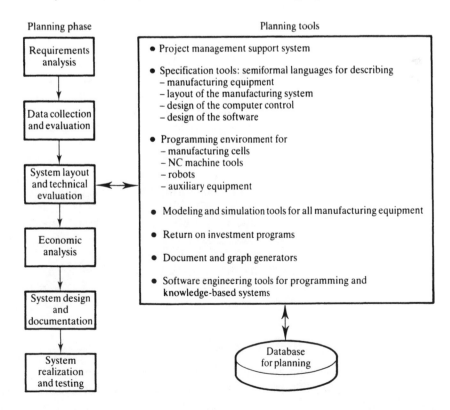

Figure 3.2

The design phase of a planning cycle for a manufacturing system and the supporting planning tools.

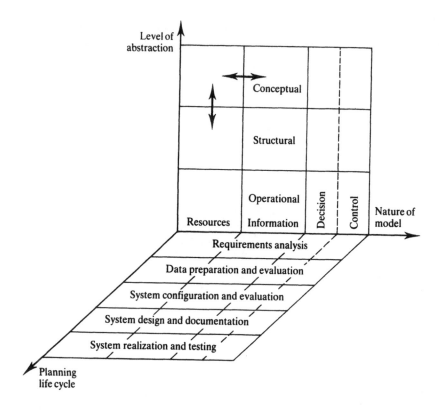

Figure 3.3
Various models of
manufacturing systems.

results. Product design and factory equipment layout is done with geometric models and the same objects are frequently presented. For this reason, there should be access provided to the engineering database. Operational data is also input information to planning and must be presented in a useful format.

The three basic components which make up the manufacturing system are the physical plant, the information database and the flow of decisions; the latter is initiated by the customer order. Each of these three components must be considered by the planning cycle and there will be several levels of abstraction for each phase which guide the designer from his concept to the implementation. The basic tools of the designer are models which help him to visualize and write down his ideas. Figure 3.3 shows a three-dimensional system of the various models of the planning phases, see Doumeingts (1989). The models must be linked by a suitable interface. The transformation from one model to another can be greatly assisted by planning tools for the hardware, software and control of the manufacturing system. Several of such tools are listed in Figure 3.2. We will be discussing the various planning phases and planning tools in the following sections.

3.2.2 The requirements analysis

In this activity the basic concept of the manufacturing process is drafted. If a product of the type to be made has already been produced previously, a careful review of the former manufacturing process will be made to find its strengths and weaknesses. Probably, many parts of this process can be taken over and will serve as the basic components of the new system. In the case of a new facility, an in-depth investigation has to be made of available manufacturing methods and processes, and several competing alternatives will be considered. To evaluate the alternatives, a mathematical model will be set up for the simulation of the manufacturing process. It may also be necessary to build up a physical model, however, this is very expensive.

The study will proceed in a hierarchical fashion. Intuitively, planning from the top down will be the most logical way of structuring the facility. Because of practical considerations, this, however, is not always possible or even advisable. When existing machines and processes are incorporated in the new facility, bottom-up planning is also necessary.

It is most advantageous to use a well-structured planning aid to define all the functions and interfaces of the manufacturing system and to verify the consistency of all information which describes the factory. For our discussion we will be using tools of the GRAI method to show how a complex production system can be systematically structured as discussed in Doumeingts (1989). Figure 3.4 shows how the GRAI method is structured. It uses decision centers and resources and refines

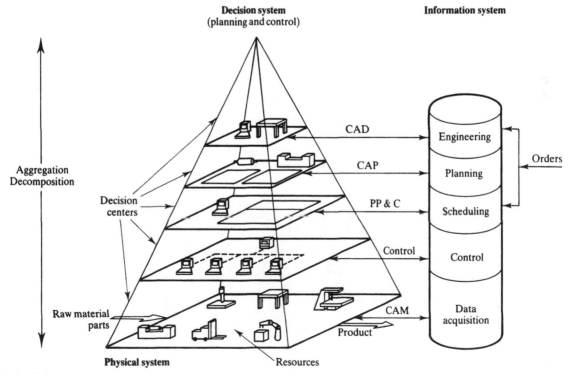

Figure 3.4
Macro structure of a manufacturing system.

the planning process by several stages of decomposition. A manufacturing operation consists of the physical system and an information system. Both of these systems are planned and controlled by a hierarchical decision architecture. The decisions are responsible for controlling the flow of information and resources. The impact and temporal value of a decision depends on the tier on which it is being made. At a higher level, a decision will affect a large part of the manufacturing system and its value is maintained for a long period. At a lower level, only a smaller part of the plant will be influenced by a decision and its value is of short duration. The GRAI method conceptually starts with a global decision center at the highest level where, for example, a decision is made to build a certain model of car. The results of this decision are one or several decision frames, the contents of which is sent to decision centers at lower levels, as shown in Figure 3.5. A decision frame can be a budget, subcontract, order, load schedule, quality test procedure, and so on. It contains information about the allocation of resources, performance criteria and responsibility assignments. The decision making is supported by various aids such

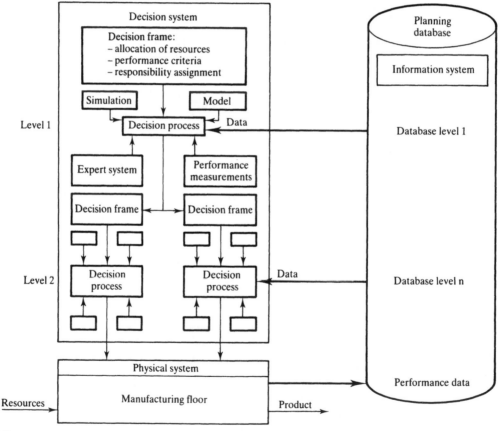

Figure 3.5
Hierarchical
decomposition of a
decision.

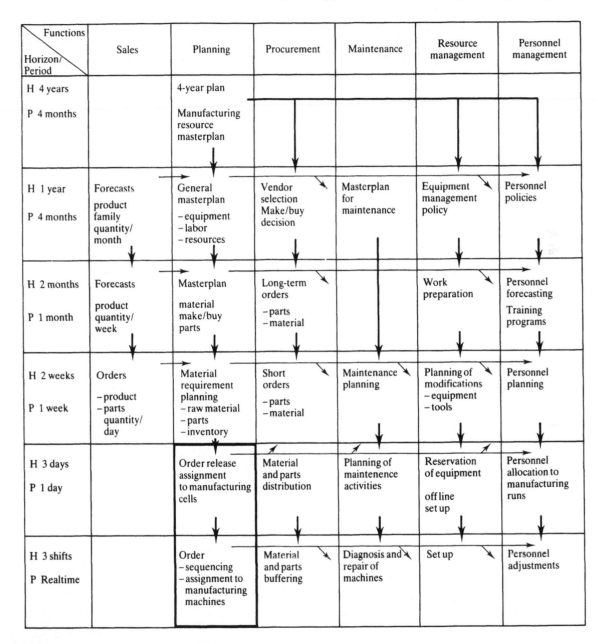

Horizon/Period \ Functions	Sales	Planning	Procurement	Maintenance	Resource management	Personnel management
H 4 years P 4 months		4-year plan Manufacturing resource masterplan				
H 1 year P 4 months	Forecasts product family quantity/month	General masterplan –equipment –labor –resources	Vendor selection Make/buy decision	Masterplan for maintenance	Equipment management policy	Personnel policies
H 2 months P 1 month	Forecasts product quantity/week	Masterplan material make/buy parts	Long-term orders –parts –material		Work preparation	Personnel forecasting Training programs
H 2 weeks P 1 week	Orders –product –parts quantity/day	Material requirement planning –raw material –parts –inventory	Short orders –parts –material	Maintenance planning	Planning of modifications –equipment –tools	Personnel planning
H 3 days P 1 day		Order release assignment to manufacturing cells	Material and parts distribution	Planning of maintenence activities	Reservation of equipment off line set up	Personnel allocation to manufacturing runs
H 3 shifts P Realtime		Order –sequencing –assignment to manufacturing machines	Material and parts buffering	Diagnosis and repair of machines	Set up	Personnel adjustments

Figure 3.6
Typical GRAI grid describing a manufacturing system. Bold arrows represent decision frames; fine arrows are decision paths.

as models, simulation, knowledge based systems, performance measurements and planning data; whereby, each level is provided with its own aids and database.

For the description of the manufacturing system the GRAI method uses a GRAI grid, as shown in Figure 3.6, also see Meyer (1987). In this grid the manufacturing functions are shown along the x-axis. The information flow is indicated by a fine arrow line and the transfer of a decision frame between two decision centers by

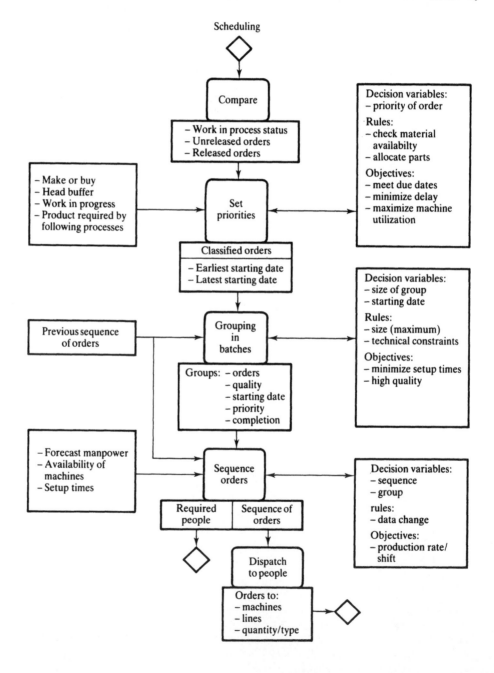

Figure 3.7

GRAI description of an activity center for scheduling and sequencing orders.

a bold arrow line. In the left hand column time frames are shown; H specifies the planning horizon and P the planning period. The individual activities of a decision center can be represented by a detailed net; a sample of a scheduling activity is shown in Figure 3.7, see also Meyer (1987). The figure contains five decision processes, shown as rounded boxes. The decision frames are the rectangular boxes which immediately follow the decision processes. The other boxes indicate the various inputs to the decision processes. The contents of this figure could actually represent the activities of the heavy outlined boxes of Figure 3.6.

Tools like the GRAI method have numerous advantages if they are used consistently for planning. First, they force the designer of the system to structure the problem in a precise manner. Second, they provide a textual and graphic framework for describing the application including input and output forms, grids, flow charts and so on. Third, they help the user to define the problem with specification languages for hardware and software; with these tools the user can check the completeness and consistency of his input and output variables. Fourth, the components of the system being investigated can be described in a modular fashion and it is possible to use them again.

The data which goes into the requirements analysis is as follows:

- workpiece data: product models, product variants, part families, parts, material,
- technology: manufacturing methods, processes and sequences,
- customer data: number of models to be sold, temporal information,
- manufacturing skills: processing know-how, available craftsmen,
- make or buy policies.

3.2.3 Data preparation and analysis

During this phase the basic data needed for the production system is collected and prepared. The principal sources of the data are the part drawings, bills of materials and the various master files of the manufacturing resources. For a new plant, much of the data has to be obtained from machine tool suppliers, books, publications and experiments. The information must be brought into a suitable form to be useful for all planning phases. For this purpose, careful attention must be paid to the design of the database and the possibility of updating it. The design of the database should take into account the following considerations:

- transparency of the database for the various planning phases,
- ease of accessing the data,
- feedback of the planning data,
- ease of maintaining the data,
- accessibility of product design data.

In this phase much manual work will be performed because there is no way of automating the tedious search for data. The most important data to be gathered is as follows:

- process planning: raw material, machine tools, machining sequences, fixtures, tools and machining parameters;
- assembly planning: main and sub-assemblies, assembly machines, tools and fixtures;
- test planning: test methods, tools and equipment;
- process capabilities: piece rate, productivity, accuracy and reliability;
- scheduling: machine tools and processes, load profile, manufacturing sequences and routes, processing alternatives and so on;
- materials: types, machinability, source, inventory cost and so on;
- labor: skills, labor rate;
- operating costs: of various machines and processes, fixed cost and variable costs.

3.2.4 System configuration and technical evaluation

The planning methods used in this phase are heuristics, operations research tools, simulation and, as a newcomer, the knowledge-based system, see Figure 3.8. The aim of the phase is to find for the product family an optimal machine configuration. The system designer will start with a rough layout of the facility and look at manufacturing alternatives, as shown in Figure 3.9. For each configuration, the system designer will perform a simulation run to investigate the dynamic behaviour of the system and to compare the alternatives. He will probably look at various optimization criteria, such as manufacturing time, utilization of the equipment and cost. The output of the system is an evaluation form containing numeric or graphic data. If the system does not perform as desired, design changes are made and, if necessary, selective test runs are performed with real production machines to eliminate bottlenecks. These changes may require that the planning cycle be repeated to obtain further refinements.

A typical production facility for discrete parts manufacturing will consist of machine tools, robots, transportation vehicles, tools, fixtures and so on. It is important that a description of this equipment is located in the planning database. There will be two kinds of simulation software, one for numerical and one for graphical simulation methods. For this purpose, numerous simulation systems have been developed which are commercially available. When the planner has obtained a workable solution for his process he performs a graphical simulation to observe the actual operation of the equipment. To do this, he needs programs for describing the cell, the workpiece contour and the robot trajectories. Good simulation systems allow the viewing of the activities of the overall systems as well as of the detailed operations, such as inserting a peg into a hole.

The planning of a production facility will be done in a hierarchical fashion as shown in Figure 3.5. This figure shows two hierarchical levels, however, in a real situation there will be several levels. Planning a manufacturing cell, typically, is

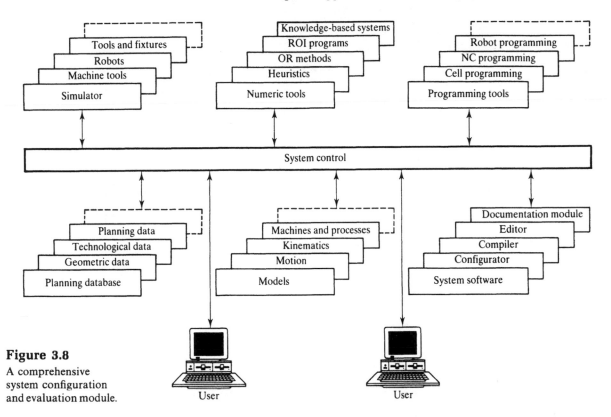

Figure 3.8
A comprehensive
system configuration
and evaluation module.

done in four or five stages as shown in Figure 3.10. Every planning box will be equivalent to a decision process and needs its own model, simulator, knowledge-based system and database, as seen in Figure 3.5. The data is obtained from the planning database in the data evaluation and preparation phases.

The outputs of the system configuration and evaluation phases are:

- the physical layout of the facility,
- production capacity of the facility,
- the manufacturing cost of the workpieces,
- the processing time,
- the machine programs,
- the process, test and assembly plans.

3.2.5 Return on investment analysis

The return on investment (ROI) analysis for a computer-integrated manufacturing system is very difficult for several reasons. First, with a new installation there is no experience available which will support the analysis. Second, the computer control system has numerous intangible benefits which cannot be anticipated before

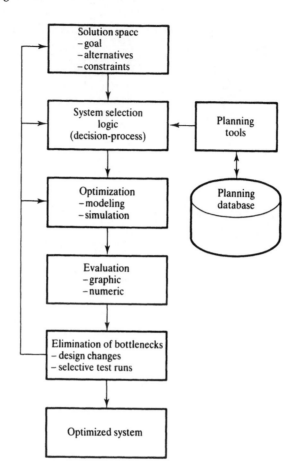

Figure 3.9
System configuration
and technical
evaluation.

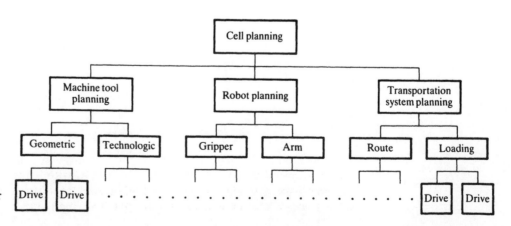

Figure 3.10
A hierarchical planner
for a work cell.

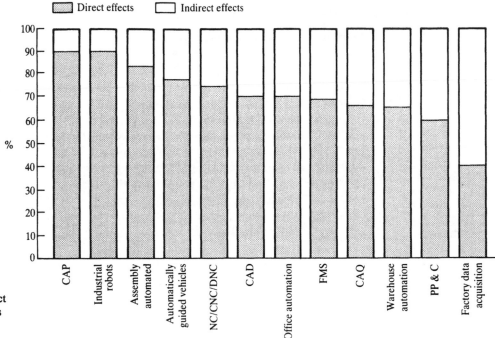

Figure 3.11
Comparison of direct
and indirect benefits
of computer
automation.

it is installed and operated. Third, a computer-controlled module may show no benefits if it is justified by itself; however, it may bring substantial benefits when it interacts with other modules of the system. For example, a computer in engineering may only be justifiable if it is also used for process planning, see Figure 3.11 and Rembold (1990). Fourth, a new technology brings structural changes to an organization; and it is not apparent how they will affect the operating cost. Small and medium-sized companies, which have had no or little experience with the CIM technology, will have the greatest difficulty calculating a return on investment. They are used to making simple cost comparisons, without considering benefits obtained by system integration and the use of various manufacturing strategies.

The return on investment analysis should use a dynamic method, considering the time value of money. For the analysis, fixed and variable costs are required. Usually, the fixed costs do not impose major problems, however, the variable costs are difficult to get, in particular for computer automation. Numerous ROI algorithms and programs are commercially available and they can substantially reduce the time for performing the calculations.

The output of the return on investment calculations is the basis on which the economic feasibility of the entire project will be determined. In most cases, changes to the plan have to be made to reduce the cost to a justifiable level.

3.2.6 Final system design and documentation

During this phase, the physical production system is configured. The most difficult and time consuming task will be the design of the computer control and the software. The system designer will be using well-established software engineering methods to conceive, write and test the software. There is not only software needed for controlling the production process, but also for servicing and trouble shooting the computers and the manufacturing processes. In most installations this additional software is far more numerous than the control software. The ease of operating a computer-controlled process depends heavily on the man/machine interface and the available service software. Knowledge-based systems may be used for trouble shooting and for simplifying the communication with the computer hardware and the manufacturing equipment.

An important, often neglected, task is the documentation of all hardware and software components. For the software, this can be done during programming where ample comments are inserted into the program. For the hardware, instruction manuals must be written to explain the start up, shut down and normal operation. If knowledge-based systems are used for trouble shooting they should be capable of diagnosing all defects or problems which may occur. It is also important to secure the source code for all software which may have to be changed during the life of the plant due to additions, deletions or corrections to the process.

3.2.7 System realization and testing

This is the final phase where all modules are installed and tested. It is necessary to check all system components separately and, if they operate properly, to test the entire process. Particular attention must be given to the hard and software interfaces to be certain that proper synchronization takes place between the system modules and the manufacturing process.

Usually, this phase is the most frustrating part of the entire planning cycle. Rarely will the installation be online after the first start-up trail. In most cases, considerable difficulties are encountered with hardware and software. Control elements may interact with each other, or the computer will suddenly enter an unexpected path which causes the collapse of the entire system or part of it. Often, there are numerous controllers, sensors, software or machine elements which do not function according to specification.

The problems and their remedy should be recorded carefully to pinpoint the source of the malfunction and to accumulate statistics for recurring breakdowns. Graphical and numerical simulation tools are excellent means for simulating the operation of a faulty system and for interpreting its behavior.

After the system has been operated over a longer period of time, a post-audit should be performed. Management must assess the performance of the entire installation and its components in order to determine its benefits. Valuable information will be obtained for justifying future installations and for improving the present system. The performance of each component must be measured against its specification. After post-evaluation, the user should be aware of the fact that it often takes a year or more until a system has reached steady state condition. The importance of the evaluation are its findings and the recommendations for actions. As the result of this post-audit many of the planning and programming tools may have to be revised.

3.3 Simulation

3.3.1 Simulation tools

A very successful aid for planning a manufacturing system and its operations is simulation. For a long time, industry viewed simulation as a very exotic tool only useful for academic studies, because simulation systems were difficult to apply and needed a large computer. However, in recent years, the situation has changed drastically. Simulation systems were augmented with user-oriented interfaces, and powerful desktop workstations were developed. In addition, a network was provided, allowing the interconnection of workstations and providing ties to larger computers.

The engineer builds a mathematical or graphical model of the process he wants to study with the computer-supported simulation and observes its behavior. The output is either a report or animated pictorials on a graphic display. Thus, the engineer is able to investigate various manufacturing operations without a physical setup of machines and processes. The model usually contains only the essential parts of the process which are to be investigated. Thus, the building and execution of the model can be greatly accelerated. It is possible to investigate quickly various manufacturing runs and manufacturing alternatives with the model. It is also possible to simplify the speed up or slow down of a manufacturing run and to take a macroscopic or microscopic view of the process. Furthermore, disturbances can be introduced and their effects observed.

A simulation can also be used for controlling a manufacturing process. In this case, the model is operated in parallel to the production system, and the plant parameters are identified and used by the model according to an algorithmic or search-oriented optimization strategy to produce the set point parameters of the process.

In the following discussion, the various phases of a simulation will be explained and thereafter, the components of a simulation system for manufacturing applications will be explained. The reader who is interested in studying simulation languages and methods will find ample literature on this subject.

3.3.2 Methods of modeling

There are various types of models which can be used for the representation of a manufacturing facility. Figure 3.12 shows a classification scheme of models.

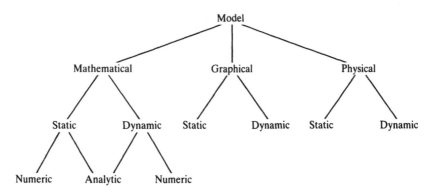

Figure 3.12

Types of models.

Mathematical models

In a mathematical model, the objects of a manufacturing system and their attributes are represented by mathematical variables. The operations and actions are described by mathematical functions which define the interrelationships between the manufacturing variables. Mathematical models can be of the static type or the dynamic type. With the static type, the modeled system is in equilibrium; and with the dynamic type, the system may be in continous change. In practice, both types of models are used to represent manufacturing operations.

Operations research (OR) methods provide various tools for the construction of manufacturing models. These tools can mainly be found in modeling of planning, scheduling and control functions of manufacturing processes. The methods of OR include linear programming, dynamic programming, network analysis, queuing models, inventory models and numeric simulation. It is very typical for OR methods to be applied to solving problems of various unrelated areas, for example, a transportation algorithm for scheduling parts to various machine tools can also be applied to planning the sequence of punching holes in a complex sheet metal part. In both cases, an attempt is made to travel to a sequence of destinations at minimal cost. OR methods are most useful for offline planning; once a plan has been drafted, it will be applied to a known manufacturing situation. If, for example, a machine tool fails, the planning has to be done over again in the offline mode. This is a weak point of the OR methods. They may also encounter difficulties if a manufacturing process has numerous disturbances and a model cannot adequately represent them.

Graphic models

Graphic models are important tools for representing the behavior of a manufacturing system. The manufacturing operation may be visualized by either icons or symbols. Icons are ideal tools for a simulation if they pictorially resemble the manufacturing equipment. Building a library of icons is, however, a very tedious and expensive endeavor. There are various ways of symbolizing a manufacturing operation. New emerging tools use Petri nets. They will be discussed later in this chapter.

Physical models

Manufacturing operations are often modeled by mock-ups or pilot operations. An attempt is made to build a physical image of a factory or a component of it with these methods in a simplified manner. It is important that these models include all functional entities which are necessary for representing the real manufacturing operation. This may become a long and expensive endeavor. Also, the extrapolation of the behavior of the model to that of the real system may lead to unrealistic results. Physical models are being replaced by mathematical or graphical models.

3.3.3 Simulation methods

When a manufacturing process is to be simulated, first a model has to be built of the object to be simulated. In the ideal case, for a continuous process the model is constructed from mathematical equations, describing the physical behavior of the object. This, however, can very seldom by done for a manufacturing process because it is virtually impossible to describe its underlying physical principles in an exact form. When experiments are performed with a process, it may be treated as a 'black box' and the behavior of the output parameters are determined as functions of variable input parameters. This type of modeling is frequently applied to continuous or semi-continuous processes such as metal treatment operations or plastic extrusion.

The principle simulation methods used for manufacturing applications are the discrete event simulation and the graphic simulation. There are various simulation systems available which combine both methods. The discrete event simulation system is suitable for studying the flow through a manufacturing operation for determining its capacity, locating bottlenecks, observing the distribution of parts to workstations, looking at manufacturing alternatives and many others. Objects are represented either by symbols or icons with graphical simulation.

When using icons, the manufacturing system can be animated directly on a screen and viewed in various levels of detail. They are very useful for observing the flow of objects, machining of parts and the assembly of products. In the following sections we will be discussing some of the languages available.

Simulation of discrete events

The majority of manufacturing processes are discrete events and must be described as such. Originally, for the simulation of these types of activities, high-level programming languages such as Fortran, Pascal and Modula were frequently used. The following features are needed for a simulation:

- random number generator,
- probability distribution functions of the events to be investigated,
- statistical data and functions,
- time-control mechanism,
- protocol facilities.

These functions can be added to the language in the form of subroutine calls, if not already available.

In a discrete parts manufacturing process, the flow of a product or information occurs in a stochastic fashion and the events, for example, the arrival of parts, behave according to a probability distribution function. This function must be known and entered into the simulation system. It is difficult to obtain the function of a manufacturing process which is to be simulated and does not yet exist. In many cases, sufficient experience is available from similar processes and an assumed probability distribution can be taken. In other cases, it may be necessary to assume the function or to get experimental data from pilot production runs.

A simulation performed with a high-level programming language can be quite involved when the flow of objects has to be described, and for this reason special languages were developed for studying discrete events. Typical discrete event simulation systems are GPSS (general purpose simulation system), SIMSCRIPT (simulation scripture), SIMULA (simulation language), SLAN and SIMAN.

There are specific manufacturing simulators emerging which simplify the simulation of factory operations. These include Witness, Promod, STARCELL, and so on. The reader is advised whenever he wants to perform a factory simulation to look at the practicability and availability of these special tools.

Graphic simulation

The graphic simulation can actually be classified by three methods:

Simulation using icons.

With this type of simulation, objects of a manufacturing system are presented as simplified images (icons) and stored in a well-structured graphical database as classes of objects. The objects can be machine tools, automatically guided vehicles, robots, workpieces, etc. When the manufacturing operation is to be simulated, the objects are retrieved from the database and configured to an operable installation. The operation of the system can then be viewed in real time and it is possible to interact with the displayed objects to try out various functions. This type of simulation is most useful for debugging, fine manipulation and material flow observances. Figure 3.13 shows a simulated robot operated assembly. Iconic modeling is also suitable for selecting components of a system to be configured. In this case, icons are provided for the various objects to be simulated. The icons can be displayed onscreen and the system will provide a template which has to be filled in by the user to define the actions of the object. The definitions can be done with user entries or pre-assembled functions.

Simulation using logic symbols.

With this method, graphical symbols are assigned to a selected vocabulary of system functions, for example, advance a part, assign a part to a machine, tabulate a production count and so on. The arguments needed for invoking the functions are provided by the user. He can display the operation of a facility with the help

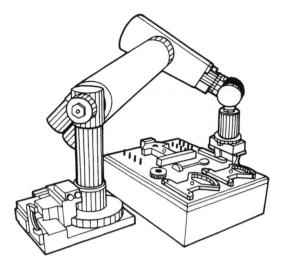

Figure 3.13
Robot operated
assembly station.

of symbols onscreen and observe it. With advanced simulation aids the object can
be colored.

Simulation using symbolic graphics.
With this method, an object is presented by a symbolic icon, for example a
temperature by a thermometer, a load profile by a bar chart, an applied pressure
by an arrow and so on. This graphic is usually associated with an output report
and can be part of a documentation. If displayed onscreen, some systems will
allow animation of the symbols for example, the increase of a pressure shown by
a pressure gauge. With advanced systems, coloring is possible.

3.3.4 The simulation cycle

When a process model is being built, it is essential to include all components
which are necessary to describe the functions under investigation. Any under-
specification will lead to wrong results, and any overspecification makes the model
too complex and may result in unacceptable computer runs. To find the right level
of abstraction is not an easy task and modeling experience is necessary. Another
problem, which is usually associated with simulation, is the time and manpower
which should be spent. Modeling may become very complex and time consuming
and manufacturing personnel may become impatient waiting for the results.

A simulation has its defined life cycle similar to any planning activity. Figure
3.14 shows the various phases. The content of this figure does not have to be
explained in detail, it is self-explanatory. The execution of the phases may have
to be repeated several times until an acceptable solution has been obtained. In
many cases, the basic data which goes into the simulation is not sufficient, and
experimental data may have to be obtained from a pilot production run. This,

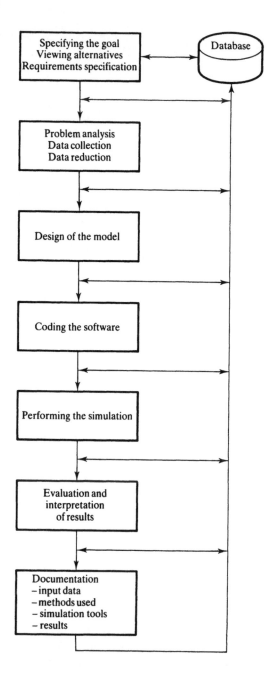

Figure 3.14
Phases of a simulation.

usually, will complicate the simulation. It is important that the documentation of the simulation is clearly formulated so that the results can be easily used. The simulation must be done in close cooperation with the factory personnel to ensure its correctness and to make the results acceptable.

3.3.5 A comprehensive simulation tool

A simulation tool for a factory operation will consist of numerous modules for the components and the entire system. There is no ready made all purpose simulation system available, and the user has to assemble a comprehensive simulation tool from numerous sources. Usually, basic simulation systems like GPSS or SLAM can be purchased, however, many processes in the factory are application-specific, so the user has to build his own modules. A comprehensive simulation system will be very complex because it is not easy to tie the modules together to one unit, and the problems will be long computer runs, interfaces and common data formats. There is no guarantee that purchased and in-house modules can ever be integrated into an efficient system. In many cases, the simulation system will consist of several modules, with its own input/output structure and maybe its own database.

In Figure 3.15 the architecture of a simulation system is depicted for a discrete manufacturing operation. The production units may consist of machine tools, robots and material transportation vehicles. The system is constructed from five modules; modeling, emulation, programming, simulation and graphics. The latter contains the tools for the user/system dialog. The data is stored in a simulation database which, preferably, has a relational data structure. We will be discussing the most important components of the system.

The modeling module

This module contains all the models needed for the simulation, including those of the product, machine tools, robots and transportation vehicles. The models describe the kinematic as well as the dynamic properties and behavior of the object under investigation. The factory model is used for the configuration of an entire manufacturing operation. There is one module for sensors and one for control loops for investigating the basic behavior of these components.

The emulation module

The emulation module is used for showing the operation of the manufacturing equipment and its components. Motion planning is done for all moving equipment like robots, machine tools and material handling devices. The navigation component emulates the travel of the transportation vehicles through the plant. The sensor and control modules allow the emulation of measuring strategies and control operations. There are also modules provide for investigating various decision and planning strategies.

The programming module

The cell operation, robots and machine tools are programmed with the help of this module. Programming can be done via teach-in, textual, graphic or task-oriented methods. A special module allows programming via experimental setup.

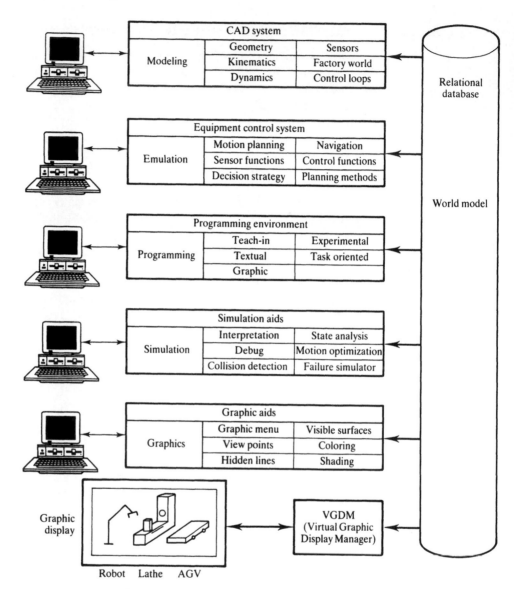

Figure 3.15

A comprehensive simulation system for factory applications.

Simulation aids

The interpreter is the interface between the programming system and the simulation. The code from the programs is accepted, interpreted and then displayed as moving icons onscreen. The debug facility is used for locating and correcting failures in the programs. The collision detection module investigates all motions of the equipment and locates collision problems. Typically, collisions which can be detected are those of robot arms, interaction of tools with a workpiece to be cut and problems with moving transport vehicles. The failure simulator allows the introduction of defects into the system to study the reaction. Motion optimization is necessary for metal cutting, navigation and assembly. A special model is provided for investigating and displaying the state of the manufacturing operation at any time.

Graphic aids

This module contains all computer tools necessary for modeling and simulation. The contents of this box are self-explanatory.

The last component is the virtual graphic display manager. It provides an equipment independent graphic output. Thereby, it is possible to connect various types of displays to the simulation system.

3.4 Petri Nets

3.4.1 Introduction

In a manufacturing process, objects such as workpieces flow in a sequential, nonsequential or parallel order through a plant. At various workstations, operations are performed with the objects, whereby material is removed, heat treating operations are done and parts are assembled. Each operation, which is called an event, is done by manufacturing resources also known as objects. Typical objects are machine tools, robots or manpower. Petri nets are a simple tool for modeling workpiece flow through a manufacturing system and for representing the orderly execution of the individual operations. Particular features of Petri nets are their ability to handle parallelism and concurrency. Their weak point is that they cannot represent time easily.

Figure 3.16 shows a simple Petri net in which a robot picks up an object and loads it into a machine tool. An object is represented in a Petri net by a token (dot) which is located in a place (circle). An operation is performed with an object when all resources are available by firing a transition (heavy bar). In the upper part of the figure all resources are available. With a token in every place, leading to a transition, the operation 'loading' is triggered and a new token appears in the place after the transition. All input tokens are removed. The lower part of the figure shows that the machine tool is loaded by the presence of a token in the lower place. With this simple example in mind, we will explain the most important properties of Petri nets.

3.4.2 Fundamentals of Petri nets

A special class of Petri nets are condition/event Petri nets, which are often used for modeling manufacturing systems. A condition/event Petri net (CEP) is a four-tuple $N = (P, T, F, M_0)$, where

(1) P is the set of places

(2) T is the set of transitions

(3) $S \cap T = \{\}$

(4) A flow relation $F \subseteq (P \times T) \cup (T \times P)$

(5) M_0 is an initial marking.

p_i is called an *input place* of the transition t_j if $(p_i, t_j) \in F$

p_0 is called an *output place* of the transition t_j if $(t_j, p_0) \in F$

$P_I(t_j)$ denotes the set of all input places of t_j

$P_0(t_j)$ denotes the set of all output places of t_j

Places are interpreted as conditions and transitions are interpreted as events depending on the conditions.

A marking M of a CEP N is a mapping of places to the set $\{0, 1\}$:

$$M: P \rightarrow \{0, 1\}$$

A place p_i is *marked* if $M(p_i) = 1$, otherwise p_i is said to be *unmarked*.

A marked place is indicated by a token in the corresponding net graph. If a place p_i is marked the *corresponding condition* $C(p_i)$ is said to be true, otherwise $C(p_i)$ is said to be false.

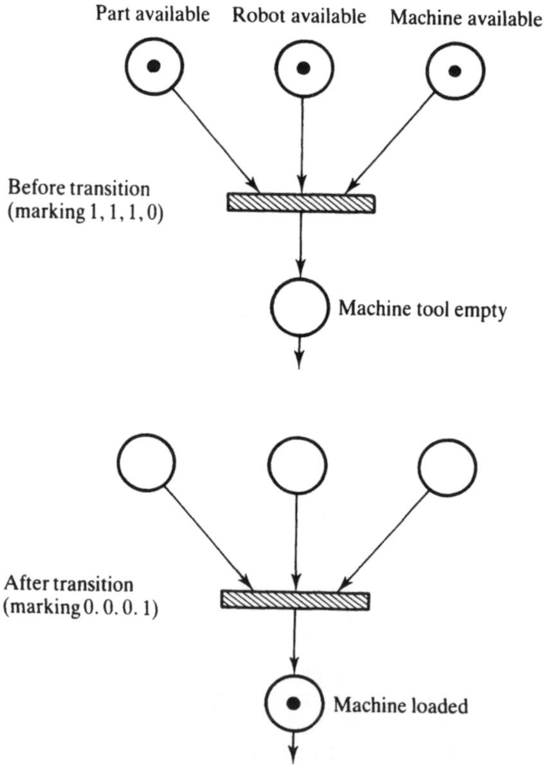

Figure 3.16

Loading a machine tool with a part by a robot; shown with Petri net notation.

A Petri net is executed by firing transitions. A transition may fire if it is enabled with respect to a marking M.

A transition t_j is *enabled with respect to a marking* M if

$$M(p_i) = 1 \text{ for all } p_i \in P_1(t_j)$$

(that is, if a token resides in each of its input places).

When t_j fires, the marking M is updated to marking M' according to the following rules:

$$M'(p_i) = O \text{ if } p_i \in P_1(t_j)$$
$$M'(p_i) = 1 \text{ if } p_i \in P_0(t_j)$$
$$M'(p_i) = M(p_i) \text{ otherwise.}$$

If a transition t fires the corresponding event E(t) is said to occur.

The reachability set R(N, M) of a CEP N with respect to an initial marking M is the set of all markings which can be generated by a valid firing sequence starting with marking M.

A condition event Petri net (CEP) can be designed from the following four basic modules, as shown in Figure 3.17, see also Zhon *et al.* (1989) and Normann (1989).

(1) The **sequence Petri net** consists of $n + 1$ places and n transitions. This module represents a sequence of manufacturing operations, see Figure 3.17(a).

(2) The **parallel Petri net** has one input place, n parallel places and one output place, Figure 3.17(b). The parallel places are separated by one transition from the input place and one transition from the output place. In this model parallel operations are executed simultaneously. They may, however, not be completed at the same time.

(3) The **conflict Petri net** has one input place, n parallel places and one output place, as in Figure 3.17(c). The n parallel places are separated from the input places by n parallel transitions and from the output places by n parallel transitions. In this module operations are in conflict with one another. There is a choice by which competing succeeding operations are performed.

(4) The **mutual exclusion Petri net**. The Petri net shown in Figure 3.17(d) has nine places and six transitions. One designated place is occupied by a token. From this firing place the Petri net is controlled. Either the left side or right side of the Petri net can be fired. For example, an autonomous vehicle designated by P_{-3} may carry parts either to P_{-5} or P_{-6}.

Various manufacturing processes can be modeled, with these four basic modules. It is possible to start with a simple Petri net by describing the operation of a system in a rather abstract manner and then by further refining it to include more and more detail. The refining is limited by the increasing complexity of the Petri net.

In the original Petri net definition, time is not considered. Time, however, plays an important role in dynamic planning and control. Thus, it is necessary to implement time delays for places or transitions of a net. Depending on the

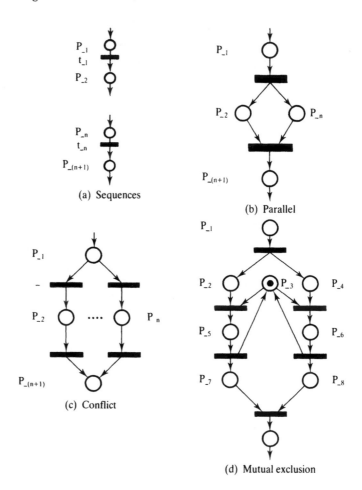

Figure 3.17
The four basic Petri
net modules.

presentation of time. Petri nets may be deterministically or stochastically timed. To introduce the notation of time, two different approaches exist to CEPs.

The system operations can be represented by places with the presence of a token at a place indicating the activity of the operation, as shown in Figure 3.18. Transitions are used to indicate the starting and termination of an operation respectively. This preserves the classical Petri net notion of transitions as instantaneous events, whereas a place remains marked until its associated operation is completed, for example the marked place of Figure 3.18 shows an operation in progress.

If, on the other hand, the system operations are represented by transitions, the execution of an operation is related to the firing of the transition, as in Figure 3.19. Each transition is associated with one of the three states (enabled, firing, inactive) shown in Figure 3.19(a). An enabled transition starts firing by removing the tokens of its input places and then activating the associated operation, shown in Figure 3.19(b). While the operation is executed the transition is kept in the firing state, shown in Figure 3.19(c). In this case, neither its input places nor

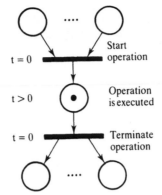

Figure 3.18
Modeling of a timed
Petri net using places
as operations.

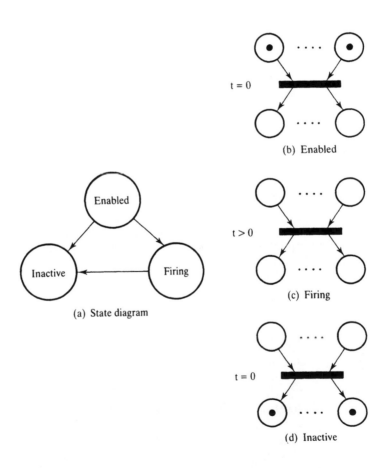

Figure 3.19
Modeling of a timed
Petri net using
transitions as
operations.

its output places are marked. Finally, if the operation has been terminated, the output places of the transition are marked and the transition becomes inactive, as in Figure 3.19(d). This notion of time emphasizes the dynamic character of transitions as active elements causing state changes.

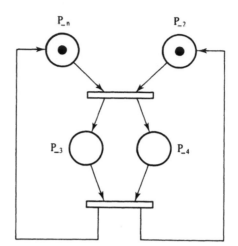

Figure 3.20

Example of a safe, live and reversible Petri net.

For many applications a Petri net must be safe, live and reversible.

(1) A transition t is live with respect to an initial marking M_0 if from every marking in $R(N, M_0)$ there exists a firing sequence which leads to a marking M' for which transition t is enabled. A Petri net is live if all of its transitions are live. If a Petri net model of a system is live, the operation of the system is free of deadlock.

(2) A Petri net is *safe* if not more than one token can ever be in any place of the net at the same time which can occur in special types of Petri nets. A CEP is by definition always safe.

(3) A Petri net is reversible if from every marking in $R(N, M)$ there exists a firing sequence which leads to the marking M. To a manufacturing operation this means that the system must be ready for a new operation after the preceding one has been completed. This may also imply that a system has recovered from a failure.

Figure 3.20 shows a safe, live and reversible Petri net.

3.4.3 A planning example using a Petri net

A robot action plan for an assembly

In this example we show how the assembly operations of a workpiece can be modeled by a Petri net. For this purpose an action plan is drafted.

An **action plan** is the description of a task a robot has to perform. Normally, this description consists of a partially ordered set of simpler subtasks. Two subtasks T_A and T_B are said to be in order $T_A < T_B$ if T_A must be performed before T_B. The sequence constraints of the subtasks are often represented by using a **precedence graph** (PG). Figure 3.21 shows on the top left corner the product to be assembled and on the right the final assembly. The corresponding PG is depicted in the lower part of the figure.

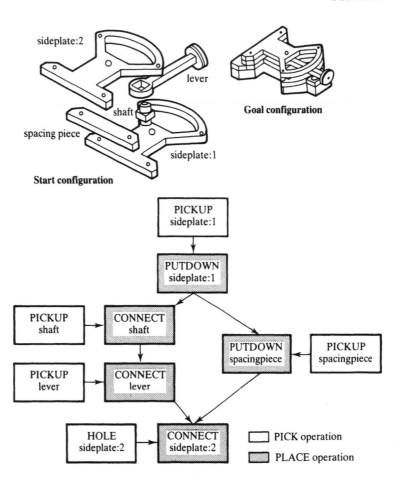

Figure 3.21
Assembly job.

The PG is a directed graph with each node corresponding to a certain subtask. Two nodes A and B are connected by an arc A → B if the corresponding subtasks are in order $T_A \langle T_B$. Thus, a PG represents all possible sequences of subtasks by which a given task can be performed.

An action plan is called an **implicit representation** of a robot task if the subtasks in the plan describe the desired goal (for example, 'pickup object') but not the way this goal can actually be achieved (for example, 'find object, approach, grasp object, depart'). A complex plan execution system is needed for decomposing an implicit subtask description into a sequence of explicit robot control commands, depending on the actual situation at execution time. To achieve this goal, the plan execution system must have knowledge about the online execution environment and the ability to schedule subtasks. The objective of scheduling is the coordination of the subtasks and the allocation of the resources (for example, manipulators, sensors) so that the overall task is completed in a timely and effective manner.

The execution of a PG, that is, the performance of a robot task, is closely related to the interpretation of a Petri net. CEPs are well suited for representing the dynamic behavior of systems which consist of asynchronous concurrent events,

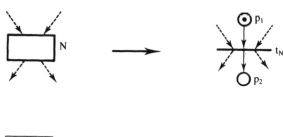

Figure 3.22
Translation of a precedence graph into a condition/event Petri net.

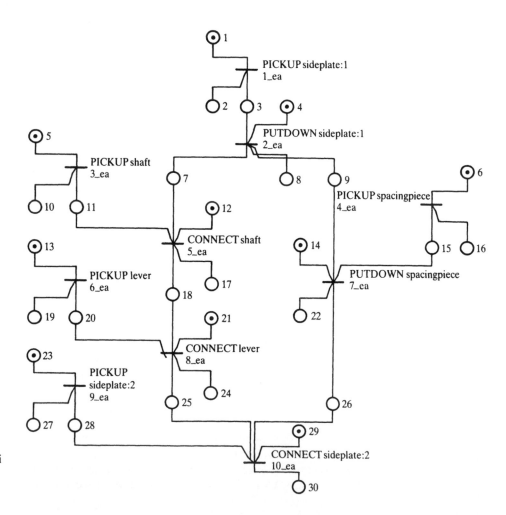

Figure 3.23
Condition/event Petri net graph of a manipulation task.

partially ordered over time. Although the PG describes only the static structure of an action plan, it can be easily translated into a CEP according to the following rules, and shown in Figure 3.22.

- Each node N of the PG is replaced by a transition t_N and two places p_1 and p_2, respectively. p_1 is a marked input place of t_N and indicates that the subtask of node N is not executed yet. p_2 is an unmarked output place of t_N and is used to indicate the termination of the subtask of node N. Thus, the execution of the subtasks of the PG is related to the firing of the transitions of the CEP.
- Each arc $A \rightarrow B$ connecting two nodes A and B is replaced by an unmarked place p and two flow relations (t_A, p) and (p, t_B).

The mapping of the assembly precedence graph of Figure 3.21 onto a Petri net presentation is shown in Figure 3.23. The task is to assemble the five main pieces of the physical pendulum shown in Figure 3.21.

3.5 Artificial Intelligence Methods for Manufacturing

3.5.1 Introduction

The power and potential of artificial intelligence (AI) tools for planning and control of manufacturing processes has been proven by many research projects and actual implementations. The most important AI techniques are quantitative reasoning and simulation, and the use of deep models. The operation of present manufacturing systems is based on human experts who have learned their skills over many years. However, there are two basic problems with human experts. First, when an expert leaves the company or retires, the manufacturing know-how goes with him. Second, many manufacturing chores become routine work after the skill has been acquired, and a planning and control job may become very repetitive and boring. For this reason, it is not surprising that the use of expert systems in manufacturing has become a favorite subject. The activities in manufacturing cover a wide spectrum of product development such as process planning, factory floor layout, scheduling, control and maintenance. Since a vast amount of know-how is needed for all activities, it is obvious that there will not be an all-knowing knowledge-based system which can be used as a central source of manufacturing knowledge. There are numerous knowledge-based systems being developed for almost any manufacturing activity. Typical applications are for design, manufacturing planning, machine diagnosis, machine layout, system configuration, task-oriented programming, man/machine communication, vision, sensor data interpretation and so on. At the present time, however, it is not known how many of these systems are practicable and are being used. A conservative estimate is that the results of only 5% of all research endeavors have found their place in the factory. This may be a very discouraging reality; but there are numerous reasons for this problem, including:

- The tools for building knowledge-based systems are not sufficiently developed and are difficult to apply. The methods for acquiring knowledge from experts are not very well understood. There are too few qualified people available who really know how to apply AI tools.

- Good knowledge-based systems usually contain several thousand rules and are huge software systems which are difficult to use on conventional computer systems. In spite of these sobering facts, it is believed that such systems will have a major impact on manufacturing planning and control. It will just be a matter of time until the right tools and methods have been developed.

In this presentation, an overview will be given on the use of a knowledge-based system in manufacturing. A knowledge-based system is also known as an expert system. The most important developments will be discussed and it will be shown where the tools can be applied and how larger systems can be integrated.

3.5.2 Basic components of an expert system

Structure of an expert system

The builder of an expert system must have a model of the system for which a solution is being sought. The model describes the properties and behavior of the system. Usually, an attempt is made to keep the model simple and to include only the important features of a process. The various models used for knowledge engineering are:

- **Informal symbolic model.** It contains an informal textual description of the process.
- **Diagram.** It may show the flow of information or material through a process.
- **Formal mathematical model.** It describes the behavior of the process with a set of mathematical equations.
- **Heuristic model.** It describes the process with a set of rules.
- **Pictorial model.** It describes the process with symbols or pictures.

Numerous programs and programming packages have been developed to solve manufacturing problems. The heart of a program consists of the algorithm and data, as shown in Figure 3.24. The execution of a program is done in a concise manner laid down by the rules of the algorithm, which may include branching and looping. Usually, the algorithm is traversed in a hierarchical manner, starting with an abstract presentation and then going into detail. Upon completion, the

Figure 3.24

Comparison of the traditional program structure with that of an expert system.

Traditional program = Data + Algorithm
Expert system = Knowledge base + Reasoning engine

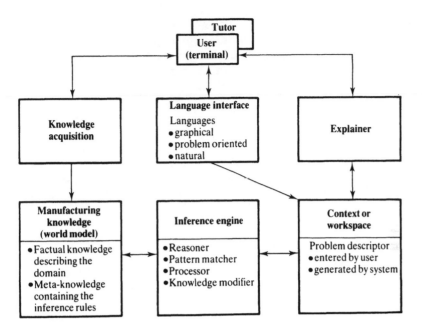

Figure 3.25
Schematic of an expert system.

algorithm may present one or several results. In principle, any algorithm and its associated data contains expert knowledge about a limited domain.

The basic difference between the program and the expert system is the way knowledge is presented and processed. The expert system consists of a knowledge database containing explicit knowledge of a human expert in a specialized domain, and a reasoning or interface engine which can access the knowledge base to come to a decision for a described problem, as shown in Figure 3.24. The description of the problem and the context are entered by the user or by the system. Figure 3.25 shows a schema of an expert system which consists of the following six components:

(1) **User or tutor interface.** There are two groups of persons who must have access to the expert system. First, the tutor who sets up the system and who prepares the knowledge to be entered into the knowledge database. The tutor will also maintain the expert system. Second, the user who tries to find a solution to a problem. He must be able to describe the context of his problem to the system. Both the tutor and user communicate with the expert system via a graphics terminal.

(2) **Knowledge acquisition module.** This module processes the data entered by the expert and transforms it into a data presentation understood by the system.

(3) **Language interface module.** There are various ways of communicating with the system; by natural, graphical or problem-oriented language. The level of abstraction of the user language will be much higher than that of the tutor language. The language used must be understood by the interface, and a special expert system may be needed to extract from the user input the semantics which describes the problem.

(4) **Manufacturing knowledge module.** This module can be understood as a world model of the domain for which the expert system was developed. It is like a huge database which contains all factual knowledge and rules needed for the operation of the expert system. There are also empirical rules in the form of meta-knowledge stored in the database, which knows how the factual knowledge has to be processed to find an answer to an inquiry.

There are two ways of operating this module. First, the user enters into the system the description (context) of his problem. It is stored for use in the system. Second, the system constructs the description by interrogating the user in a question and answer session. For this operation, there must be a description of the problem available to the system. This module can be considered as a temporary storage needed for the solution of the problem.

(5) **Inference engine.** Essentially, this module is the knowledge processor which looks at the problem description and tries to find a solution with the help of both the factual and meta-knowledge. First, all rules of the factual knowledge are investigated by the reasoner, and, with the help of the pattern matcher, the ones to be used are selected. Thus, a set of candidate rules are obtained. Second, one of the rules is selected and applied to the problem description by the processor. Third, with the result obtained, the original context is changed by the knowledge modifier. This completes the first step of finding a solution. The whole process is started over again until all selected rules have been processed and a global decision has been reached.

(6) **Explainer.** The user can communicate with the explainer to obtain a report about the operation of the expert system. He can find out how a solution was obtained and which individual steps were taken. If the user desires, he can obtain intermediate data and information on how the knowledge was used.

Some fundamentals on expert systems

Representation of knowledge.

There is a wide spectrum of knowledge needed to operate a manufacturing system. Usually, knowledge contains a physical or abstract description of the problem domain. Typical for a system operated by a human is that many decisions are made on uncertain and intuitive knowledge. When the knowledge is written down it is usually with symbolic or numeric symbols. Depending on its context, knowledge may be shallow, if it is commonly known, or deep, if it describes in detail a specific domain. Knowledge used for building expert systems may be defined as follows:

- **Factual knowledge.** For example, it is a fact that $10 + 20 = 30$.
- **Heuristic knowledge derived from experience.** For example, IF < the workpiece is rotational > AND < has a length/diameter ratio of x > THEN < use a standard horizontal lathe >.
- **Declarative knowledge.** For example, all screw machines produce threads.
- **Inferred knowledge.** This is usually knowledge inferred from applying the first three types of knowledge.

- **Meta-knowledge.** This is the knowledge needed for processing the information contained in the knowledge base.

Much of the information of a manufacturing operation is in descriptive form and is, therefore, difficult or unnatural to represent by conventional arrays or sets of numbers. For this reason, several special presentation mechanisms have been developed by which knowledge is presented in a knowledge base. The most important ones used in manufacturing are semantic nets, object-attribute-value presentation, rules and frames. We will be using the same example to demonstrate all four methods, see also Ochs (1988).

One important mechanism in a presentation scheme is the so called property inheritance. With this mechanism it is possible for one object to automatically inherit properties of a more generic object.

Semantic nets

A semantic net is constructed from nodes and arcs. Nodes present physical or conceptual objects or properties of a domain. Arcs describe binary relationships between the objects.

Semantic nets are well structured and powerful representation mechanisms for knowledge. The inheritance mechanism makes them suitable for presenting very complex problems. Figure 3.26 shows the description of a computer configuration where a host computer controls a communication bus to which a CNC system is connected. The reader should have no problem in understanding the context of this figure. When we look more closely at the presentation, we will discover that both the CPU and the memory board have the European card format and need an identical single slot to be plugged in. This means the CPU can inherit both of these attributes (card format and single slot) from the memory board.

Object-attribute-value presentation (OAV triple)

The structured domain knowledge about the objects of the previous semantic net (Figure 3.26) can be presented with the OAV method by a description of the object, an attribute the object has, and a value of the attribute. A section of the semantic net describing the computer configuration is shown in Figure 3.27 as an OAV presentation. The reader can see that this is also a very systematic presentation of knowledge. The value of an attribute can also represent another object. Thus, the inheritance mechanism is used in this presentation, which also makes it very powerful.

Rules

With this method, the domain knowledge is represented as a collection of rules which have the form IF < condition > THEN < conclusion >. Again, using part of our example, we obtain the representation in Figure 3.28. From this figure, it can be seen that a presentation based on rules is easy to configure and the semantics easy to understand. Rules are very popular for describing procedural manufacturing knowledge. However, rules can also be contradictory, and, therefore, it is often difficult to construct large knowledge bases with them.

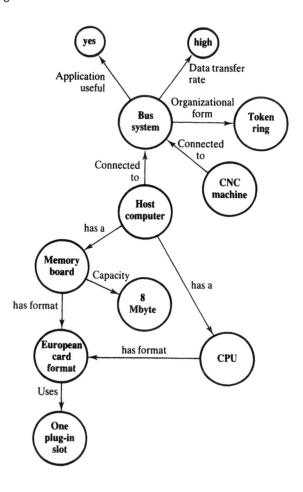

Figure 3.26
Semantic net
describing a computer
configuration.

Figure 3.27
Object-attribute-value
triple.

Object	Attribute	Value
CNC	Connected to	Bus system
Bus system	Data transfer rate	High
CPU	Uses	One plug-in slot

Frames

Frames represent a generalization of the OAV method. Figure 3.29 shows an example of a frame. They provide a well-structured method for representing objects and classes of objects for a given domain. Frames can have inheritance mechanisms where knowledge is passed from a generalized object to a specific object. By a procedural attachment, it is possible to specify procedures, called 'demons'. Another possibility is the inclusion of rules, whereby a rule is represented as a frame.

The decision on which of the four methods should be used for representing knowledge depends mainly on the application, but also on the available tools.

Object	Attribute	Value
The CPU has the European card format The European card needs one plug-in slot		Facts
x has format y and y needs z		Rules
x needs y		Conclusions
The CPU needs one plug-in slot		New facts

Figure 3.28
Principle of rule-based presentation of knowledge.

Object: Bus system	
Attribute	**Value**
Data transfer rate	High
Organization form	Token ring
Number of users	7
Maximum number of users	24

Figure 3.29
The frame concept.

The reasoning mechanism.

When a problem has been described to the expert system and a result is being sought, the inference engine activates the rules and tries to find (along with the pattern matcher) those rules which can be applied. For example, for a process plan, the cutting tip for machining a rotational stainless steel blank is to be selected. The user enters the instructions: (CUTTING MATL (BLANK?x)?y).

The system searching for the answer finds in the knowledge base the pattern (CUTTING MATL (BLANK STAINLESS) TUNGSTEN CARBIDE).

Thus, the answer will be:

x = STAINLESS

y = TUNGSTEN CARBIDE

Usually, the searching process is not as simple as shown, and the system has to go through a hierarchy of rules. After firing the pertinent rule, the knowledge is updated and additional rules are processed until a solution is obtained.

The knowledge contained in the knowledge base may be used in any problem solving phase, and the inference machine must be capable of inferring from the known facts and conditions which rules are to be applied to reach a solution. This is a special feature of the expert system to set up the rules, to test and modify the obtained knowledge, and to find a solution.

Most inference mechanisms use the so called 'modus ponens' rules from predicate calculus; IF $< x >$ is true AND $< x$ implies $y >$ THEN $< y >$ is true. Of course, it is desirable to find the most efficient path through the contained knowledge. This can only be done with very powerful control strategies.

Various control strategies have been developed for expert systems. As with many technical schemes, there are no general purpose mechanisms suitable for solving any application. Also, the solution finding process of an expert system is usually very computing intensive. For this reason, the user should select the control strategy which is most suitable for his problem. Persons unfamiliar with this technology should seek the help of experts to get started.

The most important techniques will be explained in the following sections. Often, the most convenient way of discussing the techniques is with the help of a state space graph. Each node in such a graph contains a solution process and the inference rules are applied to create and describe the arcs between the rules. Predicate calculus is most often used as the formal specification language for defining the contents of the nodes and for mapping the nodes of the graph onto the state of the problem. With this method, a certain proposition may be inferred from a set of assertions and the modus ponens inference rules. A problem may be stated with the help of modus ponens rules. The relations between the initial assertions and the inferences are expressed as a directed graph.

Many problems in knowledge processing could also be presented by other data structures such as arrays. However, the predicate calculus description and the inference rule methods are very powerful. Other knowledge presentation methods such as semantic nets, rules, and frames use similar search strategies and structures as discussed below.

Forward reasoning.
With this method, the state space search, also called data driven mode, is done in a forward chaining mode from the top to the bottom, as shown in Figure 3.30. The process of finding a solution starts with the data describing the problem and a set of valid rules. The application of a rule to the data will result in new facts, which in turn are used by new rules to produce new facts. The search is finished when a goal condition has been met. This method has advantages in the following situations. First, if the problem can easily be formulated by a given goal or hypothesis. Second, if pruning of the search space is possible in a situation where a large number of rules produce a large number of goals. Third, if the problem solver has to be guided by new data needed for finding a solution. This method is used for the solution of many planning problems in manufacturing.

Backward reasoning.
This method starts with the goal to be reached and tries to apply the rules which will lead to the goal, as shown in Figure 3.30. Thus a subgoal is determined from which the goal can be reached. Now the process is applied again to find a sub-subgoal. The search is continued upward until all facts of the problem are

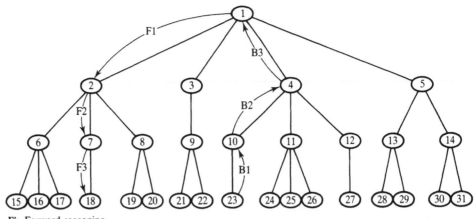

Figure 3.30

Forward reasoning and backward reasoning for a state space.

Fi: Forward-reasoning
Bi: Backward-reasoning

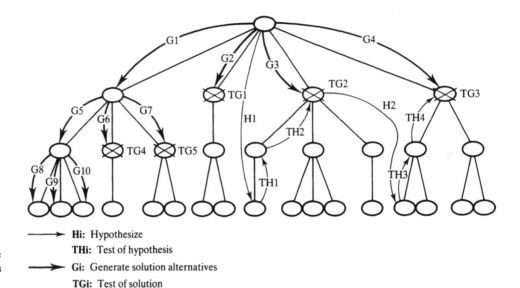

Figure 3.31

Hypothesize and test method, and generate and test method for a state space.

Hi: Hypothesize
THi: Test of hypothesis
Gi: Generate solution alternatives
TGi: Test of solution

obtained. Thus, the chain of values from the data to the goal are found in a backward fashion. This method is also called goal driven; it has the following characteristics. First, a complete set of data is available for the problem description. Second, the state space has a large number of goals. However, the use of the facts may be limited. Third, the formulation of a hypothesis is very difficult. This method is often used for action planning problems.

Hypothesis and test method.

This method is predominantly used for expert systems designed for solving diagnostic problems, as shown in Figure 3.31. In the first step, H1, the existing symptoms are analyzed and a probable diagnosis is made. In a second step, an attempt is made to check if this proposal is plausible; steps TH1, TH2. If it is not possible to find symptoms which are necessary to confirm the assumed diagnosis, a new probable diagnosis H2 will be set up. Then the system tries again to check this diagnosis; steps TH3, TH4. The system continues to set up diagnoses by forward reasoning and to check them by backward reasoning until a solution is found or no further probable diagnoses are available.

Generate and test method.

This method is often applied to design problems. First, an attempt is made to find all possible solution alternatives, steps G1, G2, G3, and G4, as shown in Figure 3.31. Second, in a following test phase, the solutions which are not feasible are rejected, steps TG1, TG2, and TG3. This search may lead to intensive computing efforts because of the numerous combinations to be considered. It is only useful if early pruning is possible to reduce alternatives.

Constraint propagation.

This method has its application in design and equipment scheduling. It assumes that any process has certain constraints which must be considered when planning is done. Different constraints are interconnected to a constraint net via common parameters. The constraint propagation tries to restrict the range of parameters as much as possible.

3.5.3 Building an expert system

People involved in building an expert system

An expert system is a software product, the development of which has a defined life cycle. People involved in building an expert system are the user, the expert and the knowledge engineer. Each person in this group has a specific role to play when an expert system is designed.

The manufacturing engineer.

The user is the manufacturing engineer who has to solve a task in his production environment, and who needs an effective planning and control tool to aid him in the fulfillment of his assignment. The user must not be burdened with the intricacies of the expert system. He is a specialist who knows how to run a manufacturing operation, and the tools he applies must be user-friendly and easy to understand. The man/machine interface for him must be designed in a simple fashion and he should be able to converse with the system in a dialogue using his accustomed manufacturing vocabulary. To enhance the ease of operating the expert system, a well-designed tutoring system may be necessary. The manufacturing engineer is instructed in a question and answer dialog session by the system to explain how to use it. This tutor may itself be using a small expert system to interpret the questions and to produce the answers. A well-designed system may even go a step further and assist the user in locating problems during the operation of the expert system.

The expert of the domain.

The expert provides the domain knowledge for the system. To be precise, there are two types of experts to be distinguished. First, knowledge may come from books, catalogs, and operating instructions. This information is put on paper by experts who are usually not directly involved in building the expert system. Their knowledge is extracted from the written material by a third person and presented in a suitable form in the knowledge base. Second, knowledge is obtained directly from an expert who is familiar with the specific manufacturing process to be automated. This expert must be interrogated about his domain, in such a manner that all important information for the proper operation of the manufacturing process is obtained. This can be a difficult job. On the one hand, the knowledge engineer, who interrogates the expert, might not be able to assess the importance of the information he obtains. On the other hand, the expert may not be willing to render the right information, or he may have forgotten what is important and what is not, because he has lived with the problem for too long.

The knowledge engineer.

He is the person who designs the architecture of the expert system. Being involved with the user and the expert, he really should have basic knowledge on all aspects of the expert system. His main knowledge will come from the fields of artificial intelligence, software engineering, and ergonomics. He must be capable of selecting and structuring the right expert knowledge, of designing the user interface, and of configuring everything into one workable system.

The life cycle of an expert system

It is the task of the knowledge engineer to build the expert system. Usually, he is not an expert in the domain for which the system is being developed. Knowledge may come from various sources, such as reports, books, experiments, and experts. Usually, the best source is the expert because his knowledge is very deep and up to date. It requires much skill to extract from all sources the information needed and to combine it in the knowledge base. There are various tools available which can help the engineer to perform the knowledge acquisition. They are machine learning methods, psychological techniques, graphical interactive acquisition methods, and hybrid tools. The transformation of the knowledge to a suitable presentation in the knowledge base can be done either by the knowledge engineer or a program.

The acquisition of the knowledge is a learning process for the engineer and the expert. One of the best aids to construct a knowledge base is to use an interactive graphical editor, which offers a systematic way to take the usually less structured knowledge repertoire of the expert and to transform it into a structured database. The contents of a book can be transformed into a computer internal presentation with the help of induction programs. In general, books have better structured knowledge than an expert.

There are five phases which make up the life cycle of the expert system, see also Figure 3.32. These are identification, conception, formalization, structuring and

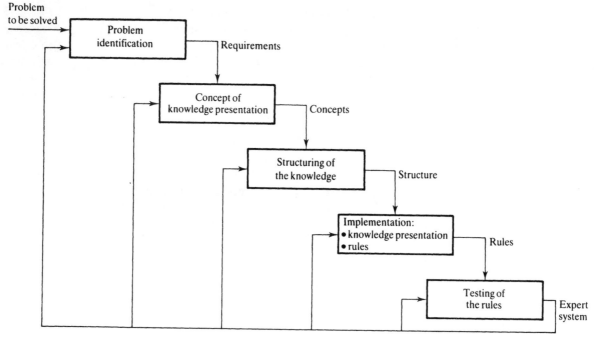

Figure 3.32

The life cycle of an expert system.

testing. In practice, the border lines between the phases are not so well defined. As with all software developments, the phases are not entered in a successive order. Development errors or new facts may lead the system architect back to previous phases. As a matter of fact, there will be feedback loops from every succeeding phase to its preceding phase. This is not shown in the figure.

(1) **Identification phase.** When the knowledge engineer and the expert are familiar with each other's field, the problems and the structure of the knowledge base are defined. This phase has to be prepared with the utmost care, since errors in this phase will be transmitted through all succeeding phases and become increasingly difficult and expensive to correct as they migrate deeper into the system. The following typical questions must be resolved.

- What are the aims of the system?
- What is the scope of the knowledge domain?
- What does the user expect from the system?
- How is the user going to apply the system
- What is the required ergonomics from the various machine interfaces?
- How can the system be hierarchically structured?
- How can the system be expanded in the future?
- What do the interfaces to other manufacturing system components look like?
- What are the sources of knowledge (expert, books, manuals, and so on)?
- How much data is needed, and how should it be structured and stored?

- What is the computer configuration for the developer and user?
- What are the languages and other development aids needed to build the system?

This phase will go through several iterations until the knowledge engineer and expert have agreed on the problem requirements.

(2) **Conception phase.** In this phase, the conceptual interdependencies of the components of the expert system are defined. Usually, graphical aids for problem description and documentation will support this work. The following typical questions will have to be answered:

- What strategies and methods are best suited for the solution of the problem?
- What are the components of the system and their causal relationships?
- How will the procedural and declarative parts be constructed?
- What does the information flow look like?
- What are the limits of the system?
- What components and methods of the system can be incorporated from previously designed systems?

This phase is also an iterative process until an agreement exists on a concept.

(3) **Structuring phase.** During this phase, the knowledge engineer will formalize the knowledge and will solve the problem with the available tools. The three most important factors of formalization are the definition of the hypothesis space, the model of the processes and the structuring of the data.

To define the hypothesis space, the concept is formalized and a decision is made on the detail and structure of the presentation. The relationship between the components is to be established and represented in a suitable form. Finally, the hypothesis may be structured in a hierarchical form, the levels of abstraction are determined and an investigation is made on the uncertainty of the system's answer.

Models are used to formulate the knowledge about the process. There are mathematical and behavioral models. In many cases, the mathematical model can be made very accurate and its knowledge can be entered directly into the expert system. The behavioral model is not so easy to design. For this reason, it should be as simple as possible.

The structuring of the data requires special skills in order to ensure that sufficient and only pertinent data is present. The type and source of the data is presented. It has a decisive influence on the formulation of the knowledge and the conception of the hypothesis. When preparing the data, the following factors must be considered.

- How is it possible to obtain relevant data?
- How can the most important features be presented with the available data?
- Is the data consistent and reliable to formulate the required knowledge?
- Is the planned level of effort justifiable to obtain the data?

When the concept of the system, the required data, and the dataflow are defined, a prototype of the knowledge base can be built. During this phase, the programming language will be selected. The prototype is usually programmed in LISP, Prolog or another typical knowledge representation language. The language to be selected depends on the problem and available programming aids.

(4) **Implementation phase.** During implementation, the development components are configured to a complete system. If this is the final implementation, it may be necessary to reprogram the system with a more efficient language to reduce computing time and memory requirements. Usually, the final system is a component of a global planning and control system. In this case, all the required interfaces have to be combined to make the knowledge-based system part of the total system. The implementation phase may require building of new development tools and the design of new languages. The latter is necessary if a special user interface is required.

(5) **Test phase.** This final phase is necessary to assure that the system provides the specified requirements. Points of particular interest are efficiency, user-friendliness, extendability and maintainability. A test will include the following questions:

- Is the system user-friendly?
- Can the user maintain the system?
- Are the inference rules consistent and complete?
- Are the answers of the system ambiguous or satisfactory?
- Are the test examples realistic?

If the system gives wrong answers, the inference rules must be investigated and it might be necessary to trace the operation of the inference process, step by step, to locate the problem. For this purpose, the system should be provided with an explanation component.

Supporting tools for building expert systems

Expert systems are very complex software structures and need various support tools for their development. It was shown earlier that there are two people who must communicate with the expert system. They are the user (manufacturing engineer) and the knowledge engineer. The user is only interested in a tool to solve his problem. For this reason, all information which he does not need for operating the system should be hidden from him. He must assume, when he queries the system, that somebody has configured the software in such a manner that the system is capable of producing an answer automatically. However, as it is often the case with software systems, this can only be applied efficiently when the structure and logic is understood. In addition, updating and smaller changes often have to be made by the user. Thus, there is a need for the user to have some knowledge of the system and its development tools.

Level	
6	Knowledge acquisition tools
5	Expert system development support
4	Expert system shell
3	Expert system shell development language
2	Implementation language
1	Operating system
0	Computer architecture

Figure 3.33
Virtual hierarchy of a knowledge-based system.

The knowledge engineer, of course, must know all development tools and their use. He will be trying to design and configure the expert system as well as possible.

If we integrate the software and hardware to one expert system architecture, we are talking about a virtual hierarchical knowledge-based system, as shown in Figure 3.33. There are seven levels of abstraction shown. The problem is defined on the upper level; to be implemented and operable, it is gradually refined by the lower levels until it can be executed on the selected machine architecture. The individual levels will now be explained in more detail.

(1) **Knowledge acquisition tools.** We have discussed knowledge acquisition in the previous section. The problem consists of extracting knowledge from the experts and text, and combining the obtained procedural and declarative knowledge in the knowledge base. An automatic acquisition tool should be capable of extracting and structuring the pertinent knowledge and of translating it into a suitable formalism for knowledge presentation. The expert will be questioned during an interview to extract the important objects and concepts of his domain. The results will be embedded in a natural language concept which can be automatically interpreted. Text processing will be simpler, because text is usually more structured. The processing of written material will start with a linguistic analysis to locate the relevant concepts and text. It can then be scanned with a menu or window technique to locate key words and phrases for further processing. The knowledge extracted from the expert and the text will be represented by an intermediate language for the generation of rules and frames. There are numerous tools being developed to support the knowledge acquisition phase.

(2) **Expert system support environment.** The support tools needed on this level are typical development aids used in software engineering. The most important equipment is the interactive graphical terminal with bit mapping capabilities in conjunction with windowing and menu techniques. The entire knowledge can be entered interactively into the system by the use of menus and forms. It should be possible to represent the knowledge base and relations between objects in the form of understandable graphics, and to observe the inference

process. Editing and debugging functions must be provided to perform corrections, additions, and deletions.

(3) **Expert system shell.** There are various shells available to assist the development of an expert system. They provide the knowledge engineer with a set of domain independent tools which are commonly used for the design of expert systems. They are:

- A component for the presentation of knowledge.
- A component for developing the control structure of the expert system.
- A component containing one or more inference mechanisms.
- A module for finding and correcting errors.
- An explanation facility to inform the user of how the decision was obtained and why certain rules were applied and others not.
- An interface to databases and programming languages.
- A comfortable user interface.

The user may find that some shells will handle this problem very well and others not so well. Often, much patience is needed to understand the functions and capabilities of a shell. A basic problem of a shell is its generality. For this reason, the developed expert systems are usually computing intensive and very slow.

(4) **Expert system development language.** In general, expert systems must make their decision with declarative knowledge. Thus, a language must be provided which presents the knowledge efficiently and facilitates the inference process. Typical programming aids are Prolog, OPS5 and object-oriented languages.

(5) **Implementation language.** This is the actual language in which the expert system is written and which is compiled for the target machine. The run time of an expert system is highly dependent on its structure and the implementation language. For this reason, expert systems are often developed and debugged with a shell and then rewritten in a more efficient language such as C or Pascal. There are many systems written in LISP, however, the programs usually result in inefficient computer runs.

(6) **Operating system.** The operating system really is part of the system software of the underlying computer structure used. It must provide the user with a good interface to the hardware, database, communication system, and other programming packages. Good operating environments for knowledge-based systems are offered by UNIX, VMS, Aegis and others.

(7) **The computer architecture.** In the early days of knowledge engineering, special machines were conceived for accommodating LISP and Prolog programs. In the meantime, powerful workstations have been developed which are capable of handling most expert system applications.

3.5.4 Expert systems for manufacturing

Defining the application

The manufacturing engineer who wants to utilize the power of artificial intelligence tools for planning and controlling of factory activities will find it difficult to decide where to start. He can be certain that all manufacturing problems cannot be solved by one large knowledge-based system. We have learned in the previous chapter that planning and control of a factory is done in a hierarchical fashion; where central activities are defined at higher levels and gradually broken down to subactivities at lower levels. For this purpose, the manufacturing engineer must conceive a model of his factory, as shown in Figure 3.34. Various manufacturing models will be discussed in the next chapter.

For our discussion, we will take a rather abstract model suggested by Meyer (1987). This model is placed on a GRAI grid to show the hierarchical levels and their planning horizons, see Doumeingts (1989). In the model, each controller activity is based, in a broad sense, on a planning, quality control, and maintenance function (PQM). As a controller is hierarchically decomposed into smaller control modules, the PQM structure is maintained. For each of these functions, expert knowledge is necessary which may be provided by man, conventional control algorithms, or expert systems. Figure 3.35 shows the typical structure of a controller. Here, each module contains an expert system and has access to a knowledge base which contains the static and dynamic world model of the control activity of this module.

When a control command is sent from an upper level to the planner, it decides with the information of the knowledge base the course of action which has to be taken. For this purpose, the status of both the current actions and of the resources has to be known. The planner interrogates the diagnosis module of the same hierarchical level. When the diagnosis has been made, the results are sent for interpretation. The results of the diagnosis are communicated to the planner to be incorporated into its actions. The planner decomposes its task into subtasks

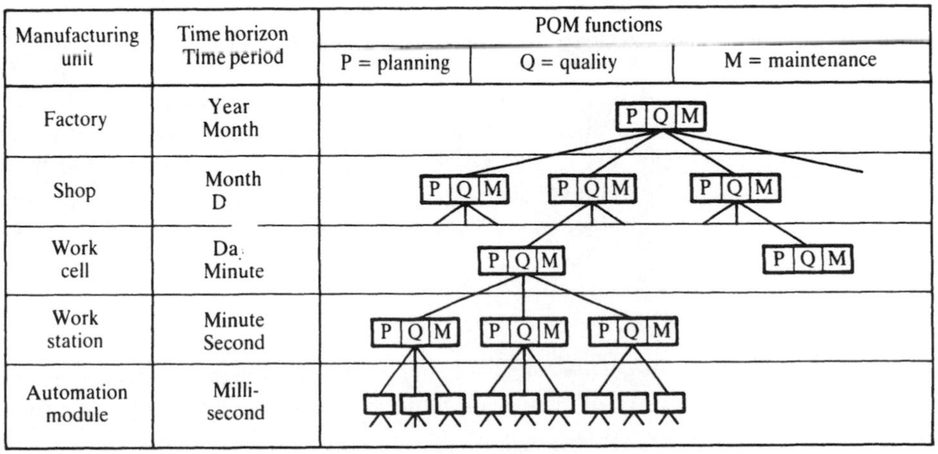

Figure 3.34
Factory reference model.

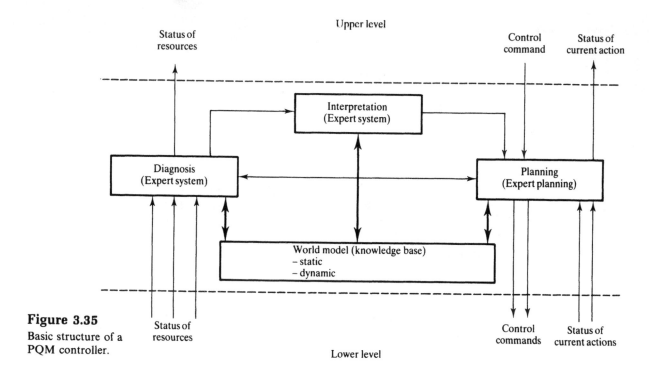

Figure 3.35

Basic structure of a PQM controller.

and sends control commands to the lower level. At the same time, it monitors the status of the execution of current actions of the lower level. Also, the upper level receives information on the status of the current actions of the controller as well as of its resources. The world model of the controller is divided into a static and dynamic part. The static model contains all information on the controller environment which is of a more permanent nature, for example, the available machine tools and their layout. The dynamic world model is used to store information and it changes rapidly as data is entered constantly from the manufacturing floor, for example, measured quality parameters.

A basic problem with the controller, shown in Figure 3.35, is the design of the world model and its interface with the planning, diagnosis, and interpretation functions. Proper control structures must be implemented so that all functions can share the knowledge base without interfering with each other's work. Also, there must be realtime capabilities in the controller to update the dynamic model online when status data is sent to the controller.

Design

When a product is being developed, the design process passes through the definition stage of the functions, definition stage of the physical principles, design stage of the shape, and the detailing stage. At the beginning of the design, the idea of the product is rather abstract and with each succeeding stage the solution becomes increasingly concrete until it is completely described by the drawing. In principle, it is possible to build an expert system which will capture the designer's know-how for conceiving and shaping the product. However, all attempts which have been made to date show that the design of even a very simple product is not an easy task for an expert system. The main problems are in the definition of the functions and that of the physical principles. It is believed that for more complex products, these phases will remain the domain of the designer for a long time to come. The expert system will help where the designer is concerned with details, such as finding available similar designs, standard components, tolerances and so on.

The designer is used to obtaining a solution for his problem with formulas and algorithms. However, in many cases, the description of the solution becomes very complex and proves to be impractical. Here, the knowledge-based systems can be of great help to show the designer alternative solutions and to significantly reduce trial and error searches.

There are three types of designs which lead to the conception of a new product. They are the new design, variant design, and modified design. It will be extremely difficult to produce a new design with the help of an expert system. There is no way of finding with this tool a solution to a problem for which no prior knowledge is stored. Man has an immense fantasy and intuition to deduce, from often remote areas, solutions which will lead him to functional designs. The variant design is based on existing functional and physical principles of a similar product. Dimensions and other physical parameters, however, may be different. Here, expert systems will be of help to propose to the designer a solution based on a variant. The modified design usually has only a few alterations of an existing product. Here, expert systems could also lead the way to a solution.

The design of the product determines, to a great extent, its manufacturing process. However, the designer is often not very familiar with the details of the manufacturing methods. Expert systems can be of assistance to consult the designer about the manufacturability of the product. One could even go a step further and obtain the process plan from the description of the parts to be manufactured. This could be done according to generative or variant principles.

The acquisition of the data for the knowledge base must be conducted very carefully to catch the most important information of the product. Figure 3.36 shows how this could be done for the design of fixtures, see also Eversheim and Neitzel (1989). First, literature, drawings, and experts are consulted to assess the scope of the problem. Second, the fixtures are partitioned into functions. Third, general functions are determined. Fourth, the functions are structured. And finally, the functional constraints are set. The information thus obtained is transferred into the knowledge base by a suitable internal presentation.

In addition, the rules are designed to use the knowledge base. The formulas needed for calculating the physical properties of the machine elements are established and entered into the knowledge base. A pictorial on the presentation of the knowledge is shown in Figure 3.37, see also Eversheim and Neitzel (1989).

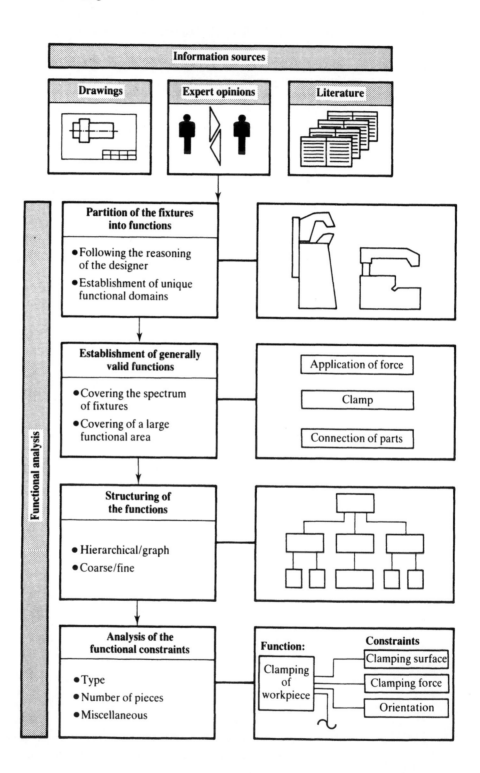

Figure 3.36
Knowledge acquisition
by functional analysis.

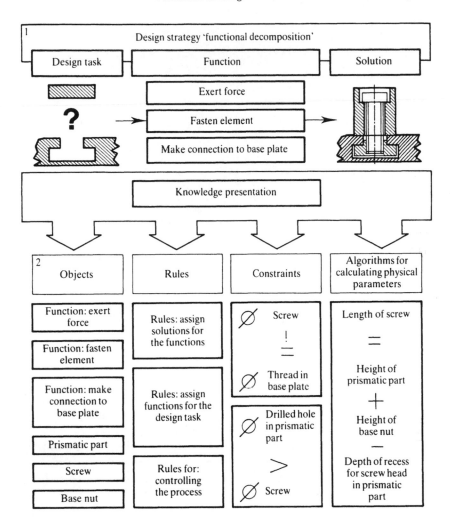

Figure 3.37
Knowledge
presentation of
fixing devices.

Process planning

When the design of the product is completed, process planning can start. Figure 3.38 shows the functions of the process planning activity, also explained in Rembold *et al.* (1985). There are two different methods used for process planning: they are the generative and variant methods. With the generative method, the part surface to be created has to be related to a manufacturing operation. The planning method must contain the modular machining primitives which are capable of selecting from the dimensional and physical parameters of the workpiece, the manufacturing processes and the sequence of processes. With variant process planning, the manufacturing processes of a part variant must be known and stored in the computer as a nonparameterized variant. During process planning, the designer queries a variant catalog in the computer and searches for the variant which is

- **Selection of raw parts or stock**
  Shape
  Dimension
  Weight
  Material

- **Selection of process and sequence of machining operations**
  Global operations
  Local operations at a given workplace

- **Machine tool selection**

- **Auxiliary functions**
  Fixtures
  Tools
  Manufacturing specifications
  Measuring instruments

- **Manufacturing parameters**
  Feeds and speeds
  Setup time
  Lead time
  Processing time

- **Text generation**

- **Process plan output**
  Process plan header
  Process plan parameters

Figure 3.38
Functions of the
process plan.

similar to the part to be made. When the variant has been found, the parameters for the new part are entered and inserted by the computer into the process model. With this information, the computer automatically generates the process plan.

Numerous variant process planning systems have been designed and are in service. They use conventional planning methods and work best with rotational, prismatic, and simple sheet metal parts. However, most of these systems are rather rigid and are difficult to handle if a variant has slight exceptions and if there is a need to add new variants. Here, expert systems could considerably improve the flexibility of process planing.

Since it is very difficult to relate the part surface to a manufacturing process, there are few systems which use the generative process planning method. For example, if a tool has to be found for machining a flat surface; under normal conditions, the process planning system will assume that the surface can be machined by a universal tool. However, if the surface is bound by a ledge, a tool has to be selected which will not interfere with the ledge.

In most cases, the variant process planning method or a hybrid method is being applied. These methods have the disadvantage that a variant must be available for every part to be manufactured.

Generative process planning is an activity where expert systems can be usefully applied. The strategies for machining a part are limited, since there is usually a limited number of machine tools, cutting tools and fixtures available. The

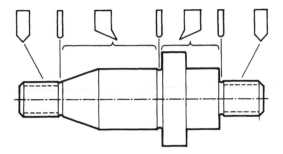

Figure 3.39
The selection of
cutting bits for turning.

time-consuming and labor-intensive work of modeling each and every process alternative can be alleviated by an expert system which knows how to deal in generalities. The system will plan the details with the help of rules which recognize patterns and, if necessary, make assumptions. The knowledge and rules have to be provided by process planning experts.

Expert systems will mainly be of help for the first four phases of process planning, as shown in Figure 3.38. There must be a master knowledge base for each phase and rules to derive a solution.

Manufacturing scheduling

Scheduling is an activity which starts after the product has been released by engineering. The information going into scheduling are the type and number of parts to be manufactured, the process plans, the bill of materials, order delivery dates, available machine tools, other resources, factory monitoring data on resource utilization and so on. There are basically three scheduling problems to be solved, as follows:

- Capacity planning, to provide for manufacturing resources,
- Order release planning, to ensure that the parts to be manufactured meet due dates,
- Operations sequencing, to utilize production equipment at its maximum capacity, if possible.

Until now, scheduling has mainly been done with operations research and related tools. Typical tools are linear and dynamic programming, branch and bound method, heuristic methods and so on. In general for scheduling, a great number of alternatives have to be considered. Since solutions are discrete, optimization by differentiation is not possible. For this reason, an attempt is usually made to obtain only a sub-optimum. However, the most severe difficulty with conventional scheduling methods is their inability to handle schedule exceptions.

The methods usually give good results for a normal manufacturing run, but fail when a sudden event such as an urgent deadline, a machine tool breakdown, or a part shortage is incurred. Knowledge-based systems should be capable of assisting the conventional methods in such emergency situations.

The two artificial intelligence tools most suitable for scheduling are the generate and test, and constraint propagation methods. Neither method will eliminate the combinatorial explosion problem. However, they have the definite advantage that they can be conceived as open systems, permitting easy extension and changes.

With the generate and test method, the scheduling problem is first considered in a longer time frame, and it is then decomposed in various steps into shorter and shorter time intervals until the required detailed short-range plan is obtained. This method is used for solving capacity and schedule problems.

Constraints imposed on a manufacturing system are considered with the constraint propagation method. Typical constraints are as follows and discussed in Fox (1983).

- Organizational constraints
  - uniform production level
  - low inventory
  - meeting delivery schedule
  - quality
  - cost
- Physical constraints
  - dimensions of machine tool
  - accuracy of machine tool
  - production rate of machine tool
- Causal constraints
  - precedence of machining sequences
  - available work space in connection with tools and fixtures
- Availability of resources
  - a machine must be available
  - a machine must be set up
  - the material must be available
- Preferences
  - preferred machine tools, tools and fixtures
  - preferred raw stock

A manufacturing schedule will be set up by considering selected constraints. The influence of the constraints on one another are investigated, and the parameter restricted the most will be selected for further investigation, and the effect of this constraint on the manufacturing system is observed. Possible questions are:

- Will the constraint lead to a solution?
- Are there other alternatives?
- Can the constraint be relaxed to solve the problem?

The best known system using the constraint-based approach is ISIS. The system tries to schedule production equipment for a factory order to meet the delivery date. The planning is done on several levels; level one selects the scheduling order, level two selects the manufacturing equipment needed, and level three schedules the resources to produce the order.

AI methods will not be competing with conventional scheduling algorithms. Their main area of application are:

- Handling of exceptions,
- Supporting simulation systems used for scheduling,
- Improvement of the man/machine interface of scheduling systems,
- Assisting machine learning of scheduling procedures.

Quality control

Quality control is an important function of the manufacturing process. If the quality of the product deviates from quality objectives, immediate correction of the manufacturing process is necessary. For a complex product, there is a variety of quality functions to be performed which may include measuring principles, procedures, and techniques. A quality control operation is done in several phases. First, a test plan is drafted. Here, the parameters to be tested are determined, the test procedure is selected, the hardware is specified, and the data deduction and evaluation programs are defined. Second, the test system is configured and programmed for the individual products to be tested. Third, the measurements are performed and the data is recorded. Fourth, the data is evaluated and test protocols are prepared. When inspecting these functions, it becomes obvious that much expert know-how goes into a quality control procedure. There are two areas where expert systems may be of considerable help to improve a quality control operation.

(1) **Test planning.** A generative or variant test planning method may be used, similar to process planning. The generative method is extremely difficult, in particular, when performance or visual inspection parameters are involved. Variant test planning is being used successfully where testing of a product family is done.

(2) **Data evaluation and interpretation.** An attempt is made to obtain from the test data the information on how well the product is doing. If only upper and lower limits of the test parameters are investigated, this is a straightforward method and no expert system is needed. If, however, an attempt is made to diagnose the problem, then expert knowledge must be available to find the cause of the malfunction. Similar situations do occur in machine diagnosis. The methods and tools used for diagnosis are discussed in a later section.

Usually, the evaluation of visual multisensor data is a very difficult and involved process, and in many cases can only be done well with expert systems.

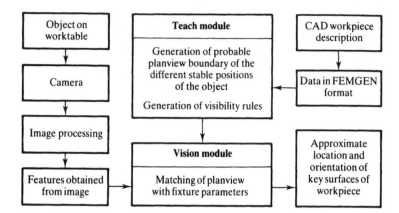

Figure 3.40
Schematic for matching camera images of a workpiece with CAD data.

Visual data.

Visual inspection has four phases. First, a picture is taken of the object. Second, the iconic features are extracted from the picture. Third, the iconic features are used to reconstruct an abstract image. Fourth, symbolic picture identification is applied to the abstract image, involving a matching technique with a template. The template can be obtained directly from the part CAD database via a FEMGEN format. Since the part may be presented to the camera in any one of its stable positions, a template must be available for all of these positions. In addition, the picture may be taken from any viewpoint. Thus, a template would be needed for all viewpoints. In reality, it is impossible to store data on an indefinite number of viewpoints. For this reason, usually an attempt is made to recognize special features of a part whose recognition is rather invariant to the position of the object and that of the camera. Rules can be applied to infer from the features and their relations to each other the identity of the object. A schematic diagram showing this method is illustrated in Figure 3.40, see also Majumdar and Rembold (1989).

Processing of multisensor data.

Testing and evaluation of a complex system or scene often require several sensors using various physical principles. The sensor data must be fused to obtain information about the object under investigation. A typical problem would be the recognition of a 3D workpiece by a robot, employing a 2D camera, and a sonic distance sensor, see Figure 3.41 and Raczkowsky (1989). The method shown employs a blackboard architecture which is a common tool for solving artificial intelligence problems. In our case, there are two sensor ports, the world model, the blackboard and control structure, and the control and communication unit shown. When the 2D camera has taken a picture of the object, it tries to get from the world model all information needed to identify the 2D projection of the workpiece. This information is sent to the blackboard and logic is used to identify the workpiece. If this is not possible, an instruction will be given to the distance sensor to take a measurement in the third dimensions. This information is sent via the matching unit to the blackboard and a new attempt is made to identify the object.

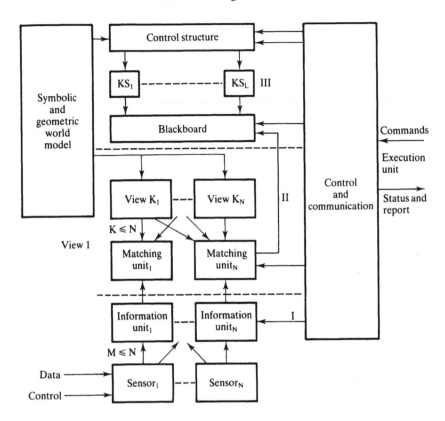

Figure 3.41
Control structure of a
multisensor
architecture for robots.

In case the workpiece still cannot be recognized, an additional sensor or another sensor combination may be suggested. This process will proceed until positive identification is made.

Diagnosis

Expert systems for diagnosis are the most advanced AI tools used in manufacturing. They play an important role in supervising complex production equipment and locating problems as soon as they arise. Thus, it is possible to reduce the average search time for a defective machine component from an hour to a few minutes. The principles of most of the machine diagnostic systems in service have been patterned after medical expert systems for diagnosing patients. The diagnostic procedure usually involves taking measurements, evaluating parameters, matching of patterns with a known hypothesis, and initiation of corrective actions. The measurement principles used are often similar to those applied by quality control.

Figure 3.42 depicts a typical diagnostic problem encountered with production equipment. A faulty operation may originate from various sources, for example, the structural components of the machine, the controller, or the interaction of the tool with the workpiece. For this reason, the goal of the diagnostic system must be clearly defined and the experiment to obtain a hypothesis must be done under

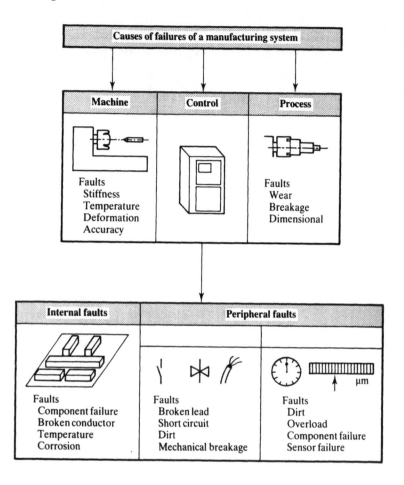

Figure 3.42
Types of failure
occurring in a
manufacturing system.

known conditions. Expert systems for locating problems with a machine or a controller are usually more reliable than those for diagnosing problems with the process. It is virtually impossible to obtain a valid hypothesis for every tool/ workpiece combination which may be encountered during the service life of a machine. Hypotheses for the machines or controllers may be derived from various statistical, heuristic, or causal knowledge sources, including:

- expert opinions
- service manuals
- mathematical calculations
- operating statistics of the machine
- experiments.

An experiment usually involves the deliberate introduction of a fault in the machine, and then observing and recording its effects. This may be very expensive if the effect of many components has to be investigated or if numerous experiments have to be run to obtain statistically significant data. Presently available diagnostic systems can be defined by three categories.

(1) **Statistic diagnostic systems.** With this method the diagnosis is obtained for a set of diagnosed cases for a given symptom. The statistical significance of the diagnosis is asserted by Baye's Theorem. In addition, the Dempster-Shafer theorem may be used to evaluate partial knowledge which can be presented by a diagnosis hierarchy. Both theorems assume that:

- The case data are representative and complete.
- The symptoms may not correlate with one another.
- The diagnosis is unequivocal.

Since the inference mechanism is based on statistical data, causal explanations are difficult to obtain.

(2) **Heuristic diagnostic systems.** These systems are built with heuristic knowledge derived from experts and simulations; they are usually not as exact as the statistic systems. The expert has to structure the relationships between symptoms and diagnosis in a formal manner by using frames and rules. Frames are a good tool to structure and represent sets of symptoms; whereas rules are used to represent heuristic causal knowledge. Well-known inference mechanisms are applied, such as forward chaining, backward-chaining, establish and refine, and hypothesize and test.

The disadvantages of the heuristic diagnosis systems are:

- It is not always possible to obtain sufficient expert knowledge.
- It is difficult to ensure that the knowledge is complete and free of contradiction.
- The explanation of the conclusions is difficult; often, they are only protocolled.
- The results may be inaccurate.

(3) **Model-based diagnostic systems.** With these methods, a causal model is built of the system to be diagnosed, see also Grimm (1987). The model contains deep knowledge of causal relationships between the diagnosis and the symptoms. When a manufacturing machine is diagnosed, the model simulates its operation at the same time. The results of the system are compared with those of the model, see Figure 3.43. The greatest drawback of this method is the problem of building an accurate model of a machine tool or of a manufacturing process.

Figure 3.43
The principle of a model-based diagnosis system.

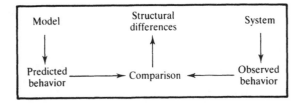

Implicit programming of robots and other manufacturing equipment

With a conventional programming language, a robot or other production equipment is instructed explicitly step by step of the actions it has to perform. This type of programming is usually very cumbersome and results in many instructions. For this reason, task-oriented or so called implicit programming languages are being developed. Here, the system is programmed by instructions which resemble sentences of the spoken or written language, for example, 'take part one and insert it into part two'. The system must have thorough expert knowledge of the task to be performed, the assembly operation, robot capabilities, the workplace, and of how to generate execution instructions. Figure 3.44 shows the various planning levels of an implicit programming system for robots, see also Frommherz (1989).

First, a specification of the assembly task is created by modeling the spatial relations between the workpieces involved in the goal state. This is done using a graphical editor, which enables the user to interactively arrange geometrical models of the objects. On the next level, the assembly task is analyzed. With respect to

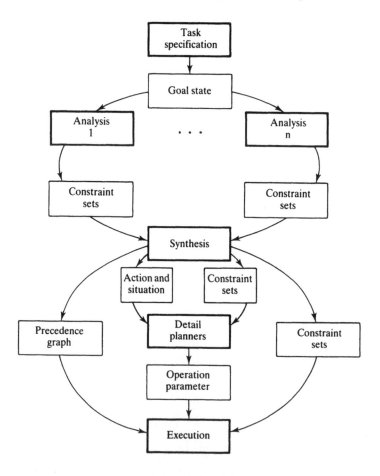

Figure 3.44

Different planning levels of an implicit programming system for robots.

the geometry of the workpiece and the geometrical goal relations, the priority relations of the various assembly operations are deduced. The analysis can be performed under the following criteria; during assembly no penetration of the parts is permitted and stability and robustness must be assured.

With the resulting constraint sets as input, the synthesis module generates a precedence graph, which is an efficient representation of a task description. The synthesis can also include constraints produced and fed back by subsequent planning modules. A second output of the synthesis module is an action and situation list of the assembly task, which describes the possible topology of the environment for each assembly operation.

The final assembly sequence is planned on the level of the detail planners. The planning process includes the planning of a sequence of elementary subtasks for each assembly operation and the assignment of the available resources (that is, robots, grippers and so on) to these subtasks. Moreover, optimal positions for the robots are calculated. In order to plan the detailed geometric parameters of the particular robot operation, this information is passed on to the motion planner, which is a module of the detail planners.

The task of the motion planner is to plan the motion parameters for the elementary operations. For example, one elementary operation is the pick and place operation. The underlying action sequence is shown in Figure 3.45. A pick

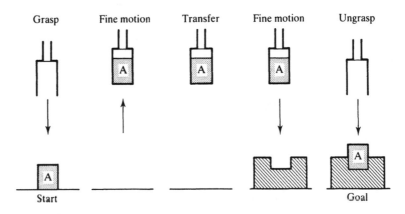

Figure 3.45
Operations necessary to perform a pick-and-place motion.

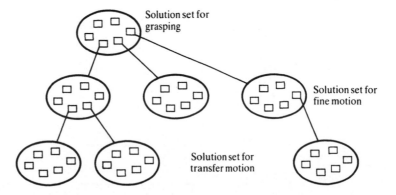

Figure 3.46
The solution space of the motion planner.

and place operation consists of five single motions:

(1) a grasp operation,
(2) a fine motion to separate the part from its support,
(3) a transfer motion,
(4) a fine motion to connect the part with its goal environment,
(5) an ungrasp operation.

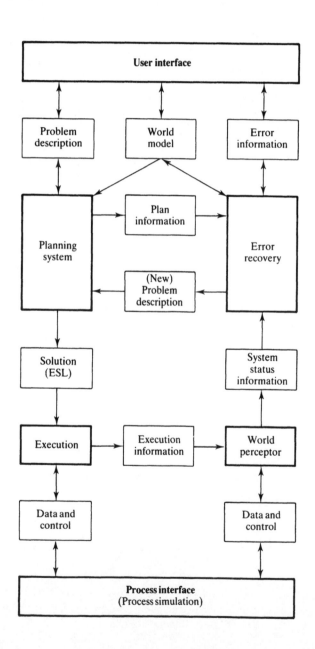

Figure 3.47
Overall implicit
programming and
control for intelligent
manufacturing
equipment.

There are three geometric sub-planners to plan this combined motion sequence. First, both the grasp and the ungrasp operation are planned by the grasp planner. Depending on the grip selected, the two fine motions both for separating and connecting the object are planned by the fine motion planner.

The tree-shaped solution space resulting from this planning process is shown in Figure 3.46. Going down the solution space, a grasping task may lead to several sets of fine motions, and a fine motion may lead to several sets of transfer motions. Thus the motion planner has to coordinate the three sub-planners in order to find a solution to solve the combined motion problem. Backtracking is necessary to select an alternative for the next higher level if one sub-planner fails to find a solution. Once a solution is found, it can be translated into an executable code such as VAL, IRDATA or MCL language (MCL is the motion command language).

When the implicit programming method is used to program an intelligent manufacturing system, a very complex control system will be needed, see Figure 3.47. In this figure, the programming module of Figure 3.44 is implemented as the planning module. The program generated by this module is sent in an intermediate code ESL (elementary solution language) to the execution module which is located in the machine controller. Every line of code will be examined if it is executable, before it is actually executed. For this purpose, the world perception module activates its sensors to obtain an actual status of the work environment. If it finds out that the instruction can be executed as programmed, the machine is told to perform the task. If, however, the world perception module finds that an instruction cannot be executed, because the state of the world does not conform with the state for which the instruction was written, the problem will be signalled to the error recovery module. This module investigates the cause of error and sends to the planning system the new state of the world. Thus, a new plan is configured and sent for execution. The cycle may have to be repeated until the world perceptor asserts that the instructions can be executed.

3.5.5 Knowledge-based configuration of a manufacturing system

Large manufacturing systems

Complex technical systems are usually designed in a hierarchical fashion as explained in Chapter 1. For such systems, a large amount of knowledge has to be gathered, structured, evaluated, and stored. Numerous planning and configuration tools have been designed to aid the system engineer in building large technical structures. Many of these tools were developed for software engineering. Recently, various artificial intelligence tools have been investigated for planning and configuring of the control levels of a manufacturing facility. The results are very encouraging and it can be expected that these tools will make system configuration

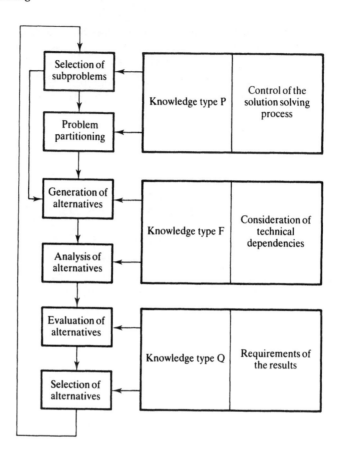

Figure 3.48
Problem solving cycle and knowledge type.

easier and more responsive to marketing changes. We will be explaining one approach in more detail, see also Ochs (1988) and Ochs *et al.* (1989).

Within this scheme, it is assumed that the planning is done according to a hierarchical decomposition concept. The planner starts with a rather abstract view of the manufacturing operation on the upper level. Thereafter, the factory activities are refined on different levels until on the lowest level are detailed the operations on the factory floor. The planning on every level is done in a similar manner; thereby the controller functions, data communication systems and the equipment are layed out and determined. On each level there are three phases of the planning cycle. First, the problem is decomposed into subproblems. Second, manufacturing alternatives are investigated. Third, the best manufacturing sequence is determined. A phase may have to be repeated iteratively until a satisfactory solution is reached for a level.

For every phase, a specific type of knowledge must be stored and processed in a comprehensive expert system. Figure 3.48 shows the solution cycle and the types of knowledge. The three types of knowledge are:

(1) functional knowledge of Type F to obtain a workable solution,

(2) quality knowledge of Type Q to assure the integrity of the solution,

(3) procedural knowledge of Type P to select the optimal solution.

The knowledge can be processed with the aid of an AND/OR graph, as in Figure 3.49. There are three dependencies to be distinguished. The first dependency is in the vertical direction; there are predecessor and successor relationships which represent the decomposition of the problem to parallel activities (I) or which describe alternative solutions (II). In case an element of the graph can be divided into several sub-elements, the corresponding edges leading to the sub-elements are combined with a circular arc. If, however, the sub-elements are alternatives then there is no arc between the edges.

The second dependency is in the horizontal direction where relationships between alternatives are described. There are three types of relationships. With an implication, a preceding decision of an alternative enforces the execution of a succeeding decision of an alternative of another partial problem (III). A restriction, on the other hand, prevents a decision of a preceding alternative from influencing a decision of an alternative of another partial problem (IV). The suspension allows the evaluation of alternative solutions (V).

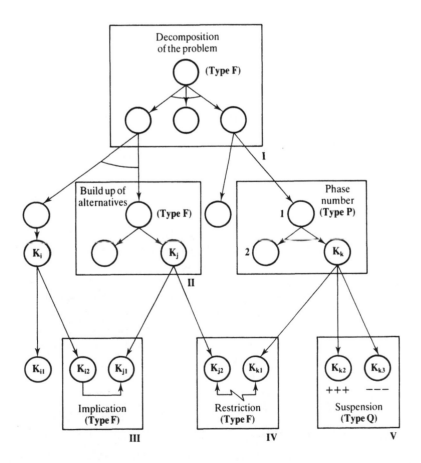

Figure 3.49

The presentation of dependencies in the AND/OR graph.

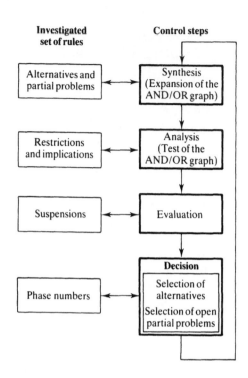

Figure 3.50
The control strategy.

The third dependency is the ordering of the partial processes which is done with the help of phase numbers. This actually represents an optimization process.

The hierarchical decomposition of a problem in subproblems or alternatives is a very powerful aid to reduce the problem's complexity. The more it is possible to isolate the subproblems into independent units, the better the decomposition will be. The graph presentation has the further advantage that various alternatives can be depicted in compact form.

The various types of P, F and Q knowledges depicted in Figure 3.48 have also been entered in Figure 3.49 and show which knowledge will be processed in which activity of a planning cycle.

When all knowledge of a manufacturing system has been classified and structured, a problem solver can be developed which uses the control strategy shown in Figure 3.50. On the left side of the figure, the types of rules are entered and on the right side, the processing steps which operate on the rules are shown. The structure of an entire expert system, for planning of a manufacturing system looks similar to the one shown in Figure 3.25.

For smaller manufacturing systems, for example a machining cell, the preparation of the process plan and the equipment layout can be highly automated. The workpiece spectrum is defined in the CAD database and the piece rate is entered. The system will draft a process plan as discussed earlier in this chapter. For the layout operation of the cell the following parameters are needed.

- number and types of machines selected,
- machine dimensions,

- operating range of machines (robots),
- available floor space,
- material handling equipment,
- part buffers,
- material flow pattern and so on.

With these parameters and the layout rules, the expert system will select the machine tools and material handling equipment, and arrange these to a workable system.

Figure 3.51 shows a typical workpiece spectrum of stampings and Figure 3.52 the user interface to an expert system for configuring sheet metal manufacturing cells. From the input description of the parts, the user can obtain in a dialog session the specification and layout of the cell. The output of the expert system is displayed on the graphical screen via a CAD interface.

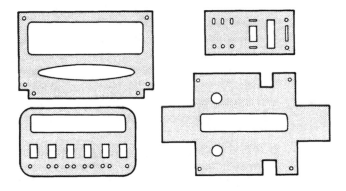

Figure 3.51
A workpiece spectrum as input to an expert system.

- Visual control of the layout
- Correction with CAD commands

User

- Input of the workpiece spectrum
- Modification of restrictions and requirements

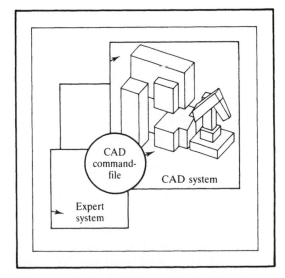

Figure 3.52
User dialog of an expert system for configuring a manufacturing cell for sheet metal parts.

3.5.6 Operation of a knowledge-based planning and control system

The success of a knowledge-based system in manufacturing depends on many factors, including the man/machine interface, the integration of knowledge-based systems in present factory planning and control activities, the use of knowledge-based systems in a realtime environment, and the development of learning capabilities of knowledge-based systems. We will be discussing important topics concerning the operation of knowledge-based systems in this section.

The man/machine interface

Until now, the development of knowledge-based systems has been the domain of the knowledge engineer. He may be a good computer scientist and an expert in artificial intelligence; however, he is usually ignorant about the domain for which a knowledge-based system is being developed. When looking back at the history of data processing, a similar situation was observed. The early data processing applications were mainly the work of computer specialists who did not know the manufacturing process. For this reason, many of the early applications did not work. We appear to have forgotten this experience and are repeating the painful learning process again. To make an application workable, it must be conceived and implemented by the manufacturing engineer. He is the person who knows the manufacturing process and its problems. The knowledge engineer should only assist him.

When the knowledge engineer and manufacturing engineer get together, there is immediately a problem of communication. Unfortunately, artificial intelligence has developed a vocabulary which does not belong in the user's language repertoire. To make things worse, there is no standard terminology and, therefore, experts often have communication problems among themselves. Here, it would be of advantage if the scientific community develops a more down to earth attitude and becomes more user oriented. A strong effort must be made to develop tools for the user so that he will be able to develop his own knowledge-based systems. This entails interfaces being developed which tutor and guide the application engineer to the use of an knowledge-based system. Languages like Prolog, LISP, Smalltalk and so on, are a deterrent to keep the user away from artificial intelligence tools. Since such languages are needed to build knowledge-based systems, they should be hidden from the user.

There is an urgent need for the conception and development of acquisition tools to extract knowledge from written text, protocols, and verbal communication. Presently, there is a considerable amount of research work in this area, however, most tools developed so far aid the knowledge engineer and not the user. Again, a similar trend can be seen here as has happened with software engineering tools. In order to apply them correctly, one must be an expert in the field and be able to read and interpret often big and confusing instruction manuals.

Good man/machine interfaces should also be developed for the user, to be able to communicate with the process. One of the greatest potentials of the knowledge-based system actually lies in this area. The man on the factory floor needs tools to understand his process to guide him during a start-up or shut-down, and to help him to locate and interpret problems. Such capability should be built into the operating console of the machine tool.

Since there is a considerable amount of production machine programming for the manufacturing system, better programming facilities for cell, machine, test equipment, and robot programming must be developed. An important aspect is shop-floor programming for job shop operations so that the machinist himself can program and set up the machine. Attempts made in Germany in this area were highly successful for NC programming; however, cell and robot programming are more difficult and need the aid of knowledge-based systems. One way is to go the task-oriented or implicit programming route which is presently being pursued, mainly in Europe. The results indicate that implicit programming is not a cure-all to solve the problem. To build the knowledge base for the world model is a formidable task, even for a narrow domain. It is believed that graphic interactive programming in conjunction with implicit programming will render more viable solutions.

Integration of expert systems in a realtime environment

In the factory of the future, there will be a flurry of knowledge-based system activities. It will become necessary to tie various knowledge-based systems together and operate them in real time. Until today, time was not an important consideration with early planning and control systems. Actually, all manufacturing operations are time dependent and many applications cannot be handled well without a time scale.

Another problem in this connection is the realtime capability of knowledge-based systems. Applications involving sensors, usually, must have a control system which can take actions in real time. To date, this problem has not been resolved satisfactorily and much research must be done to find ways of making a knowledge-based system an integral part of realtime control.

Often in a complex manufacturing system, various knowledge-based systems have to work together to solve a problem. As a common communication medium for knowledge-based systems, the so-called blackboard was developed as shown in Figure 3.53. Blackboard-based systems are often called knowledge-based systems of the second generation. The blackboard architecture consists of the various knowledge sources containing different types of knowledge. The blackboard serves as a common communication medium and working memory, and the scheduler controls the strategy to apply the knowledge. The blackboard is hierarchically structured; the objects are input data, partial solutions, alternatives, or the final solutions. During its operation, the blackboard is monitored by each knowledge base to determine if it can help to find a result. If yes, it becomes active and changes the situation in the blackboard. The results may activate other knowledge sources to contribute to the final solution of the problem.

The blackboard technology is actually being used in many applications; however,

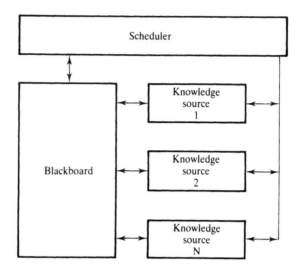

Figure 3.53
Blackboard
architecture.

its use has not been very successful for complex problems involving numerous knowledge-based systems.

Learning capabilities of knowledge-based systems

An added, desirable feature of a knowledge-based system would be learning capabilities. The system should be able to learn from the ongoing manufacturing process to update its knowledge, and to be able to find a solution for a new situation. The learning ability, of course, will be limited, for example, a process planner may learn from the machining operation of one part variant how another is to be machined. However, the planner will not be able to determine how an unknown part is to be machined. The learning strategies can be classified as:

- rote learning,
- learning from instruction,
- learning by analogy,
- learning from example,
- learning from observation.

Learning from observation will probably be the most important method to be used for manufacturing problems. Several experiments have been made with learning, however, the technology is still in its infancy.

3.6 Problems

(1) Why is it necessary to consider all manufacturing activities when planning a CIM system? Are the planning activities for a new computer-controlled plant different from those needed to retrofit an old plant for computer control?

(2) Why is the estimation of the cost of software so difficult for a CIM system? Does the degree of automation play a role when estimating the software cost?

(3) Explain the planning cycle of a manufacturing system and the tasks which have to be performed in each phase. Discuss the flow of information between the various phases.

(4) How can the planning cycle be supported with well-structured planning aids? Give an example of such a planning aid and discuss the details.

(5) Discuss the various simulation tools and explain the advantages and disadvantages of them. For what kind of problems can they be used?

(6) What are the principal simulation methods and how do they work?

(7) Explain the simulation cycle and which tasks have to be performed in each phase.

(8) Show how a robot action plan can be derived from a precedence graph. Conceive a small flexible manufacturing system and show how the flow of a product can be simulated by a Petri net.

(9) What are the components of an expert system, their function and interfaces to other components?

(10) How can knowledge be represented in a knowledge base? What are the advantages and disadvantages of the different knowledge presentations?

(11) Explain how the various search strategies for processing of knowledge work?

(12) Explain the life cycle of an expert system and the tasks to be performed in each phase.

(13) What are the three types of product design? How can design be supported by an expert system? In which design phase is an expert system most useful?

(14) What are the phases of process planning and which tasks are to be performed in each phase? How can an expert system aid process planning?

(15) What are the basic problems of using expert systems for manufacturing scheduling? How can online rescheduling be done and how would you handle sudden manufacturing disturbances?

(16) Explain the three categories of diagnosis systems. Where would you use them?

(17) Explain Figure 3.45.

(18) Devise a general scheme of configuring a manufacturing system with the help of an expert system. Can such a scheme be used for both batch-type production and mass production?

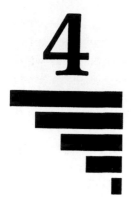

Flexible manufacturing and assembly equipment

CHAPTER CONTENTS

4.1 Introduction   176
4.2 Numerical Control and Design Features of NC Machines   177
4.3 Flexible Manufacturing Equipment   181
4.4 Planning and Introduction of Interconnected Manufacturing
   Systems   198
4.5 Flexible Assembly Systems   206
4.6 Economic Considerations   210

4.1 Introduction

For many manufacturers of consumer goods, flexible manufacturing and assembly is the only way in which they can efficiently compete in the market-place with a range of product variants. This is because clients are increasingly looking for product variants tailored to their own needs rather than mass-produced products. Such product variants can no longer be produced using classical mass-production methods and for this reason flexible production systems are required. Flexible manufacturing and assembly systems usually consist of several programmable individual machines, where, according to demand, in order to manufacture a specific product variant, several machines may be linked (communications link) to one production unit with the aid of planning and control programs.

Manufacturing and assembly equipment, together with transport and handling equipment, form the ingredients for machine support of integrated production. Today, flexible manufacturing and assembly equipment is characterized by the principle of numerical control. Thus, important components of a numerically-controlled machine include:

- computer control (for example, computerized numerical control (CNC) or robot control (RC)) as a hardware ingredient with associated functional software,
- speed-controlled drives, and
- position-controlled measurement systems.

Technological parameters such as:

- cutting speed
- feed rate
- cutting depth
- tool selection
- forces and moments

may be automatically set using computer control. Moreover, the geometric paths and positioning required in machining and joining processes may be flexibly automated via axial movements. This chapter describes flexible manufacturing and assembly equipment. The chapter begins with a brief consideration of numerical control and particular design features of NC machines; it ends with a discussion of the basic structure of various flexible manufacturing machines.

4.2 Numerical Control and Design Features of NC Machines

The NC machine is the basic building block required for flexible production. The term 'numerical control' (NC) means that an NC machine is controlled by numbers obtained from the description of the part. This permits flexible adaptation to the changing nature of tasks, particularly in small- and medium-scale batch production. Geometric and technological instructions for manufacturing a workpiece are coded in terms of numbers and held in data storage (see Section 5.6.1).

The generation of the data for the automatic machining of a workpiece is called NC programming. Programming produces a series of NC records for a workpiece (see Section 5.6). Every NC record contains geometric and technological instructions, in other words, dimensional data for generating a part and switching information for operating the machine tool.

The manufacturing information for a workpiece is fed to the NC system. It is divided into geometrical data for the tool or workpiece (G, X, Y, ...) and technological information (F, S, T, M) for the control of the machine, and stored as such, as shown in Figure 4.1.

The dimensional data contains the target positions for the generation of the workpiece contour; according to the machine tool used, this may consist of instructions for XZ, XYZ, XYZC and possibly additional machine axes. The geometry is generated using point-to-point, straight-line or continuous-path control. The individual axis positioning instructions are generated by an interpolator

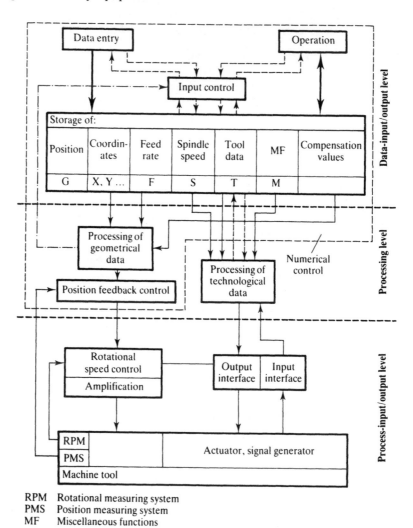

Figure 4.1
Functions of an NC
controller.

RPM   Rotational measuring system
PMS   Position measuring system
MF   Miscellaneous functions

and forwarded via an amplifier to the actuator. The three typical types of control are shown in Figure 4.2.

- Point-to-point control permits movement from point-to-point with rapid positioning (drilling cells). The path of motion between points is determined by the controller.

- Straight-line control permits movement along an axis at a defined speed. It is often possible to couple several axes in variable speed drives; thus, movement at 45° is possible. Straight-line control is now only found in exceptional cases.

- Continuous path control may be used to travel through a number of selected paths (mostly lines and circles) at defined speeds. Arbitrary contours may be composed from the available motion elements such as lines and circles.

The technological control influences various technological parameters such as

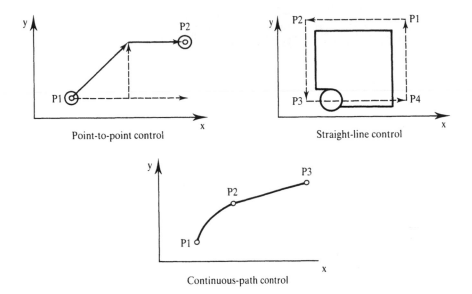

Figure 4.2

Various positioning control methods for machine tools.

the switching of the main spindle speeds, feed rates, tool-changing equipment, supply of coolants and so on.

The NC processing level was originally implemented in a hard-wired circuit. Subsequently, arithmetic components in the form of soft-wired processing algorithms were introduced into the numerical control. This type of implementation is called computerized numerical control (CNC). Nowadays, numerical control systems are manufactured as CNC systems. Chapter 5 discusses the development from NC to CNC and also to DNC.

The design of numerically controlled machines must follow certain requirements, which are a prerequisite to positioning and travel on exact paths. These requirements may be subdivided into two groups:

- General requirements (valid not only for NC machines), such as high static and dynamic rigidity of all machine-tool elements and low thermal distortion.
- Requirements to minimize the effects of forces, such as machining or joining forces together with the effects of the transmission characteristics of transfer elements or machine elements in the position control loop, see Figure 4.3.

Additional requirements for the mechanical elements are:

(1) Low moments of inertia of the machine elements which are to be accelerated and decelerated.

(2) High rigidity, so that the mechanical natural frequencies are sufficiently higher than the natural frequency of the drive.

(3) High absolute damping of forces.

(4) Avoidance of nonlinear transfer elements in the controlled system. Nonlinearities result from, for example, play and friction which is not proportional to velocity.

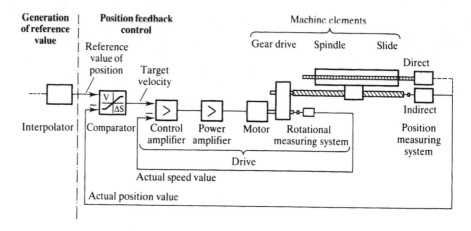

Figure 4.3

Position control loop
of a machine tool.

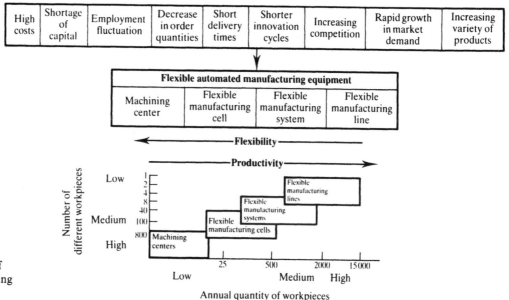

Figure 4.4

Application areas of
various manufacturing
methods.

Figure 4.3 shows a linear axis controller; the same basic mechanisms are used for rotational axes. The interpolator forwards a reference value to the numerical control system. Here, after comparison with the actual position, a target speed for the drive, with its control and power amplifiers and motor, is generated. The machine-tool table is adjusted via the gear drive and spindle. The figure shows both a position control loop and a speed control loop (see also Figure 4.1). The underlying speed control loop leads to improved damping. The control loop also contains direct and indirect measurements of position. Nowadays, more complex control structures, such as state control, are sometimes used, for example, to perform dimensional adjustments caused by machine distortion or high speed operation.

In addition to the automatic setting of technological and geometric values, NC machines may contain other automated functions such as tool and workpiece handling and buffering, use of multiple tools, measurement in the machine, and so on. Individual machines may also be linked together. When such functions are taken into account, it is possible to distinguish between different manufacturing principles such as machining centers, machining cells, flexible manufacturing systems and production lines. Figure 4.4 (see also Figure 1.6) links these concepts to flexibility and productivity. Flexibility and productivity are contradictory objectives. In the figure, the numerical values for quantities and numbers of parts are solely examples. It is clear from the overlapping sections shown in the figure that precise delimitation of the areas of application of the different methods is impossible. The top line in the figure describes current demands on manufacturing technology which may be met by flexible manufacturing methods. The classification of Figure 4.4 and the requirements may be applied to assembly equipment. In the following section, the methods listed are described in more detail and examples are given.

4.3 Flexible Manufacturing Equipment

Flexible manufacturing equipment (such as machining centers, manufacturing cells, linked systems) is used for automatic machining of different workpieces. It permits fast adaptation to changing machining tasks. Reasons for the need for flexible manufacturing include small batch sizes with short production runs, reduction in stock and short-term delivery dates, under the constraints of high-availability and high-level-of-use of production equipment (see also Figure 4.4). Because of these technical and economic requirements, flexible manufacturing requires thorough planning and effective organizational tying-in to the factory operation.

4.3.1 Machining centers

A numerically-controlled machining center is capable of performing several machining operations and of carrying out automatic tool changes from a magazine or other storage device, according to the machining program. The manufacturing method is batch production; batch changes usually imply manual retooling.

Basically, two types of machining centers are used, one for rotational and one for prismatic parts, see Figure 4.5. Both types use CNC controllers and programmable logic controllers (PLC). The handling of workpieces may vary, since clamping is often carried out manually on the machine. Turning centers are equipped with one or more turrets for storing tools. Turrets are less common in boring-and-milling centers, usually tools are changed from a magazine into the main spindle. Figure 4.6 shows the working space of an NC turning center with automatic tool change to a multiple turret and with the facility for numerically-controlled positioning and movement of the main spindle. The simultaneously-controllable axes are shown. Both turrets operate along the linear axes X and Z (see also Figure 4.3). One turret and the main spindle are arranged along the numerically-

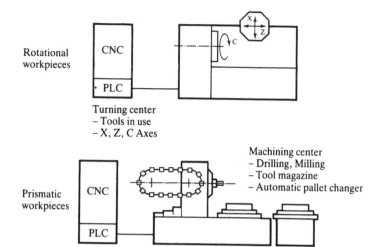

Rotational
workpieces

Turning center
– Tools in use
– X, Z, C Axes

Machining center
– Drilling, Milling
– Tool magazine
– Automatic pallet changer

Prismatic
workpieces

Figure 4.5
Machining centers, the
basic elements of a
manufacturing cell.

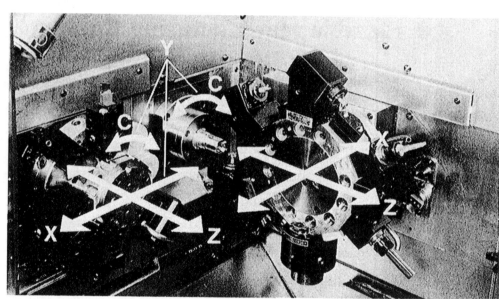

Figure 4.6
Axes of a CNC lathe.

controlled rotating axis *C*. The movement along the *Y* axis is given by the combination of two rotations, namely that of the main spindle and that of the left turret. Thus, it is possible to perform various boring-and-milling operations on the turning center. Figure 4.7 shows the performance of drilling and milling operations using both the *X* and *C* axes. These centers have a high versatility and can completely machine entire workpieces and thus, in particular, reduce the throughput time. A relief drilling station also permits drilling on the opposing sides of a part.

In many cases, a machining center for prismatic parts (Figure 4.8), in addition

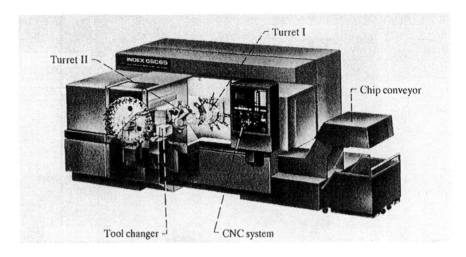

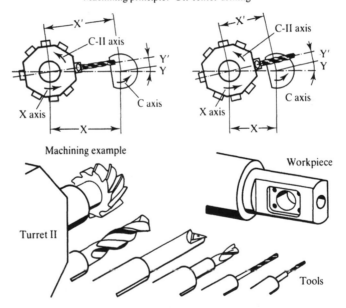

Machining principle: Off-center drilling

Machining example

Figure 4.7
Machining center
index GSC 65 –
GSC 100 with *Y*-axis
and tool changer.

to a tool magazine, also has a facility for automatic pallet changing (Figures 4.5 and 4.9). Here, workpieces are clamped to a pallet on which they may also be transported through a manufacturing system. The pallet is used almost like a machining table in each machining center; this requires a reference position. During the machining of a workpiece the next workpiece is clamped to a new pallet and prepared for machining. Thus, the machining-time component of the center is high since clamping and machining are carried out in parallel. Workpieces may be changed manually from storage during the production process. Turning and boring-and-milling centers are unquestionably the centers which are most frequently

Figure 4.8
Typical complex
workpieces produced
on a flexible
machining center.

Figure 4.9
MC 50 machining center with a magazine rack for 103 tools; free spindle tip and chip conveyor.

used and which have been available longest. However, centers are increasingly designed for other manufacturing processes, for example, punching, nibbling and metal-bending.

Figure 4.10 shows a sheet metal-bending center. This permits consistent manufacturing of complicated cabinet-shaped products from flat sheet metal as input material. The handling of the workpieces and the use of bending tools must be coordinated by programming. Here, mention should also be made of measuring centers, which facilitate both automatic workpiece changes and automatic probe changes.

We also note the importance of planning the selection of appropriate workpieces for a manufacturing system. Figure 4.11 illustrates possible coarse criteria for evaluating the selection of workpieces for machining in a center with various machine tools. It lists a number of partial objectives, divided into three groups. These groups relate to geometric, economic and manufacturing-oriented requirements. Every entry has a weighting denoted by W. These groups comprise partial objectives, which again have weightings. The weighting involves objective and subjective criteria. The lower part of the figure includes an assessment based on the overall value V. When using this evaluation technique, the geometric and technological complexity and workpiece families must be considered.

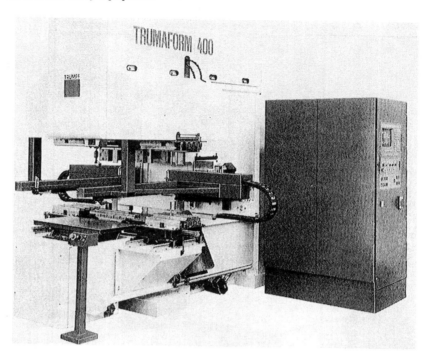

Figure 4.10
Flexible sheet metal
bending center.

4.3.2 Flexible manufacturing cells

A flexible manufacturing cell is an organizational unit and may be a single- or multi-machine system which, preferably, machines workpiece families automatically. In addition, the peripheral functions such as workpiece and tool handling together with the execution of measurement tasks and process monitoring are highly automated in a cell. Flexible manufacturing cells (FMC) may be elements of flexible manufacturing systems (FMS) and support the incremental build up of such a production facility in a sensible fashion.

FMSs facilitate manufacturing of small batch sizes with low staffing. With automatic tool and workpiece handling and storage, the unattended use of FMSs in breaks or in the second and third shift is possible. Batch changes do not always necessitate retooling, a simple change of program may be all that is required here.

Figure 4.12 shows a manufacturing cell consisting of a two-machine system comprising a machining center and a lathe linked together by a system for transporting material. The figure also shows the machine-oriented control level containing the NC and robot controls. This level is supervised by the master cell control system (see Chapter 5). Material and tool transport is carried out via vehicles and the handling via robots. The figure shows the typical machine components of a flexible manufacturing cell together with their material handling interfaces. These components include the following:

- Manufacturing equipments with robots for part and tool handling and buffering for parts and tools.

Rough selection of tools for manufacturing of workpieces in a machining center					
Workpiece: Gear housing _____	Drawing number: 0 247 474 _____				
Name: _____	Date: _____				

Geometric requirements .... (W = 50)	Y	N	W	V
1 Are more than two faces to be machined?	(*)	( )	(10)	(500)
2 Includes machining of non-parallel axes?	(*)	( )	(9)	(450)
3 Wall-height/Wall-thickness ratio ≥ 5?	(*)	( )	(3)	(150)
4 Inner radius ≤ 4mm?	(*)	( )	(3)	(150)
5 More than twelve holes?	( )	(*)	(9)	( )
6 Drill hole diameter ≥ 20mm?	(*)	( )	(3)	(150)
7 Drill holes with steps and/or threads?	(*)	( )	(3)	(150)
8 Drill holes with depth/diameter ratio ≥ 3?	(*)	( )	(3)	(150)
9 Circular milling possible for large boreholes?	( )	(*)	(9)	( )
10 Turning operations needed?	(*)	( )	(3)	(150)
11 More than three perforations or notches?	( )	(*)	(3)	( )
12 Spacing tolerance ≤ 0.050mm?	(*)	( )	(10)	(500)
13 Tolerance on fit ≤ IT7?	(*)	( )	(10)	(500)
14 Parallelism ≤ 0.1mm over 100mm?	( )	(*)	(6)	( )
15 Angularity ≤ 0.1mm over 100mm?	(*)	( )	(6)	(300)
16 Depth tolerance for offset drill holes?	(*)	( )	(6)	(300)
17 Surface quality at least RMS 6.3μm	(*)	( )	(4)	(200)
Total value of the geometric requirements				(3650)

Economic requirements .... (W = 20)	Y	N	W	V
18 Machining time/Downtime ratio ≤ 20?	( )	(*)	(15)	( )
19 Batch sizes up to 200 ... 300 pieces?	(*)	( )	(25)	(500)
20 At least four jobs during the year?	(*)	( )	(15)	(300)
21 At least one design modification per year?	( )	(*)	(25)	( )
22 Duration of workpiece load and clamping time?	( )	(*)	(10)	( )
23 Duration of set-up time per job	(*)	( )	(10)	(200)
Total value of the economic requirements				(1000)

Manufacturing requirements .... (W = 30)	Y	N	W	V
24 Are there similar workpieces (part families)?	(*)	( )	(25)	(750)
25 Spare parts including after termination of manufacture?	(*)	( )	(15)	(450)
26 Are the current fixtures expensive?	( )	(*)	(5)	( )
27 Is the space requirement for fixtures important?	(*)	( )	(5)	(150)
28 Must every workpiece be checked?	(*)	( )	(15)	(450)
29 Is uniform quality necessary?	(*)	( )	(25)	(750)
30 Is a short job time important for sales?	(*)	( )	(5)	(150)
31 Are workpiece transport and storage costs high?	(*)	( )	(5)	(150)
Total value of manufacturing requirements				(2850)

Total value of all requirements	(6500)

Assessment: Part is probably suitable.
Perform economic study.

Y = Yes, N = No, W = Weighting (%), V = Value = Group weighting × Individual weighting

Figure 4.11
Rough selection scheme of tools for manufacturing of workpieces in a machining center.

- Transport devices or vehicles for parts and tools.
- Facilities for storing tools and parts.

This two-machine system contains complementary machines which execute different machining tasks and are not capable of replacing one another. In contrast, substituting machines provide the same machining facilities and may be replaced by one another. Usually, cells consist of similar substituting machines. This greatly reduces the risk of total breakdown in bottleneck situations and means that the use of each machine can be kept uniformly high. Figure 4.12 shows a master cell control system based on a control computer or a microcomputer. This control system is superordinate to the CNC and PLC layers and carries out coordinating planning and control functions. The control layers and the control functions of the cell layers are discussed in Chapter 5.

Figure 4.13 shows a turning cell and a boring-and-milling cell. The turning cell (upper half of the figure) is a one-machine system, while the boring-and-milling cell comprises two machines (two machining centers). In this case too, the nature of the part has a crucial influence on the structure of the machining system. The method of handling and interconnecting machines for rotational parts is essentially different to that for prismatic parts. Prismatic parts are handled by pallets which represent a specific mechanical interface. Rotational parts are transported directly by gripper systems. As far as turning is concerned, surface or inline gantry operated handling devices have replaced the original solutions with standalone robots for part, tool and sometimes even clamp handling. A tool change for retooling is carried out via a central storage in the turret. Tools needed to manufacture a part of a current job should be available in the turret(s) so as to keep changeover times short. For components which are gripped in different ways (workpieces, tools, clamps), grippers may be changed automatically.

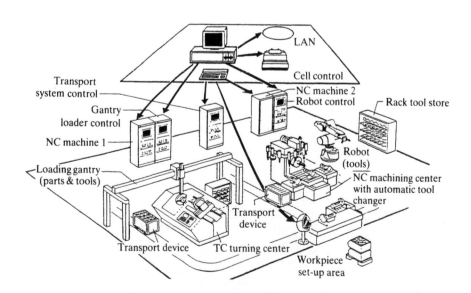

Figure 4.12

Structure of a flexible manufacturing cell.

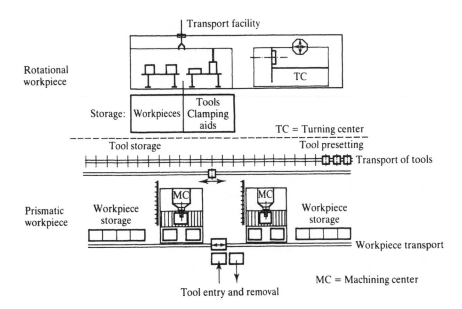

Figure 4.13
Flexible
manufacturing cells.

Figure 4.14
Flexible turning cell
INDEX GSC65 with
gantry material
handler INDEX WHZ
160.

Figure 4.15
MC 50 machining
center with pallet
storage for 8 630 × 630
pallets.

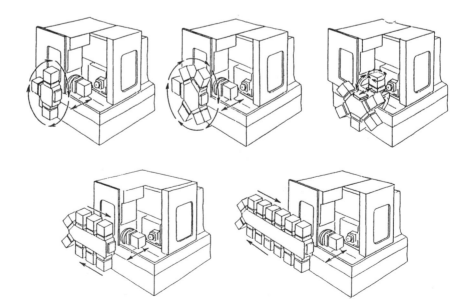

Figure 4.16
A flexible
manufacturing cell
with an extensible
storage for workpieces.

Figure 4.14 shows a turning cell with a surface gantry handling system and various grippers which permit automatic change of raw and finished parts, tools and clamping devices.

Boring-and-milling cells often have a pallet buffer and a large tool storage (Figure 4.13 – lower part and Figure 4.15). This helps to meet requirements for long uninterrupted periods of service, for example during breaks and during the third shift. Figure 4.15 shows the storage of eight pallets with various workpieces clamped to them. The clamping area for workpieces is also clearly visible. Figure 4.16 shows a machining cell for small workpieces. Various modular storage extensions permit the machining of a number of different workpieces; which may be carried out within a job mix. Machining within a job mix requires the availability of the whole range of tools. A clamp may be positioned horizontally or vertically in the work area; efficient chip disposal is also desirable.

4.3.3 Flexible manufacturing systems

Flexible transfer lines (Figure 4.17) are designed for the manufacture of variants, which are workpieces with a high degree of similarity. Manufacturing follows a fixed time cycle. Flexible manufacturing systems (FMS) consist of several interconnected individual machines which machine various workpieces simultaneously in a sequence uninterrupted by retooling. They are characterized by sequentially arranged manufacturing equipments (including manufacturing cells) linked together by a common control and transport system. The linkage obtained through the material flow is not paced. The manufacturing equipment may be accessed randomly.

Figure 4.18 shows device-oriented functions of a flexible manufacturing system. The material-flow functions are workpiece and tool storage and handling. The machining is symbolized by the machine tools in the centre of the figure. The information flow and the control system are shown in an abstract form on the left (for details, see Chapter 5). Other possible components of linked manufacturing systems not illustrated in the figure include washing stations, test stations, NC multi-coordinate measuring devices and chip-disposal equipment.

In FMS, the structure of the transport and storage equipment is also largely determined by the nature of the workpieces (rotational or prismatic). Also, with this manufacturing approach prismatic workpieces are usually clamped to a pallet, with one or more workpieces attached to each pallet. Rotational parts are transported individually or on pallets to the machines via the handling system (the gantry and gripper type system). The transport equipment may be used to transport both workpieces and tools. This depends on the time requirements which are determined by the number and the duration of the individual machining operations and the specification as to whether single- or multi-step machining is required. Single-step machining means that only one machine is provided for machining, while in multi-step manufacturing the overall machining takes place on at least two machines.

The interconnection structure of a manufacturing system has an important influence on the transportation of parts and tools in a large FMS. Figure 4.19

Figure 4.17
Flexible transfer line.

shows three basic structures of possible realizations of material flow systems, and the control principles used. In the networked structure any material-flow point may be linked to any other, which is not possible in the loop structure. In the radial structure, all links must start at the center. The choice of a specific structure depends on the time requirements, the layout of the manufacturing system, the equipment present and so on.

Figure 4.20 shows alternative workpiece supply systems and gives corresponding criteria for assessing such systems. As in Figure 4.19, solutions are described in device terms. There are transport systems with and without rails, surface gantries and overhead transport systems including both indirect (with pallets) and direct handling. Numerically-controlled guided vehicles are very reliable and are often used nowadays. One disadvantage of these is the reduced access to manufacturing equipment. In flexible manufacturing systems, which cover large spatial areas and

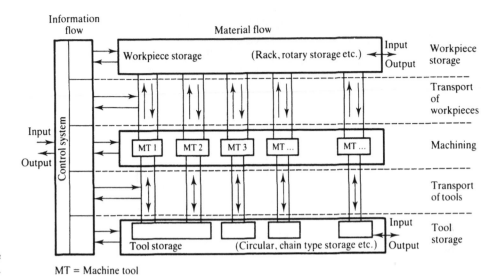

Figure 4.18
Functions of a flexible manufacturing system.

MT = Machine tool

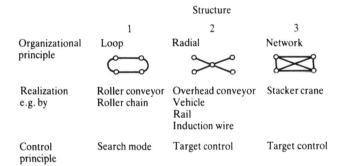

Figure 4.19
Interconnection structures of typical FMS for machining prismatic workpieces.

	Structure		
	1	2	3
Organizational principle	Loop	Radial	Network
Realization e.g. by	Roller conveyor Roller chain	Overhead conveyor Vehicle Rail Induction wire	Stacker crane
Control principle	Search mode	Target control	Target control

in which the time requirements are critical (long machining times), it is advantageous to use automated guided vehicles. The routes of these devices may be flexibly programmed (see Chapter 10). Figure 4.21 shows devices of this type used in the manufacturing of aircraft parts with long machining times. In this example the workpieces to be machined are placed in interim buffers. Thus, there is a decoupling in time of the transport and the machining processes. This example also shows an automatic supply of tools to the manufacturing machines. The tool transport system is installed on the supports of the building structure.

Figure 4.22 illustrates the principles of various tool supply systems and gives selection criteria. The various structural approaches to solutions are divided into two groups; tool change on the machine and tool storage on the machine. The selection criterion here is whether tools can be exchanged manually or automatically from the magazine on the machine tool. The figure also gives a rough assessment of tool supply systems. When preparing tools, it is important to disturb the productive time as little as possible. There are two ways of doing this. Either all

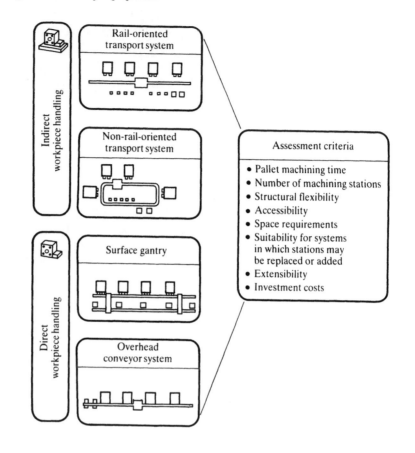

Figure 4.20
Alternative workpiece
supply systems.

the tools required during a manufacturing period may be placed in a machine-integrated storage, or tools are exchanged in parallel with production as required when changing jobs or during maintenance (see below, four approaches to these solutions). The 'additional costs' column of Figure 4.22 merits special attention. Figure 4.23 shows an implementation of an installation in which individual tools or whole tool sets may be retooled in parallel with production. All the tools in the magazine (maximum 203) may be directly accessed at any time via the tool feed track. The interconnection structure together with the transport facilities influence the control functions of the controller, as we shall see in Chapter 5.

Briefly, we now give additional examples of flexible manufacturing systems. Figure 4.24 shows the flexible manufacturing and assembly system of the ISW (Institute for Control Technology for Machine Tools and Manufacturing Systems of the University of Stuttgart). This installation is the basis for diverse, extensive research and development programmes. Its control structure is described in Chapter 5. On the right side of the figure is the manufacturing system with a machining center (10) and a turning center (8). A multi-coordinate measurement device (9), together with storage (7) and clamping areas (14) are visible in the layout. On the left is a flexible assembly system with a set-up robot (12), handling

Figure 4.21
A flexible manufacturing system for the production of aircraft parts.

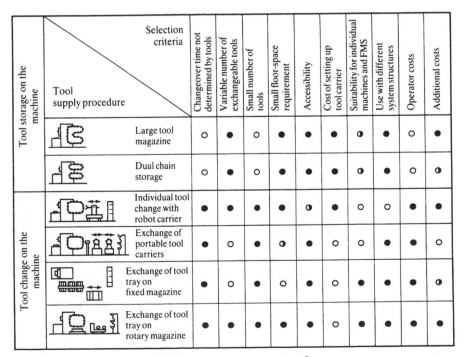

Tool storage on the machine / Tool supply procedure	Changeover time not determined by tools	Variable number of exchangeable tools	Small number of tools	Small floor-space requirement	Accessibility	Cost of setting up tool carrier	Suitability for individual machines and FMS	Use with different system structures	Operator costs	Additional costs
Tool storage on the machine — Large tool magazine	○	●	○	●	●	●	◐	●	○	●
Tool storage on the machine — Dual chain storage	○	●	○	●	●	●	◐	●	○	◐
Tool change on the machine — Individual tool change with robot carrier	●	●	●	◐	●	○	○	●	●	●
Tool change on the machine — Exchange of portable tool carriers	●	○	●	◐	●	○	○	●	●	○
Tool change on the machine — Exchange of tool tray on fixed magazine	●	○	●	○	●	○	●	●	●	◐
Tool change on the machine — Exchange of tool tray on rotary magazine	●	●	●	●	●	○	●	●	●	●

● Good   ◐ Average   ○ Poor

Figure 4.22
Selection criteria for tool supply.

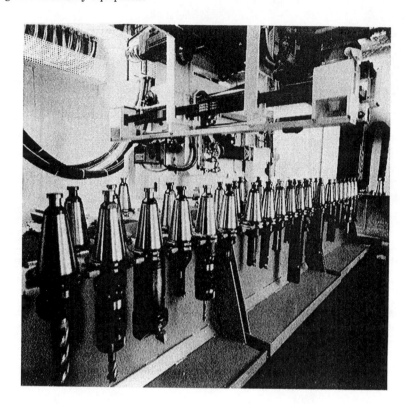

Figure 4.23
Gantry type
automatic tool
changer.

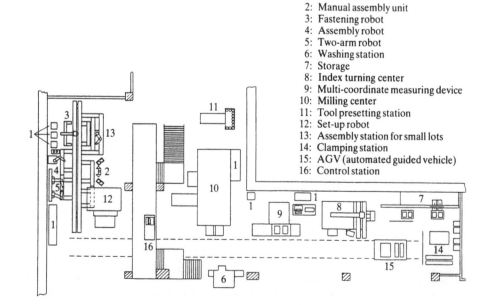

1: Control cabinet
2: Manual assembly unit
3: Fastening robot
4: Assembly robot
5: Two-arm robot
6: Washing station
7: Storage
8: Index turning center
9: Multi-coordinate measuring device
10: Milling center
11: Tool presetting station
12: Set-up robot
13: Assembly station for small lots
14: Clamping station
15: AGV (automated guided vehicle)
16: Control station

Figure 4.24
Layout of the
experimental FMS of
the University of
Stuttgart.

Figure 4.25
Flexible manufacturing system consisting of: 2 MC 50 machining centers, 2 assembly stations, 15 positions for storing 630 × 630 mm pallets and a passive magazine for 300 tools.

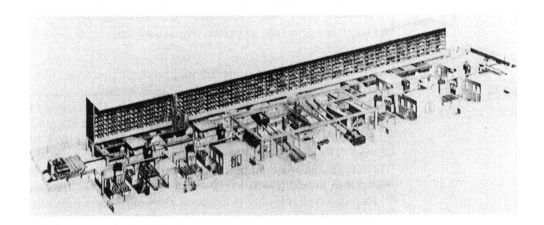

Figure 4.26
Highly automated FMS with many machining centers, an interconnecting transportation system and a high-rise storage area.

robots (3, 4, 13), a two-arm robot (5), a manual workstation (2) and conveyors. The two systems are linked together by a transport system (15, floor-type conveyor without tracks). The measuring cell is integrated into the quality control loop. The automated guided vehicle is a new development.

Figure 4.25 shows an FMS with two machining cells, two assembly areas, 15 pallet deposit points and a magazine with 300 tools. The material and tool transport is via a conveyor. Figure 4.26 outlines a highly-automated FMS consisting of four machines with a tool supply built as a high-rise rack storage. This system may be automatically operated with few people.

4.4 Planning and Introduction of Interconnected Manufacturing Systems

Complex manufacturing equipment requires integral planning, covering all business functions of a company including manufacturing planning and design (see Chapters 1 and 2). In addition, questions of training, job evaluation, working-time arrangements, wages and so on, must be considered. Extensive planning is required, which must be supported by equipment suppliers and users. Figure 4.27 shows a typical planning sequence which describes the tasks of the manufacturer and the user of the manufacturing system. This cooperation must be led by a project management. The figure shows that individual tasks may be executed alternatively by the manufacture and user sides. One important feature is the repeated coordination of interim results at various stages of the planning process.

The setting of goals involves the formulation of organizational, technical and economic targets, while feasibility studies identify alternative schemes for achieving the objectives. Project planning then leads to a proposal of a specific system for realization.

In general, planning must take account of the following:

- Flexibly-automated manufacturing is capital intensive and is in practice only suitable for a limited part spectrum.
- The task is highly-critical and important and requires large expenditure; a badly planned facility may lead to high restructuring costs when put into operation.
- The basic principle 'as productive as possible – as flexible as necessary' should be applied.

The objectives of planning are of a strategic and operational nature. Strategic objectives include, for example, the reduction of throughput times and of stock, and the formation of a basis for a CIM strategy. Operational planning objectives include a high utilization rate of a multistage operation, low staffing, high flexibility, high return on investment and gradual introduction and integration into the factory. The high utilization rate required by capital-intensive facilities must be in the foreground. Automation should take priority over flexibility. However, this utilization should not be at the cost of lack of operational flexibility.

Detailed cost planning is based on the hourly operating cost of interconnected manufacturing systems, which is very high in comparison with conventional manufacturing. Figure 4.28 shows an assessment of the flexibility of a production facility against the costs of automation. The increase in cost of the various activities due to the interconnection of CNC base machines is shown as a percentage.

Planning may be summarized by the following three important tasks:

(1) selection of appropriate workpieces and their adaptation for the purposes of automation,

(2) specification of the layout (including system components) and of the dimensions of the material flow and storage equipment,

(3) specification of the control functions and the control structure.

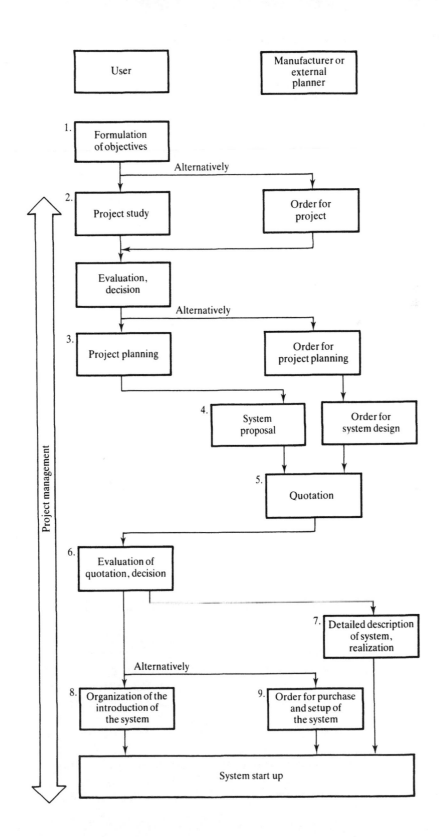

Figure 4.27
Flow chart showing the sequence of work and the distribution of assignments between the user and the manufacturer for the introduction of a flexible manufacturing system.

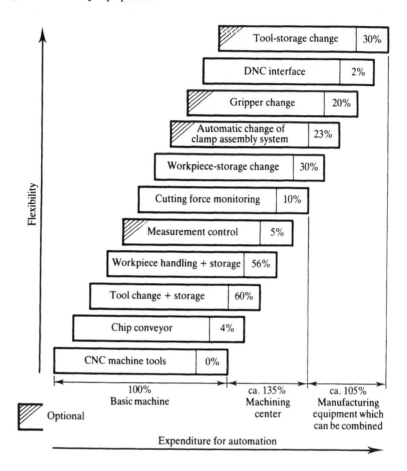

Figure 4.28

The expenditure for flexible automation of machining operations.

A study should precede these planning stages (see Figure 4.27) in order to analyze the task to be performed and the present state of the manufacturing system. This involves an analysis of the range of workpieces which may be manufactured in a manufacturing system. The selection of appropriate workpieces may be based first, on the overall range of parts within a production shop and second, on workpieces which have already been produced by NC machines, see Figure 4.29. The figure shows that from an overall range of the parts spectrum a subset of parts is particularly well suited for NC machining and a second subset may be machined by the FMS system. The figure lists the criteria for selecting the subsets together with the documents and methods needed. Geometrically- and technologically-oriented classification systems are important aids to the selection of workpieces. It may also be advantageous to introduce a code system for the clamping facilities for prismatic workpieces, for the machining and for the capacity requirements. While codes for the workpiece geometry are usually based on well-known classification systems, company-specific systems are introduced for the other classification tasks. Figure 4.30 summarizes the features of a classification of workpiece geometry, machining and clamping for an actual application. The

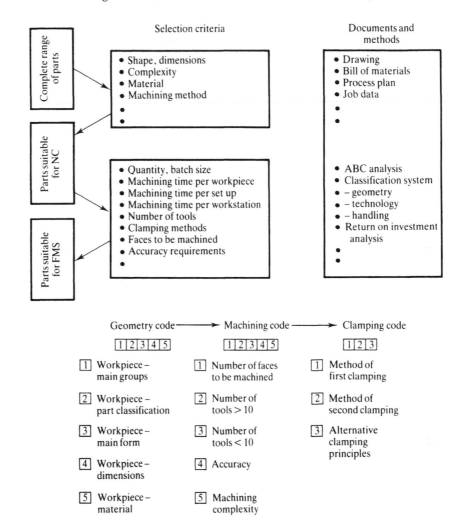

Figure 4.29

Selection of suitable workpieces for FMS.

Figure 4.30

Structure of a workpiece classification system.

classification systems form the basis for computer-assisted selection of appropriate workpieces (see Chaapter 8). Programs have been developed to evaluate classification results. Figure 4.31 is a rough illustration of the structure of such a program. The program analyzes the classification features and determines handling parameters, and necessary tools and clamping facilities. This data together with the required manufacturing capacity form the basis for designing the manufacturing system and its material handling equipment. The program also determines the requirements of various workpieces in terms of machines, tools and linkage. Finally, a suitable family of parts is chosen.

It is important to specify families of parts suitable for FMC and FMS which essentially leads to an increase in batch size and which may be manufactured in a fixed, uninterrupted order. A possible change to the production program should be borne in mind during the planning to ensure long-term flexibility. The first batch size of the first job will scarcely provide for favourable economic

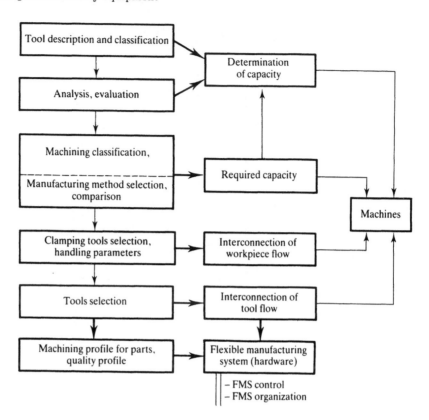

Figure 4.31

Analysis of a
workpiece spectrum.

manufacturing, since, amongst other things, tested, error-free NC programs are
required; obtaining these is costly and time consuming. Figure 4.32 shows an
example of a workpiece which was analyzed by a method similar to that shown
in Figures 4.30 and 4.31 and classified as being suitable for machining in an
FMS.

For a fixed range of workpieces, the tools, manufacturing equipment, material-
flow components and auxiliary equipment (in other words, the layout and the
control system) are selected by rough and detailed planning (Figure 4.33). The
figure distinguishes between the planning of workstation, material-flow equipment
and control- and auxiliary-equipment components. In addition, planning objectives
for the rough and the detailed planning are given. The detailed planning
specifications lead to the final layout of the manufacturing system. Planning is an
iterative procedure involving frequent revisions of earlier planning steps. Thorough
planning requires task-oriented working groups of users and manufacturers, with
effective coordination, see Figure 4.27.

Figure 4.34 illustrates the sequential planning of the machine components of a
flexible manufacturing equipment in detail, including an analysis of the range of
tools at the beginning of the planning activities. The figure clearly shows the
iterative procedure. The planning subtasks are shown (in blocks), together with
the starting points for the planning (drawings, process plans, quantities and limiting
conditions such as clamping facilities). On the right side of the figure is a list of

Figure 4.32
Typical workpieces for
FMS machining.
(Courtesy of
Heidelberger
Druckmaschinen,
Heidelberg, Germany)

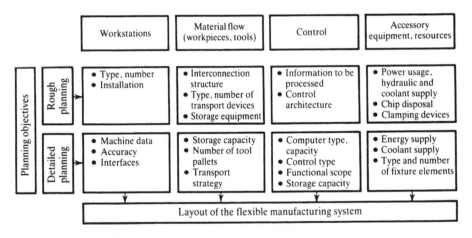

Figure 4.33
Planning of flexible
manufacturing systems.

		Workstations	Material flow (workpieces, tools)	Control	Accessory equipment, resources
Planning objectives	Rough planning	• Type, number • Installation	• Interconnection structure • Type, number of transport devices • Storage equipment	• Information to be processed • Control architecture	• Power usage, hydraulic and coolant supply • Chip disposal • Clamping devices
	Detailed planning	• Machine data • Accuracy • Interfaces	• Storage capacity • Number of tool pallets • Transport strategy	• Computer type, capacity • Control type • Functional scope • Storage capacity	• Energy supply • Coolant supply • Type and number of fixture elements

Layout of the flexible manufacturing system

the various criteria and parameters which influence the planning and its subtasks; in each instance, cost and economic objectives must be taken into account.

Planning should also investigate the time-dependent behavior of the functions of the interconnected manufacturing systems. Applied methods (Figure 4.35) include analytical procedures and numerical simulation (see Chapter 3). Analytical mathematical methods are usually only applicable to simple systems and rough planning. Simulation is based on discrete-time dynamic models. Because of interactive graphics and the possible speed up of the creation of the model, and the generation of control strategies, simulation is increasing in importance (computer developments also contribute to this). Figure 4.36 illustrates a simulation with the associated sequential animation. It shows the simulation of a manufacturing system with two transport devices on a common track (1, 2), 13 machine tools,

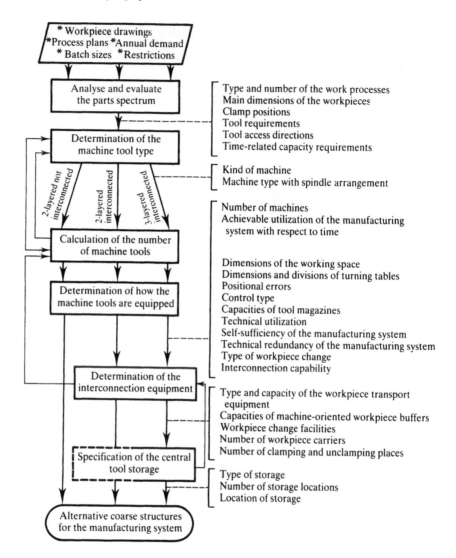

Figure 4.34

Technical planning of
an FMS (source:
Behrindt, 1986).

(02 ... 07 and 09 ... 15) and an input/output point (I/O). For improved visualization, colors are assigned to the workpiece conditions (raw, partially machined, ready). Times are given in the series of boxes above and below the manufacturing system. Collisions may be detected and loads may be assessed from the representation.

Figure 4.33 contained information on the planning of the control of a flexible manufacturing system; corresponding control structures and control functions are discussed in Chapter 5.

The organizational measures associated with the introduction of a flexible manufacturing system include various preparation activities.

- Monitoring of the planning sequence to observe results and dates of every planning phase (Figure 4.34).

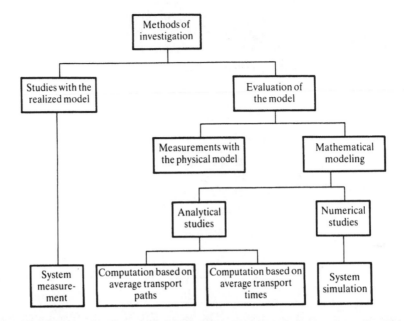

Figure 4.35
Methods of studying the time behavior of FMSs.

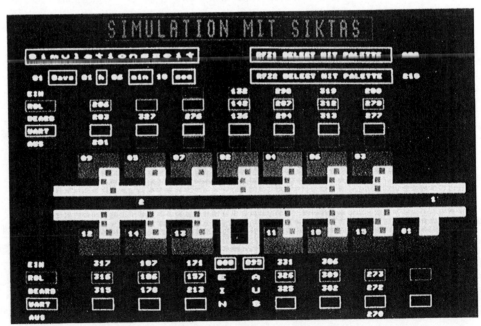

Figure 4.36
Detail of a simulation shown on a CRT.

- Necessary modifications to production control and planning. This means the inclusion of the manufacturing system in the production control whereby are considered batch sizes, maintenance, and quality assurance.

  In addition, NC programs (work plans) must be adapted to the new manufacturing structures; similarly, work stock, clamping equipment and NC programs must be prepared with the corresponding logistics. Also, the tool supply and the use of tools, together with the associated data organization, must be determined.

- The design should take account of automatic workpiece handling and should strive towards standardization of the geometric elements of the workpieces to be machined.
- It is very important to control personnel matters such as participation in works' committees, and questions of wages, shifts and training.

Staff qualifications and the introductory and ongoing staff training are of central importance. This is crucial to an economical and successful use of complex manufacturing systems with high availability. In addition, the motivation of people is of great importance. Training measures are required in the following areas:

- basic NC technology,
- the programming of numerically-controlled machines,
- the organization and operation of a manufacturing system (system aspects),
- monitoring and diagnosis associated with maintenance.

It is also essential to provide a graded training course starting from the level of the individual NC machine and leading to the linked manufacturing system.

4.5 Flexible Assembly Systems

In flexible assembly systems, different assembly tasks must be executed for small batches. Until today, flexible automation has been introduced for many manufacturing tasks. However, this does not apply to assembly. Currently, a large number of developments are concerned with flexible assembly, with the goal of making use of the existing automation potential.

Assembly facilities may be essentially represented by analogy to manufacturing systems (Figure 4.37). It is clear that there is a correspondence between the basic equipment for manufacturing and assembly, see Figure 4.4. Thus, in the case of assembly, there exist assembly stations, cells and systems (Figure 4.37). Corresponding concepts for manufacturing have been discussed in the previous sections and are listed in the bottom line of the figure. In Figure 4.38 the concepts of assembly stations, cells and systems are defined in more detail with reference to the control structures (see Chapter 5), the underlying machine-support components, and the workpiece and tool supply. It is clear (as in the previous discussions for manufacturing systems) that a comprehensive view should be taken when specifying an assembly system. Components of such a system are often assembly cells (as for example, in Figure 4.24).

The control costs, the assembly depth and the material-flow costs are important and often interrelated parameters of assembly systems. Figure 4.39 illustrates the influence of these parameters for various types of assembly equipment. The schema

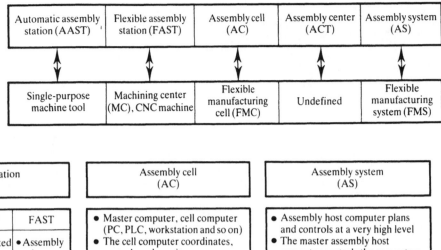

Figure 4.37

Analogy between assembly units and flexible manufacturing units.

Automatic assembly station (AAST)	Flexible assembly station (FAST)	Assembly cell (AC)	Assembly center (ACT)	Assembly system (AS)
Single-purpose machine tool	Machining center (MC), CNC machine	Flexible manufacturing cell (FMC)	Undefined	Flexible manufacturing system (FMS)

Assembly station (AST)			Assembly cell (AC)	Assembly system (AS)

MAST	AAST	FAST
• Assembler	• Automated assembly machine	• Assembly robot
• Arbitrarily flexible	• Not programmable	• Programmable

Assembly cell (AC)
- Master computer, cell computer (PC, PLC, workstation and so on)
- The cell computer coordinates, controls and monitors all subsystems
- Core of the ACT consists of 1 to 5 FAST
- May also contain MAST and AAST integrated into the material and information flow
- Integrated automatic workpiece and tool supply

  - An additional materials storage (tools) is managed by the cell computer

ACT

Assembly system (AS)
- Assembly host computer plans and controls at a very high level
- The master assembly host computer controls the computers of the individual assembly units and coordinates the job processing
- AS consist of arbitrary combinations of AAST, MAST, FAST, ACT and AC

ACT = Assembly center
MAST = Manual assembly station
AAST = Automatic assembly station
FAST = Flexible assembly station

Figure 4.38

Definition of assembly systems.

of this figure was constructed by analogy with that of a manufacturing system. At the same time, the figure illustrates the detailed planning necessary for highly-complicated equipment having an intricate material-flow and control system.

A number of examples are presented below. Figure 4.40 shows an assembly cell which also contains integrated manual workstations (manual assembly stations). The total number of stations in this cell is small and the transport system does not have a buffer facility for pallets. Manual workstations, as used in this cell, are still needed in assembly systems because certain joining tasks cannot be satisfactorily automated; the required sensor technology is frequently unable to meet practical tasks. This cell contains a workpiece loader. In addition, tool change is shown at individual stations.

The flexible assembly system of Figure 4.41 has a few more automatic stations and also a manual workstation; this figure actually shows the right part of Figure 4.21. The individual assembly components are linked together by an interconnected material-flow structure. The transport system is the same as that of the cell in Figure 4.40, however, there are now transport tracks for buffering pallets.

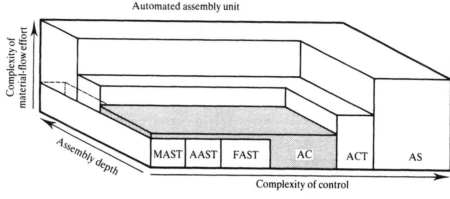

Figure 4.39

Functional relationship between control, assembly depth and material flow for flexible assembly installations.

MAST = Manual assembly station   AC = Assembly cell
AAST = Automatic assembly station   ACT = Assembly center
FAST = Flexible assembly station   AS = Assembly system

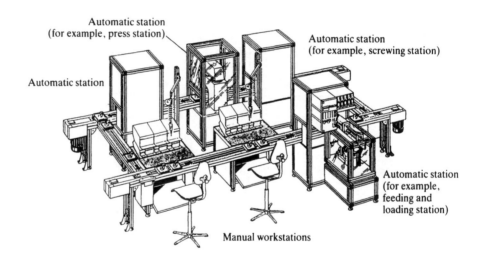

Figure 4.40

A semi-automatic assembly system.

Machining and assembly jobs are coordinated by a master computer. The parts to be assembled may pass through the individual stations in any order. With a corresponding organization, several product variants may be assembled simultaneously. The workpieces for the individual assembly jobs are prepared by the set-up robot. The two-arm robot in the figure is a new development and used for complex assembly tasks: two robots are attached to a crossbeam in a mobile arrangement and work independently or in a complementary way. For example, they may interchange joining tools.

The statements made for FMSs on planning and control largely apply to assembly systems. However, as far as the organization and material flow are concerned there is a fundamentally different way of looking at the problem. Machining is concerned with individual parts, whereas in assembly several parts are assembled into a product. It is no longer a matter of specifying and controlling what happens to the individual workpiece. Instead, there are now several

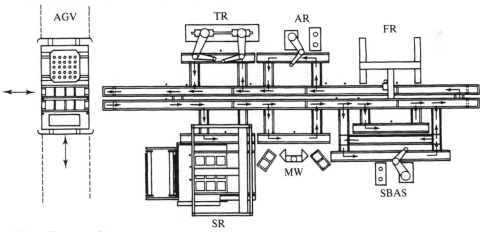

Figure 4.41

Layout of the
assembly station of
Figure 4.21.

TR	Two-arm robot		
AR	Assembly robot	MW	Manual workstation
FR	Fastening robot	SR	Set-up robot
SBAS	Small batch assembly station	AGV	Automated guided vehicle

workpieces processed and it is a matter of controlling their subsequent joining to
assemblies or devices. Conveyor belts with switches (as shown in Figures 4.40 and
4.41) permit flexible transport and buffer structures. Another feature is the
distinctive processing of sensor signals. All these features impose high demands
on the control system.

4.6 Economic Considerations

Every flexible manufacturing system must be economically justified, which is usually very difficult. Both quantifiable and unquantifiable factors must be taken into consideration. Often, it is not possible to predict the future development of the processing tasks or that of the workpieces to be machined (including their quantities); thus, no suitable parameters may be available for economic evaluation. It is also true that manufacturing during breaks and during the third shift is an important factor for inclusion in the return on investment analysis. Other factors include the reduction of set-up and throughput times. Figure 4.42 shows the main parameters which influence calculations of the return on investment. Profitability of the investments is achieved by cost reductions and increases in turnover and returns, which in turn are made possible by automation efforts, market advantages and flexibility of manufacturing. These three objectives are listed in the figure together with other sub-objectives. Parameters for justifying flexible manufacturing systems were previously given in Figure 4.4. Important prerequisites for the achievement of the objectives include thorough planning (as already mentioned) together with an effective logistics, particularly in the area of automated manufacturing. This involves ensuring that workpieces, manufacturing aids, NC data and other information are correctly distributed and are ready on time at the

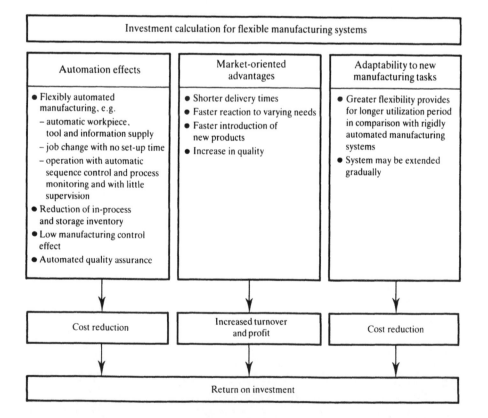

Figure 4.42
Factors influencing
the investment
calculation for flexible
manufacturing systems.

manufacturing equipment at the beginning of each job. This is discussed further in Section 5.5.

Currently, calculations use mainly static procedures such as cost-comparison calculations and economic efficiency analysis. Both these procedures require the

Economic efficiency analysis

Goal criteria	Weighting		FFS 830-4 and FFS 800-4		12 individual machines	
	Share	Factor	Points	Partial efficiency	Points	Partial efficiency
System structure	15%					
– High flexibility with machines with same functions	30%	0.045	10	0.45	7	0.315
– Extensibility	30%	0.045	7	0.315	4	0.315
– Integration of extra processes (e.g. washing, deburring, checking)	15%	0.0225	8	0.18	4	0.09
– Centralized operator facilities	25%	0.0375	8	0.3	3	0.1125
System control	15%					
– Control hierarchy	20%	0.03	8	0.24	4	0.12
– Integration possibilities of CIM structures	15%	0.0225	8	0.18	3	0.0675
– Job- and tool-data management	15%	0.0225	9	0.2025	3	0.0675
– Tool management	25%	0.0375	9	0.3375	3	0.1125
– Prompting	25%	0.0375	8	0.3	3	0.1125
Workpiece and tool flow	15%					
– Automation of workpiece supply	25%	0.0375	10	0.375	8	0.3
– Automation of tool supply	25%	0.0375	10	0.375	4	0.15
– Utilization of workpiece storage space	10%	0.015	10	0.15	4	0.06
– Tool change parallel to cutting time	25%	0.0375	8	0.3	2	0.075
– Specific single tool change	15%	0.0225	10	0.225	3	0.0675
Manufacturing organization	25%					
– Integration into master CAP system	25%	0.0625	8	0.5	5	0.3125
– Integration into master material-flow system	25%	0.0625	8	0.5	6	0.375
– Assembly-oriented execution with simultaneous manufacturing of small batches	25%	0.0625	8	0.5	4	0.25
– Adherence to schedules	25%	0.0625	9	0.5625	6	0.375
Company-specific aspects	30%					
– Short run times	20%	0.06	10	0.6	7	0.42
– Low operating capital	20%	0.06	8	0.48	5	0.3
– Short delivery times	20%	0.06	9	0.54	7	0.42
– Future-oriented technology	15%	0.045	10	0.45	5	0.225
– Long-term flexibility in regard to part spectrum	15%	0.045	10	0.45	10	0.45
– Ergonomic workplace design	10%	0.03	9	0.27	5	0.15
Sum		1		8.78		5.2425
Absolute efficiency				87.8%		52.4%
Relative efficiency				100%		61.8%

Figure 4.43

Example of a detailed economic analysis for flexible manufacturing cells.

consideration of alternative manufacturing systems. Figure 4.43 shows a comparative economic efficiency analysis of two different systems which produce the same product. One of these systems is an FMS, the other comprises 12 individual machines. The figure lists five goal criteria: the system structure, the system control, the workpiece and tool flow, the manufacturing organization and company-strategy (with a 30% share, this latter is deemed to be the most important aspect). For each of these five goal criteria, sub-goals are also weighted relative to a weighting of the goal critiera. The weighting factor is obtained by multiplying the absolute share of a goal criterion by the relative share of a sub-goal. If preference points are allocated to each sub-goal, it is possible to calculate a partial efficiency by multiplying the points by the corresponding weighting factor. The individual partial efficiencies for the two solutions may then be added and a comparison made. The figure lists the subjective, absolute and relative weights and the resulting weighting factors. The allocation of preference points is also subjective; in order to obtain largely objective values, this allocation should be agreed by several persons. The results of the economic efficiency analysis are the absolute and relative efficiencies. This concluding discussion brings together all the objectives of automated flexible manufacturing discussed in this chapter. It is also clear from the figure that economic values or goal criteria are mainly represented quantitatively.

The goal criteria for system control listed in Figure 4.43, namely control hierarchy, CIM integration, tool data management, tool organization and operator prompting, are important keywords for the next chapter.

4.7 Problems

(1) How are instructions for control commands entered into an NC controller? What kind of control information is there?

(2) What is an interpolator for an NC machine? How does the interpolator process control information for the axis controllers?

(3) What kind of measuring systems (direct or indirect) are needed to obtain high positional accuracy?

(4) Why is a rotational speed controller used in position controllers for NC axes?

(5) What effect has play and friction on a position controller?

(6) Why are flexibility and productivity counter-acting?

(7) What do we mean by complete machining of a workpiece? Why do we have to distinguish between rotational and cubic workpieces for machining?

(8) What do we understand by complementing and substituting machine tools?

(9) What is the difference between animation and simulation? What are the aims of a simulation?

(10) What aspect of planning a manufacturing system is of particular importance? How is planning started?

(11) What is it that makes the automation of flexible manufacturing systems difficult?

(12) What are the basic differences between a machining cell and an assembly cell?

5

Control structures for manufacturing systems in the CAM area

CHAPTER CONTENTS

5.1 Introduction   213

5.2 Function-oriented Structure   214

5.3 Hardware Structures   217

5.4 Software-oriented Structures (Program Building Blocks
and Files)   224

5.5 Computer-aided Organization of Manufacturing Resources   231

5.6 Programming NC Equipment   242

5.1 Introduction

Chapter 1 was concerned with the basic concept of hierarchical planning and control and Chapter 2 dealt with CIM structures (see for example, Figure 2.4). The process- or machining-oriented area is called CAM. This area predominantly involves the execution of control tasks; however, data monitoring, entry and evaluation together with control may also be functions in this area. The corresponding control systems may be structured according to various points of view.

A distinction is made between functional, hardware and software (program building blocks and files) control structures. Often these three structures are not the only critiera which determine the control architecture. The representation of controls and control systems in terms of structures helps in their design and promotes an understanding of these facilities. In what follows, structures will be presented according to the three criteria.

5.2 Function-oriented Structure

The control tasks for flexible manufacturing and assembly equipment may be divided into three levels (Figure 5.1). The middle level is called the central control or the supervisory control system. It generally incorporates the host computer and cell-controller levels and until now has usually had an offline link to production planning and control (PP&C and CAP). Online interfacing is now being realized with the gradual implementation of the CIM concept.

The vertical dimension of the data processing structure (Figure 5.1) distinguishes between technical control, organizational control and machine and manufacturing data acquisition and processing. Technical and organizational control involve the processing of predominantly technical or organizational data, respectively. The bidirectional dataflow is shown in the figure by the functional block 'Production data acquisition and processing'. There is also a horizontal dataflow. Increasing use is being made of closed control loops, in which state-data entries (for example, quality data) have a direct effect on the technical or organizational control data within the control system.

The main tasks of the central control system, according to Figure 5.1 are:

- Technical control
  - NC program management and distribution which is called direct numerical control (DNC).
- Organizational control
  - Detailed planning and job execution
  - Generation of control data for workpiece and tool flow.
- Production data entry (PDE)
  - Acquisition and processing of manufacturing, machine and state data.
  - Processing of display data, documentation and operational data feedback are tasks for the control system. Data may be entered from control data processing and via sensors or pickup units.

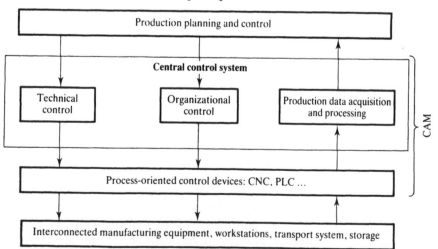

Figure 5.1

Control levels for flexible manufacturing.

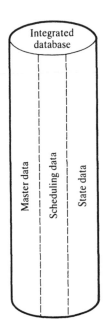

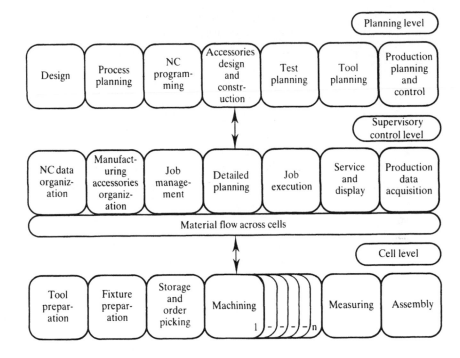

Figure 5.2
Hierarchical structuring of a manufacturing system with information processing levels.

— Management of the global system state in which the local system states are represented.

Figure 5.2 provides for a detailed consideration. This figure shows the planning, supervisory control and cell levels. The planning level contains tasks of areas discussed elsewhere in the book such as CAD, CAP and PP&C. The planning tasks are executed on a longer-term basis and include the generation of manufacturing drawings, and work, test and tool plans. These documents are used in the supervisory-control level to specify data for the process sequence.

Some of the above subfunctions of the process control level are described in more detail.

- Job management is the interface to the high-level planning system. It converts the specification data into a system-internal format, manages the jobs and returns corresponding actual data once the latter has been produced.

- Detailed planning processes the specification data of the level above, taking into account system-specific circumstances such as the size of a tool magazine storage or criteria for optimizing the set up. The main tasks to be executed include:

  - detailed scheduling of manufacturing steps,
  - determination of the manufacturing accessory (MA) requirements,
  - ordering the preparation of MA.

The result of detailed planning is the so-called job schedule, stating which jobs should be machined and on which unit. The job schedule serves as a specification for the job execution.

- Job execution, which is also known as sequence control, permits fully automatic operation of the manufacturing system. Based on the feedback from the underlying level and on the current state of the system, it decides which process should be executed next. It is governed by the primary objective of complying as closely as possible with the specification data from the detailed planning in the form of the job schedule.

- Organization of tools, fixtures and test accessories (MA organization). The tasks of the organizational building blocks for the various manufacturing resources are basically identical. The most important tasks are:

  - management of representative actual manufacturing accessories data

  - management of the description of each type of accessory,

  - setting up of process-oriented control devices according to the specification data in the work and supply plans,

  - support for detailed planning involving the determination of gross and net requirements, control of availability and verification of preparation.

  The tasks of the MA organization are summarized in Section 5.5.

- NC data organization (DNC system). NC data is taken to include not only NC programs in the conventional sense but also all data files which are needed to execute a specific production task. Thus, this definition covers NC handling programs, NC measurement programs, the correction files assigned to NC programs, robot center (RC) programs and operational instructions for manual workstations or tool assembly. NC data is also known as manufacturing accessory data; the corresponding organizational building block is listed separately here, since, unlike other MA building blocks, there is no material flow to control in this case. The tasks are:

  - the management of NC data,

  - the transmission of data files to and from the machine control systems,

  - the management of the CNC working memory,

  - availability tests before jobs are loaded,

  - setting up manufacturing equipment for a process according to the manufacturing accessory plan.

- Material-flow control manages the transport orders generated by the automatic job execution or by the operator within the manufacturing system, forwards these to the transport device control systems and coordinates the transport processes.

- Service and display. The service and display system represents the man–machine interface of a control system and is crucial as far as acceptance by operators is concerned. The user interface permits active intervention in the system and manual or semi-automatic control of the plant. In addition to current system state data, the display system can also show more detailed information (raw data, statistics, and so on) which supports operator-based control of the plant.

Current preparatory activities are arranged by the cell level (Figure 5.2), for example, tools and fixtures are made available before a job starts and control data is transmitted to the machine CNC system on time.

With the integration into the operational data flow, the interfaces between the central control of the supervisory control system and the planning layer are increasing in importance. There must be interfaces to:

- Production planning and control (PP&C)
  - receipt of manufacturing job data.
  - feedback of progress in manufacturing.
  - notification of long-term faults.
- Job preparation (job planning and NC programming) (CAP)
  - receipt of NC and process-plan data,
  - transfer of correction data.

Likewise, the interfaces to the cell level are important for a well-defined planning structure. This includes interfaces to:

- Tool preparation and presetting
  - receipt of tool data from the tool presetting operations,
  - transfer of the current tool-state data after use of the tools.

5.3 Hardware Structures

Figure 5.3 shows the functional and hardware organization of a control structure which was typical in the early 1970s. The figure shows the data distribution (DNC), program control, functional control and servo levels. The first practical implementations of CNC systems contained data distribution and program management levels, the tasks of which correspond to the basic functions of a DNC system. In the late 1960s they were the first control functions to be realized by using a realtime computer. Program control involves the removal of a program line from a program and the issuing of parallel function commands. Functional control associates function commands with output commands to actuators, taking feedback into account. As the manual entries to the levels illustrate, automatic association (interconnection) of these levels is not necessary in principle. Characteristics of this structure in Figure 5.3 include the functional and hardware correspondence of the levels together with the hard-wired implementation of the program and functional control levels (NC and PLC). Every level has input, processing and output sublevels. This structure was typical for the first DNC solutions.

With the development of computer technology based on progress in microelectronics, the DNC structure was introduced from the beginning of the 1980s; it is shown in Figure 5.4. This figure also shows data distribution, program control,

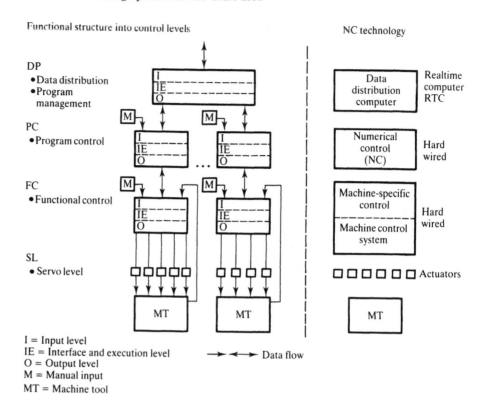

Functional structure into control levels

NC technology

Figure 5.3

Structure of a machine
control system.

I = Input level
IE = Interface and execution level
O = Output level
M = Manual input
MT = Machine tool

→ ← → Data flow

functional control and servo levels. However, the boundary between functional
and hardware levels is no longer defined. For example, the data distribution level
(now also called the supervisory-control level) contains both hierarchical and
parallel computer structures. A hierarchical computer structure is, for example,
exhibited by the previously-mentioned host computer and cell-controller levels
(Figure 5.2). The hardware levels cover more than one functional level. Moreover,
functional extensions are now possible and soft-wired hardware solutions (CNC,
PLC) are the rule. This flexible architecture for assigning functions to devices calls
for a structured view.

The hardware components of the control systems are, according to Figures 5.3
and 5.4, control computers, process- or machine-oriented control devices and
communications systems. Central control tasks or supervisory control functions
(which are similar) are assigned to control computers. Because the development
of computers and storage building blocks has resulted in cost-effective advanced
technology, the functional scope and structure of control systems have changed
greatly over recent years, for example, with the use of personal computers (PCs)
with realtime features. However, we note that, proportionally, software costs have
risen sharply (this is often not fully recognized).

Figure 5.5 illustrates more recent developments of the underlying hardware
structure with host computers and cell controllers. Cell controllers are not only
used for machining but also for other functions such as transportation, measurement

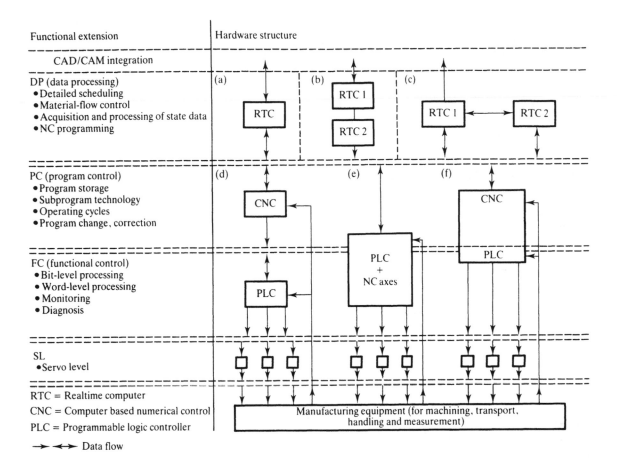

Functional extension | Hardware structure

CAD/CAM integration

DP (data processing)
• Detailed scheduling
• Material-flow control
• Acquisition and processing of state data
• NC programming

PC (program control)
• Program storage
• Subprogram technology
• Operating cycles
• Program change, correction

FC (functional control)
• Bit-level processing
• Word-level processing
• Monitoring
• Diagnosis

SL
• Servo level

(a) RTC (b) RTC 1 / RTC 2 (c) RTC 1 ↔ RTC 2

(d) CNC (e) PLC + NC axes (f) CNC / PLC

PLC

Manufacturing equipment (for machining, transport, handling and measurement)

RTC = Realtime computer
CNC = Computer based numerical control
PLC = Programmable logic controller

→ ↔ Data flow

Figure 5.4
Structure of a control system.

and tool management. Advantages include an incremental construction, a modular structure, a high availability and an optional assignment of control functions to the host computer or cell controller levels. On the right of Figure 5.5, the order in which the functions are listed is only given as an example; it may vary depending on the applications.

Figure 5.5 (like Figure 5.4) shows parallel and hierarchical arrangements of control computers. Host computers and cell controllers form a hierarchy whereby parallel systems undertake task sharing; this is particularly clear in the cell controller level.

We again note that functional and task areas have been described. Thus, nothing has been said about the actual tasks to be performed via a supervisory control system and the sharing of these tasks between the host computer and cell controller levels. Multiple use of the same functional building blocks in the supervisory control and cell levels is technically possible. Even manual workstations may be included (see assembly systems in particular). Two examples are given as an addition to the general structure.

Figure 5.6 shows a flexible manufacturing system for the production of precision machined prismatic workpieces. The components of this system include four

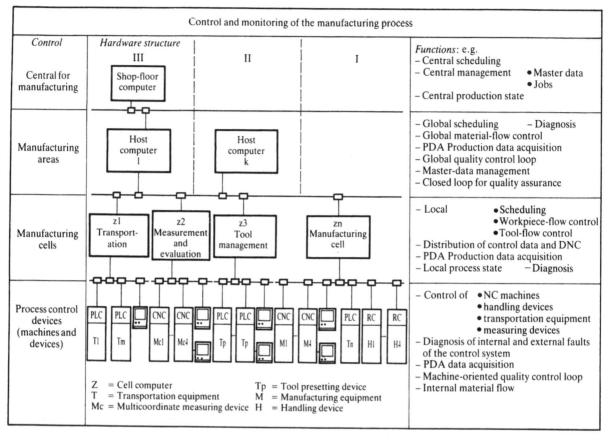

Control	Hardware structure III	II	I	Functions: e.g.

Control and monitoring of the manufacturing process

Control	Hardware structure	Functions: e.g.

Central for manufacturing — III: Shop-floor computer

Functions: e.g.
- Central scheduling
- Central management • Master data
  • Jobs
- Central production state

Manufacturing areas — III: Host computer 1 / II: Host computer k

- Global scheduling – Diagnosis
- Global material-flow control
- PDA Production data acquisition
- Global quality control loop
- Master-data management
- Closed loop for quality assurance

Manufacturing cells — z1 Transportation / z2 Measurement and evaluation / z3 Tool management / zn Manufacturing cell

- Local • Scheduling
  • Workpiece-flow control
  • Tool-flow control
- Distribution of control data and DNC
- PDA Production data acquisition
- Local process state – Diagnosis

Process control devices (machines and devices) — PLC Tl / PLC Tm / CNC Mc1 / CNC Mc4 / PLC Tp / PLC Tp / CNC M1 / CNC M4 / PLC Tn / RC Hl / RC H4

- Control of • NC machines
  • handling devices
  • transportation equipment
  • measuring devices
- Diagnosis of internal and external faults of the control system
- PDA data acquisition
- Machine-oriented quality control loop
- Internal material flow

Z = Cell computer   Tp = Tool presetting device
T = Transportation equipment M = Manufacturing equipment
Mc = Multicoordinate measuring device H = Handling device

Figure 5.5
Hierarchical and automation levels of CIM.

Figure 5.6
Flexible manufacturing system (courtesy of Carl Zeiss, OberKochen, Germany).

1 Operation panel
2 Multi-coordinate measuring device
3 Central storage for workpieces
4 Workpiece transport system
5 Manufacturing centers

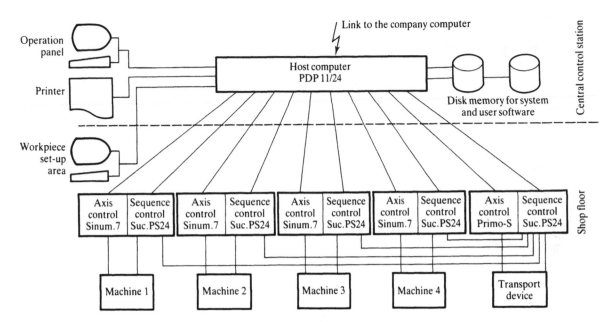

Figure 5.7

Hardware
configuration of the
Zeiss FMS.

machining centers with tool storage and a multi-coordinate measuring device. Its central control is executed by a manufacturing host computer. A control computer is used (Figure 5.7). The main memory contains 512 kbytes and the storage capacity of the disk drives is 2×10 Mbytes. Two display terminals are attached to serve the control systems. One of these terminals is in the workpiece clamping area and a printer is attached for output of system and operational messages.

The underlying control level which consists of the control systems for the machining stations and the transport device is linked via serial interfaces (RS232C standard) with the manufacturing or host computer in a star arrangement. This solution is advantageous as far as expenditure, test facilities and costs are concerned. Standard controls with DNC interfaces are used to control the four machining stations, programmable logic controllers (PLC) are used for the sequence control. Both NC and PLC are linked over serial interfaces with the host computer. The control system for the transport device includes axis control to position the vehicle and sequence control for the pallet handling. Like the machining centers, both are attached to the manufacturing computer. In addition, the transport device control system is responsible for controlling the workpiece clamping position and the pallet changer of the multi-coordinate measuring device.

In order to synchronize transport devices and machining center pallet changers during pallet transfer, the PLC of machines and transport devices are also interlinked. Thus, collisions of the transport device and the machines may be avoided even during manual operation when the manufacturing computer is switched off.

Figure 5.8 shows the hardware control and network structure for the CIM pilot facility of the Institute for Control Technology for Machine Tools and Manufacturing

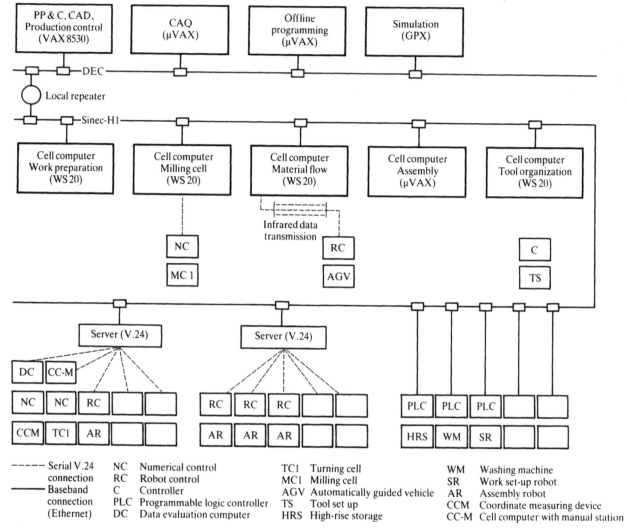

Figure 5.8

Computer and communication architecture of the ISW CIM installation.

Systems in Stuttgart (see Figure 4.21). Important features include the bus structures, the manufacturing control and planning tasks (PP&C and CAP) (see Figure 5.1) and the partially pragmatic approach to coupling peripheral devices with given characteristics. For the latter, RS232C interfaces are also provided. This structure permits experimentation with hardware and software control and communication tasks. (Chapter 6 is concerned with more recent developments in communications technology). The pragmatic networking approach is necessary to be able to handle devices which are not designed for the new communications technologies.

In Figure 5.8 the control structure for the flexible assembly equipment is also shown (by the reference to assembly robots (AR)).

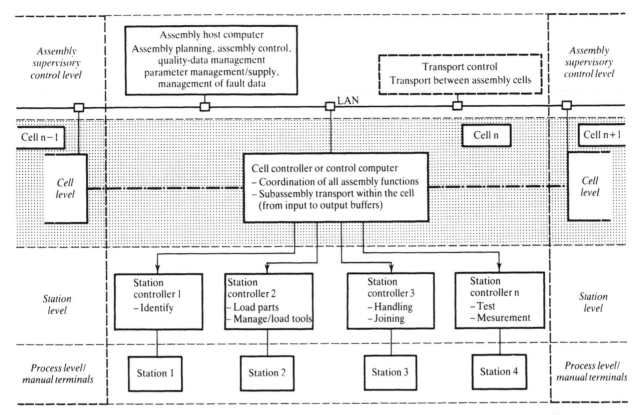

Figure 5.9

Control architecture
of an assembly cell.

For assembly systems (in particular, linked systems for flexible assembly tasks) functional and device-oriented structural considerations analogous to those for manufacturing equipment are used, as shown in Figure 5.9. Several decentralized station controllers, and the cell and supervisory control levels form the hierarchical structure. The figure also shows tasks in these levels. Functional and device-oriented considerations are combined in this figure. A typical sharing of tasks between supervisory control, cell and station control systems can be seen. Assembly planning specifies sequences, quantities, cell and station occupancy and loads, while the assembly control system using this data executes overall coordination. In the cell level PLC or realtime computers are used. The stations should be able to function self-sufficiently. Devices used may include PLC, CNC, RC and/or computer systems.

The dataflow through all the levels of the control hierarchy in an assembly cell is shown in Figure 5.10. This dataflow-oriented figure shows the data entry functions under the headings assembly state, quality data (Q data) and fault data. Operator control involves local parameter changes allowing an adaptation to current behavior. Basically, as in Figure 5.1, the figure shows the levels and the bidirectional dataflows. We also note the use of an identification code carrier (see also Figure 10.16) on the pallet and the feature of data compression in the higher levels.

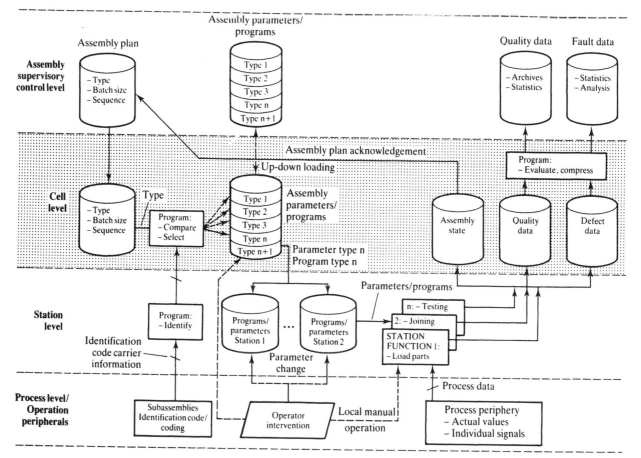

Figure 5.10
Information flow in an
assembly cell.

5.4 Software-oriented Structures (Program Building Blocks and Files)

The software implementation of control systems requires special operating system functions, user programs, files and master, temporary and state data (see Figure 5.2) together with documentation. In total, the complete software architecture determines the reliability and availability of the manufacturing system. In engineering terms, it is to be constructed in a quality oriented and economical way.

Even when flexible manufacturing systems were first developed, attempts were made to construct modular software, in which every module had a precisely defined, self-contained function to execute. The user modules to control and monitor the data and material flows in the plant obtain their instructions from feedback from the manufacturing process (see Figure 5.1). In addition to this planned feedback, various alarm messages from the manufacturing equipment and operator entries must be collected.

These tasks can only be executed by realtime computers. To support the modular construction, the functions are implemented in self-contained functional building blocks which are executed as standalone programs (tasks). This results in computer requirements such as multitasking and the facility for task synchronization. Special-purpose programming languages such as Process and Experiment Automation Realtime Language (PEARL) or Process Fortran have been developed to permit general implementation of these requirements. In addition to the constructs to describe the execution, these languages also contain constructs for data exchange between tasks and with process-oriented control devices. Owing to inadequate support from computer manufacturers (software licenses) these languages are now only rarely used. Commonly-used languages for control systems include Fortran and, more recently, C. In these languages, the above requirements must be met via operating system routines. Most operating systems provide:

- shared memory, global sections (common data areas in the working memory)
- so-called mailboxes (for messages)

for fast data exchange between tasks (Figure 5.11).

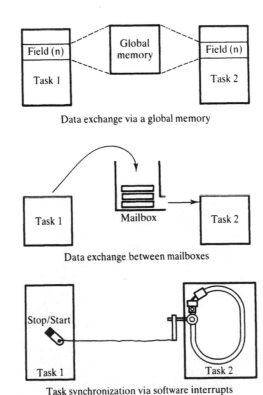

Data exchange via a global memory

Data exchange between mailboxes

Task synchronization via software interrupts

Figure 5.11
Possibilities for data exchange and task synchronization.

The synchronization of tasks and that of tasks and the process-oriented control devices may involve:

- software interrupts, or
- mailboxes with waiting points.

The principle of task communications was discussed in Section 2.3.1.

With the use of operating-system routines the tasks of the control system become in part operating-system specific and thus dependent on the type of computer used.

In the first control systems the functional structure of the control system was mapped into the control programs (Figure 5.12). In this construction, the functional modules have several interfaces. For example, the internal scheduling (detailed planning with job execution) is based on interfaces to the workpiece-flow building block, the DNC, the production data acquisition system (PDA), the master data management and the operator building blocks, and to the files of temporary (scheduling) and state data. The implementation of each of these individual interfaces requires the programming of several operating system calls, so that individual functional building blocks consisting of more than 50% operating system-specific commands are not rare. The testing of the functional modules and the putting into operation of the control system are hampered by this construction. In addition, with the strongly networked program structure, exchange of functional modules or adaptation to another plant is also impossible and extensions are difficult. However, the design of flexible production structures with computer-aided facility control has had a beneficial effect on both the developer and the user whereby increased attempts have been made to produce adaptable control software.

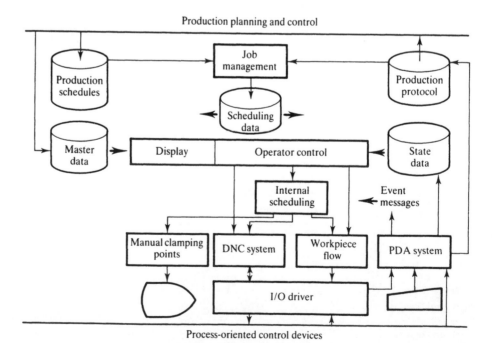

Figure 5.12

Components of a functional and software-oriented control structure.

Open control systems should be developed with the following characteristics (see also Chapter 2):

- Plant independence. The control system should be usable in different system configurations in different hierarchical levels (for example supervisory control or cell level).
- Computer independence
  - operating-system independence. The control system should run under different operating systems.
  - hardware-structure independence. Not all functions of a control system must run on a single computer. Individual functions may, for example, be relocated to other computers on performance reasons.
- Procedure independence. In addition to control functions in the manufacturing area those associated with the assembly or with the preparation of manufacturing accessories should be covered.

In order to achieve plant independence, the functions of a control system must be specified; one should strive for task sharing by plant-independent and plant-specific subfunctions. Figure 5.13 shows such a subdivision for the example of DNC, where several interface levels with different degrees of plant dependence are specified.

Figure 5.13
Obtaining modular independence by standard interfaces.

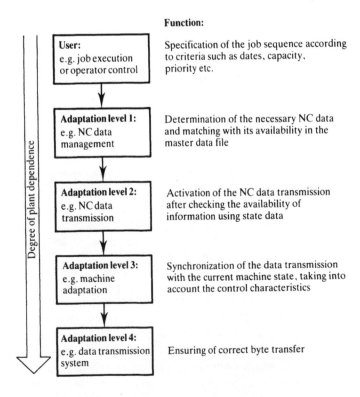

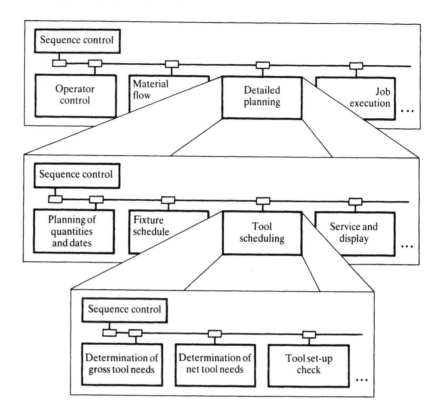

Figure 5.14
Hierarchical
description of the
control function
'detailed planning'.

The uniform structuring of all functional units of the supervisory control level in Figure 5.14 (see Figure 5.2) supports this subdivision and facilitates the exchange of individual functional modules together with a subsequent extension.

However, the problem of adaptability is only partly solved by this structuring since every individual functional building block, in particular in the lower levels of Figure 5.13, must process plant-specific information in order to execute its function. Thus, as far as possible, plant-specific parameters, strategies or sequences should *not* be stored implicitly in the program code but explicitly in editable (easy to alter) files. In Figure 5.15 these files are called a configuration database. The functional building blocks should be programmed to be generally valid and should only carry out their plant-specific tasks after interpreting this database at run time. The configuration database is generated as required using graphically supported dialogue masks or formal graphically supported description languages such as Petri nets or state graphs.

Computer indepence is largely supported by the use of a software communications building block (Figure 5.16). In this way all operating system calls are concentrated in this building block and in linkable interactive task modules. With the consistent subdivision of the individual functions into self-contained functional building blocks, each building block requires only one interface (subprogram with operating system specific instructions) to forward and receive command messages and for realtime connections (interrupts, program synchronization) of tasks among them-

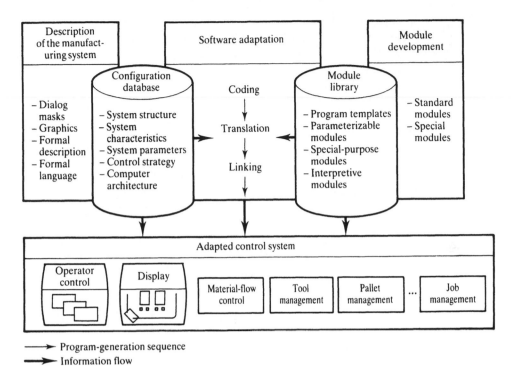

Figure 5.15
Adaptable control system for flexible manufacturing and assembly.

selves and with the process. In addition to the possibility for gradual design and consolidation, this procedure also offers the following advantages:

- Functional building blocks of the control program may be arranged on a number of computers. In Figure 5.2 the various functions are assigned to the supervisory control and cell levels. According to this scheme the assignment to computers is almost arbitrary.

- Easier exchange of functional building blocks via standardized messages between functional building blocks and process components.

- Good facilities for testing the functional modules.

- The control system is more modular and the adaptation to particular computers is simplified.

- Openness in respect of standardization activities such as manufacturing message specification (MMS) or automation protocol (AP) (see Chapter 6).

The data required for the control system may be divided into three main groups (Figures 5.2 and 5.16), namely:

- master data
- scheduling data (temporary data)
- state data

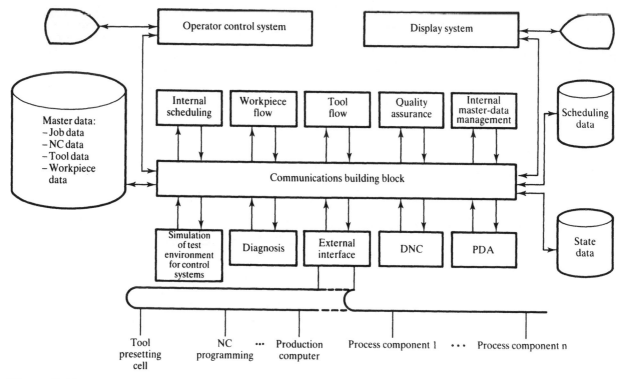

Figure 5.16

The software structure of the ISW control system.

Master data does not vary with time (see Section 2.3.8). This includes, for example, the description of a manufacturing accessory and also planning data which is only installed once such as plans generated during process planning. The master data mainly consists of more or less permanent data sets which are less frequently accessed. Thus, it is sufficient to store this data in a peripheral store, like disk or tape units. Simultaneous access to this data is available to a number of programs (tasks). When changing master data, and in particular plaintext lists, using the editor, the corresponding basic data area is protected to avoid data inconsistencies which usually occur during changes. Master data is stored until it is explicitly deleted. A relational database should be used to manage the master data.

Temporary (scheduling) order related data is data which relates to a particular planning period such as a shift, a day or a manufacturing period (Section 2.3.8). Typical examples include the job schedule or the tool-supply list which states which tools should be prepared for a given planning period. Temporary data is invalid after the planning period to which it refers. After all statistical evaluations for the given period have been completed, this data is implicitly deleted.

State data includes all time-varying data, for example entries describing the current location and state of manufacturing accessories within the system. One particular feature here is the definition of the system boundary, since the data is deleted as soon as a manufacturing accessory leaves the system. The state data (system image) includes all data which changes often and rapidly. In particular, the material flow control and the internal scheduling or detailed planning systems often read the state data while the factory data acquisition system updates (alters) the data. Thus, this data cannot be stored in a peripheral store. In this case, the read access would be too slow and the constant locking to prevent inconsistencies would be intolerable. To meet these requirements the system state (image) is stored directly in the working memory of the computer. A common storage area is used to permit simultaneous access to the data by several tasks (for DEC computers with the VMS operating system this is called the global section; for computers with RSX it is known as the shared common). For security reasons, since the data in the global section is lost after breakdowns, the complete system state is cyclically stored in the peripheral store (updated).

This section has described the analogy between the software-oriented control structure of manufacturing systems and assembly systems. Undoubtedly, in the future, the software structure and the software engineering of control systems will be of paramount importance. Criteria relating to hierarchical and modular structuring are important, as are questions of software maintenance, upkeep and reusability. These determine the availability and the reliability, and thus the economy, of complex cost-intensive manufacturing and assembly systems.

5.5 Computer-aided Organization of Manufacturing Accessories

In the previous sections, many references have been made to data. Data is a basic building block of CIM structures. One important feature is the product model (Chapter 7). However, data about manufacturing resources must also be managed and prepared for the areas of CAD, PP&C, CAQ and CAM. In the following sections we shall discuss the organization of manufacturing accessories data.

When high organizational availability of computer-controlled manufacturing and assembly is desired, it is essential that uniquely identifiable data and materials (workpieces, tools, test devices, clamping devices) be prepared on time. The organization of computer-supported manufacturing accessories serves this purpose, as can be seen from the previous sections.

Manufacturing accessories include tools, test devices, fixtures and clamping devices (Figure 5.17). The organization of manufacturing accessories within a company may be divided into five functional areas (Figure 5.18), namely planning, management and control, scheduling, supply and use. The term 'supply' arose in connection with the cell level (Figure 5.2). Figure 5.18 lists individual subtasks of the five functional areas. This classification is generally applicable. It is also related to time: while planning tasks are executed over long periods, as far as time is concerned, use is closely linked with the manufacturing process.

The following features of the organization of manufacturing accessories follow

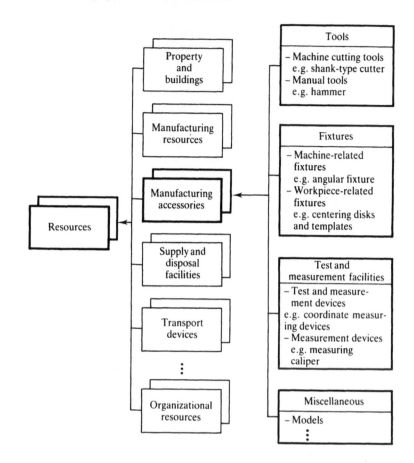

Figure 5.17
Structuring of
manufacturing
accessories as part of
the resources.

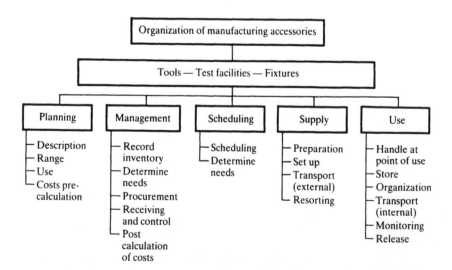

Figure 5.18
Manufacturing
accessories.

from this classification:

- The execution of various subfunctions requires various links between individual data items.
- The data varies in its temporal relevance and includes master data (long term), temporary data (medium term) and current state data (see Section 5.4); however, attention is often only paid to master data.
- Finally, it is clear that the construction of a computer-assisted organization requires a graduated scheme which takes account of operation-specific starting positions and objectives and initially only considers sub-areas.

In what follows, prominence is given to manufacturing tools. Substantial rationalization effects may be obtained through a computer-assisted tool organization. Here, organization is taken to include data entry, upkeep and management together with task-oriented access to this data.

The functional classification of a tool organization must be integrated into an operational structure. Figure 5.19 illustrates this for the CIM concept using the

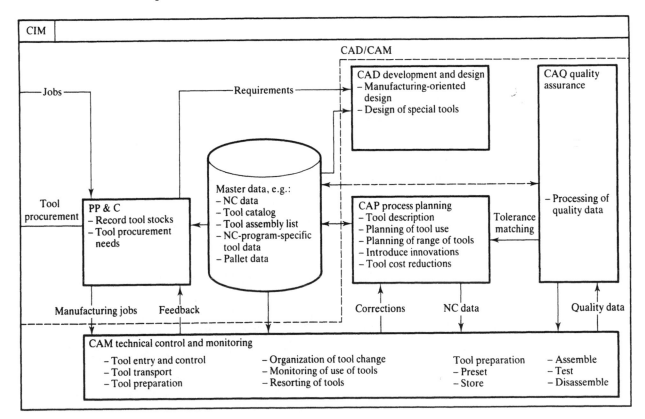

Figure 5.19

The task of tool management within a CIM concept.

tool management as an example. This figure shows the familiar CIM structure with the CAD, CAP, PP&C, CAQ and CAM components and their links. This structure only contains data and task entries referring to tools. The figure shows the assignment of subfunctions of the given classification to operational areas such as manufacturing control (PP&C), job planning and NC programming together with control technology and machine-oriented control (CAM). CIM is provided for in an operationally-specific manner and thus represents a comprehensive organization of manufacturing accessories. This does not depend on whether the tasks of the operational areas involve computer-assisted machining in so-called island solutions or conventional machining. Interfaces are very important in this scheme, as is the question of data maintenance. This is illustrated centrally in Figure 5.19 for tool master data. Realizations often involve distributed and consistent data maintenance; a logically-central data maintenance must be guaranteed.

In addition to the specification of a sharing of the functions across the operational areas and a consideration of the resulting tool flow, data maintenance is also important for efficient tool organization. This includes the specification of the nature of the data to be administered and determines how the data should be managed.

A major factor for efficient tool organization is the ability to retrieve stored data quickly. Usually simple identification keys are not sufficient, since, because of limiting conditions from the machining task, individual operational areas such as job planning require classification or search criteria to find tools. Tool data is not classified according to temporal characteristics and conditions such as a period of validity. Tools are distinguished by the following types of characteristics:

- geometric characteristics, in conjunction with graphical characteristics,
- technological characteristics, including the tool type,
- organizational characteristics, for example parts lists,
- statistical characteristics.

Geometric and technological characteristics may determine the classification of a tool. The temporal classification comprises the following categories:

- constant, master data,
- constant over a period, temporary data,
- variable, or state data.

Most of the tool data sets consists of master data. A classification into complete tools (CT) and individual tool components (ITC) appears advantageous (Figure 5.20). Figures 5.21 and 5.22 show typical master data for lathes. Organizational, geometric and technological data is also shown. NC programming (see Section 5.6) with simulation and collision analysis together with tool preparation and set up may be supported by task-oriented data preparation.

From the user's point of view, the tool organization must be based on a database with solid, comprehensive, cross-domain data (Figure 5.23). This figure again shows the five functional areas depicted in Figure 5.18. Each of these functional areas requires a different combination of data items which are stored in the central

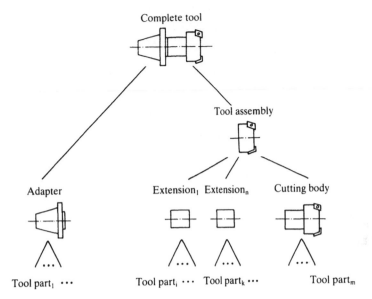

Figure 5.20
Tool parts and
sub-assemblies.

Identifying data		
Field name	Format	Description
TTN	N, 8	Tool type number (primary key for master data) contains no classification part The value range will be parameterized
TTI	N, 3	Identification of tool type
DPTN	A, 20	Number of the tool type
CUTEDG	A, 20	Number of the cutting base
Organizational data		
TTYID	A, 20	Tool identification in plaintext
MAGTY	N, 2	Identification of different magazine types for large/ small tools, boring heads ...
WEIGH	I	Identification of weight for handling devices (in Newtons (N))
Technological data		
CMO	A, 6	Cutting material code
*CHIP	R	Greatest permissible chip thickness
*MINC	R	Min. relative cutting depth (relative to the entry in the material file of a material/cut combination)
*THRAT	R	Thickness ratio of the chip (to adapt the feed for this tool to the feed in the material file)
*MinSpanR	R	Minimum chip thickness
⋮		

Figure 5.21
Tool master data,
Part 1.

Geometric data:

TAD	R	Tool adjusting dimension L between tool cutting point and adjustment reference point
PTAD	R	Positive tolerance
NTAD	R	Negative tolerance
QADJ	R	Tool adjusting dimension Q
QPTOL	R	Positive tolerance

⋮

From this data, task-oriented combinations are required. Two examples illustrate this.

TOOL supply plan

TTN	N,8	Tool type number
JNO	A,20	Job number
MCNO	I	Machine/cell number
DOU	D	Date of use

NC-program-specific data

NCPNO	A, 12	NC program number
TNO	N, 8	Tool type number
AJOB	A, 32	Assembly job number
OPLIF	I	Required operating life, standardized in min/10
LIFF	R	Operating life factor
BRM	R	Breaking-risk multiplier for special uses (to be multiplied by the breaking-risk factor given in the technological master data)

⋮

Figure 5.22
Tool master data,
Part 2.

database. The figure shows a number of subareas of the five functional areas together with relevant tool-related terms. In reality such a logical view of the database is an ideal which is almost beyond the technical capabilities of data processing today.

Database systems provide the necessary prerequisites for a uniform tool database (Section 2.3.8). As Figure 5.24 shows, these consist of the database and database management components. The database represents the total of all the data stored in a database system. The database management system is responsible for executing all the operations needed to manage the database. As the number of different users accessing a database increases, the usefulness of the database improves. This is also of advantage to the integration of the CIM system.

The data maintenance of a tool organization system is designed to represent all the tool data in a logically central database to be managed by a database management system. We shall consider the main aspects of a database system together with possible ways of adapting the database maintenance to the requirements of the real operational environment.

Before a large database is planned and designed a thorough analysis and structuring of the data is required. The most important steps are:

- analysis of the existing tool data,
- analysis and description of the relevant objects,

- analysis and description of identifying and defining object attributes,
- analysis and description of the relevant relationships between the objects.

An object is a real entity which may be described by attributes and which has a specific significance. The following example should help readers to understand this. It concerns a sample of tool master data and is illustrated in Figure 5.25.

Complete tools (see Figure 5.20) are uniquely characterized by an identification number. They have a designation and belong to a specific tool type. The complete tools are constructed from tool components which are also uniquely identified by a number. Complete tools and tool components are stored in various places and supplied by various suppliers. Storage locations and suppliers are identified by a name. A complete tool or a tool component is stored in a single position but may be supplied by several suppliers. As far as ordering is concerned, the order number and the price are important.

It must be possible to map the description of tools onto the underlying data model

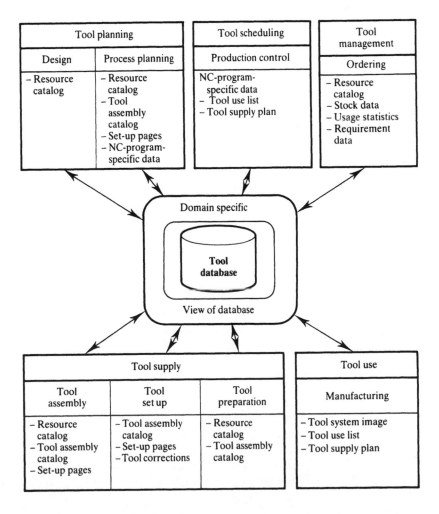

Figure 5.23
A tool database.

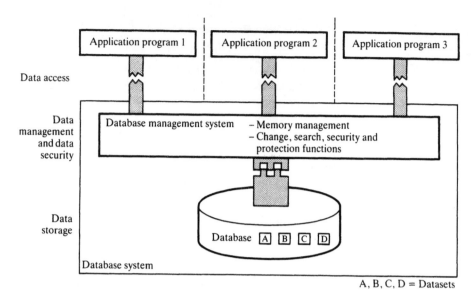

Figure 5.24
An integrated
database.

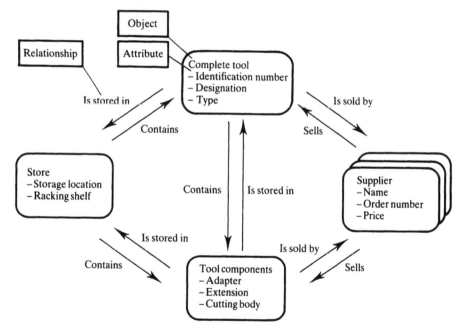

Figure 5.25
A tool data model.

of a database system. In recent years, the relational data model has gained in importance. Thus, it has become possible to configure database systems with sufficient flexibility in an application-independent manner. A database system conceived as a relational model consists of a collection of mathematical relations (see Figure 2.42). As Figure 5.26 shows, relations may be represented graphically as tables. Every database has a name and contains a finite number of attributes. The primary key of a relation is the attribute or set of attributes which characterizes

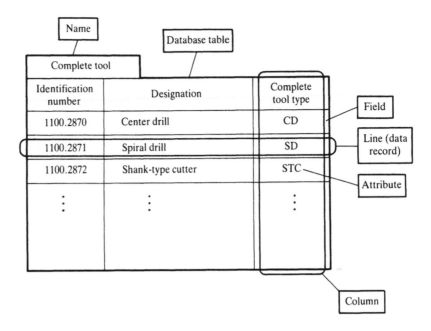

Figure 5.26

Database tables for a
complete tool.

a data record (line) uniquely. Manipulative operations on data may be described using the relational algebra. The data model for the example is that shown in Figure 5.27. The parts list is important for tool preparation and tool presetting.

Commonly managed data must be comprehensively protected and secured. Data protection includes all measures which prevent unauthorized access. Data security on the other hand refers to the avoidance of errors associated with authorized access.

Section 5.2 contained a detailed discussion of the assignment of tool organization functions to the control technology (and in particular the tool scheduling). Figure 5.28 shows the functions which are assigned to the tool-scheduling building blocks in the supervisory control and cell levels. It shows the software-oriented structuring, as in Figure 5.14. The manufacturing accessories-related functions are individually named. It is important to determine the gross and net requirements of a manufacturing task. It is clear that it is advantageous to base the control technology, and thus the control system, on data from a tool-organization system.

An economic evaluation of a computer-aided organization of manufacturing accessories must involve non-quantifiable values. It is not easy to obtain quantifiable values from a comparison with conventional forms of organization; and future developments and their effects are difficult to assess.

A few brief words of advice regarding economic considerations are appropriate. First, we discuss the designer's responsibility for costs and the costs generated by his design. A computer-assisted tool management system may reduce production-specific construction and thus also the generation of manufacturing costs. Thus, Figure 5.19 rightly shows the association of CAD with a tool database. Figure 5.29 illustrates the effects arising from faults in the preparation of manufacturing accessories and from missing job documents. Tools, fixtures, NC programs and set-up data almost exclusively determine the number and duration of faults which

Complete tool		
Identification number	Designation	Complete tool type
1100.2870	Center drill	CD
1100.2871	Spiral drill	SD
1100.2872	Shank-type cutter	STC

Storage		
Identification number	Storage location	Shelf rack number
1100.2870	Storage A	12
1100.2871	Storage B	17
1100.2872	Storage C	8

Supplier			
Identification number	Name	Order number	Price
1100.2870	L1	A12.736	246.36
1100.2870	L2	1348	241.36
1100.2871	L1	A13.924	299.99
1100.2872	L1	A12.878	235.35
1100.2872	L2	1435	230.35
1100.2871	L3	K132.465	220.00
1100.2870	L3	K234.465	231.81

Parts list			
Identification number	Adapter	Extension	Cutting body
1100.2870	SK.1100	SH.1020	CD.1005
1100.2871	SK.1200	SH.2050	SD.5010
1100.2872	SK.1100	MK.1000	STC.2020

Figure 5.27

Database model for the example.

may affect the manufacturing sequence. This is based on a monitoring period of 24 work days of two shifts each.

The data in Figure 5.29 points to the need to provide for supporting dataflow between manufacturing control and preparation (the supply and use of manufacturing accessories). A prerequisite for this is a knowledge of the time expended on preparation. It is known that a company can justify investment on an organizational system by slight increases in the availability of NC manufacturing equipment on job change. Possible stock reductions as a result of a tight organization also leads to economic advantages.

It follows from these few examples that there is an overall potential for cost savings which remains to be exploited. Important objectives here include the synchronization of the manufacturing accessory flow (Figure 5.30) with the workpiece flow at various manufacturing equipments, together with the preparation of accurate data for preparatory planning and controlling areas. Three of the five functional areas of Figure 5.18 are listed vertically in Figure 5.30. The corresponding workflow oriented subtasks associated with these three functional areas are shown. The figure shows what happens to the tool in each case. In the figure, it is assumed that the manufacturing system in question consists of several machines together with both a central tool store (CTS) and machine-related tool stores (MTS).

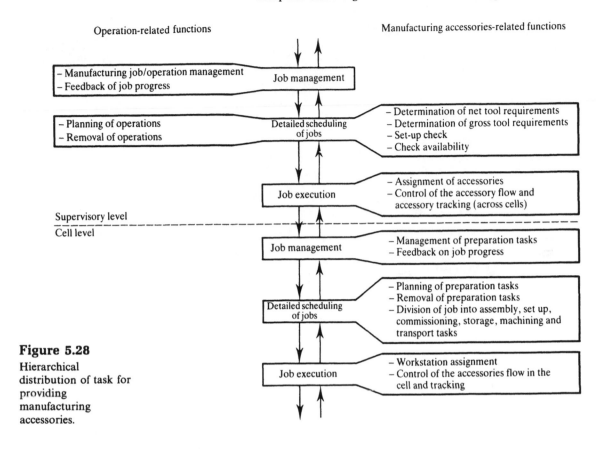

Operation-related functions   Manufacturing accessories-related functions

- Manufacturing job/operation management
- Feedback of job progress

Job management

- Planning of operations
- Removal of operations

Detailed scheduling of jobs

- Determination of net tool requirements
- Determination of gross tool requirements
- Set-up check
- Check availability

Job execution

- Assignment of accessories
- Control of the accessory flow and accessory tracking (across cells)

Supervisory level
Cell level

Job management

- Management of preparation tasks
- Feedback on job progress

Detailed scheduling of jobs

- Planning of preparation tasks
- Removal of preparation tasks
- Division of job into assembly, set up, commissioning, storage, machining and transport tasks

Job execution

- Workstation assignment
- Control of the accessories flow in the cell and tracking

Figure 5.28
Hierarchical distribution of task for providing manufacturing accessories.

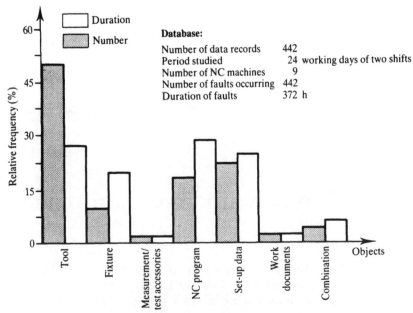

Relative frequency (%)

Duration
Number

Database:
Number of data records   442
Period studied   24 working days of two shifts
Number of NC machines   9
Number of faults occurring   442
Duration of faults   372 h

Objects: Tool, Fixture, Measurement/test accessories, NC program, Set-up data, Work documents, Combination

Figure 5.29
Overview of number and duration of faults with the supply of manufacturing accessories.

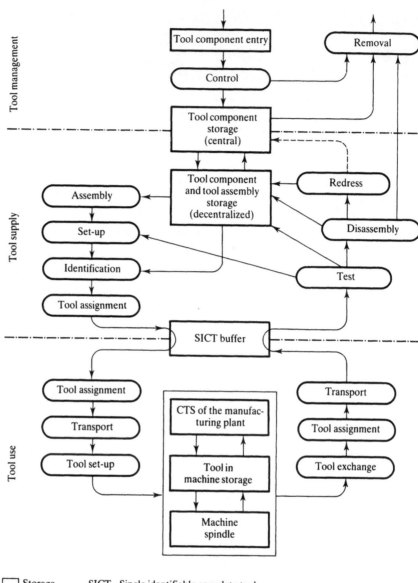

Figure 5.30
Tool flow.

☐ Storage   SICT   Single identifiable complete tool
◯ Function   CTS   Central tool storage

5.6 Programming NC Equipment

5.6.1 NC machines

NC programming is an important planning task (see Figure 5.2) which in computer-aided machining falls in the CAP area. It supplies data for the technical control system, in other words for process-oriented control devices (Figure 5.1).

As in Section 5.5, the control data is considered to be a manufacturing accessory where the control system has the task of preparing the data on time for the individual machine using DNC. We will now be looking at the programming aspects in more detail.

By the programming of a numerically-controlled manufacturing equipment we mean the specification of the data required for the machining (handling, measurement, jointing) of a workpiece in a task-specific (for example, machining-specific) sequence and its transfer to a data carrier which can be read automatically by the control system. In this respect, according to the usual definition, a program is a sequence of instructions, for example, for machining a workpiece. A punch tape, the back-up storage of a control computer in DNC systems and the program memory of a numerical control device may be data carriers. DNC means that the distribution of control data via a realtime computer to several numerically controlled manufacturing equipments (for example, machine tools) is as described above.

As a result of developments in computer technology and in numerical control systems with associated peripheral devices, the number of programming facilities and procedures has steadily increased. For programming, the following criteria should be considered:

- programming with/without computer-support,
- organizational schemes,
- hardware assignments,
- area of application to manufacturing or production (for example, turning, boring and milling, handling, measuring or programming of several integrated manufacturing tasks).

The first criterion refers to manual and computer-aided programming (also known as automatic programming). Figure 5.31 contrasts both methods.

In manual, machine-oriented programming, the programmer generates all the instructions in a form which is directly readable by the numerical control device. Usually, the sequence of tasks steps is specified in records, the structure of which corresponds to ISO 6983 or DIN 66025. Figure 5.32 shows a section of a program manuscript (also known as the coding sheet) for a turning task. Starting points for manual programing include not only the workpiece drawings but also card files which contain data about the machine and the control system and about the workpiece and the tool. The address characters in Figure 5.32 have a standardized meaning, as does the associated coding (see Figure 5.33). Figure 5.34 (right) shows a section of a program in ISO or DIN notation.

Automatic or computer-aided programming (Figure 5.31, right) is done largely independent of the manufacturing equipment, and directly solves the task, which is defined in a part program (Figure 5.34, left). In programming systems it is possible to process part programs written in a problem-oriented language and have access to computer files containing detailed application oriented information, in order to generate the control data needed for machining (Figure 5.35). Figure 5.36 contains possible geometric definitions in program-oriented languages, while

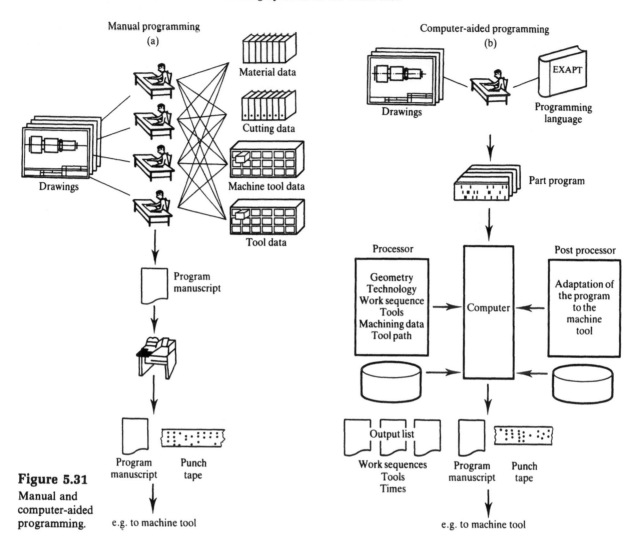

Figure 5.31
Manual and
computer-aided
programming.

Figure 5.37 shows a complete part program for a simple workpiece. The part program is structured; it contains header data, geometric and technologic definitions, execution instructions and the end symbol. The execution of a program is supported by processors and post processors. The processor processes the part program (this involves access to files) and generates the machine-independent CLDATA interface (cutter location data). The post processor which accesses the machine data adapts the CLDATA to the particular controller of the manufacturing equipment. Thus, the components of a programming system are:

- part programs, written in an application language
- processing programs, to translate the part program
- files, to store one or several part programs

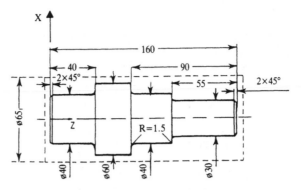

Coding sheet (short form)

Record number	Tool path functions	Tool path commands		Distance from center of circle		Feed command	Rotational speed command	Tool command	Miscell-aneous functions
N	G	X	Z	I	K	F	S	T	M
⋮									⋮
(Finishing)									
N410	G00	X13	Z161				S200		M03
N420	G01		Z160			F12			
N430		X15	Z158			F18			
N440			Z106.5						
N450	G02	X16.5	Z105	I1.5					
⋮									⋮

G00	Path functions	(= Fast feed)
G01	Path functions	(= Straight line path)
G02	Path functions	(– Circular interpolation)
X30	Path command	(= Travel in X direction)
F12	Feed command	(= Feed rate 0.1 mm/hr)
S200	Rotational speed command	(= Rotational speed n = 2000 revs/min)
M03	Misc. function	(= Clockwise motion of spindle)

Figure 5.32
A typical coding sheet.

One important objective of programming systems is the cost-effective generation of error-free NC control data, thereby providing to the user an affordable way of writing programs for machining.

As far as the organizational scheme is concerned, we note that, until a few years ago, both manual and computer-aided programming was carried out centrally as part of the job planning and manufacturing planning. The introduction and development of computerized numerical control (CNC) systems with program memories and the increasing capability of small robust computers (for example,

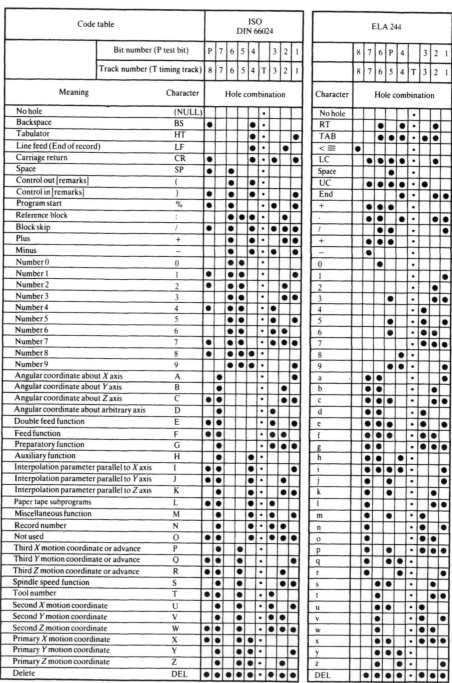

Figure 5.33
Standard code for NC tape.

/WR = space character

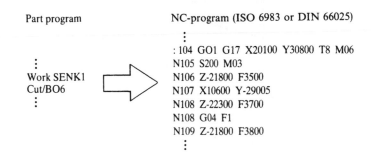

Machining task

Countersinking (symbol: SENK1) of a borehole (symbol: BO6) with automatic selection of tools and technological control commands

Part program	NC-program (ISO 6983 or DIN 66025)
⋮ Work SENK1 Cut/BO6 ⋮	⋮ : 104 GO1 G17 X20100 Y30800 T8 M06 N105 S200 M03 N106 Z-21800 F3500 N107 X10600 Y-29005 N108 Z-22300 F3700 N108 G04 F1 N109 Z-21800 F3800 ⋮

Figure 5.34
Example of a part program and an NC program for a machining task.

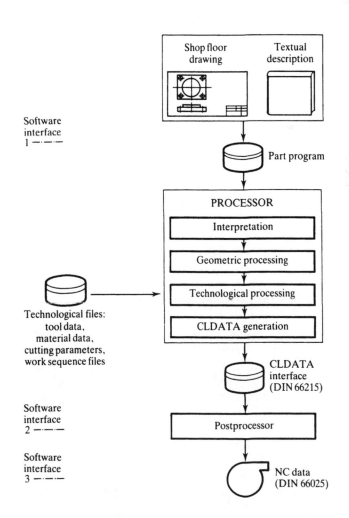

Shop floor drawing
Textual description
Software interface 1
Part program
PROCESSOR
Interpretation
Geometric processing
Technological processing
CLDATA generation
Technological files: tool data, material data, cutting parameters, work sequence files
CLDATA interface (DIN 66215)
Software interface 2
Postprocessor
Software interface 3
NC data (DIN 66025)

Figure 5.35
Processing of a part program.

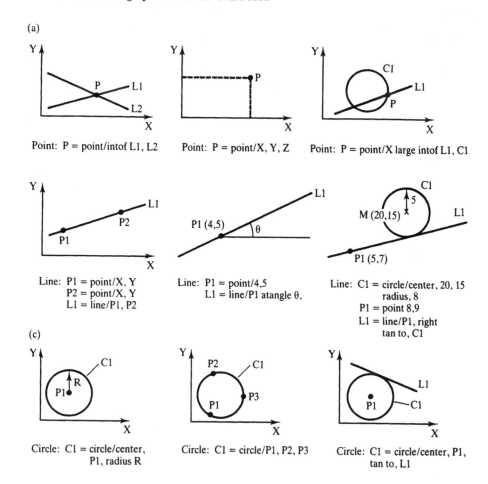

(a)

Point: P = point/intof L1, L2   Point: P = point/X, Y, Z   Point: P = point/X large intof L1, C1

Line: P1 = point/X, Y
P2 = point/X, Y
L1 = line/P1, P2

Line: P1 = point/4,5
L1 = line/P1 atangle θ,

Line: C1 = circle/center, 20, 15
radius, 8
P1 = point 8,9
L1 = line/P1, right
tan to, C1

(c)

Figure 5.36
·Possible descriptions
of points, lines and
circles with APT.

Circle: C1 = circle/center,
P1, radius R

Circle: C1 = circle/P1, P2, P3

Circle: C1 = circle/center, P1,
tan to, L1

16- or 32-bit office computers) have brought the question of decentralized programming more to the fore. This decentralized programming may take the form of shop-floor programming (which may or may not be linked to a machine). A fine distinction is not always possible; it depends on which hardware the programming capability is provided for (Figure 5.38), for example, a control computer or separately controlled PCs. The once common acronym WOP (workstation-oriented programming) refers to machine-linked, interactive, graphically-supported programming. An important objective of workstation-oriented programming is to increase the work volume of NC machines.

In addition to the execution of the basic functions, the powerful and cost-effective computers and control systems of CNC permit the inclusion of new functions for NC-systems. Thus, program changes and corrections in the control system are easily implemented based on the NC program and correction memory, and one-time entering of the control data, without the need for central or machine-oriented programming. The re-use of an optimized and corrected program is supported by read out to an external data memory (whether a punch tape or, in DNC operation, a central disk memory).

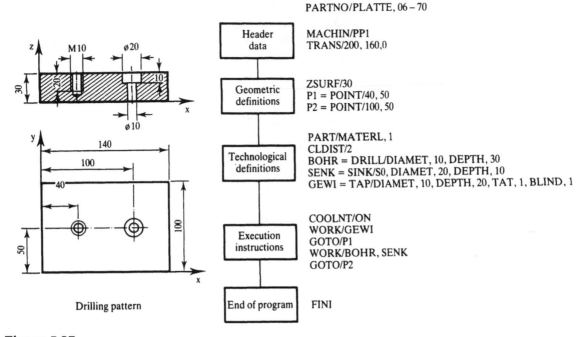

Figure 5.37
Example of an
EXAPT program.

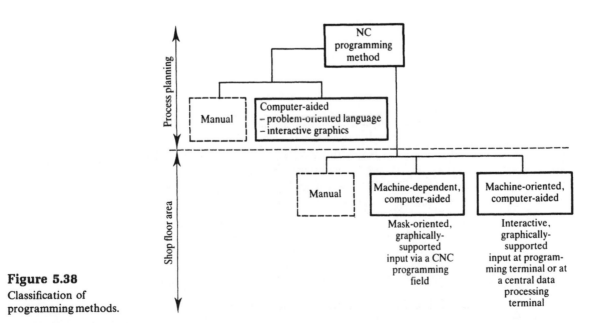

Figure 5.38
Classification of
programming methods.

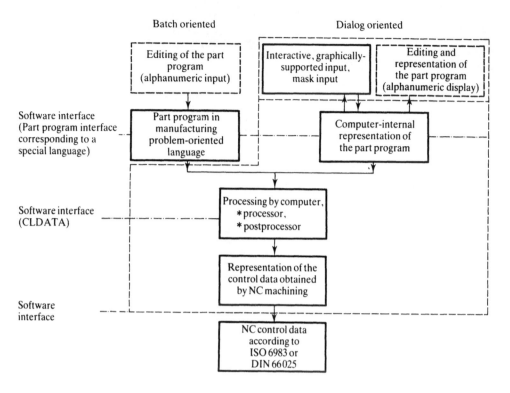

Figure 5.39
Possible
representations of the
part program and
interfaces.

In computer-aided programming, as shown in Figure 5.39, the part program represents the input interface to processing in the computer. While the direct generation of a part program is illustrated in alphanumeric form on the left of the figure, on the right the interactive graphics support is shown as an indirect path. This is made possible by the development of convenient peripheral devices and software tools such as window techniques. In the following, we shall consider the part program in more depth.

Figure 5.37 shows a simple part program. Figure 5.40 shows the interactive graphically-supported input using mask techniques. The mask-supported entry describes a complex NC grinding task. A conical-helix full-radius milling cutter is manufactured by several grinding operations in a single setting.

Figure 5.35 shows the characteristic structuring of NC programming which is typical of many programming systems (and in particular of APT-like languages including EXAPT (see Figure 5.37)). Figures 5.35 and 5.39 also emphasize the interfaces. These are very important for the structuring, as are the module interfaces within the processor and postprocessor. Programming systems are subject to constant ongoing development, which is supported by interface definitions and by upkeep and maintenance of the software. The figures do not show the interface to CAD systems; this will be discussed later.

The purpose of the processor and the postprocessor is:

(1) To produce machine-independent data, known as the CLDATA (cutter

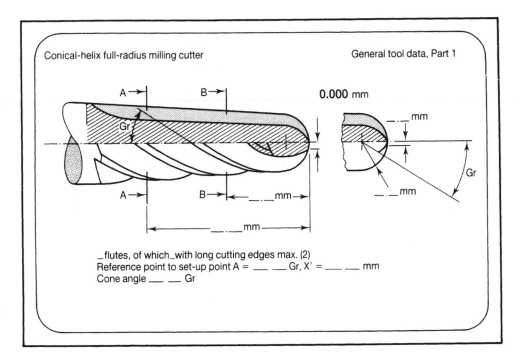

Figure 5.40
Interactive graphics
input for NC tool
grinding.

location data) (ISO 4343 DIN 66215), using the geometric and technical program building blocks of the part program.

The corresponding processing program is called a compiler or a processor. While the processor is a functional unit of a computer installation which includes an arithmetic unit and a control unit (= hardware), an NC processor is a 'program which describes compiler, translator and related functions for a programming language for numerically-controlled machines'.

(2) To adapt the part program to the particular NC manufacturing equipment using the interface program or the postprocessor.

'An NC postprocessor is a program which describes the conversion from CLDATA texts into specific control data for numerically controlled machines'.

One important aspect is the writing of collision-free programs which should lead to short machine run-in times. A collision analysis and graphical simulation support this. The quality of a programming system must be expressed in terms of the shortest possible manufacturing times for the programmed workpiece. Expert systems are increasingly being developed to plan the machining steps and the tools to be used.

5.6.2 Industrial robots

We will now give an overview of methods used to program industrial robots (Figure 5.41).

Direct (online) programming methods, often used in connection with robots,

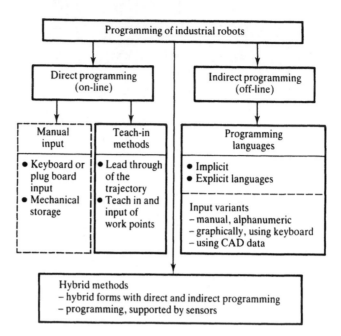

Figure 5.41
Programming
methods for
programmable
manipulators.

include the following motion-oriented procedures:

- Manual programming via keyboard, plug board input or mechanical storage.
- Programming by demonstration (play-back method), for example, by leading the robot effector by hand through its work motions and subsequent storage of the coordinate values.
- Programming by sequencing the robot effector through its work motions with the help of a teach pendant and storing the trajectory data (teach-in programming).
- Hybrid procedures. The program sequence is generated using a programming language (if necessary offline) and afterwards the work points are entered by the teach-in method.

In indirect (off line) programming, each task is described symbolically without an industrial robot using an application oriented language. Both explicit (based on a model of the environment) and implicit programming languages for industrial robots have been developed.

In explicit (motion-oriented) programming the path of motion of the handling device between different positions is described by the programmer, taking account of the freedom from collisions. This places high demands on the programmer and his ability to visualize space.

Implicit (task-oriented) languages are based on a known environment for the industrial robot which must be described in a model where the spatial conditions (travel range, collision zone) of the robot construction and the relationships between objects are specified in full. The programmer only describes the handling sequence.

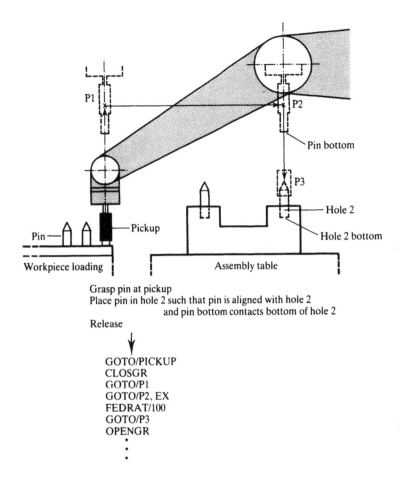

Grasp pin at pickup
Place pin in hole 2 such that pin is aligned with hole 2
                         and pin bottom contacts bottom of hole 2

Release

GOTO/PICKUP
CLOSGR
GOTO/P1
GOTO/P2, EX
FEDRAT/100
GOTO/P3
OPENGR
.
.
.

Figure 5.42
ROBEX programming
example.

All other data needed for the operation is determined by the processing program itself in order to generate collision-free paths of motion.

Figure 5.42 is a self-explanatory example of the use of an implicit language. The figure shows an assembly task. The sequence of motion is shown; except for the point P3, the pickup points are not entered precisely, they will be determined by teach-in. This is a hybrid method.

5.6.3 CAD/NC integration

The interconnection of CAD and NC programming systems is an important step towards CIM (Section 2.3.4). Until now, batch and dialog-oriented generation of part programs has been highlighted, in which both the technical drawing and the language description are basic ingredients (Figure 5.43, left). The drawing may be generated conventionally or by CAD. The term CAD/NC-programming is used to describe the generation of workpiece data for NC using the CAD database

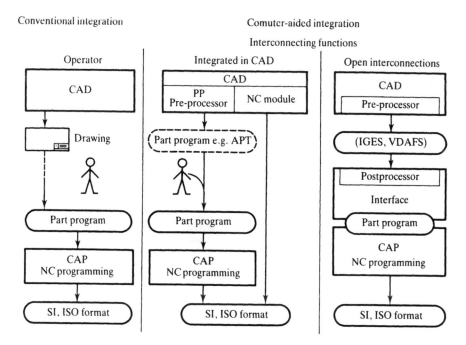

Figure 5.43
Various methods of
integrating CAD with
NC.

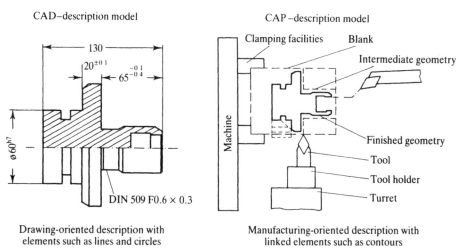

Figure 5.44
CAD and CAP
models describing
turned parts.

implemented by the designer. CAD/NC-coupling is a CAD/CAP coupling in the narrower sense.

In CAD/NC integration, the conception and design of the proper facilities and procedures are an important task of software engineering (see also Chapter 7). Here, we distinguish between:

- The program building blocks for the integration itself and for man–machine communication.

- The description model (Figure 5.44).

- The data interfaces ('SI' in Figure 5.43), for example, part-program and IGES interfaces.

It follows that description and interpretation rules together with interface specifications are particularly important. When integrating CAD into CAM, two tasks should be considered:

- modification of the geometry,
- extension of the technology.

Both of these tasks are usually carried out via an interface building block involving interactive graphics (see Figure 5.39).

The modification of the geometry results from the fact that a CAD (drawing-oriented) geometry is not suitable for manufacturing. Neither is it directly suitable for computation or for quality assurance. Thus, a product model suitable for CIM requires a number of slightly different views of a workpiece (see Section 2.3.3).

Integration may be expected to lead to the following advantages:

- increase in quality in the product development,
- guaranteed data and model consistency,
- minimization of faults,
- reduction of costs associated with the acquisition and generation of data,
- faster job throughput,
- time savings in planning tasks,
- improved quality of NC programs and work plans.

5.6.4 Cells

In the longer term, uniform programming of different control devices is desirable. Assembly or machining cells equipped with robots (see for example Figure 5.5) contain various control devices such as CNC, RC or PLC controllers. The concept of uniform programming involves three complementary ways of generating programs (see Figure 5.45):

(1) explicit language programming for all the control types used in the robot cell (RC, NC, PLC), for example, based on a motion and sequence-oriented programming system (MSPS),

(2) product-oriented implicit programming,

(3) sensor-assisted programming.

It must be possible to combine these programming methods. The interface question is of major importance in uniform programming of manufacturing equipment (Figure 5.46). One way is the use of standardized input formats, and another is the use of an intermediate language. The standardized input format (in the figure

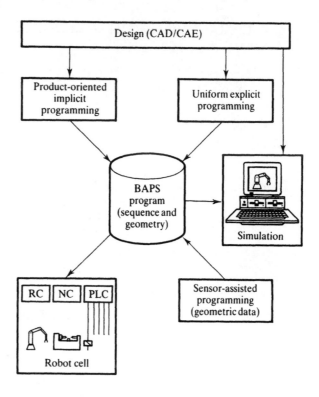

Figure 5.45

Overall concept of programming robots cells.

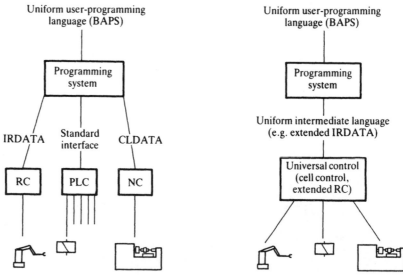

Figure 5.46

Possible structure of programming languages.

on the left) provides a solution which permits the incorporation of many of the control devices now on the market. The second solution produces a uniform code generated by the various programming languages.

At the beginning of this subsection (5.6.4), we referred to implicit product-oriented programming. This involves the definition of a target state. Programs determine how this state is attained. The method involves changing initial data, describing the target state, into output data (control programs). This comprises the 'implicit' concept. The 'product-oriented' concept implies a specialization on specific problem classes (for example, the manufacturing of printed circuit boards, the assembly of cable harnesses, surface handling), also known as products.

Thus, for an assembly task, in product-oriented implicit programming, data about the following components is required:

- assembly equipment: robots, loading equipment, tools;
- assembly tasks: products, individual parts, interrelationships between these components and to limiting technological conditions.

In implicit product-oriented programming the automatic generation of control data is based on these specifications and associated classifications by the programming system. The programmer is relieved of many routine tasks.

This section on uniform programming brings up new development tasks. Experience has been gained from small cells with only a small number of integrated manufacturing or production processes. However, we note other tools such as testing and quality monitoring together with error monitoring which could also be integrated into automated systems. Thus, this gives rise to the need for further in-depth consideration of a uniform programming system.

5.7 Problems

(1) What are the two types of data needed for detailed planning in order to schedule manufacturing equipment?

(2) When does time-variant data lose its value?

(3) How is horizontal and vertical structuring of control systems done?

(4) Why does cell control software depend on operating system functions? How can this problem be avoided?

(5) Why is configuration data needed for adaptive cell control software?

(6) Are software modules for cell control dependent on the manufacturing process?

(7) Discuss the tool supply activities.

(8) What do you understand by gross tool supply and by net tool supply?

(9) What does the term logic central data management mean to you?

(10) Why do NC programming systems to which extended interactive graphic input features have been added maintain the part program interface?

(11) What data files are used by the processor and post processor to run a part program?

(12) What is the purpose of using an interface for interconnecting CAD with NC?

(13) List the parameters which are characteristic for a high-quality NC program.

6 ⎯ Communication nets and protocol standards

CHAPTER CONTENTS

6.1 Introduction 258

6.2 Communication Topologies 259

6.3 Access Procedures 261

6.4 The ISO/OSI Reference Model 263

6.5 Communications Profiles 280

6.6 Field Bus 283

6.7 CNMA Pilot Installation as an Example of the Use of Open Communications 287

6.1 Introduction

Communication plays a central role in computer integrated manufacturing (CIM). The choice of a communication system largely determines the capability and productivity of a factory as a whole. Moreover, in the implementation of CIM systems, the costs associated with the interconnection of the individual CIM components are very important.

The various device technologies used in CIM and the different demands in the individual areas of computer-integrated manufacturing necessitate different communications networks to meet these requirements. In addition, office communication networks impose different requirements than factory communication networks. In the office area, communication is primarily used for inter-computer file access and transfer. In such a communication system there are also high data-protection requirements. In computer-integrated manufacturing, communication is largely used to control programmable manufacturing equipment. Here, the time requirements are high, and error-free data transmission is a necessity.

In the 1970s, the broad range of communication devices used and the numerous manufacture- and system-specific implementations (mostly incompatible) led to the cooperation of various international standardization bodies with the goal of a systematic analysis of the requirements for open communication systems and suggestions for standardization.

Open communication tries to provide standard data links between both manufacturer-specific and technology-specific data terminals. The functions to be implemented by an open communication system are specified by the International Organization for Standards (ISO) with the Open Systems Interconnection (OSI) reference model.

In this chapter, we shall first discuss the topological structures of communication architectures and the access protocols and then we shall explain the ISO/OSI reference model and various specifications.

6.2 Communication Topologies

The type of the connection and the logical configuration of the participants in the communications network (in other words, the topology) have a significant effect on the operational security, the capability and, not least, on the economics of automated manufacturing installations (see also Section 2.3.7). This also includes the topology and the geometric arrangement of the participants in the local network (Figure 6.1).

Figure 6.1
Various computer communication topologies.

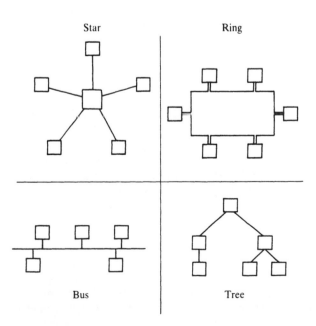

6.2.1 Star topology

The star topology represents the conventional centrally-oriented information-processing structure. It is the oldest and simplest implementation of a network (for example, a telephone network PBX). It is characterized by a number of participants connected to a central station (switching computer) which communicate with one another via this computer. Each station can be accessed directly by the central computer; all stations must communicate with one another by this computer. The extensibility of this structure is limited by the power and speed of the central computer. Other disadvantages include the high cabling expense together with the total dependence on the reliability of the central computer. The main advantages are that this topology allows very fast communication and that a fault in one station does not affect the operability of the system as a whole.

6.2.2 Ring topology

The ring topology is currently gaining in importance since this scheme is favored by major computer manufacturers (Token Ring). Every station always has two fixed neighbors on a ring highway. With this arrangement, all stations are connected to a ring. They communicate via a protocol consisting of addresses, control information and data placed on the ring by a participant. There are various methods of operating the ring. The cabling expense is low and only simple protocols are required. Disadvantages include the fact that with high data transfer rates this topology may get congested and when an individual station or a section of the cable breaks down, the ring is no longer operable (since a closed ring is a prerequisite for operability of the network). Moreover, the ring structure is not suited for covering a wide manufacturing area since branching (as for example in the star configuration) is impossible with one ring.

6.2.3 Bus topology

In the bus topology, all stations are connected to a central highway. Communication is done via a master or a method by which each participant works as an independent agent. The cabling requirements of the bus are similar to those of the ring; however, its operation is more fail-safe, particularly against the breakdown of individual components. This allows easy addition or deletion of stations and has led to very widespread use of this network topology.

6.2.4 Tree topology

The tree topology is similar to the star topology; however, it permits connection of sub-branches. This is its main advantage. Branching is associated with increased installation costs, in particular when fiber-optic or coaxial cables are used.

6.3 Access Procedures

It is understandable that, in view of the wealth of possibilities within the network structure, specific access procedures governing the participation in the network traffic together with protocols controlling the actual data exchange are needed. For network access, suppliers of such systems have reached different solutions, which are usually mutually incompatible. In what follows we shall describe procedures which are of importance to manufacturing; these are polling, time division multiplexing, carrier sense multiple access and token passing (Figure 6.2).

6.3.1 Polling procedure

The polling procedure is now rarely used for local networks. It uses a star topology with a master station in the center. The master station controls access rights by polling all stations in turn to see if they are trying to access the network. The exchange of polling information results in a loss of network transmission capacity. Moreover, most polling procedures are not directed towards handling the available amount of data from the individual stations because fixed data blocks are specified for the polling.

6.3.2 Time-division multiplexing procedure

With this method, the existing transmission capacity of the network is divided into time segments by time-division multiplexing. The time segments are allocated to the individual stations as available time slots. Thus, every terminal is able to transmit data over the network during a specific time interval.

The disadvantage of this procedure is that it has a rigid pattern of time allocation; thus, stations with nothing to send are still assigned the authorization to send, even if there is no data available. In addition, when individual terminals are introduced or removed the time slots must be altered.

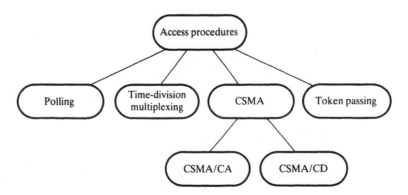

Figure 6.2
Various bus-access procedures.

6.3.3 Carrier sense multiple access

Carrier sense multiple access (CSMA) is used for networks with bus topology. Before a data packet is sent, the transmission medium is sensed. A station only accesses the bus when it has ascertained that nothing else is being sent on the bus. If the transmission medium is busy, the station which wishes to send may behave in one of three possible ways:

- Non-persistent CSMA. The send request is deferred for a random waiting time. However, under certain circumstances this reduces the transmission capacity since the medium may become free while the station is waiting. Moreover, when the access to the bus is re-attempted, the bus may again be busy.

- 1-persistent CSMA. The bus is sensed, and as soon as it becomes free the station sends its data. In this version there is no loss of transmission capacity but collisions are preprogrammed when several stations are waiting for a transmission to end. In this case, all stations may try to send their data at the same time.

- p-persistent CSMA. The bus is continuously sensed and the data is sent with an access probability p when the bus becomes free. Otherwise (with probability $1 - p$) there is a random waiting time before the transmission is repeated. This is a compromise between non-persistent and 1-persistent CSMA.

In CSMA, data packets which collide are nevertheless transmitted in full. This leads to a decrease in data capacity since transmitted data packets which have collided are unusable. This may be avoided with the procedures described below.

In the CSMA CD (collision detection) procedure a station constantly senses the bus while sending its own message and compares the data which it has sent with that on the bus. If the two are not the same, a collision is assumed and the transmission is interrupted. In this way, the loss of transmission capacity is limited to the time taken to detect the collision.

When a collision occurs, the station detecting a collision sends a JAM signal (it gives up and complains) so that the other stations also notice the collision and end the emission. After a random waiting time, a new attempt is made to transmit the data packet.

In the CSMA CA (Collision Avoidance) procedure all participants attached to the medium are allocated different retreat or delay times. This ensures that following a collision the participants in the network will (re-)attempt to access the medium at different times; thus, two stations which wish to send will not access the medium at the same time again. Thus, there is at most one collision possible on the medium. The allocation of retreat and waiting times is similar to a prioritized allocation of the right to send to the individual stations; stations with small delay times have a high priority to send. This procedure is less flexible than the CSMA CD access procedure if stations are added to or deleted from the medium.

6.3.4 Token passing

Access procedures using the token passing method are among the best-known and oldest procedures in local networks. A token (like a baton) is an entrance key or voucher for access to the network which is forwarded from participant to participant. A token may have a free or a busy state. A participant receiving a free token may send to the network.

When a station, which wishes to send, has received a free token it alters it to a busy token and attaches the data to be sent to the token. Another free token is then attached to the end of the data block. The message recipient returns the data with an acknowledgement to the sender, which may then monitor the error-free transmission of the data emitted by it.

The token passing procedure may be used on both ring and bus topologies. In the token ring the token is sent to the next neighbor according to the topological configuration. In the token bus, a logical sequence of stations is specified and the token must be forwarded with an address for the next recipient. This of course means that every station must be known to its successor and that when a station drops out of the network the other participants in the network must notice this and a new sequence must be specified.

6.4 The ISO/OSI Reference Model

6.4.1 Introduction

A communication system is characterized by its protocol architecture, which includes a specification of the topology, the access procedure and the messages to be exchanged. There are currently very many incompatible manufacturer-specific protocol architectures for general computer communications, for example: DECNET, SNA, TCP/IP and SINEC. When participants supported by two different architectures wish to communicate, expensive network interfaces are required which lead to a decreased communications performance and mapping losses. Manufacturer-independent, internationally standardized communication systems are needed to overcome these disadvantages in the future. The standardized ISO/OSI base reference model forms the fundamental structure for such an open system; it is described below. More information about manufacturer-specific systems may be found in Kühn *et al.* (in prep.).

6.4.2 Fundamentals

The base reference model is a general model for digital communication between two or more participants. It provides the framework for the elaboration of standards for open communication systems.

The reference model (Figure 6.3) is based on the following four principles:

(1) The communication functions are divided into layers.

(2) The services to be provided by each layer are specified.

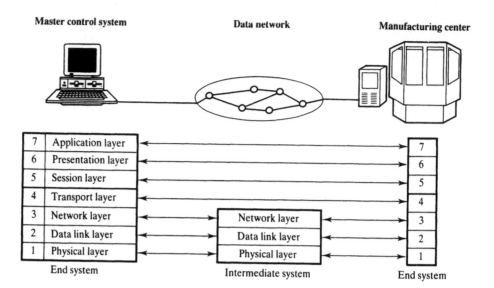

Figure 6.3
The ISO/OSI
reference model for
open communication.

(3) Layer $N + 1$ which is immediately above layer N uses the services of the latter to implement its functions.

(4) The communication between the layer N and the participating terminals is specified by the ISO protocols.

The OSI reference model comprises seven layers, where each layer may consist of several sublayers. The four lowest layers form the transport system which transports the data transparently and thus provides end-to-end communication. There may also be a transit system between the two end systems for traffic routing and switching. The three highest layers form the application services which, unlike the transport system, provide the transport of information rather than that of data.

Figure 6.4 shows the relationships between consecutive layers in the base reference model. The layer N is the (N)-service provider for the layer $N + 1$ and makes its services available to the latter. The layer $N + 1$ (the (N)-service user) has access to the services of layer N via the (N)-service access point (SAP) using the (N)-service protocol (adjacent layer protocol). The layer N of the end systems involves communication via a (peer-to-peer) protocol and exchange data in the form of (N)-protocol data units (PDUs). To transmit the (N)-PDUs the layer N uses the services of the layer $N - 1$. Each (N)-PDU contains the ($N + 1$)-PDUs of the layer $N + 1$ which are to be transmitted and the protocol control information (PCI) of layer N. This continues downwards until the data is transmitted over the physical transmission medium to the attached destination end system. In the destination end system, the data follows the reverse path up to layer $N + 1$. This is illustrated in Figure 6.4.

A brief description of the individual layers of the OSI reference model is given below.

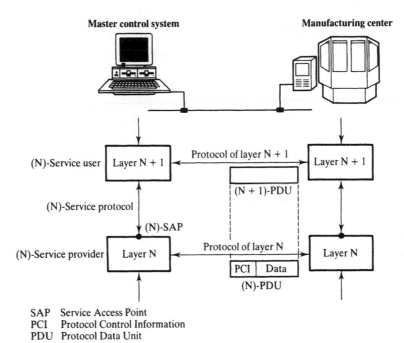

Figure 6.4
Relationships between
the layers of the
reference model.

SAP   Service Access Point
PCI   Protocol Control Information
PDU   Protocol Data Unit

Layer 1: physical layer

Layer 1 defines the functional, electrical and mechanical characteristics of the transmission medium necessary for transparent physical transmission.

The most important tasks include:

- parallel/serial conversion, multiplexing (frequency, time division),
- physical interfacing to the transmission medium (fibre-optics, coaxial cables and so on),
- synchronization at the bit level,
- definition of the valid signals and connecting lines.

Layer 2: data link layer

This layer provides for access to the transmission medium and for secure transmission of individual data blocks. This involves:

- activation/deactivation of access to the transmission medium,
- block synchronization,
- monitoring of the data-block sequence,
- dataflow control,
- error detection and, if necessary, error correction.

Functionally, layer 2 may be divided into layer 2a (Medium Access Control (MAC), for access to the transmission medium) and layer 2b (Logical Link Control (LLC), for the security tasks).

Layer 3: network layer

The main task of the network layer is to switch the data to be transmitted between the end systems by selecting the best path. By end systems, we mean the sender and the recipient of a message; under certain circumstances, these may be physically interconnected via a transit system. The control of the passage (switching and transmission) of the information packets through the network involves the following functions:

- packet assembly, disassembly,
- control of the exchange of packets between data terminals and the network,
- establishment and release of virtual connections between data terminals,
- transport of the data packets between two data terminals,
- routing within the network.

Layer 4: transport layer

The task of the transport layer is to give the user a reliable logical transport connection. The layer-4 services provided to the user enable him to communicate with the target end system in a network-independent fashion, without having to be concerned with the detailed physical characteristics of the transmission medium in question. Thus, this layer represents the dividing line between the transport-oriented and the application-oriented parts of the OSI reference model. The most important functions of the transport layer include:

- establishment and release of transport connections,
- multiplexing of layer-3 transport connections and virtual connections (connection multiplexing),
- end-to-end flow control,
- fragmentation (assembly/disassembly).

Layer 5: session layer

Layer 5 is responsible for synchronizing the execution of the message exchange between two terminals or end users. It provides:

- conversion of symbolic addresses into real addresses for the transport connection,
- declaration of the session parameters (half/full duplex, flow-control characteristic quantities, and so on),
- dialog control (synchronization points, token passing, and so on),
- re-establishment of interrupted transport connections.

Layer 6: presentation layer

The main task of this layer is to convert the system-internal data presentation into a uniform network presentation, which can be negotiated between the communications partners. A formal description of the data is called an abstract syntax. ASN.1 (Abstract Syntax Notation One) is defined by ISO as a formal data description language for layer 6. ISO also specifies 'Basic Encoding Rules for ASN.1'. These functions are needed because of the various computer-internal presentations (for example, of integers) and the various coding methods (for example, ASCII, EBCDIC).

Layer 7: application layer

Layer 7 is the most important layer of the OSI reference model from the user's point of view. The definition of uniform protocols for the application layer is made extremely complicated by the large variety of possible applications. For this reason, layer 7 is subdivided. Layer 7a contains general functions which are used by the most application-specific protocols of layer 7b. One important element of layer 7a is ACSE (Association Control Service Element) which contains services for connection and release of users and for connection monitoring. Service elements of layer 7b provide the user with services for a specific application area.

As far as the user is concerned, layer 1 (physical layer) together with layer 2 (data link layer) and layer 7 (application layer) are the most important. Thus, we shall discuss the protocols for these layers used in the MAP/TOP specification in detail.

6.4.3 Standards for the physical layer and the data link layer

Figure 6.5 lists the currently important ISO standards for the physical layer and the data link layer.

The individual standards are distinguished by the transmission medium, the transmission procedure and the access procedure used. The most important features of the individual standards are described in brief below.

Transmission medium

The main transmission media are the coaxial cable and the twisted pair. Fiber-optic cable is increasingly being used to increase data security and data rates.

Communications technology

The data is transmitted bit serially which in the computing equipment entails a parallel to serial conversion or vice versa. The bit stream to be transmitted is encoded using an appropriate method (for example, Manchester code). There are three types of communication technologies: baseband, carrierband and broadband.

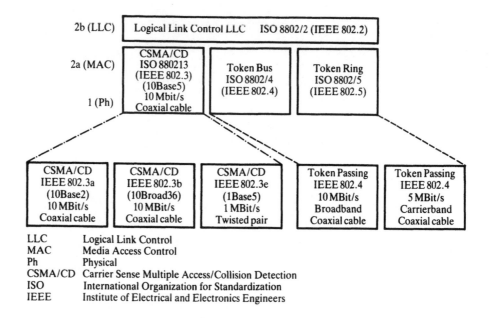

Figure 6.5
Standards for LANs.

In baseband communications the original signal is transmitted directly. When the signal is modulated using an appropriate modulation procedure, the term broadband transmission denotes the use of several carrier frequencies (several communication channels), whilst the term carrier-frequency modulation denotes the use of a single carrier frequency (only one channel).

Access procedure

The use of a LAN in the manufacturing area is almost always associated with the use of the decentralized CSMA CD procedure and token passing.

The access procedure, the transmission medium used and the transmission procedure together determine the characteristics of the local network.

Security methods

In order to guarantee the integrity of the data to be transmitted, the useful data must be supplemented with additional control data. There are two basic methods for this, namely the cyclic codes (Cyclic Redundancy Check, CRC) or the generation of block checksums. CRCs are used in OSI networks. In the CRC procedure a generator polynomial is applied to the useful data to form additional security bits.

6.4.4 Application-layer standards

This section describes the OSI application-layer standards which are most important from the user's point of view. Only standards for the MAP/TOP specification are considered. The usefulness of the standards to the user depends on the services provided.

The important protocols for computer-integrated manufacturing are FTAM (File Transfer, Access and Management) and MMS (Manufacturing Message Specification). After a brief introduction to the FTAM standard, we shall describe the MMS standard in more detail since it plays a central role in computer-integrated manufacturing. Concepts such as MMS objects, MMS services, MMS servers and MMS clients, which are important for the understanding, are defined.

File transfer access and management (FTAM)

FTAM supports file transfer, access and management within a heterogeneous computer network. This standard is mainly applied to the information exchange between systems with peripheral data systems. A typical application is, for example, the exchange of files between workstations. It is based on the definition of a virtual file storage (Figure 6.6). The implementation must map the existing physical file system onto this virtual file system. The following services are provided:

- estalishment, release and interruption of a logical connection,
- file selection and release,
- file opening and closure,
- file read and write,
- generation and deletion of a file,
- reading of file characteristics.

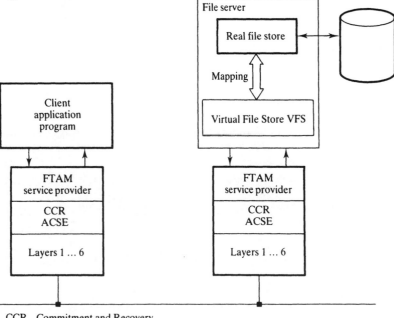

Figure 6.6
Scope of the FTAM standard.

CCR Commitment and Recovery
ACSE Application Control Service Element

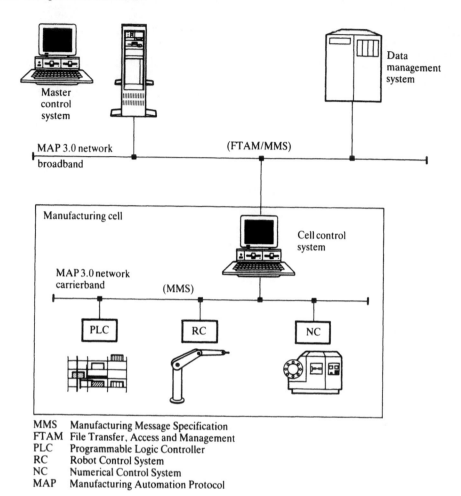

Figure 6.7

Typical application of the MMS standard.

MMS Manufacturing Message Specification
FTAM File Transfer, Access and Management
PLC Programmable Logic Controller
RC Robot Control System
NC Numerical Control System
MAP Manufacturing Automation Protocol

Manufacturing message specification (MMS)

Application area. ISO standard 9506 (Manufacturing Message Specification, MMS) is a service element of the OSI application layer (layer 7) and specifies the message exchange between programmable devices for computer-integrated manufacturing. The configuration shown in Figure 6.7 is a typical MMS application.

The parts of the MMS standard. The MMS standard consists of several parts. Parts 1 and 2 form the 'MMS core' upon which the 'companion standards to MMS' are built (Figure 6.8). The MMS core defines the syntax and semantics for an application-independent message exchange. The companion standards define the application-specific specifications relating to the general syntax and semantics (for example, for numerical or robot control).

Part 1 of the specification (ISO/IEC 9506-1, Service Definition) contains the definitions of the services. It defines the following:

- an abstract model which describes the interaction of two MMS participants,

- the visible functionality of an MMS implementation, viewed from the outside,
- the MMS services,
- the parameters which must be supplied for the services,
- the relationship between events and actions in MMS and their valid sequences.

Part 2 of the specification (ISO/IEC 9506-2 Protocol Specification) specifies the MMS protocol. This includes specifications of:

- The MMS protocol data units (Protocol Data Unit, PDU),
- The assignment of the MMS services to the MMS PDUs,
- The mapping of MMS services onto services of layer 7a (Application Control Service Element, ACSE) and services of layer 6 (presentation).

Tasks and scope. In order to facilitate modeling which is independent of the application layer, MMS defines application-independent objects which are combined into classes. MMS describes services which facilitate the creation, deletion and manipulation of the objects and specifies how they should behave for certain service requests. MMS describes the mapping of the resources of a manufacturing device onto an abstract model (Virtual Manufacturing Device, VMD). In this mapping, only resources visible or accessible from the network are considered (communications engineering view). The VMD consists of elements of the individual MMS object classes.

The main objectives of MMS are (Figure 6.9):

- MMS defines building blocks for the generation of an abstract model of a manufacturing device. The actual nature of this model is determined by the implementation.
- MMS defines services which facilitate work with these building blocks. The given requirements specify which services are actually implemented.

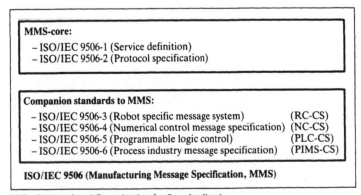

Figure 6.8
Parts of the MMS standard.

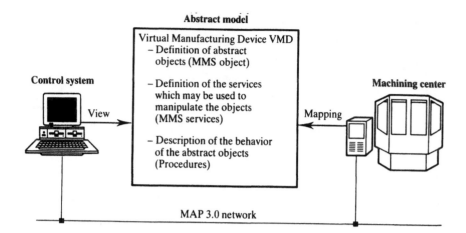

Figure 6.9
Scope of the MMS
standard.

Communication sequence. The terms are as follows:

- **MMS provider:** provides the MMS services, and exchanges corresponding PDUs with the MMS provider of the target system.
- **MMS user:** the part of the application which uses the MMS services provided by the MMS provider for the message exchange with the MMS user of the target system.

The MMS user may play two different roles within a message exchange:

- **MMS server:** the part of an application which maps the physical resources which may be accessed from the network onto MMS objects and manages these.
- **MMS client:** the part of an application which accesses the MMS objects (and the physical resources) made available by an MMS server.

A typical MMS sequence has the following form (Figure 6.10):

(1) The MMS user (MMS client) requests an MMS service from the MMS provider (MMS service request).
(2) The MMS provider receives this request and forwards the appropriate MMS-request PDU to the MMS provider of the target system.
(3) The MMS provider of the target system informs the MMS user (MMS server) that it has received a request (MMS service indication).
(4) The MMS user (MMS server) processes this request and forwards the result to the MMS provider (MMS service response).
(5) The MMS provider forwards the corresponding MMS-response PDU to the MMS provider of the source system.
(6) The MMS provider forwards the result of the request to the MMS user (MMS service confirmation).

As far as the user is concerned, the interface to the local MMS provider is of primary importance during this sequence of events.

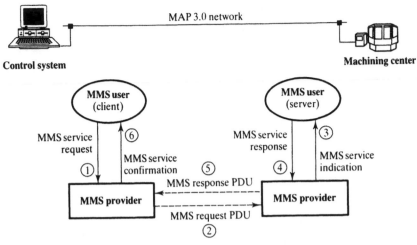

Figure 6.10
The logical
communications
sequence for MMS.

MMS   Manufacturing Message Specification
PDU   Protocol Data Unit

MMS object classes	MMS service classes	
	Environment and general management services	(5)
VMD	VMD management services	(6)
Domain	Domain management services	(12)
Program invocation	Program invocation services	(8)
Variable access manag. model (5 objects)	Variable access services	(14)
Semaphore manag. model (2 objects)	Semaphore management services	(7)
Operator station	Operator communication services	(2)
Event management model (3 objects)	Event management services	(19)
Journal management model (2 objects)	Journal management services	(6)
	File access service	(1)
	File management services	(6)

Figure 6.11
MMS object classes.

( ) Number of individual MMS services

The MMS server makes its resources available to the MMS clients with the help of MMS objects. The MMS implementor must know which MMS objects are available to him.

MMS defines the following classes of objects (Figure 6.11):

- VMD (Virtual Manufacturing Device)
  - the abstract model of the real manufacturing device.
  - consists of representatives of the following object classes.

- Domain
  - data storage: for example, NC records, tool correction data,
  - information storage: for example, a machine image.
- Program invocation
  - facilitates the management and control of programs, for example, initiation, termination, abortion, relocation and deletion of NC programs,
  - a state diagram ensures that in every state only appropriate services may be applied in the program invocation, for example, only programs in the RUNNING state may be terminated.
- Variable
  - Information storage: for example, maximum spindle speed, number of axes and so on. A number of variables may be combined into a list, in order to obtain the information content of several variables through a request. Specific access to an element of a structured variable is also possible. Moreover, data-type descriptions may be defined to describe the structure of the contents of a variable. In MMS, the object classes for variables permit modeling at all levels of complexity from simple to very complicated.
- Semaphore
  - synchronization for concurrent access to shared resources.
- Operator station
  - input/output via a terminal, for example, the control panel of an NC system.
- Event
  - allowance for events, for example, alarm messages, pallet arrival,
  - specification of when events occur (event conditions),
  - specification of action(s) to be taken,
  - coupling of event conditions and actions (event enrollment).
- Journal
  - recording of specific information/data, for example, for diagnosis, quality assurance and so on.

An implementor may use these building blocks to make the resources of a manufacturing device available to the MMS client.

Access to elements of the MMS object classes

An element of an object class may only be accessed using MMS services defined for that object class. For example, elements of the object class 'domain' may only be accessed using domain management services.

The following questions arise for the user:

- Onto which MMS objects are the resources of a programmable manufacturing device mapped?

- Which MMS services may be used to manipulate these objects?
- How can objects be created on the server?
- How can an MMS client access these objects?

MMS does not contain any specification which is specific to an application area as far as modeling an MMS server on (for example) an NC system. For example, it does not specify the MMS object onto which the spindle speed is mapped or which data type should be used. This and other specifications for modeling a control system (for example, an NC system) in a VMD are provided by the companion standards.

MMS service classes

Every object class is associated with a service class which facilitates the manipulation of the object. The service classes contain several services with different functions.
MMS defines the following services classes:

- Environment and general management services
    - for connection, establishment, release and interruption,
    - for signalling protocol errors.
- VMD management services
    - to access the VMD attributes.
- Domain management services
    - to upload and download domain contents,
    - to delete domains and to read out domain attributes.
- Program invocation services
    - to define a program invocation using domains,
    - to start and stop program invocations,
    - to delete program invocations and to read out program-invocation attributes.
- Variable access services
    - to define and delete objects describing an abstract variable type,
    - to define and delete variable objects,
    - to read from and write to variable objects,
    - to read out the attributes of a variable object.
- Semaphore management services
    - to create and delete semaphore objects,
    - to request and relinquish control over the semaphore object.
- Operator communication services
    - to input and output data to and from a terminal.
- Event management services
    - to create and delete an event object,

- to report an event,
- to confirm an event.

• Journal management services
- to create and delete a journal,
- to make entries in the journal.

Example. We give an example to illustrate the use of MMS services and in particular the communication sequence for a read access (read service) to a variable object with the name SSL. The example shows the exchange of PDUs from the point of view of the MMS user and thus illustrates the behavior of the MMS server and client (Figure 6.12).

The client sends a read request to read the variable SSL. The server then receives a read indication. The server accesses the variable object with the name SSL in its VMD to obtain the value of the variable object. Finally, the server executes a read response with the current value of the variable SSL. The client receives the read confirmation containing the value requested, and the communications procedure is closed.

Companion standards to MMS

The companion standards are based on the MMS core (which consists of the service definition (ISO/IEC 9506-1) and the protocol specification (ISO/IEC 9506-2) parts). The companion standards define the application-specific part of MMS. The

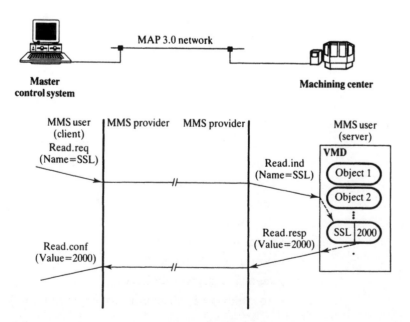

Figure 6.12

Readout of a variable object (SSL) according to the MMS standard.

MMS core specifies that if a companion standard is required then at least one of the following conditions must be satisfied:

- There is a need for an information model for a certain application area and no existing companion standard defines extensions of the MMS objects or new application-specific objects which cover the need in full.
- There is a need to extend the semantics of individual MMS services and no existing MMS companion standard defines identical semantic extensions.
- The object names standardized in the companion standard are not sufficient for the application area.

It is explicitly noted in the MMS core that the need for a companion standard is not justified if the only criterion is that there is no subset of services of the MMS core specific to the application area.

The most important tasks of a companion standard are already evident from this list. The MMS core proposes the approach, which is reproduced in the arrangement of a companion standard. Here, we distinguish between an MMS-independent and an MMS-related part.

The MMS-independent part includes:

- The description of the application area in which the companion standard may be applied (for example, robot control).
- The definition of an abstract information model of the application area.
- The description of application-specific functions (for example, to load, start and delete a program).

The MMS-related part of a companion standard describes:

- The mapping of the application-specific information model onto a virtual manufacturing device (VMD).
- The mapping of application-specific objects onto elements of the object classes defined in the MMS core (for example, mapping of an NC program onto a program-invocation object in conjunction with domain objects).
- The mapping of the application-specific functions onto MMS services.

If necessary, it defines:

- New abstract object classes or extensions of existing MMS objects so as to permit the mapping of application-specific objects.
- New MMS services or extensions of existing MMS services.
- Application-specific, predefined representatives of the available object classes.
- Application-specific conformance classes. A conformance class represents a subset of the available MMS services.

This approach is illustrated in Figure 6.13 for the example of the companion standard for the application of robot control (RC-CS).

State diagrams for some of the MMS object classes are defined in the MMS core. These may be refined using a companion standard to include new substates.

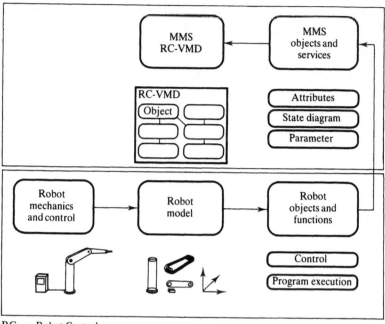

Figure 6.13

Mapping process of the companion standard for robot control.

RC	Robot Control
VMD	Virtual Manufacturing Device
MMS	Manufacturing Message Specification

When existing MMS object classes are extended, the guidelines specified in the MMS core should be adhered to. Here, we note that not all MMS object classes may be extended with additional attributes (for example, the MMS object class named Variable cannot be extended). Extensions are possible for the following MMS object classes: VMD, domain, program invocation, operator station, event condition, event action, even enrollment and journal entry.

In addition, each companion standard contains an informative appendix describing an illustrative application.

Standardization activities for companion standards

National and international standardization bodies are working in close cooperation to generate appropriate companion standards. Currently, they are concerned with companion standards for the application areas of numerical control (NC), robot control (RC), programmable logic control (PLC) and process-control technology (PCT). Figure 6.14 shows the path from a draft proposal (DP) to an international standard (IS). The times shown represent average values.

Harmonization of the companion standards

The area of computer-integrated manufacturing is currently characterized by the development and integration of flexible manufacturing cells consisting of a number

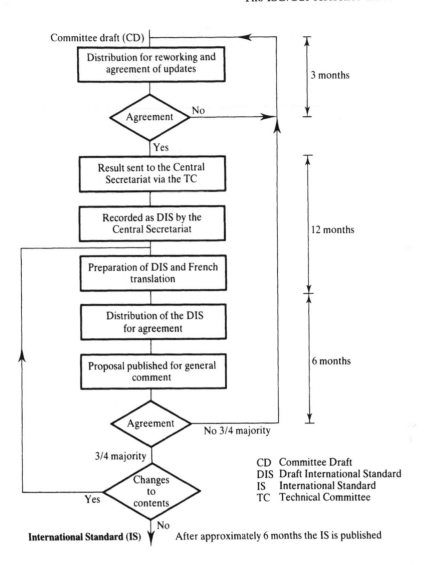

Figure 6.14
Evolution of an international standard.

of different hardware devices (for example, robot control systems and numerically controlled machine tools and so on). An economic implementation of flexible manufacturing cells requires the development of functionally-delimitable, reusable functional building blocks. This requires the mapping of the device-oriented approach onto a functionally-oriented approach.

This mapping is currently complicated by the various approaches to the generation of companion standards. This has led to attempts to harmonize companion standards.

Companion standards may be used to provide specific object classes in the MMS core with additional attributes. As far as harmonization of the companion standards is concerned, it is desirable that this facility should only be rarely used.

The examples of robot control systems (RC-CS) and numerically-controlled machine tools (NC-CS) illustrate two different approaches to the companion standards.

In the companion standard for robot control systems (RC-CS) the object class VMD is extended by the following attributes

- Attribute: Safety interlocks violated (TRUE, FALSE).
- Attribute: Robot VMD state (ROBOT IDLE, ROBOT LOADED, ROBOT READY, ROBOT EXECUTING, ROBOT PAUSED, MANUAL INTER-VENTION REQUIRED).
- Attribute: Any physical resource power on (TRUE, FALSE).
- Attribute: All physical resources calibrated (TRUE, FALSE).
- Attribute: Local control (TRUE, FALSE).
- Attribute: Metric measure (TRUE, FALSE).
- Attribute: Reference to selected controlling program invocation.

RC-specific information is stored in these additional attributes. The current operational mode is made available (for example, to the RC-CS client) in the 'local control' attribute. The RC-CS client may query the contents of the additional attributes using the MMS service 'status'. For this, the service must be extended by the necessary parameters.

Moreover, the RC-CS server may forward this information to the RC-CS client unsolicited using the service 'unsolicited status'. In addition, this RC-specific VMD-related information is stored in predefined representatives of the object class 'named variable', the contents of which the RC client may query using the 'read' service.

In the NC-CS on the other hand, no additional attributes are defined for the object class VMD (NC-CS-VMD). The NC-specific, VMD-related information is only made available to the NC-CS client in predefined representatives of the object class 'named variable'. The current operational mode, for example, is stored in the element 'N_REMOTE' of the object class 'named variable'. The NC client may request this information from the NC server using the 'read' service. In addition, the NC server may send this information unrequested to the NC client (information report).

The approach in the case of NC-CS does not require extensions of object classes or services and thus supports the harmonization of the companion standards.

6.5 Communications Profiles

6.5.1 MAP/TOP

The international efforts to specify a communication system which provides for data linkage between heterogeneous computer and control systems for CIM at a reasonable price are currently engaged in the MAP/TOP specification. Version

3.0 of the MAP/TOP specification has been available since October 1988 and will be valid until 1994. This should provide both the suppliers and the users with the necessary stable time to use the standard.

MAP

MAP stands for manufacturing automation protocol. It is a protocol which was specified for open communication in manufacturing automation.

MAP selects, from existing standards for each layer of the OSI reference model, the ones appropriate for computer-integrated manufacturing. A selection of existing standards is called a profile. Thus, MAP is a profile for the area of computer-integrated manufacturing.

TOP

TOP (technical and office protocol) like MAP is a profile, but in this case for office communications.

Figure 6.15 lists typical MAP/TOP applications.

The ISO standards used in MAP/TOP

The functions of the individual layers may be realized in different ways. Thus, for an application, there exist a number of protocols for the individual layers (ISO standards, CCITT recommendations and so on). The implementation of an OSI communication system requires the selection of the appropriate existing standards to meet requirements. The protocols selected must also be mutually compatible.

Figure 6.16 lists the ISO standards currently used in the MAP/TOP specification.

The MAP/TOP specification is the result of the initiative of General Motors (USA) and Boeing (USA), supported by international MAP/TOP user groups, to

MAP	TOP
• Factory monitoring and control system	• Electronic mail (MHS/X.400)
• Factory data collection system	• File transfer, access and management (FTAM)
• Quality control	• Electronic data interchange (EDI)
• Distributed numerical control systems (DNC)	• Distributed word processing
• Flexible manufacturing systems (FMS)	• Office compound documents (graphic and text)
• Manufacturing cells and lines	• Product data exchange
• Robot assembly cells and lines	• Manufacturing resource planning (MRP)
• Automatic storage and retrieval systems (AS/RS)	• Computer-aided design (CAD)
• Computer-aided manufacturing (CAM)	• Computer-aided engineering (CAE)
• Computer-integrated manufacturing (CIM)	• Just-in-time (JIT)
• Process control	

Figure 6.15
Typical MAP/TOP applications (Source: EMUG).

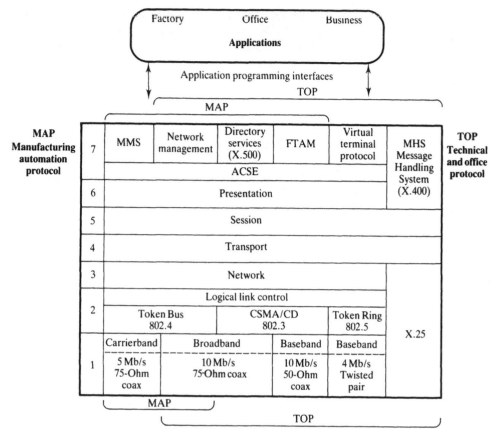

Figure 6.16
The protocols used in
MAP/TOP (Source:
EMUG).

achieve a cost-effective implementation of a communications link between individual
CIM components. We note that the MAP/TOP specification is not an international
standard and does not lie in the area of responsibility of any national or
international standardization organization. The North American MAP/TOP users
group (NAMTUG) has taken on the role of the patron of the MAP/TOP
specification, and future changes to the specification will be released by the users
group.

6.5.2 Mini MAP

Since it is so complex, MAP (also called Full MAP) requires large amounts of
memory and for some applications exhibits undesirably long run times. For this
reason, the Mini MAP communications profile was developed, which only uses
the three layers 1, 2 and 7. As in Full MAP, Mini MAP uses the token passing
procedure on baseband or carrierband buses. Thus, layers 1 and 2a of the ISO
reference model remain unchanged. Layer 2b had to be modified from that of

Full MAP, since layers 3 to 6 above it are missing. PDUs from MMS (which is unchanged) are mapped directly onto layer 2.

Since layer 4 (the transport layer) is missing, the communication is packet-oriented and several users may be addressed with a single request.

Despite the large increase in speed, Mini MAP is primarily suitable for cell buses. However, it is not suitable for typical field-bus applications since the connection costs due to the use of coaxial cable are too high.

The MAP-enhanced protocol architecture (MAP/EPA) is an architecture which supports both Full MAP and Mini MAP.

6.5.3 CNMA

The Commission of the European Community is supporting the MAP/TOP users group in the framework of the ESPRIT programme (European Strategic Programme for Research and Development in Information Technology) via the CNMA project (Communications Network for Manufacturing Applications). Work within this project is concentrated on the following projects:

- updating the international standards used in the CNMA communications profile;
- definition and implementation of the function directory service;
- development of standardized application interfaces for FTAM and MMS;
- extension of the network-management components;
- development of network-management applications and optimization of performance, for network configuration and for the diagnosis of faults;
- implementation of four pilot installations.

For the user, the differences between CNMA and MAP are as follows:

- CNMA supports ISO 8802/3 which corresponds to the IEEE 802.3 (Ethernet).
- CNMA uses ISO/IEC IS 9506 whereby MMS is an international standard (IS). MAP uses ISO/IEC DIS 9506 whereby MMS is a draft international standard (DIS).

We note that while TOP supports ISO 8802/3 (Ethernet) as transmission medium, MAP does not.

The partners in the CNMA project are directly connected with the national and international standardization institutions and other organizations such as EMUG, MAP/TOP users group and the OSI/NM forum.

6.6 Field Bus

The term field bus was coined by the IEC. It is the digital replacement of the analog 4–20 mA interface which is very widely used in industrial process control. This interface has the serious disadvantage that it can only transmit analog values in one direction. With the introduction of intelligent sensors and actuators for the field use in process control, there arose a need for a bidirectional digital communication system to link field devices to the master control system. This communication system should meet requirements for increased accuracy and transmission security, reduced cabling costs and remote calibration together with diagnosis and maintenance functions.

A number of manufacturers of automation systems for process control developed company-specific digital communications protocols, which users have only installed hesitantly, since they introduce a dependence on the supplier. Thus, national and international standardization activities have also been initiated in this area, with the aim of developing a uniform communications standard for the field area, which would guarantee integration into the communications structures of the master automation system.

6.6.1 Application areas

There are four different overlapping areas of manufacturing technology in which computer networks may be used (Figure 6.17).

At the higher levels of the control hierarchy there exist factory buses for factory-wide networking and cell buses for interconnection of manufacturing units (such as a manufacturing cell). Broadband, carrierband and baseband networks based on MAP are mainly used for this purpose.

Field buses are used at the lower levels of the control hierarchy, where sensors, actuators, regulators, programmable control systems and I/O modules are interconnected. MAP networks are not used here for cost reasons and because of the particular requirement profile. As previously mentioned, the idea of a field bus system for interconnecting simple sensor and actuator systems stemmed from industrial process control. Here, the applications are characterized by large distances between the devices so that reduction of cabling cost for sensors and actuators is particularly important.

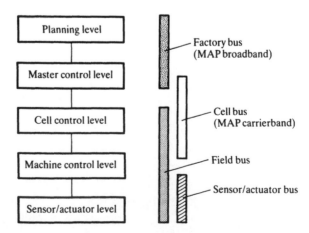

Figure 6.17

Application areas of communication systems.

The areas in which field buses and cell buses (which are based on carrierband or Ethernet) are used have a relatively large overlap. Most of all, at the cell control level, field buses or cell buses based on MAP may be used as alternatives. The separation line must be drawn according to the specific application and the type of devices used.

However, the lower border line cannot be determined uniquely. In addition to their use as cell buses, field buses are also typically used to interconnect intelligent sensors, actuators and decentralized I/O devices to the automation system. Thus, the requirements for realtime capability are very different. In particular, in manufacturing technology there are also applications which are characterized by such special requirements that are difficult or impossible to cover by a general, open field bus. This is true, for example, for highly dynamic control loops, as they are found in a machine tool. Here, in the future, special sensor/actuator buses will be used which meet the specific requirements (in particular, the requirement for a realtime capability) with a reduced and specially adapted range of functions.

6.6.2 Requirements

The requirements for a field bus for manufacturing technology and industrial process engineering are very similar (Figure 6.18). The most important differences are the short response times which are required in most manufacturing technology applications, together with the topological requirements of industrial process systems where the distances between participants may be large. Moreover, in certain process applications additional features such as the supply of power via the bus and intrinsic safety may be required.

Although the features of a field bus largely match the possible application areas, the development of a uniform international standard will require that the user be

Manufacturing engineering	Chemical engineering
Realtime behavior Simple protocols Short messages Favorable coding efficiency	
Response time 1 ... 50 ms	Response time 50 ms
Cost effective	
High transmission security	
High availability (redundancy)	
Bus length 10 m – 100 m	Bus length > 1000 m
Galvanic decoupling	
	May be used in hazardous areas. Auxiliary power supply via the bus
Easy linkage to MAP networks (MMS)	

Figure 6.18
Requirements for a field bus.

offered various alternatives (particularly for layers 1 and 7), from which he may select the most suitable option for his particular application.

6.6.3 Incorporation in the ISO/OSI reference model

The ISO/OSI reference model divides the individual functions of a digital transmission into seven layers. In an implementation of a communication system, not all layers of the ISO/OSI reference model are necessarily realized. As more layers are implemented, the system becomes more complicated from the coding point of view and the time behavior of the system is degraded.

In the case of the field bus, only layers 1 (physical), 2 (data link) and 7 (application) are usually specified. Layers 3 to 6 are not explicitly present, since their functionality is not necessarily required for field buses (Figure 6.19). The functions which are required are added to layers 2 and 7. This assures a high message efficiency, short response times, high data throughput and low hardware and software costs. In particular, this also applies to the cost of the individual interfaces.

In layer 1, according to the above requirements, a number of variants must be made available from individual application areas.

Because of the realtime requirements, only a deterministic solution is considered for the layer 2 bus access procedure. However, under this constraint, both central and decentralized access procedures (or even a hybrid) may be used.

The demand for universal mutually compatible bus schemes for nearly all applications greatly influences layer 7 and the nature of the services made available to the user. In addition to the request for effective information transmission in the lower layers of the control hierarchy, in which the field bus is used, the demand for a uniform design at all levels also plays a major role.

At the higher levels of the control hierarchy, MMS is predominantly used in layer 7. Thus, a field bus standard should always be oriented towards the concepts described in MMS if it is a universal device.

Figure 6.19

Incorporation of the field bus in the ISO/OSI reference model.

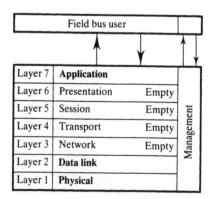

6.7 CNMA Pilot Installation as an Example of the Use of Open Communications

The ESPRIT project 'Communications Network for Manufacturing Applications (CNMA)', involves a consortium of 17 European companies and universities and is aimed towards the specification, implementation and validation of international communications standards for computer integrated manufacturing. The objective of a CNMA pilot installation is to demonstrate the communication of devices from the various companies involved, based on international standards for open systems.

In what follows, we shall describe the CNMA pilot installation, which is installed at the Institute for Control Technology for Machine Tools and Manufacturing Systems (ISW) of the University of Stuttgart. Figure 6.20 shows the computer and network configuration used in the pilot installation.

6.7.1 Description of the pilot installation

The pilot installation consists of two parts. The first part is a five-axis milling machine for single-part production. The production process is integrated with

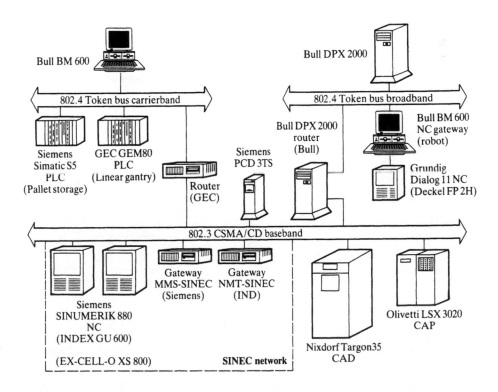

Figure 6.20

Network configuration of the pilot installation. of the ISW.

workpiece design. It demonstrates interfacing of computer-aided design (CAD) and computer-aided process planning (CAP) with an NC machine.

The second part of the installation demonstrates the manufacturing control of a fully-automatic flexible manufacturing cell. The realtime data exchange between computers and control systems is demonstrated here.

Part 1: workpiece design and single-part production

A five-axis milling machine (Deckel FP2H), a model typically used to make molds, is used for the single-part production. It cannot be integrated into an automatic workpiece flow, therefore the machining program must be started by a machine operator. However, the machine is linked to the CAD and CAP system via a MAP

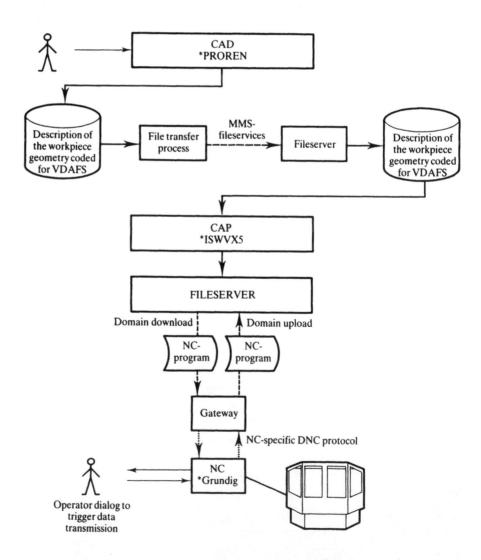

Figure 6.21

Data link between the CAD/CAP line and the NC system.

carrierband network. Figure 6.21 shows the data link between the CAD/CAP line and the NC system.

The workpiece geometry is generated using a CAD system. The data describing the workpiece geometry (coded in VDAFS) is transmitted to the CAP system (ISWVX5) using the MMS file management and the MMS file access services. In the framework of this data link between CAD and CAP, MMS is used to transmit files. The CAP system receives the geometry data and uses it together with other technological data to generate the NC data needed for the production on a five-axis milling machine. This may be requested by the operator of the NC system. The NC data is transmitted using the MMS domain management services. The data link with the NC system requires a conversion of the manufacturer-specific protocol stack into the CNMA protocol stack. The operator is also able to upload altered or newly generated NC data. MMS domain upload services are used for this.

Part 2: manufacturing control

The manufacturing cell operated by the manufacturing control system consists of a turning center, a boring and milling center, a pallet storage and a linear gantry system (Figure 6.22).

The manufacturing control system consists of the production planning, supervisory control and cell control functional units, together with the individual control

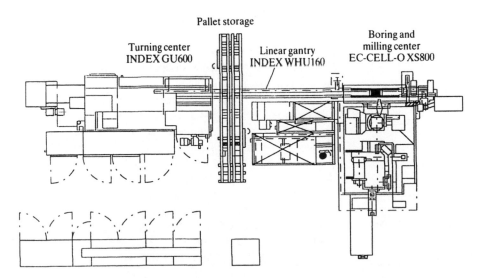

Figure 6.22

Machine layout of the ISW flexible manufacturing cell.

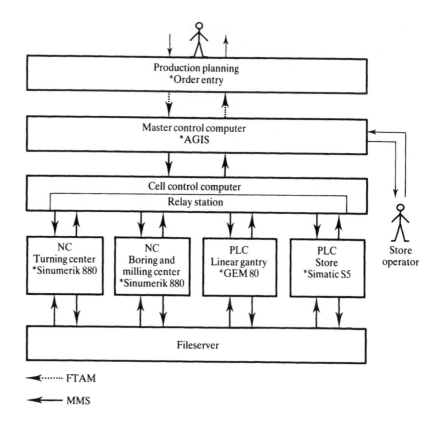

Figure 6.23

Data links between
the functional units
for the ISW
manufacturing control
system.

systems (NC, PLC) for the turning center, the machining center, the linear gantry
system and the storage.

The functional unit for production planning performs the elementary functions
to create, modify and remove manufacturing tasks. This primarily involves
emulation of a complete production planning system. The manufacturing jobs are
passed to the supervisory control system (AGIS) using the FTAM file copy service.
The supervisory control system monitors the individual manufacturing steps of
the manufacturing machines involved.

The cell control system receives manufacturing instructions and the necessary
information via the MMS variable access services. The individual control systems
(NC, PLC) are tasked and supervised by the cell control system via the MMS
services. This involves a protocol stack conversion between the NC system and
the cell control system. Thus, the control systems (NC, PLC) have access to a
fileserver which is responsible for the central data organization.

The cell control system facilitates (as a special relay station function) the data
exchange between the individual control systems (NC, PLC). This is intended to
demonstrate the use of the MMS services for the time-critical information exchange
when the individual control systems are synchronized (Figure 6.23).

Figure 6.24 shows details of the information exchange between the individual

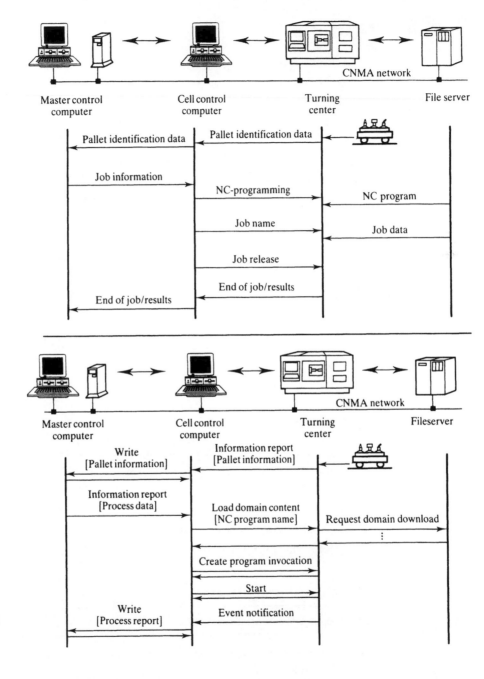

Figure 6.24
Information flow and MMS services of the ISW system.

functional units (except production planning) together with the MMS services used. The figure shows only those MMS services which are crucial to an understanding of the process. The sequence begins with the arrival of a pallet in the turning center and thus the manufacturing process is started.

6.8 Problems

(1) What are the two main parts of the OSI reference model?

(2) Name the advantages of bus topology over those of ring topology.

(3) Using the CSMA/CD or token passing methods, what is the maximum waiting time for a permission to send?

(4) What layers of the OSI reference model may not have to be used in a network?

(5) Are MAP and TOP international standards?

(6) Discuss the principal difference between TCP and PDU.

(7) Which OSI reference model protocol layer has a task similar to the FTP protocol?

(8) For which applications is MMS specified?

(9) To which layer of the OSI reference model does MMS belong? What are the constituent parts of MMS?

(10) Which of the following definitions of the MMS server are correct?
  (a) The MMS server is the MMS user that only processes MMS requests of an MMS client which itself is not permitted to use an MMS request.
  (b) The MMS server is the MMS user that provides objects and administers them.
  (c) The MMS server administers MMS objects and can request services from an MMS client.

(11) Which statements are correct?
  (a) Companion standards to MMS can define object classes and services.
  (b) Companion standards to MMS are only allowed to use object classes and services defined by the MMS core.
  (c) Companion standards of MMS can extend any object class defined by the MMS core.

(12) Why is it not possible to connect a router between two segments of a field bus?

(13) Which delay times have an influence on the realtime behavior of the named communication models?

	Producer–consumer	Client/server
Waiting time for bus access		
Response time of an addressed peripheral		

(14) A controller sends setpoints to an actuator in cyclic mode (for example, an intelligent valve).
  (a) The controller wants to supervise the functions of the actuator in order, if necessary, to pursue an alternative control strategy. Is this possible for the following cases?
    —Producer–consumer model (the controller is the producer)
    —Client/server model (the controller is the client)

(b) The actuator wants to check the functions of the controller in order to initiate an alternative strategy. How can it be done for the following cases?
—Producer–consumer model (the actuator is the consumer)
—Client/server model (the actuator is the server)

(15) Discuss why the CSMA CD bus cannot be used for field bus applications.

7

CAD: Its role in manufacturing

CHAPTER CONTENTS

7.1 Introduction   295

7.2 Historical Perspective   295

7.3 The Design Process   297

7.4 Design Hierarchy   298

7.5 The Role of the Computer in the Design Process   299

7.6 Methods of Constructing Geometric Elements in CAD   301

7.7 Transformation in 2-Dimensions   304

7.8 Transformations in 3-Dimensions   306

7.9 Computer Graphic Aids   307

7.10 CAD Modeling and Database   310

7.11 Solid Representation Schemes   313

7.12 Representation Schemes   318

7.13 Bill of Materials   321

7.14 Interfaces for CAD/CAM   324

7.15 Types of Interfaces   326

7.16 Description of Various Interfaces   329

7.17 Requirements of a Product Model   342

7.18 Design Features   343

7.19 Feature Classification by Application   349

7.20 Product Level Classification   356

7.21 Concurrent Engineering   356

7.22 The Product Modeler   361

7.23 Quality Methods in Design   366

7.24 Life Cycle Costs in Design   367

7.25 Conclusions   369

7.26 Problems   369

7.1 Introduction

The purpose of this chapter is to show how CAD data is used for the planning and control of the manufacturing process. We will not be discussing the CAD process as such; it is very complex and the reader is advised to obtain fundamental knowledge of CAD to understand the interaction between CAD and CAM.

The main goal of this chapter is to describe the methods used to produce manufacturing documents, drawings, bills of material, process planning and so on. We will also show the procedures necessary to automate these activities and how they relate to each other in order to achieve more coordinated, flexible and automatic manufacturing systems.

7.2 Historical Perspective

Computer-aided design (CAD) is a relatively new technology and only became a prevailing engineering tool in the 1980s. The use of this technology has transformed the normal working practices of designers in industry.

The origin of CAD can be traced back to a series of independent projects (1956–1959) which started in the 1950s with the APT project at the Massachusetts Institute of Technology (MIT). APT stands for automatically programmed tools and was intended to be used for representing the geometric shape of workpieces to numerically controlled machines performing high-precision operations. But APT was not interactive in nature. Another related project was the development of the light-pen which came out of a radar project called SAGE (semi-automated ground environment system). The objective of the project was to develop a system which can be used to analyze radar data and present possible aircraft positions on the CRT screen, see also Groover and Zimmers (1984).

The systems culminated in the development of the Sketchpad at MIT in 1962–1963, which was the first interactive computer graphics (ICG) system. Prior to the development of the Sketchpad, computers had been used in performing engineering calculations, but the advent of this system meant that a designer could now interact with the computer graphically using the CRT and the light-pen.

This first Sketchpad was limited to two-dimensional representation of objects. But now drawings could be analyzed using programs to ensure structural validity. In 1963, T. E. Johnson, (see Johnson (1963)), extended the Sketchpad's capability to three dimensions. With this system, it became possible to do perspective drawings of objects on the CRT screen.

A special aspect of the computer is that it can make basic product data readily available for product design, various operational activities, process planning, and the generation of the programs to drive the manufacturing equipment. Thus the data only has to be entered once into the system and is accessed by various activities, as shown in Figure 7.1. Technological information such as material,

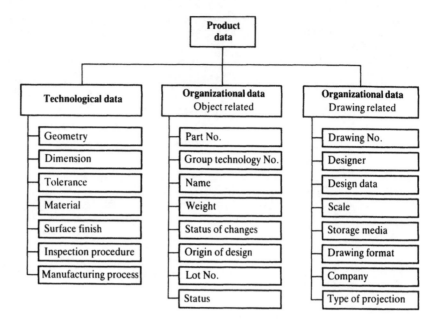

Figure 7.1
CAD database containing product and manufacturing planning data.

drawing related organizational data including drawing number and drawing format, and all the object-related organizational data are contained in the manufacturing master production file.

An interactive computer graphic system must be user-oriented, whereby the computer is used to create, transform, and display data in the form of pictures and symbols. The computer communicates to the user via the CRT screen, where images are created by entering commands to call the desired software subroutines stored in the computer. Geometric elements such as points and lines can be enlarged or reduced in size, and moved from one location to another on the CRT screen by the transformation process of rotation and translation.

The interactive computer graphic system normally includes hardware and software. The hardware can be composed of the CPU, one or more graphics display terminals, and peripheral devices such as printers, plotters, and drafting equipment. Software can include those computer programs needed for a variety of graphics processing on the system.

The advantage of the interactive computer graphics system is the synergistic effect it has on the design process. The designer is able to perform the aspect of the design most suitable for human intellectual skills including conceptualization and reasoning. The computer then performs that aspect of design most suited to its capabilities as described in Groover and Zimmers (1984). These capabilities include speed of calculations, visual display, and storage of large amounts of data.

The evolution of what is now commonly known as computer-aided design/ computer-aided manufacturing (CAD/CAM) has been strongly influenced by the APT and Sketchpad projects. The marriage of both of these systems enabled the generation of numerical control programs from the geometric model of the design which is now in the computer, whereby design and manufacturing activities can use the same CAD database, as shown in Figure 7.1.

The early CAD systems were very expensive and therefore the production and use of the machines were limited to the large industries. In the early 1960s mainly the automotive and aircraft companies experimented with several notable computer graphics systems. The late 1960s saw the emergence of many CAD system manufacturers. The 1980s saw a heavy growth in this industry. In advanced industries, most products today are designed on CAD systems and the product specification information is expected to be part of the CAD data. Many manufacturers also have been able to connect CAD with CAM for special products. Such design systems are also commercially available.

7.3 The Design Process

A conventional design process is shown in Figure 7.2. The first phase starts with a recognition that a customer has a need for a product which initiates any design activity. This recognition can take the form of a discovery of a dysfunctional system

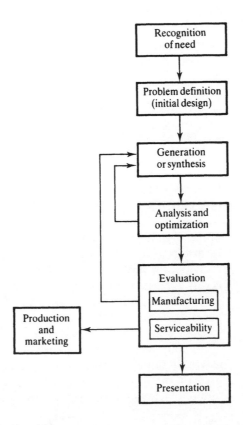

Figure 7.2
Traditional generalized design process.

which must be redesigned or a realization that the market needs a certain new product.

The second phase is the problem definition. This is the specification of the entity which will be designed. It includes the determination of the physical and functional characteristics of the product, its operating principle and service life. It can also involve the gathering of data about costs, legal requirements and standards, manufacturing, or quality and serviceability requirements.

The third phase is the conception phase of a new product generation or synthesis of alternative designs. This is a very creative activity. It is often believed that this phase of the design process is the most crucial and is where the designer's creativity is employed. The design synthesis is normally tied to the fourth phase, the analysis, since the representation of a concept would normally be subjected to analysis, which can result in improved design based on the analysis constraints. This process can be repeated several times until the design is optimized.

The fifth phase is the evaluation. It may try to use defined specifications which can be standard engineering and manufacturing practices whereby the design is evaluated in order to assure non-violation of the established constraints at the problem definition stage. The evaluation systems must include manufacturability and serviceability evaluations. In the past, evaluation could require the fabrication and testing of a prototype model to determine its performance, quality, life and so on. But with advanced computer technology, it is now often possible to model and test prototypes in the computer. This saves time and cost.

Finally, the design is presented as engineering documents, and in a modern manufacturing environment, the design data can also be directly transferred to a process planning system for product manufacture.

7.4 Design Hierarchy

Complex designs, such as aircraft, are built by many teams. Each team may yet have many other smaller teams. For instance, an aircraft is made up of thousands of components, one of which is the jet engine. The engine is normally designed by a team of engineers who would have the responsibility for individual segments of the engine. The designs produced by various teams would have to be assembled in order to test the final product. This test does not preclude the tests which individual teams must have performed on their specific components in order to certify them as completed.

It is clear that because of the component generate-and-test as well as the system generate-and-test nature of design, there are two discernable approaches to design. These are the bottom-up approach and the top-down approach. In the bottom-up approach, the designer proceeds from detailed design of the parts to the system assembly of the product. In the top-down approach, the designer makes a global decision about the product before considering the detailed parts. For the design of manufacturing processes, similar approaches are taken (see also Chapter 3).

The design process, however, needs both approaches to be effective. High-level decisions made about a product design depend on the various characteristics of the parts. Such characteristics include: the use and cost of the parts or violation of a physical law, and so on. In a similar manner, lower level decisions are concerned with the constructability, reliability, or function of the product. For instance, every subcomponent or part may not indicate its interactive behavior with other subcomponents or parts until it is considered as part of the entire system.

If, however, the problem specification phase provides adequate constraints and precise rules of manufacture (assembly), then the bottom-up approach would be a preferred alternative. This is because this approach would generate more reliable products if the strict constraints are obeyed.

7.5 The Role of the Computer in the Design Process

When the computer is used for design, many engineering activities can be automated; however, the design phases are the same as with the conventional method, see Figure 7.2. The application of the computer to design can be divided into five areas:

- problem definition
- geometric modeling
- engineering analysis
- design evaluation
- automated drafting.

Figure 7.3 shows the areas of computer application in the design process.

7.5.1 Problem definition

In the early phase of this activity, the designer must be very creative to determine the functions, performance and appearance of the product. Here the computer may not be of great help since it lacks human experience. If, however, the product had been designed before, the computer can be an invaluable tool to suggest an existing design and to search for standard components and manufacturing processes. So this phase can be partially manual and partially automatic.

7.5.2 Geometric modeling

Geometric modeling involves the use of the computer to employ a computable mathematical description of an object's geometry in the representation of the object. Usually, the object is simplified and only its essential features are represented.

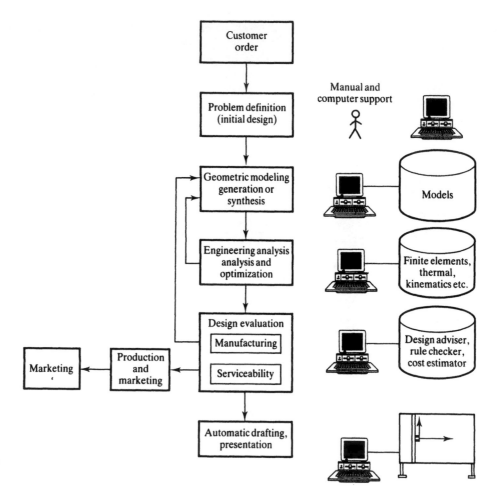

Figure 7.3
Computer applications
in the design process.

These mathematical descriptions enable the object's image to be displayed on a graphics terminal and the object may be animated to show its operating characteristics. With the help of the animation it is possible to detect problems and to suggest corrective action.

7.5.3 Engineering analysis

Designs usually require some form of analysis. This may involve finite element modeling and calculations to determine the dynamic performance of the design. Some programs can be employed to simulate the performance of the design and to collect information on such issues as power consumption, heat transfer, wear, interference and so on. Optimization is also possible if some objective measure can be represented subject to a set of performance constraints. In the traditional design approach, these performance programs are employed at a later stage of design rather than at the conceptual design stage because they normally require precise data concerning dimensions, shape, materials and so on. One important and cumbersome problem associated with the use of these programs is that CAD data cannot be used directly in their operation. For instance, geometric data still needs to be interpreted by a human in order to encode shape data. This is a problem which may be alleviated by a future design system such as ProMod, see Nnaji (1990), which captures the intentions of the design for the life cycle of the model.

7.5.4 Design evaluation

Evaluation of designs is to ensure that specific rules established for the design of a certain type of product are not violated. Some of these rules are standard operating procedures; others are costs, service rules, and so on. This phase is where the accuracy of a design is checked, manufacturability and assembly are evaluated, and the kinematics to depict spatial behavior is investigated.

7.5.5 Automated drafting

Automated drafting is concerned with the production of the detailed working drawings used to communicate design information to processing, process planning, programming of manufacturing equipment, and so on. Early application of interactive computer graphics was intended to facilitate drafting only, since drafting productivity can increase many times over with the use of CAD. With CAD, the drafting functions of automatic dimensioning is possible, along with generation of cross-hatched areas, scaling and development of sectional views. Views can be enlarged and objects can be rotated or translated to obtain oblique, isometric, or perspective views of the part.

7.6 Methods of Constructing Geometric Elements in CAD

In most engineering drawings, the geometric elements are constructed from basic primitive geometric entities. These entities can be created as shown in Figures 7.4, 7.5 and 7.6. They are entered into the computer interactively via graphic symbols or textual programming languages. The same type of primitives are also used to produce the parts program for describing the contour of a part to a machine tool (Figure 5.36). In many simpler applications, the interconnection of CAD and CAM is done by directly using the graphic description of the workpiece to produce the NC program. Other entities not shown in Figure 7.4, 7.5 or 7.6 are conics, curves and surfaces.

CAD: its role in manufacturing

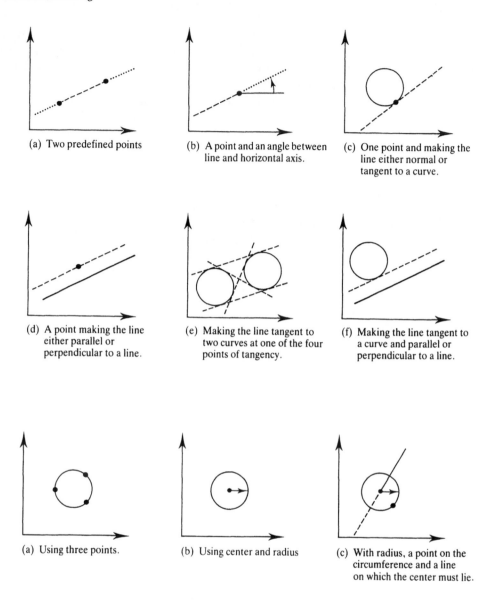

Figure 7.4
Straight line
construction.

(a) Two predefined points

(b) A point and an angle between line and horizontal axis.

(c) One point and making the line either normal or tangent to a curve.

(d) A point making the line either parallel or perpendicular to a line.

(e) Making the line tangent to two curves at one of the four points of tangency.

(f) Making the line tangent to a curve and parallel or perpendicular to a line.

(a) Using three points.

(b) Using center and radius

(c) With radius, a point on the circumference and a line on which the center must lie.

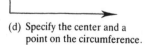

Figure 7.5
Circles construction.

(d) Specify the center and a point on the circumference.

(e) Making the circle tangent to three lines.

(f) Specifying the radius and the curve tangent to two lines or curves.

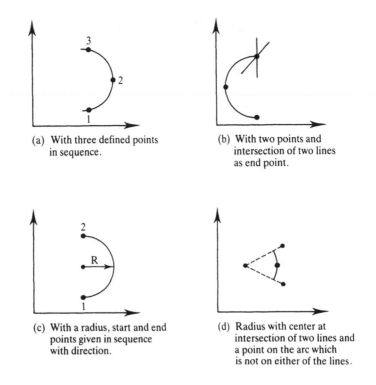

(a) With three defined points in sequence.

(b) With two points and intersection of two lines as end point.

(c) With a radius, start and end points given in sequence with direction.

(d) Radius with center at intersection of two lines and a point on the arc which is not on either of the lines.

Figure 7.6
Circular arcs.

Conics

Conics such as parabolas, hyperbolas, and ellipses can be constructed by specifying five points on the element or by two orthogonal lines and the point of intersection of the lines.

Curves and surfaces

Curves and surfaces can be described by mathematical expressions which fit through given coordinate points. There are many curve generating functions including Bezier curves and B-spline approaches. Both the Bezier and the B-spline approaches employ a blending process in order to smooth the data points.

The surfaces generated through the mathematical expressions can be multi-curved or sculpted surfaces such as ship hulls or car bodies. The surfaces are typically generated by using rotating lines and/or curves around a specific axis to produce a surface of revolution. Another approach is by using the intersecting line or surface of two intersecting surfaces. Typically, surfaces generated using blending functions are defined in certain regions of space. Each region of definition is a surface patch. An artifact will typically require several of these patches in order to be modeled. Therefore, what is obtained is usually an assembly of patches. In general, the lower the polynomial degree of the patches used, the more patches are required.

7.7 Transformation in 2-Dimensions

When modeling or programming a manufacturing or assembly process it is necessary to visualize geometric elements on a screen. After a geometric element or shape is created, it may have to be linked with another element. For that reason, the manufacturing engineer must be capable of moving or rotating the image of the part on the screen.

Since the subject of transformation plays a major role in robotics, a detailed description of how rotation and translation matrices are derived can be found in Chapter 9. The transformation operations in the CAD system can be supported by the following functions.

7.7.1 Translation

This is an operation which results in a movement of an object in the plane from one location to another, which preserves the shape of the entity and does not change the orientation. For example, let T_x and T_y be the distance which the entity is moved in the direction of the x and y axes respectively. Then the translation to new location:

(X_2, Y_2)

from current location

(X_1, Y_1)

can be stated as follows:

$$X_2 = X_1 + T_x$$
$$Y_2 = Y_1 + T_y$$

7.7.2 Rotation

Rotation results in a movement of an object in the plane, from one position and orientation to another position and orientation, which preserves the object shape. Let θ be the angle of rotation in the plane about the origin, then this can be defined as:

$$X_2 = X_1 \cos \theta - Y_1 \sin \theta$$
$$Y_2 = X_1 \sin \theta + Y_1 \cos \theta$$

Rotation can also be expressed in matrix form as follows:

$$\begin{pmatrix} X_2 \\ Y_2 \end{pmatrix} = R \cdot \begin{pmatrix} X_1 \\ Y_1 \end{pmatrix}$$

where $R = \begin{pmatrix} \cos\theta & -\sin\theta \\ \sin\theta & \cos\theta \end{pmatrix}$

Line segments can be represented in matrix form allowing for rotation or translation of the line as follows:

$$L = \begin{pmatrix} X_1 & X_2 \\ Y_1 & Y_2 \end{pmatrix}$$

A rotation of a line segment is defined as:

$$\begin{pmatrix} X_1^1 & X_1^1 \\ Y_1^1 & Y_2^1 \end{pmatrix} = \begin{pmatrix} \cos\theta & -\sin\theta \\ \sin\theta & \cos\theta \end{pmatrix} \begin{pmatrix} X_1 & X_2 \\ Y_1 & Y_2 \end{pmatrix}$$

Reflection

Reflection is an operation which results in a movement in the plane from one position and orientation to another position and orientation, which produces a congruent mirror image of the original shape. There is usually a straight line left after the transformation which may or may not be within the shape. This operation is often called a mirror operation. This line is called the **mirror line**. The following are various possible reflections:

(1) A reflection about the x-axis is obtained by reversing the sign of all the y-values without changing the sign of the x-values:

$$X_2 = X_1$$
$$Y_2 = -Y_1$$

(2) Similarly, a reflection about the y-axis is as follows:

$$X_2 = -X_1$$
$$Y_2 = Y_1$$

(3) Double reflection is obtained as follows:

$$X_2 = -X_1$$
$$Y_2 = -Y_1$$

This double reflection is equivalent to an inversion.

Inversion

The inversion operation results in a movement of an object in the plane, from one position and orientation to another position and orientation, by preserving its shape. Normally, the inversion operation leaves only one point in its original position. This point is the center of the inversion and is located at the origin.

Shearing

This operation moves the object in the plane and produces a new shape of the same area as the initial shape. Usually, this operation leaves only the points on one straight line in a fixed position. This line, called the axis of shearing, may not necessarily be within the shape and normally is located along the x- or y-axes.

Shearing can be effectively achieved by this expression:

$$X_2 = X_1$$

$$Y_2 = \propto X_1 + Y_1$$

This results in a shift of each point, in a direction parallel to the y-axis, at a distance which is proportional to the x-value of the point. The x-values remain unchanged.

Scaling

The scaling operation is a transformation in the plane of an object which results in an object of geometrically similar shape to the original shape and preserves the orientation of that original shape. Scaling in 2-dimensions can be represented as follows:

$$X_2 = X_1 S_x$$

$$Y_2 = Y_1 S_y$$

where S_x and S_y are the scaling factors associated with the x- and y-axes, respectively. The effect of scaling is that the object can become larger or smaller along the axis of scaling.

7.8 Transformations in 3-Dimensions

The transformation operations in 3-D are, in principle, the same as those of 2-D. Therefore, we will only specify the matrices of the analogous 3-D operations.

(1) Translation

The translation distances in the direction of the x-, y-, and z-axes are T_x, T_y, T_z:

$$T = (T_x, T_y, T_z)$$

(2) Scaling

The scaling factors associated with the x-, y-, and z-axes are S_x, S_y, and S_z:

$$S = \begin{pmatrix} S_x & 0 & 0 \\ 0 & S_y & 0 \\ 0 & 0 & S_z \end{pmatrix}$$

(3) Rotation

The following are the rotation matrices in 3-dimensions.

$$R_x = \begin{pmatrix} 1 & 0 & 0 \\ 0 & \cos\theta & -\sin\theta \\ 0 & \sin\theta & \cos\theta \end{pmatrix}$$

$$R_y = \begin{pmatrix} \cos\theta & 0 & \sin\theta \\ 0 & 1 & 0 \\ -\sin\theta & 0 & \cos\theta \end{pmatrix}$$

$$R_z = \begin{pmatrix} \cos\theta & -\sin\theta & 0 \\ \sin\theta & \cos\theta & 0 \\ 0 & 0 & 1 \end{pmatrix}$$

Concatenation

This is the combination of a sequence of transformations to yield a desired transformation. For instance, a model could be rotated about a fixed axis then translated from the original position to another.

7.9 Computer Graphic Aids

7.9.1 Software

The software in computer graphics is intended to enable the user to interact with graphics terminals in the process of creating or editing an image on the CRT. The software available on a given CAD system is often a function of the type of hardware used. For instance, different software may be written for vector displays as opposed to raster scan CRT. In general, image generation using graphics software is one of two types: vector graphics or raster graphics.

7.9.2 Vector graphics

Vector graphics, also known as the stroke writing technique, allows for an image to be created by constructing lines, usually straight line segments, which are defined by the coordinates of their end points. A sequence of these lines can be used to create a variety of shapes including polygons or polyhedra represented in wireframe. In this type of image generation, curved lines can be approximated using a number of straight lines as shown in Figure 7.7. The degree to which the straight lines approximate the desired curved shape is dependent on the resolution of the graphics display system, or more succinctly, the computational power of the hardware.

In a vector display, an electron beam which operates like a pen is used to draw the line segments or vectors by tracing the path of the required line segment across the screen. Each segment is defined by its end points in x- and y-axes.

7.9.3 Raster graphics

In raster graphics, the screen is divided into a large number of discrete phosphor elements called pixels, as shown in Figure 7.8. The matrix of pixels is the raster. There could be 256×256 or 1024×1024 or more depending on the power of the raster display. Each pixel element can be made to glow with a certain brightness when scanned. Here, the electron beam scans the raster from left to right and from top to bottom. When the beam reaches the right end of the screen, it is temporarily turned off as it starts on the next line. The screen is swept at a frequency of 30 to 60 cycles per second to give the appearance of a persistent image.

7.9.4 Graphics display terminals

There are three major types of display terminals: directed beam refresh, direct view storage tube (DVST), and raster scan. Information is usually written in vector graphics or in raster graphics. The main differences among these three types of

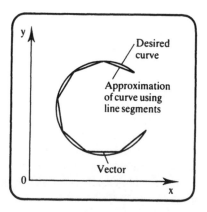

Figure 7.7
Vector graphics.

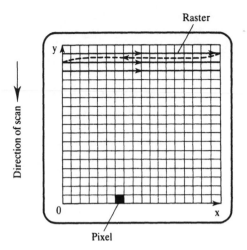

Figure 7.8
Raster graphics.

Table 7.1   Comparison of graphics display terminals. *Source*: Groover and Zimmers, 1984

	Directed-beam refresh	DVST	Raster scan
Image generation	Vector graphics	Vector graphics	Raster scan
Picture quality	Excellent	Excellent	Moderate to good
Data content	Limited	High	High
Selective erase	Yes	No	Yes
Color capability	Moderate	No	Yes
Animation capability	Yes	No	Moderate
Grayscale	Yes	No	Yes

CAD display terminals can be found in the type of screen phosphor coating, availability of color, the pixel density, and the computational memory necessary to generate and refresh the picture on the CRT (Table 7.1 contains a comparison of graphics display terminals).

Directed beam refresh

This display device uses the vector graphics, or stroke writing, approach to generate its images. The phosphor elements can only maintain their brightness for seconds when activated. This means that the screen must be 'refreshed' many times per second to ensure that there is no noticeable flickering on the screen. Sometimes, flickering can occur when sections of the screen contain many line representations of an image. This results in a higher population density of that segment of the screen. Selective erasing of segments of an image is possible, however, because of the continuous refreshing nature of the screen.

Direct view storage tube (DVST)

Unlike the directed beam refresh, DVST maintains an image which has been created on the screen without further regeneration. It uses an electronic flood gun which is directed at the phosphor elements. These phosphor elements continue to be illuminated once they are energized. Because the entire screen is energized uniformly, it is not possible to erase selectively portions of an image on the screen. In addition, this type of display device does not allow for color or animation.

Raster scan

This type of display device is becoming increasing popular with the decreasing cost of computers. The raster scan display normally requires large memory. The screen is scanned by using an electron beam to trace from left to right and from top to bottom. The scanning is done continuously thereby refreshing the brightness of the screen.

The screen is divided into a number of discrete elements called pixels. There could be 256×256 or 1024×1024 pixel elements or even more. Each bit memory or frame buffer normally has an on/off status for the given pixel element. The density of the pixels on the screen determines the resolution of the display and thus the quality of the pictures. It is possible to add color by adding gray scale. This is accomplished by varying the intensity levels which are displayed. Additional bits would normally be required for each pixel element to achieve the gray scale. For instance, two bits would be required to obtain four levels while three bits would yield eight levels. Up to twenty-four bits are normally required to obtain the three colors of red, blue, and green.

7.10 CAD Modeling and Database

The database of the CAD system contains basic graphics elements such as points, lines, and curves, and elements that define the geometry and topology of the object, giving it its shape. The topology shows a network of how the geometric elements are interconnected. The geometry shows those items which finally help to complete the description of the shape such as angularity, or dimensions. CAD systems are also expected to carry organizational and application related information such as material properties and analysis specific programs such as finite-element analysis programs, as shown in Figure 7.1.

7.10.1 Organization of database

The database is used for storing coordinates of the geometry, as well as other information needed to describe the model or to use the stored application programs. For instance, a cylinder might be created by rotating a line segment parallel to a given axis about that given axis. The cylindrical data may be stored as follows: points to describe the axis of rotation, plus points on the line segment, plus data record. In this approach, the implicit specification of the cylinder can generate a picture of the cylinder on the screen but the database lacks complete data regarding the cylinder.

Another approach is to represent graphically the information in the database. Any approach would however show the interconnection relationships between vertices, edges, and faces of an object.

Modeling

There are two basic modeling techniques – the wireframe and solid modeling approaches. The wireframe is the oldest and the easiest approach while the solid model is intended to produce the realism of the object. It is important to observe that the model selected in the CAD system may also have to be used for planning and controlling manufacturing processes, for example, a wireframe model cannot be used for planning assembly operations because no surfaces are defined (see Section 2.3.3).

Wireframe

The wireframe representation of a three dimensional entity consists of a finite set of points and their interconnecting edges (lines or curves). The result is a figure which depicts a visual representation of the real object. There are various ways of representing the wireframe data of an object. One popular way is as shown in Figure 7.9.

The wireframe modeling technique is the simplest geometric representation of the object. It offers easy computation for simple calculations about geometry. However, there are some serious disadvantages with this technique. These include the ambiguity of the models and the severe difficulty in validating the models. An example of an invalid model of the wireframe is shown in Figure 7.10. It can be seen that the shape shown cannot be produced since the geometry is ambiguous. Figure 7.11 shows the ambiguous top view of a model using wireframe geometry.

vertex list	edge list	edge type
.	.	.
.	.	.
.	.	.

Figure 7.9
Wireframe data structure.

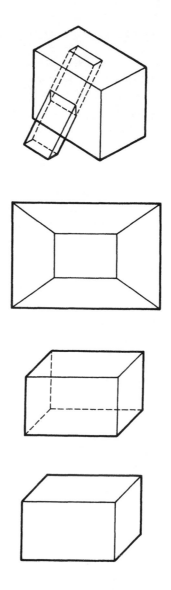

Figure 7.10
An invalid wireframe image.

Figure 7.11
The top view of a wireframe model leads to an ambiguous interpretation.

Figure 7.12
Wireframe with dotted line.

Figure 7.13
Wireframe with hidden line.

The polygonal feature could be a protrusion or a depression, but it cannot be validated in the way the wireframe is represented.

In general, objects can be represented as 2-D, $2\frac{1}{2}$, or 3-D images of the object. Wireframe models are very popular because they are easy to construct, their representation amounts to use of points, lines, and curves to represent objects. However, for complicated parts, wireframe models can be very confusing. A number of approaches can be used to enhance the visual outcome of wireframe models. One approach is to use dashed lines (Figure 7.12) and the second is to use the hidden line removal method (Figure 7.13). Yet another method to enhance the wireframe is to provide a surface representation which would enable the object to appear as solid to the viewer but maintain a wireframe object representation in the database. Because of the complex nature of the subject of geometric modeling and its relationship to the generalized concept of design and manufacturing integration, we present a more detailed description of the subject in the next section.

7.11 Solid Representation Schemes

A solid object is an object which has a finite volume in space, and which has both geometric properties like surface, volume, the center of shape and so on, and physical properties like mass, center of gravity and inertia. A solid modeling system can usually represent both geometric and physical properties of a solid object. Thus, a solid modeling system in a CAD system should not only have a representation scheme, but also have a set of processors to determine all the properties of a solid object. Additionally, with a solid model, various information for engineering applications may be obtained from a CAD system, such as NC program and finite element analysis (FEA), by attaching proper processors to the solid modeling system. A comparison of popular solid modeling techniques is presented in Table 7.2.

7.11.1 The desired CAD information and its extraction

There are a number of representation schemes used in CAD systems to represent a solid object, such as pure primitive instancing scheme, cell decomposition, constructive solid geometry (CSG) scheme, sweep representation, and boundary representation (B-rep), see also Requicha (1980).

Table 7.2 Comparison of solid modeling schemes

From \ To	Exact				Approximate	
	Simple sweep	Cell decomposition	CSG	B-rep	B-rep	Spatial enum
Simple sweep		K	K	K	K	K
Cell decomposition	I		R1	K	K	K
CSG	I	E		K	K	K
B-rep	I	E	R2		K	K
B-rep	I	I	I	I		K
Spatial enum	I	I	I	I	K	

K: Known.

E: Experimental.

I: Impossible, except perhaps in restricted domains.

R1: Cell decompositions are a restricted form of CSG ('gluing'). Algorithms to produce general CSG are not known.

R2: Known in 2-D; proposals (no known code) in 3-D.

Each representation scheme has its advantages and disadvantages. However, CSG and B-rep are two of the most popular and understandable schemes used in solid object representation.

B-rep is important because it is close to computer graphics, unambiguous, and available to computing algorithms. Disadvantages include: verbosity, complicated validity checking, and difficulty in creating objects.

CSG is preferred because it has features such as simple validity checking, unambiguity, conciseness, ease of object creation, and high descriptive power. In addition to these benefits, at least 95% of manufacturing parts do not require multi-curved surfaces found on ships' hulls, car bodies, and aircrafts and hence can have extensive capabilities for developing engineering drawings. Its disadvantages include inefficiency in producing line drawings, and inefficiency in graphics interaction operations.

We will discuss the most prominent modeling techniques available starting in the next section.

7.11.2 Pure primitive instancing scheme

The concept of group technology uses a pure primitive instancing scheme, in which solid objects are divided into many different families. Each object family is called a generic primitive, and the member within a family is a primitive instance. A generic primitive is represented by its name and finite variables. For example, **block(L, W, H)** is a generic primitive called block and the variables represent the length, width, and height respectively. A primitive instance is represented by the generic primitive name which it belongs to and values of each variable. **block(20, 10, 5)** is a primitive instance belonging to the block family with length, width, and height equal to 20, 10 and 5.

For a finite domain, such as machining tools and electronic parts, pure primitive instance is a good representation scheme. But for complex objects, the representation scheme can be inadequate. This is because the complex objects cannot be easily described using the primitives as described. A better use of primitives is described in the constructive solid geometry section.

7.11.3 Cell decomposition

An object can be represented in cell decomposition by using some cells with an arbitrary number of sides. Cells must be either disjoint or meet precisely at a common face, edge, or vertex. When the scheme is such that all of the cells are cubic and lie in a fixed grid, it is called spatial occupancy enumeration. This representation is always used in solid finite element methods for the numerical solution of differential equations, see also Requicha (1980).

7.11.4 Constructive solid geometry (CSG)

The CSG representation is an ordered binary tree as described in Requicha (1980), shown in Figure 7.14. The nonterminal nodes of the binary tree represent operators, which may be either rigid motions or regularized unions, intersections, or differences. The terminal nodes of the binary tree are either primitive leaves which represent a solid in modeling space, or transformation leaves, which contain the defining arguments of rigid motion.

Basically, CSG is a partially geometrical explicit representation scheme with topological information implied. For instance, in Figure 7.14, the solid object can be represented in CSG as:

$$\text{block}(10, 10, 10) - \text{block}(5, 20, 5)$$

This CSG representation can be modified such that the solid object can be represented as a Boolean expression as described in Nnaji (1990) and as follows:

$$(H_1 \cap H_2 \cap H_3 \cap H_4 \cap H_5 \cap H_6) \cap (H_7 \cup H_8 \cup H_9 \cup H_{10} \cup H_{11} \cup H_{12})$$

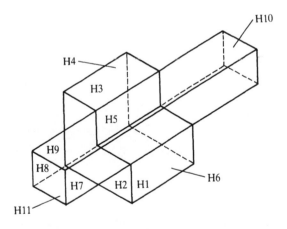

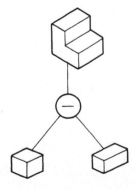

Figure 7.14
CSG representation.

Any point in the model space for which this Boolean expression is true belongs in the interior of the solid object. In the equation, all the half spaces which indicate the location of a plane in a specified coordinate system are geometric data, and all the topological relations are implied in the Boolean operator.

The advantages of CSG are as follows, see also Requicha and Voelker (1982):

(1) it is concise,

(2) it guarantees, automatically, that objects are valid,

(3) the reliable algorithms for converting CSG into boundary representation are known.

7.11.5 Boundary representation (B-Rep)

A boundary representation scheme represents a solid object by segmenting its boundary into a finite number of bounded subsets usually called faces or patches, and representing each face by its boundary edges (loop) and vertices as described in Requicha (1980). Figure 7.15 provides an example of boundary representation. The boundary representation may also be divided into exact and approximate B-rep. The exact B-rep is composed of a group of well-defined, mathematically,

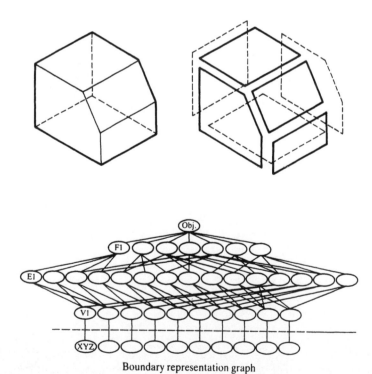

Figure 7.15
Boundary
representation.

Boundary representation graph

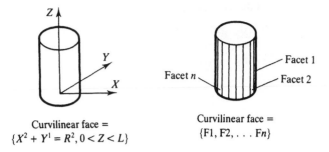

Figure 7.16

Exact B-rep and
approximate B-rep.

Curvilinear face =
$\{X^2 + Y^1 = R^2, 0 < Z < L\}$

Curvilinear face =
$\{F1, F2, \ldots Fn\}$

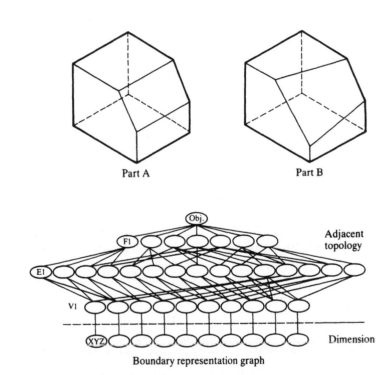

Part A

Part B

Figure 7.17

Boundary
representation
showing two different
objects with same
topology.

Boundary representation graph

curvilinear faces and plane faces. In contrast, the approximate B-rep always uses plane faces to approximate a curvilinear face, see also Requicha and Voelker (1983). An example of this is shown in Figure 7.16.

Basically, boundary representation is a topologically explicit representation. The boundary representation tree is a topological relation tree. For part A and part B shown in Figure 7.17, the topological relations among vertices, edges and faces are completely identical. The only difference between these two objects is the coordinates of the points in their boundary representation tree.

The advantages of B-rep are as follows:

(1) the information is complete and it has a long history in the CAD system community and, thus, there is plenty of software available for manipulation of the B-rep geometric data;

(2) the applications and algorithms based on boundary representation are competitive with those based on CSG;

(3) the available sculpted surface technology is surface-oriented, and hence may be easier to incorporate in boundary-based systems than in CSG-based systems.

7.12 Representation Schemes

Although each different scheme has its own characteristics and advantages, the scheme ultimately employed in a CAD system is quite application-dependent. Table 7.3 shows the capabilities and properties of each scheme for different application purposes, see also Requicha (1980). The other important factor that influences the selection of representation scheme is the conversion capability. Table 7.2 shows this ability of each scheme to be transformed into another.

In machine reasoning systems, different information for a solid object is needed to support various internal reasoning engines. For example, boundary information such as edge, loop, face, and area of face, are required for an object interpreter to determine the object envelopes, features, and the coordinate systems of objects as well as those of the features. Solid information derived from boundary representation is also used in form and function reasoning systems, as described in Nnaji (1988), in order to perform surface mapping and static interference analysis. The representation scheme used in an automatic machine reasoning system must possess two characteristics, the explicit adjacent topology and uniqueness.

7.12.1 Adjacent topology for reasoning about objects

In a task-level reasoning system such as RALPH, see Nnaji (1988), the procedure to plan a task can be divided into three steps. The first step involves understanding the world objects as well as the features of each of the objects. This recognition is performed by an object interpreter. The second step involves reasoning about the task and operations. The third step is motion planning and generation.

The adjacent topology is an important factor in the interpretation of objects since there are two types of information needed in the determination of the form features of an object: topology and geometry. In order for two objects to have identical shape, their topology must also be identical. For example, if there are six faces in one object and seven faces in the other one, then it is impossible that the shape of the two objects can be identical. If the two objects both have six faces

Table 7.3 Some three-dimensional modeling systems

Name	AD2000	BDS/GLIDE	BUILD-1	COMPAC	EUCLID(F)	EUCLID(S)	GEOMED	GEOMAP	'HOSAKA'	PADL	TIPS-1
Developer	Hanratty	Eastman et al.	Braid	Spur et al.	Brun et al.	Engeli	Baumgart	Hosaka and Kimura	Hosaka and Kimura	Voelcker et al.	Okino et al.
Institution	Mfg. & Consulting Services Inc.	Carnegie-Mellon University	Cambridge University	Technical University of Berlin	LIMSI		Stanford University	University of Tokyo	University of Tokyo	University of Rochester	Hokkaido University
Country Host/ implementation language	USA FORTRAN	USA BLISS	UK FORTRAN SAL	Germany FORTRAN	France FORTRAN FOCAL	Switzerland SYMBAL	USA ASSEMBLY	Japan FORTRAN	Japan GIL	USA FORTRAN	Japan FORTRAN
Machine	Many different machines	DEC PDP-10	TITAN PDP-7	DEC PDP-10	UNIVAC-1110	CDC6500	DEC PDP-10	TOSBAC 5600 (GE635)	TOSBAC 5600 (GE635)	PDP-11/45	FACOM machines
Interactive/ batch	Interactive	Interactive	Interactive	Batch	Either	Batch	Interactive	Batch	Batch	Interactive	Batch
Purpose/ area	Commercial: engineering and NC	Research: database for architecture and engineering	Research: engineering and NC	Commercial: engineering and NC	Commercial: engineering and architecture	Commercial: engineering and NC	Research: visual shape recognition	Research: engineering and NC	Research: engineering and NC	Commercial: engineering and NC	Commercial: engineering and NC
Shaded graphics	—	—	Yes	Yes	—	—	—	—	Yes	—	—
Curved surfaces	Splines, torus and others	Approximate cylinders and cones	Cylinder	Cylinder and others	Circle and curve segments	(Version: Bezier patches)	—	Approximate cylinders, cones	—	Cylinders	Cylinder sphere bicubic path
Shape operators	Intersection (of faces)	Union intersection difference	Union intersection difference addition	Union difference	—	Union intersection difference addition	Union intersection difference	Union intersection difference	Union intersection difference	Union intersection difference	Union difference
Solid definition method	Generating lines and surfaces	Euler operations	Primitive solids	Generating lines, primitives	Face list	Face list primitives solids	Euler operations	Primitives	Intersecting planes	Primitive solids	Primitive solids

Source: This table first appeared in Baer, A., Eastman, C. and Henrion, M. Geometric modelling: a survey. *Computer-Aided Design*, **11** Number 5 (September 1979).

but the number of adjacent faces to each face in one object is not exactly the same as the number in the other object, then the two objects do not possess identical shape either. Figure 7.18 gives an example of this case. The information about the number of faces, edges and vertices and the number of adjacent faces of each face can be regarded as boundary topology information.

If the topologies of two objects are same, and their dimensions are also the same, then they are identical. The dimension information should include the area of the faces, the length of each edge, the angle between two adjacent faces, and the coordinates of each vertex. All objects in Figure 7.19 have the same boundary topology but are not the same shape because they do not have the same dimension.

The recognition of the shape of an object, for example, with the help of a vision system, or an automatic processor of CAD generated object data, can be done in two stages. The first stage is the recognition of the topology and the second is the recognition of the dimension. Because a recognition system can involve complex

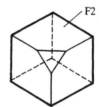

Facet   7
Edge   15
Vertex   10

The number of adjacent facets of F1 = 4

Facet   7
Edge   15
Vertex   10

The number of adjacent facets of F2 = 5

Figure 7.18

Adjacent topology of two objects.

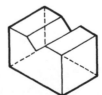

Figure 7.19

Objects with the same adjacent topology.

interpretation and reasoning, the object database should offer explicitly both topology information and geometry data.

7.12.2 Uniqueness of representation scheme

When an object extracted from a set of representation data does not allow for equivalent representation using a different set of data, the representation scheme is considered to be unique. When a set of data in a scheme represents only one object, the scheme is unambiguous. Most representation schemes in CAD systems are unambiguous but non-unique.

The representation scheme in a machine reasoning system must be both unique and unambiguous, for at least two reasons:

(1) If a set of data can represent more than one solid object (ambiguous), then it is difficult for a machine reasoning system to infer which object the set of data represents.

(2) If the representation scheme is developed to be unique, it will facilitate both the reasoning and the interpretation of data.

All representation schemes described in the previous section are unambiguous, but none of them is unique. Since a machine reasoning system requires uniqueness in interpretation, some modification is necessary. After modification, the boundary representation and pure primitive instance scheme can be unique in a certain domain. The modification to make CSG, cell decomposition, and sweep representation unique is not feasible. This is because of the various constraints imposed on the CAD user during the creation of solid objects.

Although CSG can provide all boundary informations, see also Requicha and Voelker (1983), through a complex conversion algorithm (boundary evaluation algorithm), boundary representation can offer the adjacent topology more explicitly. After modification, the boundary representation can be a unique scheme for representing a polyhedron. In dynamic robot planning, the available algorithms for static interference checking and dynamic collision detection are based on boundary representation, see also Boyse (1979). Certainly, in assembly planning, this is the case. But a variety of representations can be adopted in machining depending on the machining process of interest.

7.13 Bill of Materials

A bill of materials is a product data structure which captures the end-products, its assemblies, their quantities and relationships, The structure of a part's list determines the accessibility of the part's information by various departments in a company. It also helps to determine the level of burden put on the computational device in searching for product information. In many companies the bill of materials is structured for the convenience of individual departments. This, however, engenders problems in other departments.

In Figure 7.20, a product named product 1 is shown graphically with the summarized product structure, and the number of all items that are needed to make the parent product are enclosed in brackets. Figure 7.21 contains a bill of materials for product 1 in which the total usage of each item is collected into a single list for the product. This kind of list is convenient for the master production schedule but results in the duplication of assemblies. This implies that each product bill that uses assembly must be changed whenever there is a change in assembly. Furthermore, since lead times of intermediate assemblies cannot be detemined, parts are ordered too early the first time they are encountered in the product structure.

Another arrangement used in arranging the bill of materials is by indenting the product data as shown in Figure 7.22. One disadvantage of this method is that all components of an assembly are repeated each time the assembly is used, resulting in massive duplication of data.

One solution to the duplication problem is by holding each assembly only once in 'single-level' bill of materials as shown in Figure 7.23. In this approach, it identifies only the components used by one level and a required subassembly. This means that engineering changes can be made in only one place.

None of the file structures described above actually show when an item is used

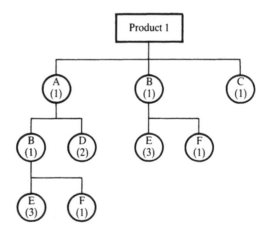

Figure 7.20

Product structure for product 1.

Part	Qty
A	1
B	2
C	1
D	2
E	6
F	2

Figure 7.21

Summarized bill of materials (BOM).

Product 1			Qty
A			1
	B		1
		E	3
		F	1
	D		2
B			1
	E		3
	F		1
C			1

Figure 7.22
Indented bill of materials.

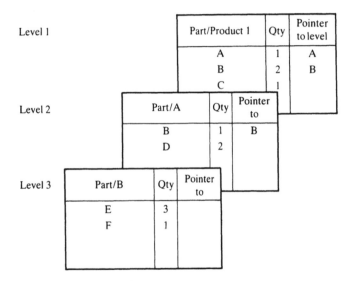

Figure 7.23
Single-level bill of materials.

on all of the assemblies and products. Some systems maintain a separate file just for this, thereby causing maintenance problems. When bills of materials are included in the product definition database in such a way that assembly data is included only once, then maintenance such as deletions, additions, and changes to the structure are simplified. In this structure, retrieval can be in any desired format described above including: summarized, included, single-level and where-used.

There are usually two kinds of bills of materials needed for a product: engineering and manufacturing bills of materials. The engineering bill of materials usually lists items according to their relationships with parent products. But this may not be sufficient to show that manufacturing constraints or tolerances may force the arrangement on the product structure to be different in order to assure manufacturability. Thus, engineering and manufacturing will usually have different valid views for the same product.

Table 7.4 shows the typical information which can be found in a bill of materials.

Table 7.4   Typical information in a bill of materials

Type of material used	Tools and dies used
Part name	Part identification number
Parent product	Quantity needed
Classification number	Type of standard part
Measuring units used	Process plan number
Part source	Part weight
Version number	Validity status
Production status	Value of part
Drawing number	Drawing format
Part description file	

7.14 Interfaces for CAD/CAM

Since the introduction of the computer as an automation tool for planning and control of manufacturing operations, the problem of interconnecting various software and hardware systems has become a major issue. The first computer-based planning and control systems were custom designed and configured for dedicated applications. It was extremely difficult and often impossible to take existing software and hardware modules developed for one plant and use them for the configuration of a planning and control system for another plant. This problem had been recognized many years ago and initiated standardization activities in almost every industrialized country. Also an attempt was made to draft standards on an international level to conceive standard interfaces for interconnecting CAD and CAM systems in order to transfer design data to the planning activities of manufacturing. Information which is communicated across the interfaces contains graphical, drawing related, geometry and product model data as in Figure 7.24, also see Schilli *et al.* (1989). This data is used for process planning, scheduling, manufacturing and quality control.

The concept of a universal data exchange system is depicted in Figure 7.25. In this figure it is shown how various CAD systems communicate with the production planning and control system. The adaptation of protocols, data formats and data transmission rates is done by the interface. The interface must provide electrical and physical compatibility and it must ensure that the semantics of the exchanged information is maintained. Data conversion may be done directly in the interface with preprocessors or postprocessors. With the help of such an interface it becomes possible to use existing software and hardware modules for various applications in a factory. In addition, a company's own manufacturing data can be easily merged with those from vendors and suppliers. It is also possible to build manufacturing planning and control systems from heterogeneous modules supplied by different vendors.

There are certain requirements which a standardized interface must be capable

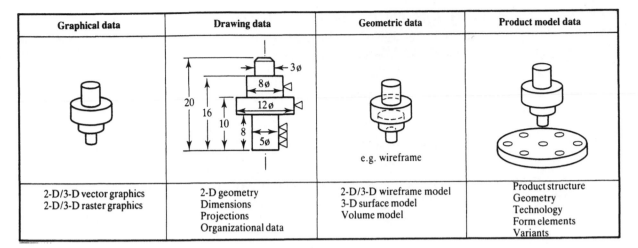

Graphical data	Drawing data	Geometric data	Product model data
2-D/3-D vector graphics 2-D/3-D raster graphics	2-D geometry Dimensions Projections Organizational data	2-D/3-D wireframe model 3-D surface model Volume model	Product structure Geometry Technology Form elements Variants

Figure 7.24
Product information communicated between CAD/CAM activities.

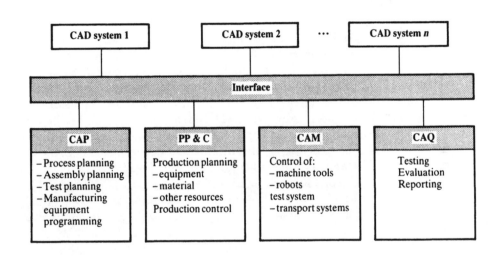

Figure 7.25
Universal interface for interconnecting the computer planning and control activities of a manufacturing system.

of providing. They are:

- The interface must be capable of handling all manufacturing data.
- There should be no information loss when data is transferred between heterogeneous systems. In other words, it must be possible to maintain the semantics during conversion.
- The system must be efficient to be capable of handling the realtime requirements of a manufacturing system.
- The system should be open-ended to permit extensions or contractions.
- The system should be adaptable to other standards.
- The system must be independent of the computer and communication architecture used.

- The number of data entities to be handled should be a minimum.

- It must be possible to form application-oriented subsets of the standard to reduce cost and overhead.

- A logic decoupling of data structure and physical memory format is necessary to ensure a separation of the specific information structure and the transmitted data units.

- The system should be capable of achieving production data.

- The interface must be upward- and downward-compatible in a hierarchical control structure.

- Test procedures must be provided to verify the efficiency and accuracy of data transmission.

7.15 Types of Interfaces

An interface can be regarded as an aggregate of conditions, rules and conventions which describes the information exchange, between two communicating objects. An object in our context can be human, software, modules, computer hardware or manufacturing processes. In principle, three types of interfaces can be distinguished including language interfaces, procedural interfaces and descriptive interfaces (Figure 7.26). The CAD system must be capable of exchanging data with the manufacturing world, and the interfaces are the windows to this world (Figure 7.27). In this figure, the interconnection of the CAD module with the various manufacturing system components are shown.

The language or user interface (Figures 7.26 and 7.27) is the window through which the user can communicate with the CAD module. The user employs the icons, pictorials, menus or textual languages to describe design and manufacturing problems. The textual languages are still the most important means of user/system communication. However, the interactive graphic languages are increasingly gaining importance, they are easier to learn and offer a more natural method of user/agent communication.

The procedural interfaces refer to language interfaces which are necessary to do the technical calculations, CAD modeling, process planning and machine programming. These interfaces may use subprograms, parameter descriptions or host languages to perform the individual CAD activities; thereby the host language concept is of particular importance. A host language is an existing language such as Fortran which has been extended to be able to handle specific constructs like the description of a circle. This is often done via subroutine calls. Procedural interfaces are also used to set up, in a generic way, the CAD configuration to be used or to manipulate data of the CAD database.

Figure 7.28 shows the concept of the Graphical Kernel System (GKS) which is a standard interface that can be used to operate different CAD hardware/software systems through an application program. GKS defines a language independent

	Language interface	Procedural interface (program interface)	Descriptive interface (data interface)
Types	Language constructs Language syntax Language grammer	Subprograms (of extended languages) Parameter description Host languages	Data structures Data formats
Representatives	Fortran, COBOL, Pascal • Data manipulation languages • Manufacturing machine programming language • Form element and variant description languages	GKS-graphic interface Data manipulation interface Fortran variant programming . . .	IGES SET VDAFS Libraries Archives . . .
Examples	P1 = POINT/0, 0, 12 L3 = LINE/P1, ATANGL, 45 C2 = CIRCLE /P1, P2, P3	CALL POINT (X, Y, Z, P1) CALL LINE (P1, 0) CALL CIRCLE (P1, P2, P3, C2)	'P1' 421, 0, 0, 12 'L3' 520, 0, 0, 12, 45 'L2' 200, 0, 0, 12, 4, 5 3, 5, 6, 10

Figure 7.26

Types of software interfaces (Courtesy of Institute RPK University of Karlsruhe, Germany).

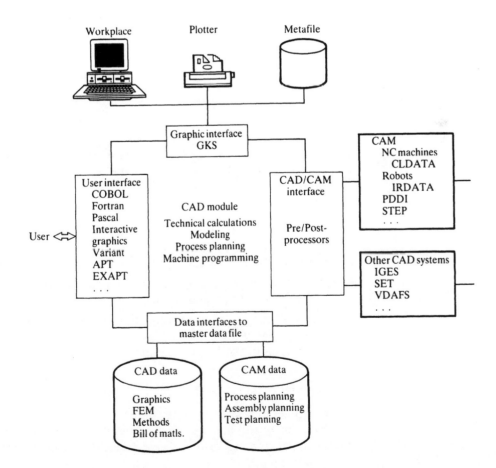

Figure 7.27

CAD/CAM interfaces.

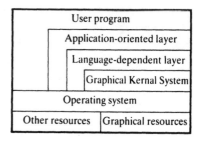

Figure 7.28

The layer model of the Graphical Kernel System (GKS).

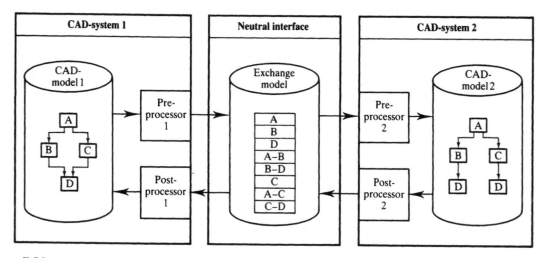

Figure 7.29

Exchange of CAD models between two different CAD systems.

nucleus for operating a graphic system. The user communicates with the graphic system via his program. The GKS system is embedded in the user program by an application-oriented layer and a language independent layer; whereby each layer calls upon the functions of the adjacent lower layer. The following services are available to the user:

- methods for the abstract definition of the peripherals which are needed for a CAD session;
- definition of basic graphic elements, polynomials, text and so on;
- performance of graphic operations such as transformations, rotations, picture segmentation, editing and so on;
- generation of a neutral picture file (meta-file) for the system independent storage of data, the exchange of data between different CAD systems and the organization of the output of graphic data;
- definition of a segment file for storing picture segments during a CAD session;
- input and output of graphic information.

Descriptive interfaces are the window to other CAD systems, to the master databases or to the manufacturing activities. The basic concept of a descriptive interface is shown in Figure 7.29, see also Grabowski and Glatz (1986). With this

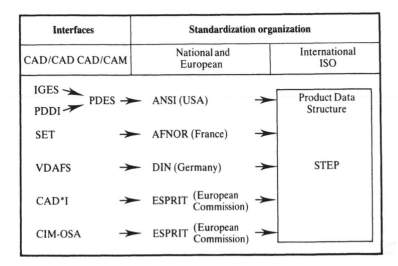

Figure 7.30
Development of
CAD/CAD and
CAD/CAM standards.

concept it is possible to exchange models between two different CAD systems. For example, model data of the CAD system A is converted via a pre-processor to a neutral presentation form and from there via post-processor to the representation scheme needed by the CAD system B. The exchange model representation is a system-independent method of describing the application. It can be accessed by any CAD system which has the proper pre- and post-processors. Typical equipment independent interfaces to manufacturing are CLDATA for NC machine programming and IRDATA for robot programming; they are explained elsewhere in the book.

In the following sections CAD and CAD/CAM interfaces will be discussed in more detail. The activities described will eventually lead to the product data structure defined by STEP (Figure 7.30).

7.16 Description of Various Interfaces

7.16.1 IGES (Initial Graphics Exchange Specification)

The basic concept of IGES is concerned with the exchange of product description data from one CAD system to another one as explained in IGES (1988). This exchange takes place via a neutral data format, see Figure 7.29. If information has to be transferred from one type of CAD system to another it is converted via a pre-processor to the neutral data format specified by IGES and then converted via a post-processor to the proper data format of the other CAD system. With this concept, it is possible to transfer CAD data between any set of CAD systems, as long as they are equipped with the corresponding pre- and post-processors.

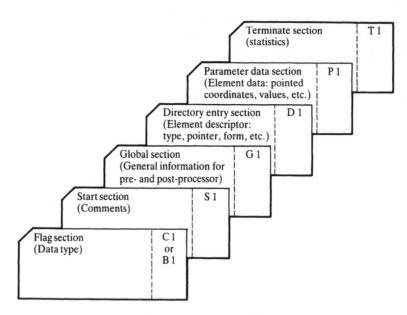

Figure 7.31
Sections of the
IGES Datafile
(Source: NIST).

IGES tries to represent technical drawings; three-dimensional wireframe, surface and volume models; FEM models; and symbolic representations in a neutral data format.

The IGES concept developed to date has the following five versions:

Version	Year
IGES – Version 1.0	(1981)
IGES – Version 2.0	(1983)
IGES – Version 3.0	(1984)
IGES – Version 4.0	(1986)
IGES – Version 5.0	(1988)

Version 1.0 was conceived for the early CAD systems and was tailored toward the technology of the 1970s. Presently, most equipment manufacturers support Versions 2.0 and 3.0.

IGES contains records of 80 character length which are presented by the 80 columns card format; where columns 1–72 hold information in ASCII code and columns 73–80 hold an alphabetic character followed by a serial number to indicate sections.

An IGES file has six parts, including a flag section, start section, global section, directory entry section, parameter data section, and terminal section, see Figure 7.31.

● The flag section indicates if the file is in binary or in a compressed ASCII format. It may not be used when the contents are represented by ASCII characters. If the binary presentation is selected, the file can be compacted by 70%.

Object-related entities		Structure entities	Predefined associations
Geometric entities	**Modeling entities**		
Circular arc	Finite element	Associativity definition	Group w/o backpointers
Composite curve	Nodal	Associativity instance	External reference file index
Conic arc	Displacement/rotation	Drawing	Views visible
Copious data	Offset surface	Line font definition	Views visible, color
Plane	Curve on a parametric surface	Macro definition	Line weight
Line	Trimmed (parametric) surface	Macro instance	Entity label display
Parametric spline curve	Nodal results	Property	Single parent association
Parametric spline surface	Element results	Subfigure definition	External reference file index
Point		Network subfigure definition	Dimensional geometry
Ruled surface	CSG-model	Singular subfigure instance	association
Surface of revolution		Rectangular array subfigure	Ordered group w/o backp.
Tabulated cylinder	Block	Instance	Planar association
Transformation matrix	Right angle wedge	Circular array subfigure instance	Flow
Flash	Right circle cylinder	Network subfigure instance	
Rational B-spline curve	Right circle cone	Text font definition	**Drafting entities (annotations)**
Rational B-spline surface	Frustum	View	
Offset curve	Sphere	External reference	Angular dimension
Connect point	Torus	Nodal load/constraint	Centerline
	Solid of revolution	Text display template	Diameter
	Solid of linear extrusion	Absolute text display template	Dimension
	Ellipsoid	Incremental text display template	Flag note
	Boolean tree	Color definition	General label
	Solid instance	Attribute table definition	General note
	Solid assembly	Attribute table instance	Leader (arrow)
			Linear dimension
			Ordinate
			Dimension
			Point dimension
			Radius dimension
			General symbol
			Sectioned area
			Section
			Witness line

Figure 7.32
Elements of IGES
Version 4.0.

- The start section contains comments; it provides human readable comments keyed in by the user and contains information for the receiving station.
- The global section holds information describing the pre-processor and data needed by the post-processor to handle the file.
- The directory entry section describes all IGES entities specified in the file (Figure 7.32). The entities are the various geometric elements including points, curves, surfaces and relations which are collections of similar structured entities; annotations like dimensions, viewing angles and notations, and structure elements like line font definitions, macro definitions, drawing specific data and so on.
- In the parameter data section, the parameters of the specified entities are stored in a free format.
- The terminate section contains a statistic indicating the size of each section which is represented by the last sequence number of a section.

Although IGES is a commonly used interface it has many problems. Version 1.0 was criticized for its huge file size and redundancy. This problem was reduced in Version 2.0 by introducing the binary format for setting up files. A problem exists of repeated definitions between the data entry section and the parameter data section, and also files are not very readable.

The pre- and post-processors are frequently of insufficient quality because companies who supply them try to maintain proprietary rights. Thus a one-to-one functional match between the transferred information of two CAD systems is often not obtained. Also the more recent versions of IGES have problems. For example, the IGES Version 3.0 was not developed to present information on solids. Thus solids must be described in an awkward method by boundary entities, and thus some information about solids cannot be transferred. In addition, there are representation schemes which are not unique and have unstructured file formats.

For the evaluation of the efficiency of interfaces a test matrix is used. Figure 7.33 shows in the upper part an input matrix for the CAD system and in the lower part the output to an NC programming system, see also Milberg and Peiker (1987). The 2-D geometry is transmitted in its entirety whereas the dimensions and text are missing.

7.16.2 PDDI (Product Definition Data Interface)

The PDDI interface was developed in connection with the ICAM project which was funded by the Air Force in order to improve the manufacturing methods in the aircraft industry, see also Weiss (1984). The essential goal of this product was the creation of an interface between CAD and CAM. PDDI is design-oriented, it provides a formal specification of a product model and its submodels. The latter represent:

(1) The 3-D geometry of the product which is defined by a set of curves, surfaces and volumes.

(2) The topology which describes the set of elements and relations for defining the product boundary.

(3) The tolerance which is inherent in the accuracy of the manufacturing processes.

(4) The future elements needed for defining manufacturing operations.

(5) The non-geometric information for organizational purposes and description of material and its properties.

Additional information includes the specification of the pre- and post-processors needed for the data exchange between CAD and CAM. The PDDI efforts were conceptual research activities and the results were used for the development of the PDES activities and are a major input to STEP. The STEP concept is described in Section 7.16.7.

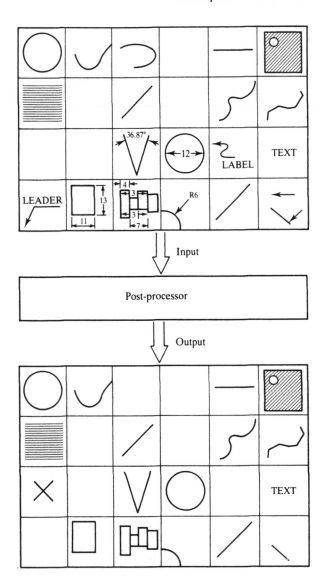

Figure 7.33
Evaluation of a
post-processor with
the help of a test
matrix (Courtesy of
Institut für Werkzeug
maschinen and
Betriebswissenschaften,
Technical University
of Munich).

7.16.3 PDES (Product Data Exchange Specification)

The aim of PDES is to provide an interface which permits the exchange of data of the entire product development cycle and production cycle, see also PDES (1985). It can be viewed as an expansion of IGES whereby organizational and technological data have been added. Functionally PDES will contain IGES Version 4.0. The physical and logical structure of both interfaces will be different. For this reason there will be a conversion program to transfer the IGES into the PDES format.

A special feature of the PDES development is the use of the formal language EXPRESS for modeling the product information. There will be capabilities for defining manufacturing features, FEM and special applications for civil engineering, electronics, ship building and so on. The results of the PDES project will be the major input to STEP.

7.16.4 SET (Standard d'Echange et de Transfert)

The interface SET is a standard for the exchange of all data which is generated for CAD/CAM. It is a French development and was first published in 1984 (SET 85). Several additions have been published since (SET F 88, SET S 88 and SET V 88).

The information transferred via the SET interface are wireframe, surface, B-representation and FEM models, as well as technical drawings and scientific data. Organizational, material property and tolerance information cannot be defined by individual models but they must be contained in the drawing for transfer.

SET allows the exchange of data between different CAD/CAM systems and also between CAD/CAM systems and central data banks. It features a much more general data structure than IGES; data is stored in a very compressed form. Thus the neutral file size is much smaller and the processing time is much shorter than for IGES.

The structure of a typical SET file is shown in Figure 7.34. The delimiters of the file are begin and end blocks. The BEGIN section contains administrative data and other information of a general type.

The END section describes the total number of blocks stored in the SET file. The file is structured from assemblies which contain a series of blocks; the blocks may be nested and hold the CAD information which has to be transferred. There

Figure 7.34
Structures of a SET
data file.

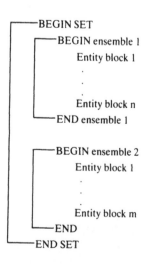

```
BEGIN SET
    BEGIN ensemble 1
        Entity block 1
            .
            .
            .
        Entity block n
    END ensemble 1

    BEGIN ensemble 2
        Entity block 1
            .
            .
            .
        Entity block m
    END
END SET
```

are different types of information units defined. They are:

Class 0. The geometric primitives needed for describing 2- and 3-dimensional objects such as points, lines, circles, parabolas, planes and surfaces.

Class 1. Complex geometric entities for defining complex curves, complex solids, shells, faces and so on.

Class 2. Graphical aids to map product information onto technical presentations. These aids include view block definitions, block definitions, line fonts, cross hatching, characters for defining symbols and so on.

Class 3. Grouping mechanisms to assist the mathematical operations needed for the manipulation of the design objects. Typical representations are coordinate transformations, scalar series, general matrices, tabulated functions, physical quantities, material properties and so on.

Class 4. Descriptive aids to represent diameters, angles, labels, text, cross-hatching, surface finishes and so on.

Class 5. Elements defining the structural relations such as calling block, group, attribute and homogeneous entities.

Class 6. Connecting elements to describe logic between the elements of any class.

Class 7. FEM elements to build the FEM model. Typical elements are finite elements, model for an element, computations, definition of a computation, loads, damping, Eigenvalue, knot renumbering and so on.

Class 80. Gives the user the possibility of defining his own elements.

Class 99. Management aids to structure SET files. They include SET header, ensemble header, ensemble header as well as 'End' specifier for an ensemble and a 'SET'.

7.16.5 VDAFS (Verband der Automobilindustrie Flächenschnittstelle)

This interface was developed by the Association of the German Automobile Industry for the transfer of 3-D curves and surfaces for which IGES only offers an unsatisfactory solution. Two versions of this interface have been developed and are discussed in Frankfurt Verband der Automobilindustrie (1983) and Mund *et al.* (1987).

Similar to IGES, the VDAFS interface contains a fixed record length of 80 characters. A file consists of a header and a data section. The header describes the data source, the project name, the origination date, the validity date, the source CAD system and the user. In the data section the geometric objects are described with the help of entities. Characteristic for the data presentation is an APT-oriented data format.

Elements of the VDAFS Version 2.0 are shown in Figure 7.35. Curves and surfaces are described with the help of points, set of points and set of vectors, (Figure 7.36).

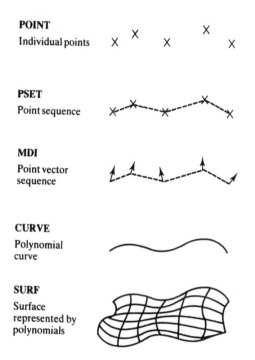

Geometric elements	Non-geometric elements
Points	Begin specifier
Point sequence	Comment
Point vector sequences	Structuring (SET groups)
Circle (curve)	Transformation matrix
Polynomial curve	Transformation table
Surface	End specifier
Curve of surface	
Trimmed surface	
Surfaces of patches	

Figure 7.35
Elements of VDAFS
Version 2.0.

POINT
Individual points

PSET
Point sequence

MDI
Point vector
sequence

CURVE
Polynomial
curve

SURF
Surface
represented by
polynomials

Figure 7.36
VDAFS geometry
elements.

With the VDAFS interface it is possible to transfer any described workpiece surface, including the orientation of the cutting tool. Problems include the transfer of text, dimensions, cross-hatching and so on. In addition, single elements are received by the user in a basic atomic format and are difficult to use for further processing. Another problem is that the interface has insufficient means of structuring data.

7.16.6 CAD*I interface

The CAD*I interface was developed within the framework of the ESPRIT Project 322 of the European Commission. The project was started in 1985 and had a duration of five years. The results are recorded in the literature of Schlechtendahl

(1988 and 1989). The emphasis of this project was placed on the data exchange between various CAD systems, the exchange of CAD and FEM models and the conception of a neutral databank for CAD data.

For the realization of the data exchange reference architecture models for 3-D wireframe, surface, CSG and B-rep representations were established, and an interface for FEM applications was specified. For the description of CAD data structures, the specification language HDSL (High-level Data Specification Language) was designed. This language allows the expression of all essential elements of a CAD data structure with the help of special language constructs.

Within the framework of this project, an attempt was to be made to improve the deficiencies of IGES, SET and VDAFS which have the following problems:

- many of the specifications are imprecise,
- the file formats are not suited for an efficient processor implementation,
- the structuring capabilities of the file format for CAD data are not adequate,
- the scope of the interfaces is very restricted.

Work on this project was done in the following areas:

- investigation of CAD data structures and the presentation of 2-D and 3-D models by using new modeling techniques;
- specifying neutral file formats and providing access mechanisms;
- development of pre- and post-processors for interfacing various CAD systems;
- development of interface test methods, and aids for performing and documenting tests.

One special goal of the CAD*I project was the interaction with the STEP activities to develop solutions which could be directly implemented in STEP.

The publications of the CAD*I project are very complete and are well suited to give the reader not familiar with CAD/CAM interfaces a good insight into the general problems of software standards.

7.16.7 STEP (Standard for External Representation of Product Data)

The STEP (Standard for External Representation of Product Data) project is supported by the ISO-workgroup TC 184/SC4. It is also sometimes called Standard for Exchange of Product Definition Data. It is an effort by this group to develop an international standard for representing product model and a data exchange format for all information needed for the life cycle of the product. In general, it is a series of standards intended to provide a common mechanism for representing product model data throughout the life cycle of a product independent of any application software that may be used to process it. The basic principles of STEP are shown in Figure 7.37. The figure shows a computer-integrated manufacturing system where CAD data is tightly coupled via the STEP interface to the main production activities such as CAP, PP&C, CAM and CAQ. The following are the predecessors of STEP (which have been discussed earlier in this chapter): IGES (USA), SET (France) and VDAFS (Germany). See Figure 7.26 for these standards. It is estimated that by 1994, STEP will be replacing the standards mentioned above. The series of standards in STEP provide a neutral representation of product model data in the form of a set of integrated resources that support a complete and unambiguous definition of a product. The resource constructs are documented in a single product data language definition as discussed in Mason (1991).

There are five components in the STEP standard.

(1) The first component discusses the purpose of the standard. It contains:

   (a) an overview describing the fundamentals of the entire standard,

   (b) EXPRESS which is a formal language for specifying the information structures which are needed for defining the product model of STEP,

   (c) framework for product modeling in which the basic methods of modeling are described.

(2) The second component defines standards for the implementation of STEP. It contains:

   (a) physical file implementations which define the rules according to which the data structures defined in EXPRESS are mapped on a segmental data file.

(3) The third component specifies the framework for conformance testing. This segment defines the rules according to which an implementation is tested for standard conformity.

(4) This component includes standards for product modeling (Figure 7.37). The models include:

   (a) a presentation model which defines the type of presentation of information. For instance how the product can be visualized on a CRT or paper. Features like color, illumination, text font, view angles and so on are determined here, see Figure 7.37(a).

   (b) Materials model: is where the characteristics of material can be described. For instance an elasticity matrix, material-specific coefficients such as a coefficient of expansion, thermal conductivity, heat transfer are specified, see Figure 7.37(b).

   (c) Tolerance model describes dimensional tolerances, geometric tolerances, form tolerances and so on, as shown in Figure 7.37(c).

   (d) Surface model describes the specification for surface finish required, as in Figure 7.37(d).

   (e) Form-feature model describes the object from the point of view of manufacturability whereby the form pattern is described or a method to produce the form element. The manufacturing relevant features can include holes, threads, keywords and so on, as shown in Figure 7.37(e).

   (f) Shape representation model is based on the elements of geometry and

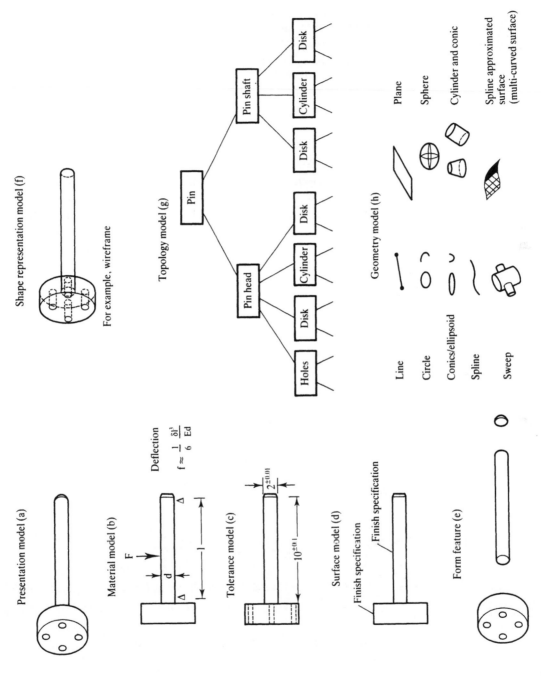

Figure 7.37
Models of STEP.

topology model. There are three methods of object representation specified: the wireframe, surface, and volume models. In the volume we can have constructive solid geometry, and boundary representation. The concept in these models is the formation of a neutral shell over the basic elements of the topology model. The intention of this concept is the separation of the operation models from the topology models as in Figure 7.37(f).

(g) Topology model defines the entities for description of neighbourhood relations. Surfaces, edges and vertices are defined. These topological features are described by association with corresponding geometric elements as seen in Figure 7.37(g).

(h) Geometry model contains all geometric information about the definition of lines, and surfaces. By specific entity structures, points, vectors, coordinate systems, transformations, curves and surfaces can be described. Free-form surfaces can be approximated to computable forms. All known mathematical forms are considered as shown in Figure 7.37(h).

(5) Standard for the specific standard technical areas. These norms are based on previously described models. Typical categories of these norms are:

(a) drafting, which describes the data type and structures for the layout of the drawing and the presentation of technical objects. Various views, cross sections, as well as dimensioning are captured.

(b) product structure configuration management defines data structure, makes possible the administration and the comparison of specified various product versions. From this data structure of the partial model, the bill of material and the product structure can be derived.

(c) finite element analysis which is represented by the partial model analysis application.

(d) kinematics where information for gears and robots is kept.

For ship building and building construction, the partial model architecture, engineering, engineering construction are described.

(6) Application protocols, which are used to give the user a certain amount of freedom, are still left open in the above standards. The application protocols are the final references to abide by:

● Version 1 of STEP: Exchange of CAD drawings.
● Version 2 of STEP: Exchange of drawing from 3-D product model.
● Version 3 of STEP: Exchange configuration controlled 3-D product definition.
● Version 4 of STEP: Exchange of B-Rep models.

In principle, the STEP development is done in three phases. In the first phase, an application-oriented information structure is provided. This structure is also called the application model; and is referred to as the application layer.

In the second phase, an information structure is provided by which all application models can be described with the help of the language EXPRESS. This phase is called the logical layer, and its output includes the information, objects, entities, attributes, associations and conditions.

In the third phase, the physical layer, the data formats and data files are defined in the Bachus-Naur-Form of the Wirth-Syntax Notation (WSN). The latter is a common language known in computer science for describing compilers.

7.16.8 EXPRESS language

EXPRESS is a definition for a formal language which was developed for use in STEP. It has syntactical similarity to Pascal and contains object-oriented features. The language can also be used in a broader sense for the specification of other information structures. The central role in EXPRESS is the entity. An entity has attributes which either constitute an entity or can derive from it. An entity can also have restrictions which define the range of values for the attributes.

With EXPRESS, a specification done for STEP can be tested formally. Furthermore, EXPRESS gives the fundamentals for software tools for the definition of product data and the description of data files.

An example of how EXPRESS is used in STEP is shown in Schlechtendahl (1991) and is as follows:

*Right Circular Cylinder
A right circular cylinder is defined by an axis point at the center of one circular cylinder face, an axis, a height, and a radius. The faces are perpendicular to the axis and are circular disks with the specified radius. The height is the distance from the first circular face center in the positive direction of the axis to the second circular face center.

```
    *)
    ENTITY right_circular_cylinder
SUBTYPE OF (primitive_with_one_axis);
radius  :  real;
position  :  axis 1_placement;
height  :  real;
    WHERE
WR1  :  radius > 0;
WR2  :  height > 0;
    END_ENTITY;
    (*
```

ATTRIBUTE DEFINITIONS:
Radius: The radius of the cylinder.
Position: The location of a point on the axis and the direction of the axis.
Height: The distance between the bases of the cylinder.

PROPOSITIONS:
(a) The radius must be positive.
(b) The height must be positive.

7.17 Requirements of a Product Model

During the design phase, about 70% or more of the cost associated with the marketing of the product and the way the product goes through the life cycle is established. The life cycle issues can include design, process planning, manufacturing scheduling, purchase of components from vendors, assembly, inspection, use, repair, modernization and scrapping, as discussed in Whitney *et al.* (1989). Of these problems, perhaps the most difficult one is the configuration of the product for manufacturability or assembly. The designer must consider the methods of manufacturing during the design process. With the computer gradually replacing the human in the generation of the design, an ideal product development system must be capable of providing all information needed to design, make and maintain the product.

The objective of a product model is to generate a scheme necessary for the development, manufacture and use of a product, whereby a product may be a complete assembly, a subassembly or a part. A model contains data, algorithms and a defined structure. The model should automatically generate the design, functions, service life, manufacturing methods and all data needed for the processing of a customer order.

The information contained in the product model is very numerous and can be classified as follows.

Administrative description

Here, all administrative data is provided for processing an order form from the request of quotation to the delivery of the product. Typical information includes part identification, part classification, part document identification, product family and so on.

Functional description

This type of information provides the functional description of the product, defines its application environment and gives instructions for repair.

Shape description

For the design of the product and its manufacture, the geometry and topology are defined, including the part of the surface and tolerances. The shape description also contains the spatial and assembly relationship between features of a part and generates the basic data for process and assembly planning.

Process information

This information describes the manufacturing process for making the product. An ideal model deduces from the functional and shape information the manufacturing and testing methods and sequences.

7.18 Design Features

When planning for manufacturing, one would like to derive from the design features the manufacturing process. In manufacturing, a feature is described as a specific geometric configuration formed on the surface, edge, or corner of a workpiece and is intended to aid in achieving a given function. This definition is sufficient for an implicit representation of form features but inadequate in automatic process planning. For instance, in assembly, features largely dictate how parts are assembled. The above definition also does not lead to a clear mathematical definition of features which is vital in reasoning about geometric interaction among features. Some definitions of features include 'any named entity with attributes of both form and function', see Dixon (1988); 'sets of product information' or 'recurring patterns of information related to a part's description', see Shah (1988) and 'set of geometric elements that forms a unit of interest', see Prinz *et al.* (1989).

These definitions actually describe the essence of features by implying function in the form of feature geometry. A definition which captures both geometrical and functional implications of a feature is 'a set of surfaces together with specifications of the bounding relationships between them and which imply an engineering function (or stereotypical entity) on an object and which may be formed on the faces, edges, or corners of an object,' see Nnaji (1990). This means that not all the surfaces of an object are feature surfaces.

Based on the definition, a feature is defined to carry the information of the elements related to its nominal geometry; thus, form features are the major concern in this description for the purpose of process planning. Form features are important for a variety of reasons:

(1) they contain geometric elements (vertices, edges, and faces), and these elements may be used as the input to mathematical definitions used for matching extracted features against ones in the database,

(2) they provide the 'actual structure or configuration' of features, and can become the basis for establishing features for other purposes,

(3) they can be interpreted by reasoning about the current output from a CAD database,

(4) they can have an application-independent classification at the highest level.

7.18.1 Genus and the Euler formula

A feature can be regarded as a connected open entity rather than a closed entity (object) located on the envelope of an object. An entity is said to be open if the entity's faces separate space into two regions, the object itself and the remaining space. A closed entity separates space into three regions; internal region, the object itself, and external region. Therefore, Euler's formula for features can be shown as follows:

$$v + f - e = (s - g) + r$$

where v, e, f are the number of vertices, edges, and faces of the open object, respectively. s is the number of the shell contained in the object. g is the number of genus in graph theory, it is greater than or equal to zero for a feature. r is the number of rings that the feature has. The number of rings can be obtained by:

$$r = N_{\text{rings}} = N_{\text{loops}} - N_{\text{faces}}$$

A ring is a closed loop with its co-face loop level equal to 1. Co-face loop level (CFLL) is the property of surface bounding loop. The co-face loop level of the outermost surface bounding loop of co-surface bounded faces is said to be 0, while that of inner loops (or rings) is said to be 1. A face is bounded exactly by one loop if its co-face loop level is 0. If $g = n$ for $n \geqslant 1$, then the feature possesses n through holes; otherwise, it is not a through hole. For application purposes, this formula is useful in determining whether the feature is a through hole or not. See Figure 7.38 for a determination of how to use the Euler formula.

7.18.2 Feature classification

Features can be properly classified based on the engineering characteristics of the application domains. Yet, even in the same application domain, features may be further classified by different users when different types of 'similarity' of features are concerned. We classify features according to three levels. These are generic, application, and product level (Figure 7.39); the results of the higher level classification strategies should be common for every lower classification level, as shown in Figure 7.40. The details of each will be explained later.

The attributes used in the first two levels are obtained or created from the boundary information of an object. The differences between the first two levels, mainly, are the generally functional characteristics of a particular application that are considered in the application level; nonetheless, in the generic level, only the topology and geometry of features are studied. Obviously, the attributes used in

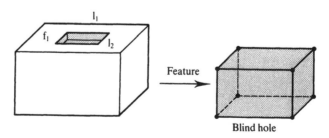

Number of feature faces = 5
Number of feature edges = 12
Number of feature vertices = 8
Number of feature rings: $r = l_1 + l_2 - f_1 = 1$
$s = 0$
$g = 0$
$v + f - e = (s - g) + r = 1$

Figure 7.38

Verification of the Euler formula.

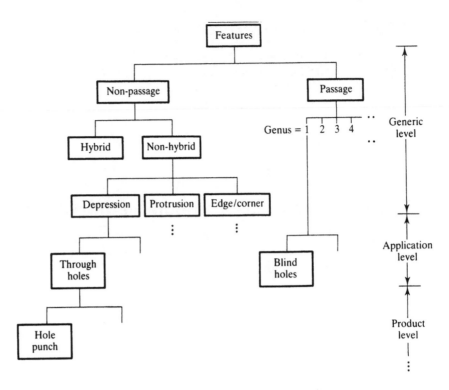

Figure 7.39
The three levels of
feature classification.

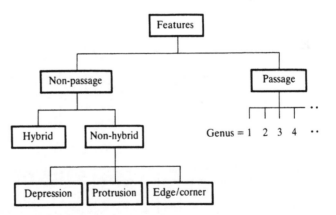

Figure 7.40
Hierarchy of generic
feature classification.

the generic level will be the same universally, but the attributes of the application level should be specific to an application.

The third level, product level, is a user-oriented feature classification scheme. The attributes should be related to the information used for process planning, such as the machine tools or the tool set to produce the features, the set-ups, the sequence of machining operations and so on. These attributes are different among various users because of the hardware and software constraint imposed by the users.

7.18.3 Generic feature classification

The intention of explicit feature classification is to classify features in a general, application independent manner. Therefore, the generic feature classification is to define features by their boundary structure. They contain geometric and topological characteristics only. Four attributes are used on this level to classify features: topology which can be captured in a spatial relationship graph (Figure 7.41) and described in Nnaji and Chen (1991), genus, convexity and concavity of edges, and locations.

7.18.4 Generic classification methodology

In generic feature classification, features can be broadly classified into passage and non-passage features (Figure 7.39 and Figure 7.42) by evaluating Euler's formula. By evaluating the convexity/concavity of their Face Bounding Loops (FBLs), the

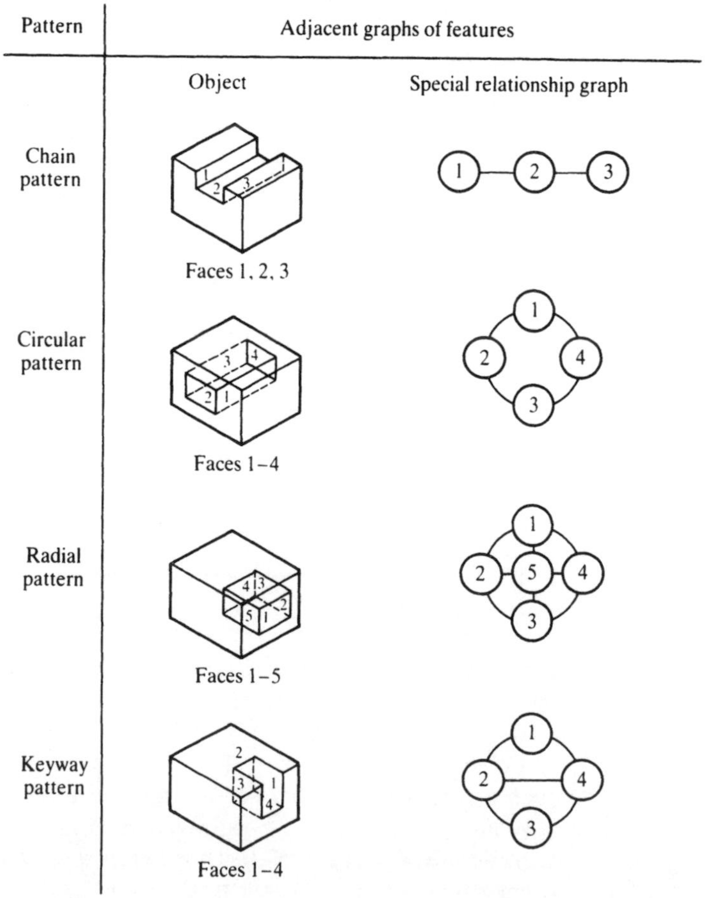

Figure 7.41
Pattern of feature adjacent graphs.

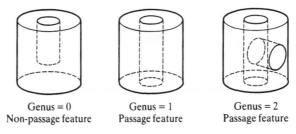

Figure 7.42
Genus determination
for objects showing
passage and
non-passage features.

Genus = 0
Non-passage feature

Genus = 1
Passage feature

Genus = 2
Passage feature

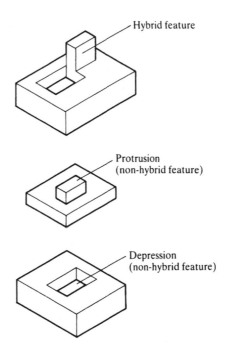

Hybrid feature

Protrusion
(non-hybrid feature)

Depression
(non-hybrid feature)

Figure 7.43
Hybrid/non-hybrid
features.

non-passage features can be categorized into hybrids and non-hybrids (Figure 7.43). The non-hybrid feature can be further classified into depressions, protrusions, and edge/corner (Figure 7.44). Similarly, passages features can be categorized by the number of genus. Finally, features are classified by their patterns. The methodology to perform the generic feature classification (Figure 7.40), is as follows:

(1) Check the validity of the boundary data of the feature by calculating the Euler formula for features. Subsequently, the number of genus, g, can be obtained. Based upon the definition, g is the number of through holes (passages) of the topological structure (or object). This process automatically classifies features into either the passage group or the non-passage group when this step is complete.

(2) If the feature belongs to the passage class, $g \geqslant 1$, then the feature can be classified according to the value of g. For example, subcategories for $g = 1$, subcategories for $g = 2$, and so on.

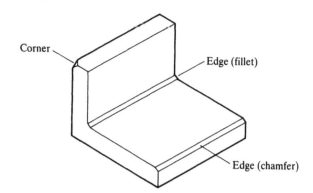

Figure 7.44
Edge/corner features.

(3) If the feature belongs to non-passage class:

    (a) The features can be categorized by evaluating the convexity/concavity of the feature bounding loop. If the convexity/concavity of all feature bounding edges are the same, then the feature is a non-hybrid feature; otherwise, it is a hybrid feature.

    (b) For a non-hybrid feature: if it is convex, then the non-hybrid feature is a depression; if concave, it is a protrusion.

    (c) For a hybrid feature: the hybrid feature can be further classified by the number of elements in the Feature Extrusive Faces Set (which is a set of all the faces which intersect the feature boundary loop with at least one and at most two vertices of their face bounded loops).

(4) All features, at this point, can be further classified according to their patterns.

7.18.5 Feature patterns

Pattern matching is an essential component of the feature recognition process in a feature-based system. The classification of features helps the recognition process in searching through a certain portion of feature instances stored in a feature database. A detailed classification scheme can make the recognition process more efficient. The feature recognition methodology can be conceptually divided into three steps: topology family checking, geometric family checking, and identical feature matching, see also Nnaji and Kang (1990).

The fundamental principle used in the methodology is that for identical objects, the same topology adjacent relationships must exist but the reverse is not true. The objects shown in Figure 7.45 have the same topology adjacent relationships. But, by checking the geometry of these three objects, one will find that they are not identical to each other. When objects have the same geometric shape, they are defined to be in a geometric family. The step of grouping features, geometrically, into families is rather tedious and less useful than grouping features using topological characteristics. One obvious reason is that the dimensional information of features, even when they are identical in shape, is often different.

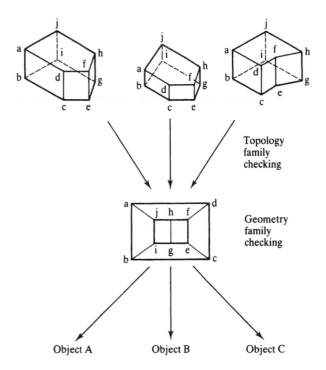

Figure 7.45
Topological and
geometric family.

By observing the 'adjacency' relationships of spatial relationship graphs (the topology of features), five adjacent patterns can be defined. Let $G_{adj}(v, e)$ be an adjacent special relationship graph of a feature, where v and e are the sets of nodes and edges, and $n(v)$ and $n(e)$ represent the number of nodes and edges of G_{adj} respectively. Additionally e_{ij} represents the adjacent relationship between v_i and v_j. Examples of patterns which have been identified using the above reasoning include chain, circular, radial and keyway, as shown in Figure 7.41 and Chen (1991). In this figure, the faces represent the nodes of the graph while the edges form the arcs which link the nodes to one another spatially.

7.19 Feature Classification by Application

Feature classification on the application level (for example, sheet metal, assembly, machining and so on) may first assume generic feature classification. The application attributes of the feature class will then be added to obtain a general (in application only) classification of the features. This is due to the fact that features are normally defined differently by applications. In this book, the assembly of machined objects and sheet metal component fabrication are used to show how features are classified.

7.19.1 Feature classification for assembly

An assembly is an operation in which two or more bodies are brought physically in contact and which reduces the degrees of freedom of each body, and preserves the resulting relationships permanently, see also Nnaji (1992). Consequently, the approach taken here is to apply the concepts described earlier which relate directly to assembly. These include:

(1) symmetry of feature,

(2) degrees of orthogonal feature freedom (the possible orthogonal approach directions for assembly with regard to the feature),

(3) complexity of a feature (degrees of feature graph).

These attributes are added to the generic classifications to obtain the complete feature classification for assembly application.

Symmetry

Lexically, symmetry is defined as 'correspondence in size, shape, and relative position of parts on opposite sides of a dividing line or median plane or about a center or axis'. Intuitively, one can observe objects as either symmetric or asymmetric, geometrically. It is also true that in performing a manufacturing operation, if the shapes of the linking features are geometrically symmetric, a tremendous amount of time can be saved in reasoning about the manufacturing task.

There are two types of symmetry, rotational and reflective symmetry, which are important in engineering applications. Rotational symmetry is essential in an assembly task as well as in a turning operation; meanwhile, reflective symmetry is important in a milling operation, for example, the NC program for machining the second side of the reflectively symmetric object may just duplicate the NC program of the first side with a transformation. Rotational symmetry means that the orientation of a feature is repeated when the part is rotated through an angle, θ, about a certain axis as explained in Boothroyd (1975), and Boothroyd and Dewhurst (1983); this axis is called the symmetric axis or the principal axis. Reflective symmetry of an object exists when the object is symmetric about a plane, line, or a point.

For all mechanical parts, a solid model is a rigid body. Thus, one could consider a model as a set of geometric elements, G. Then a symmetry or permutation of the set G is a bijection from G to itself. More specifically, a permutation is a one-on-one transformation of a finite set into itself. Because of the nature of a rigid body, the concept of group can therefore be applied to deal with symmetry.

Mathematically, a group is an algebraic system which is a non-empty set in which at least one equivalence relation (equality) and only one binary operation is defined. A symmetry of a geometrical figure is, by definition, a one-on-one transformation of the set, S, of the elements. The set, S, may be a 'space', for example a plane, sphere, etc. The bijections are the 'symmetries' of S with respect to suitable properties, the bijections of S satisfy some nontrivial laws, such as associative law, in any case. If 'o' is a binary operation, such as mapping, and $\mathcal{X}$ is the set of bijections from any set X to itself, then $(\mathcal{X}, o)$ is a group under composition. This group is called the symmetric group or permutation group of X. Consequently, the definition of a symmetry of a figure can be given as follows:

If $\mathcal{G}$ is a figure in the plane or in space, a symmetry of the figure $\mathcal{G}$ or isometry of $\mathcal{G}$ is a bijection $f: \mathcal{G} \to \mathcal{G}$, which preserves distances; that is, $\forall$ points $p, q \in \mathcal{G}$, the distance from $f(p)$ to $f(q)$ must be the same as the distance from p to q. The elements of a symmetrical group are the transformations of the feature, or object. Figures 7.46 and 7.47 provide some examples of rotational and reflective symmetry, respectively (1 denotes the identity of transformation; $rot(v, \theta)$ denotes a rotation about the vector v by θ degrees). One of the benefits of applying group theory in describing symmetry is that one can reason, mathematically, about the manufacturing operations by utilizing group theory.

Assembly classification methodology

Before classification for assembly can be developed, two issues must be reinforced. One is the definition of an assembly and the other is the type of information necessary to accomplish an assembly process.

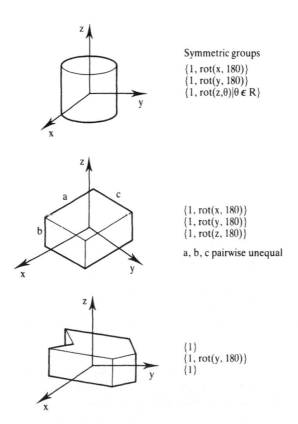

Symmetric groups
$\{1, rot(x, 180)\}$
$\{1, rot(y, 180)\}$
$\{1, rot(z, \theta) | \theta \in R\}$

$\{1, rot(x, 180)\}$
$\{1, rot(y, 180)\}$
$\{1, rot(z, 180)\}$

a, b, c pairwise unequal

$\{1\}$
$\{1, rot(y, 180)\}$
$\{1\}$

Figure 7.46
Rotational symmetry.

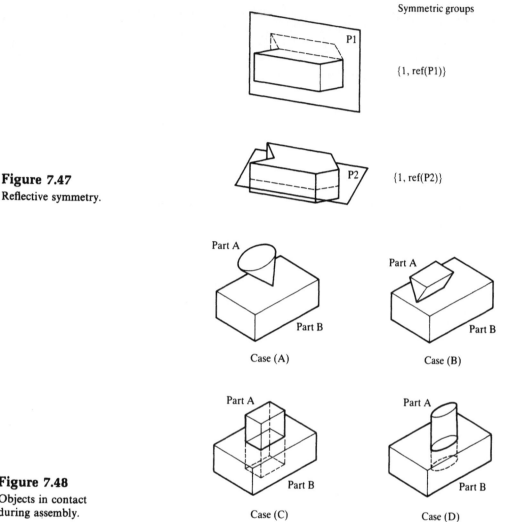

Symmetric groups

{1, ref(P1)}

{1, ref(P2)}

Figure 7.47
Reflective symmetry.

Part A

Part B

Case (A)

Part A

Part B

Case (B)

Part A

Part B

Case (C)

Part A

Part B

Case (D)

Figure 7.48
Objects in contact
during assembly.

What is an assembly process? The first requirement of an assembly is that there must be two or more objects included in each step of this process. These objects must physically come in contact with each other and keep the mutual spatial relationships after the assembly process is completed.

After the assembly process is accomplished, each object will lose at least one degree of freedom. For two objects assembled together, the number of degrees of freedom lost for each object must be equal. Figure 7.48 illustrates some examples where the objects are in contact. However, because of gravity, additional force is always acting on these objects to break the spatial relationship between them. Hence, when two objects have linked, one or both might become unstable and fall because of the force of gravity. Therefore, it is reasonable not to consider these cases in an assembly process as valid assemblies. In case C, Figure 7.48, just one degree of freedom has been left after assembly and for case D, one rotational degree of freedom is preserved in the objects and the translational degrees of freedom decrease by one for each object.

Implicit assembly features

The implicit classification of features for assembly, basically, follows the procedures of explicit feature classification. The symmetries and their relationships, multiplications, as well as the angles of feature freedom (AFF) and degrees of feature freedom (DOFF) of the features are found at the explicit stage. According to the information needed for assembly process planning, the only required information for the domain of machined object assembly is the surface's spatial relationship of the feature faces. Thus, the procedure for classifying features in this application domain is as follows:

(1) Classify features into symmetric or asymmetric categories.
(2) Find the DOFF of the feature. The features may be classified by the number of the DOFF.
(3) Check the complexity of the feature, and assign the feature to a proper family.

The hierarchy of classification is shown in Figure 7.49 constructed according to the symmetry groups and the consequent degrees of freedom in assembly. Based upon the special specifications of the products or the conventions of the designs, users may extend the classification presented in this section to include as much detail as is desired. Furthermore, the output of this classification scheme can be represented in the form of group technology codes by merely assigning numbers to the levels in terms of row and columns.

7.19.2 Sheet metal application

Metalworking processes, usually, are divided into two main categories: bulk deformation and fabrication processes. Bulk deformation process, described in Nnaji and Kang (1990), covers the processes in which the thickness and dimensions

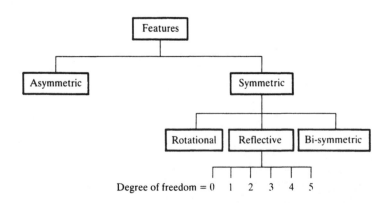

Figure 7.49
Feature classification
by symmetry groups.

of the metal are substantially changed. In sheet metal fabrication the change in thickness is only incidental. Because of the improvements of the material properties, tools, and manufacturing machines in recent years, the formability of sheet metal fabrication has been substantially improved.

A quick study of sheet metal components will easily reveal that the features of a component largely influence the method and ease of forming such a component. Furthermore, the formability is critically dependent on the manufacturing processes used to produce the component. Therefore, an ideal sheet metal features classification scheme should provide information to a user regarding how a particular feature or a set of features may be efficiently fabricated.

Before the characteristics of sheet metal fabrication are discussed, one assumption needs to be specified. The assumption is that the 'thickness' of the part which is usually very small relative to the rest of the other dimensions of the components is constant. Thereafter, one can assume that if the edges that contribute to the thickness are removed (also the faces which are bounded by the edges of thickness) then the rest of the geometric elements, edges and faces, can be divided into two groups. Each of the geometric elements are specified to belong to a certain group based on which side of the sheet they lie on (outer or inner which can also be specified as top or bottom). Based on this assumption, the surfaces of a sheet metal material will not include the faces of thickness (Figure 7.50). Furthermore, in sheet metal application, an internal feature and an external feature are distinguished by whether they intersect the contour of the sheet metal material (external features do and internal features do not include thickness faces).

7.19.3   Sheet metal feature classification

Based on the material and manufacturing characteristics of sheet metal fabrication, sheet metal features are classified into passage and depression features categories except bends. The fundamental functional attribute of a bend is to change the normal direction of a sheet metal surface; therefore, bends belong to the class edge/corner. The hierarchical relationships of a sheet metal features classification scheme are displayed in Figure 7.51. Figure 7.52 shows a coding system where group technology is applied in this application.

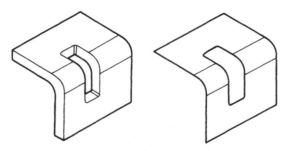

Figure 7.50
Sheet metal with and without thickness.

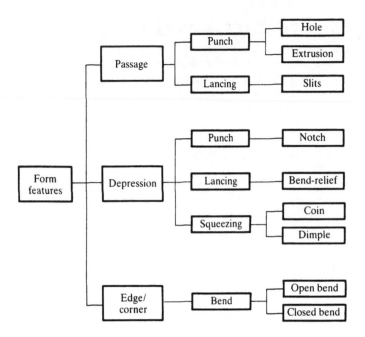

Figure 7.51
Classification scheme
for sheet metal parts.

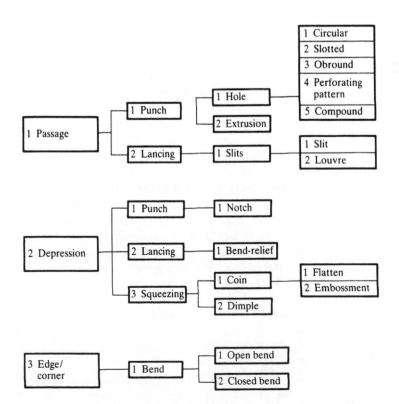

Figure 7.52
Feature codes for
sheet metal application.

7.20 Product Level Classification

Product level classification is normally determined by the user and relates the features in an application domain to their uses or functions. From the manufacturing point of view, the use of a feature is often related to what tools are used to make them. For instance, in sheet metal application, a hole can be made by a hole puncher and thus there is correspondence to the tool geometry and the feature made. In assembly, the same hole could have a functional implication of holding another object's feature in place.

7.21 Concurrent Engineering

Recent efforts in industry and research laboratories across the United States, Europe and Japan have attempted to integrate the design work with other life cycle issues, including the evaluation of the product for producibility, inspectibility, serviceability, and so on. This means that the design of the product and the manufacturing system are executed concurrently.

Nevins *et al.* (1989) stressed that engineering schools teach a fairly straightforward version of how something is designed. This approach is illustrated in Figure 7.53. It can be seen that product and process designs are sequential in this approach. Needs are determined, product specifications are done, trial designs are produced, prototypes are produced for bench testing, and then final designs are produced. After this the required manufacturing process plan is developed.

The traditional relationship between the designer and the manufacturing personnel has always been one of initiator and implementer. This relationship merely required the designer to dictate what should be done without much care about whether it could be achieved. Thus, designs are created and it is the responsibility of the manufacturing engineer to determine whether or not the product can be manufactured. If the product violates manufacturing rules, then either the manufacturing environment is redesigned to accommodate the new product or a redesign of the product by manufacturing engineers, in consultation with design engineers, will be essential. Violations can occur in several ways. Available tools and equipment might be incapable of fabricating the product to the desired tolerance, or the general cost of making the product might be beyond the budget of the company.

It is clear then that design of a relatively complex high-technology product will require a lot of analysis, investigation of basic physical processes, experimental verifications, complex trade-offs, and difficult decisions as described in Nevins *et al.* (1989).

The lack of communication between the designer and the manufacturing engineer

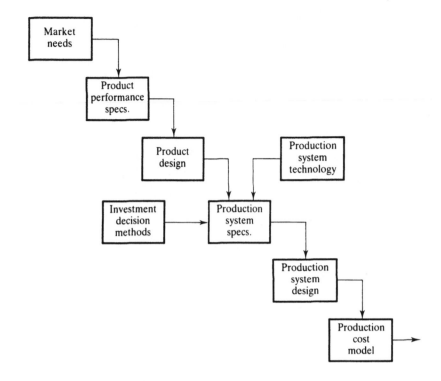

Figure 7.53
The traditional product-process design approach.

can be bridged by:

(1) training the design engineer in manufacturing principles,

(2) making the manufacturing engineer a design engineer,

(3) having manufacturing engineers work with design engineers during the design of a product.

The first two options are difficult to achieve since neither the design engineer nor the manufacturing engineer would want to take on the full responsibility of the other. It seems plausible that the manufacturing engineer and the design engineer could work together to produce a product that can be manufactured.

The state of the evolution of product and process design interaction is as shown in Figure 7.54, which is adapted from Nevins *et al.* (1989). This figure shows the depth of interaction between the product design process and the process design. Since no designer has the total knowledge necessary to evolve the concurrency in design and product, and process, a team approach is the advocated norm today. It means that people whose roles normally come later in the life of the product evolution are involved early in the design process.

The method of effecting this cooperation can vary from company to company. But in general, the manufacturing engineer would provide some manufacturing rules which become design constraints for the designer. A designer can submit a 'draft' design to a manufacturability evaluator, which can be an expert system or just an automated evaluator. Design for manufacturability requirements will vary

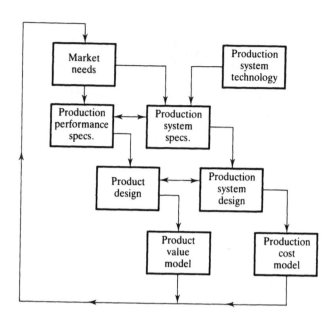

Figure 7.54
The strategy for
concurrent product
and process
development.

from one manufacturing application to another. That is, it will be different for
metal machining than for assembly processes.

Designing for manufacturability requires the fusion of the designer's intentions
for a part with the manufacturing engineer's manufacturing requirements for the
product. The designer's intentions are a set of functions which the product will
provide or require. The functions are related to the features of the product as
described in Nnaji (1990). For instance:

(1) What is the function of a given feature on the part?

(2) What operation is implied by this feature?

(3) What tool can be used to make this feature?

In actuality, the design engineer is usually totally responsible for answering
question (1). He communicates the responsibility of a feature entity on a part for
life cycle consideration. It could be that the feature is on the part only to satisfy
a temporary function, for instance ease of assembly, or that the feature is there to
serve a life-long function of the part or the product. In some cases, a surface may
only be on a part to provide support or to provide solid requirement for the part.
In the case where a surface provides support, it may be required to satisfy certain
spatial relationships with other part surfaces or with features of other parts.

Question (2) is partially answered by the design and partially by the manufacturing
engineer. Although this question derives partly from the functional intentions of
a part or a part feature with regard to its intermediate and ultimate use in some
part or product, the operational requirements in large measure derive from
geometric considerations and material properties. For instance, while knowing
that a hole is to be made may be a designer's prerogative, knowing that a drilling

operation will yield the geometric configuration which we call a hole is an operational consideration which the manufacturing engineer stipulates. While the design engineer stipulates the tolerance requirements for the part, the manufacturing engineer uses operational requirements planning to assure that the desired tolerance is achieved. The manufacturing engineer goes the next step to stipulate which machines and/or tools can be used for the drilling operation and establishes not only the operational sequence but the operational tool requirements as well, see Figure 7.55.

In answering question (3), the manufacturing engineer takes the designer's intentions, which can include the temporary functional objectives of the feature as well as the operational requirements for making the part, to generate a workable process plan which will obey the capacity constraints of the manufacturing environment of the organization.

Designing for manufacturability allows the designer to work with a set of manufacturing constraints and requirements in the design of the product.

It is clear from the above discussion that design for manufacturability is achievable by establishing some design rules in the form of knowledge engineering which can be used to guide the design to obey manufacturing requirements.

Computationally, predicate logic, semantic networks, production rules or frame theory can be used to represent the manufacturing knowledge (Chapter 3). These methods can represent data in a uniquely specified structure which can be augmented with deductive reasoning, as discussed in Ando *et al.* (1988). The more common method for representing manufacturing knowledge has been frame theory. In this application, frames can be used to capture the various manufacturing rules as well as tool requirements for various processes. Certain parameter constraints in the process planning activies such as process selection, machine selection, process sequencing, tool selection, and jig and fixture selection can be captured in the manufacturing knowledge base. It will be useful to use manufacturing application areas to illustrate the design for manufacturability concept in an application area. Then we will present the product modeler to show how future CAD and manufacturing relationships will work.

The basic rules in designing for manufacture are described in Boothroyd (1975) and are as follows:

(1) use standard components whenever possible;

(2) take advantage of the work material's geometric shape in order to design the components;

Figure 7.55
The mapping relationship of object and feature form and function.

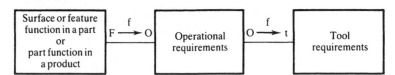

Where: F is the feature, surface or part functional responsibility
O are the operational requirements
f is an algebraic mapping function
t are the tool requirements

(3) use previous designs whenever possible;

(4) minimize required machining whenever possible;

(5) during geometric shape design, consider the ease of material handling, fixturing, machining and assembly;

(6) avoid over-specified tolerances and surface finish specifications;

(7) consider the kinematic principles during the initial steps of a design.

Whitney *et al.* (1989) outlined a guide to strategic design in the concurrent engineering approach:

(1) Determine the character of the product to understand what kind of product it is and how to develop appropriate design and production methods. The character of the product is articulated with the resulting consequences. For example, complex items, with no model mix, which are used by untrained personel, must have 100% reliability when used once and thrown away. Examples of such products are missiles, fire extinguishers and so on. The consequences of such a product character is that the product should be made of high-quality parts, glued or welded together. It should not be designed for repair after production.

On the other hand, if the product is a complex item, with a model mix, the user has options. Also, if this user is untrained and the product is intended to last for years, then the consequence is to make the product with high-quality parts, screwed together, and to provide replacement parts and accommodate field repair service.

(2) Subject the product to a product function analysis to assure that the design is made rationally.

(3) Carry out a design for producibility and usability evaluation to determine if the product's producibility and usability can be improved without impairing its desired functions.

(4) Design a fabrication and assembly process for the product that takes into account its character. This task can be accomplished by determining the appropriate assembly sequence, identifying subassemblies, integrating a quality control strategy with assembly, and designing each part so that its functional tolerances and tooling tolerances are compatible with the assembly method and sequence, and its fabrication cost is compatible with the product's cost goals.

(5) Factory design should fully involve the production workers, and ensure that inventory (including in-process inventory) is minimal, and is integrated with the procedures and capacities of the vendor.

7.22 The Product Modeler

The product modeler will be capable of creating a product in engineering and directly using the basic elements of the model throughout the life cycle of the component. Figure 7.56 contains the system architecture of the product modeler (see also Section 2.3.3). The life cycle can range from engineering to manufacturing to quality control. The product modeler is thus expected to possess an intelligent database which documents all the product information. This database supports many architectures. Figure 7.56 contains the conceptual environment of the product modeler.

The concept of concurrent engineering whereby some aspects of the process plan are evolved at the time of design is a goal of the product modeler. Some of the key development areas which will play significant roles in the evolution of the product modeler are the generalized feature characterization allowing for design with features and feature-based reasoning for process plan generation; and development of expert systems for manufacturing.

The ProMod systems, see Nnaji (1990), illustrate well the product modeling type of CAD system. The ProMod system embodies the attributes of the future CAD system as explained above. As can be seen, the CAD systems of today produce drawings on paper or geometric models on the computer as a result of a design. Since they are only capable of producing the geometrical characteristics of the part, they are merely able to assist in the final stages of the design process which is the design documentation, for example, drawings. The product modeler can capture the designer's intentions throughout the whole stage of the design process. The product modeler is concerned with the process which generates the designer's intentions to ensure preservation of the information. For instance, the physical and geometric properties of a part are normally conceptualized at the early stages

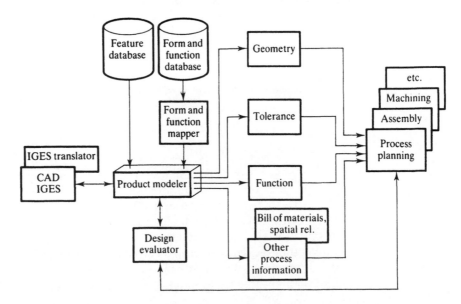

Figure 7.56

The schema of the product modeler (Source: Nnaji (1991)).

of design, the conceptual design and detail stage rather than the drawing stage. In conventional design/manufacturing systems, the production planner must supplement the raw geometric data with additional manufacturing information such as dimension tolerances, surface finish, straightness of surface, spatial relationships between surfaces and so on. In addition, the functions of features and surfaces are incorporated by the planner.

7.22.1 Design for automated assembly

In the past, designers have produced designs for assembly in which little or no attention was paid to how the product was made. As mentioned in the previous section, what is needed is a design system which accommodates the manufacturing requirements. Therefore, in the assembly application, the goal will be to design products which are not only assemblable but also cost efficient as products. In addition, they should be easy to disassemble.

It is essential for assembly to be considered at three levels, including product variety, product structure, and the product. This consideration allows for product rationalization. Productivity is influenced by product variety. If a product is not designed to accommodate the variations in product lines, then costs will increase due to non-flexibility. Also, since frequent production starts and stops increase cost and production time, fewer components should be desirable. Therefore, similarity of parts as well as minimization of the number of components needed for a product should be an objective. The product structure consists of the components which aggregate to yield the product. Figure 7.56 illustrates the method of capturing the product structure.

In designing for ease of automated assembly, two general objectives can be seen. The first one is to advise the designer of the difficulties in automated handling of the parts during assembly, as discussed in Swift (1987). The second is the estimation of handling cost for parts being designed.

In the design, four major solution stages have been identified, also discussed in Kusiak (1990):

(1) feeding of parts,
(2) orientation of parts,
(3) presentation of parts,
(4) handling system cost.

The feeding of parts necessitates a check whether the part can be fed automatically. The result of this check could be that (1) there is no suitable feeding method for the part in its current design; (2) the part could be handled by incurring some extra time, or; (3) the part could be handled without extra penalty in time or cost. In the first two cases, some redesign changes would have to be made to stay within the required design constraints. Some design rules in the form of using the production rules could be used to deal with feeding problems. These rules will deal with such issues as tangling, flexibility, weight, size, and quality.

The orientation shows the general attitude of the part or feature with reference to some other parts or features in the assembly task. In automatic assembly to date, this problem has been largely solved using the α and β symmetry approaches of Boothroyd and Dewhurst (1983). The symmetries are illustrated in Figure 7.57. In 'intelligent' CAD systems such as ProMod, see Nnaji (1990), the symmetry of a part can be automatically inferred from the spatial relationships which have been specified by the designer. In any case, the orientation problem is normally solved to determine the complexity of orientation of a part as depicted in the tooling time and efficiency or orientation.

The presentation problem is normally solved to determine the best approaches to presenting parts for the orientation area, and presenting the oriented parts to the assembly area.

Finally, handling system cost can be classified into the basic feeder cost and the cost of tooling, see Kusiak (1990). The handling device could be a vibratory feeder as shown in Boothroyd and Dewhurst (1983), in which case the cost will be for the basic feeder (including drive unit and bowl) and the cost of tooling the bowl.

7.22.2 Product assembly modeler

The objective of product assembly modeling is to illustrate the product modeling concept in the assembly domain of application. One major stumbling block in the development of the automatic process planning for assembly is the loss of parts relationships among the assembly objects. Although standard systems such as PDES may eventually provide the framework for transferring such information

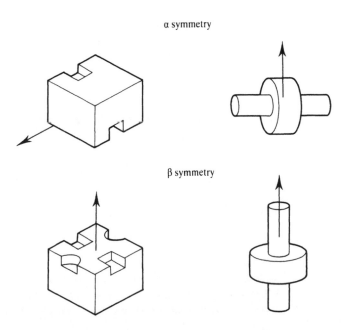

Figure 7.57
Symmetry of objects.

from the designer to the process planner, not much work has gone into this area. Whitney *et al.* (1989) have shown how to automatically generate and evaluate alternative assembly sequences coupled with economic analysis tools for designing assembly systems. It is a bold effort in addressing the shortcomings of today's design process. In addition to the assembly sequence problem, there is a need to automatically generate key process information directly by the designer. This product assembly model (PAM) serves as an interface between designer and process planner.

The product assembly model works as follows:

(1) A designer uses a modeler, for example, CATIA, GEOMOD and so on, to represent a product.

(2) The designer initiates the product assembly model which provides a list of all the components which make up the product.

(3) The product assembly model provides a menu system through which spatial relationships can be established for each pair of components. The system also automatically labels the faces and features of each component, discussed in Nnaji and Liu (1989).

(4) The system will then arbitrarily choose a part from the list and exhaustively pair the other parts against this part. Each pair is presented on the CRT screen along with a list of spatial relationships, such as against, fits, attached, coplanar, aligned and so on, which the designer can use to establish the relationship which must exist between the two objects, Figure 7.58.

(5) The designer uses the menu system (employing a mouse or the keyboard), to establish spatial relationships between each pair of components. The output of a work session is shown in Figure 7.58.

The product assembly model is also capable of building the bill of materials tree. A bill of materials can be built by using the physical relationship **part_of** to connect parts and establish inheritance attributes (where they exist) in **ancestor/ descendant** manner. When all the parts in the list are linked in the bill of materials, an upper or a lower triangular matrix will thus be established for the parts and their relationships. A bill of materials graph can be obtained from this, and the number of part *i* which directly goes into the making of part *j* will be used to complete the product. Similarly, the product assembly modeler automatically establishes feature matrices needed for linkage. The feature matrices work the same way as the above parts relationships.

One of the most significant benefits of building this product assembly model is seen in automatic assembly planning. A planner must establish these relationships in the form of expressions. For instance if part A is against part B. The expression:

against(PartA,Feature(face),PartB,Feature(face))

can be derived; where *against* is a developed spatial relationship protocol, also discussed in Nnaji and Liu (1990). With a product assembly modeler, we can now automatically generate the same expressions based on the entries into the triangular matrix. It is now easy to develop and solve the huge algebraic problems inherent

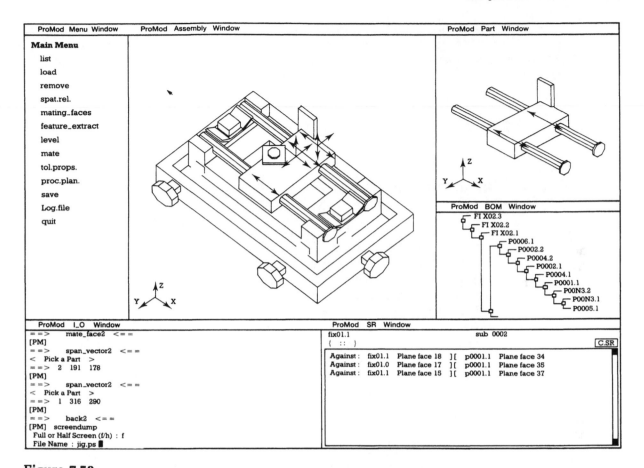

ProMod Menu Window   ProMod Assembly Window   ProMod Part Window

Main Menu
    list
    load
    remove
    spat.rel.
    mating_faces
    feature_extract
    level
    mate
    tol.props.
    proc.plan.
    save
    Log.file
    quit

ProMod BOM Window

FI X02.3
FI X02.2
FI X02.1
P0006.1
P0002.2
P0004.2
P0002.1
P0004.1
P0001.1
P00N3.2
P00N3.1
P0005.1

ProMod I_O Window

```
==>     mate_face2  <==
[PM]
==>     span_vector2  <==
< Pick a Part >
==>  2  191  178
[PM]
==>     span_vector2  <==
< Pick a Part >
==>  1  316  290
[PM]
==>     back2  <==
[PM] screendump
Full or Half Screen (f/h) : f
File Name : jig.ps ▮
```

ProMod SR Window

fix01.1   sub 0002

{ :: }   C.SR

Against :   fix01.1   Plane face 18   ][p0001.1   Plane face 34
Against :   fix01.0   Plane face 17   ][p0001.1   Plane face 35
Against :   fix01.1   Plane face 15   ][p0001.1   Plane face 37

Figure 7.58
Product assembly
modeler.

in solving the spatial relationships of an assembly manipulator with its work
environment. The latter is represented in a world model.

There is another major advantage. The establishment of relationships among
fixed objects in an assembly world will be easier. This task in world modeling has
been quite cumbersome and therefore discouraging to even the most dedicated
automatic programming workers. Once the entities in a facility plan have been
identified, the object spatial relationship approach can be used to relate them in
the world.

The product assembly model is an interactive module through which a designer
or process planner can enter the functions between assembly parts and specification
requirements. This information contains the designer's intent in carrying out
assembly process planning. The designer creates the part on a CAD system with
boundary representation or CSG as its mode of representing the CAD data.
Through the product assembly model, the geometric information combines with
linkage functions, and design specifications. The kernel of the product assembly
modeler is based on the spatial relationships among the parts being assembled.
While these spatial relationships are being assigned, attributes such as linkage

functions, assembly tolerances, required torque for screwing, the contact force exerted on each part which signals the robot to stop (using the information carried via force sensors) and so on, are requested.

7.23 Quality Methods in Design

7.23.1 Value engineering

One of the earliest methods for reducing product cost while maintaining quality and function is value engineering (VE) which originated in the 1950s. The original intention in value engineering, discussed in Salvendy (1982), was to apply the concept throughout the organization with the type of multi-discipline teams advocated in concurrent engineering. It differs from concurrent engineering in that it assumes that a design is available and the design can then be improved. Recent value engineering applications have tended to look at single components rather than the entire product. The concept uses a functional statement formulated to evaluate whether the item can perform the function at a cost, with savings derived by eliminating some nonessential functions or by redesign. Team members also contribute to deciding the actual value of a given function and functions which are too expensive and are targeted for improvement.

7.23.2 Taguchi method

Another method which is quite effective and which is targeted at quality improvement is the Taguchi method, explained in Taguchi and Wu (1986). It is a mathematical analysis method which is intended to aid the designer in creating a product that is produced within economic tolerances, on economical equipment and which still functions as desired. There are several important elements identified in product design. The first element is system design, which implies the product concept design and engineering analyses to assure functionability. The second element is parameter design which implies the selection of dimensions and tolerances so that the product can be produced. The third element is tolerance design which implies tightening performance specifications on materials, work-pieces, and processes so that the product will work. Taguchi does not recommend going as far as tolerance design since it is not smart design and implies wasting money. He states that designers often tend to perform system and tolerance designs, skipping over parameter design and the encumbered benefits it affords in providing a robust product-process design.

System design in this method involves more than product concept design. It involves the separation of design factors into controllable and uncontrollable factors – 'noise'. The controllable factors are those that are predictable variables, which can be made to conform to certain desired levels and tolerances. Examples are time and weight elements. Uncontrollable factors are often futile and expensive to correct. Instead, the method suggests that designs of product and process should be aimed at providing immunity to the noise as much as possible. The method is called 'reducing the product's sensitivity', or 'offline quality control'. This implies that the quality control is designed into the product. This contrasts with the conventional approach, online quality control, where the attempt is to keep machines running at tolerance performance, or improving their behavior to reach tighter tolerance.

Parameter design is an attempt to use statistics to deal with unpredictable factors. In this, one identifies the probable noise factors and the controllable factors associated with the product's function. A set of experiments is then designed using standard statistical methods to determine how the controllable and uncontrollable factors affect performance. It is important to note that the controllable variables must be controllable during experiments for this scheme to work. Results are then analyzed to determine which settings of the controllable variables produce the least variations in the uncontrollable variables. These settings are the ones which will be least sensitive and are therefore chosen.

7.24 Life Cycle Costs in Design

Fabrycky and Blanchard (1991) state that system engineering is a process that has recently been recognized to be essential in orderly evolution of people-made systems, and involves the application of efforts to:

(1) Transform an operational need into a description of system performance parameters, and a preferred system configuration through the use of an interactive process of functional analysis, synthesis, optimization, definition, design, test and evaluation.

(2) Incorporate related technical parameters and assure compatibility of all physical, functional and program interfaces in a way that optimizes the total system definition and design.

(3) Integrate performance, producibility, reliability, maintainability, human factors, supportability, quality and other specialties into the overall engineering effort.

The role of systems engineering is an unstructured set of goals of concurrent engineering. But frequently, products and processes are planned, designed, produced and operated with little or no attention paid to life cycle costs. In many cases when costs are considered, they are only considered in a fragmented manner. For instance, usually, the costs associated with research, design, testing, production or construction, customer use, and field service are considered in isolation. Experience now shows that the major portion of the total cost of many products derive from their operation and service. These cost commitments should actually be made at the early stages of design. Also, the various costs associated with the system are interrelated and therefore should be addressed together thereby providing the opportunity to evaluate the overall product. Figure 7.59 adapted from Fabrycky and Blanchard (1991), shows the various stages of a life cycle cost analysis process.

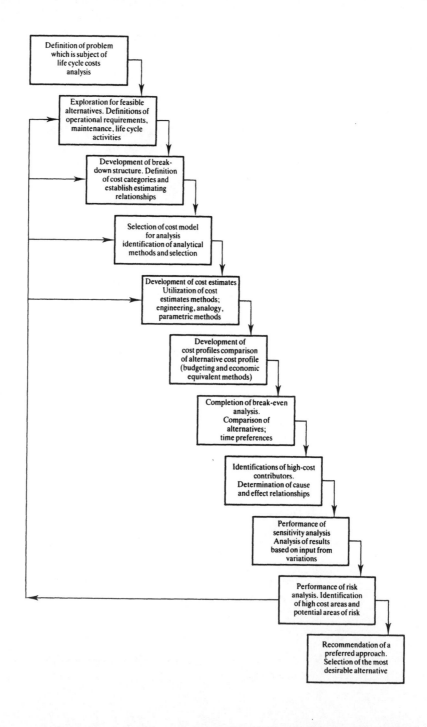

Figure 7.59

The basic life cycle cost analysis (Source: Fabrycky and Blanchard (1991)).

7.25 Conclusions

In this chapter, we have presented the role of the computer in design. We have presented a more contemporary perspective of design which embodies the advocacy of concurrency in product and process design. We presented a product modeling system which is representative of the desired futuristic designer system. Finally, we discussed the progressive issues of standards and considered a product's entire life cycle during the design process.

7.26 Problems

(1) What was the significance of the APT project in the development of today's CAD systems?

(2) Describe the traditional design process. What are the limitations of this approach?

(3) What areas of the design process can be computerized and what way do you think this is possible?

(4) Give three ways for constructing each of the following: circles, surfaces, and arcs.

(5) How can a cubic box ($3 \times 4 \times 2$ cm) be rotated an angle of $90°$ from the horizontal to the vertical in the positive direction and then translated a distance of 2 m?

(6) Derive the equations for rotating a 2-D object in the plane.

(7) Show how to scale the box above by a factor of 2 in x, and y.

(8) If this box shows an inversion, write the expression for the new reflected box.

(9) Why are raster graphics so popular in modern CAD systems? Compare this to other approaches for representing data on the CRT.

(10) Use the wireframe model to show how to organize the database of a CAD system. Can you use the wireframe model to describe assembly operations for a robot?

(11) Why is solid modeling the best technique in today's intelligent CAD systems? Describe the features of boundary representation that make it attractive for geometric reasoning.

(12) In CAD data representation in automated manufacturing, what elements are necessary for unique representation?

(13) Use a gearbox assembly to illustrate the models of a bill of materials in CIM.

(14) What are the limitations of IGES? What features of STEP make it valuable in a CIM environment?

(15) Discuss the main principles which are essential in the development of a product modeler.

(16) Develop a framework for a product modeler for sheet metal component fabrication.

(17) What is the principal function of the Euler formula in geometric reasoning? Why is this needed in automated manufacturing?

(18) What are the principal levels in feature classification hierarchy?

(19) Can concurrent engineering concepts be applied to all domains? How can you tailor concurrent engineering for the design and manufacture of electronic chips?

(20) Discuss an integrated product model architecture for the production of an automobile. How would this product model differ from, say, the manufacture of dentures?

8

Process planning and manufacturing scheduling

CHAPTER CONTENTS

8.1	Introduction	371
8.2	Process Planning	372
8.3	Automated Process Planning	373
8.4	The Operational Sheet	375
8.5	Group Technology	377
8.6	Coding Structure	379
8.7	Available Coding Systems	383
8.8	Design Data and Automated Process Planning	386
8.9	Manufacturing Resource Planning	401
8.10	Material Inventory Systems	404
8.11	Material Requirements Planning	407
8.12	Lot Sizing Techniques	411
8.13	Sequencing of Operations	418
8.14	Scheduling Systems	430
0.15	Project Scheduling	439
8.16	Assembly Line Balancing	443
8.17	Summary	447
8.18	Problems	447

8.1 Introduction

In discrete parts manufacturing, process planning involves the act of preparing a plan which outlines the operations, routes, machine tools, fixtures, tools, and parameters required to transform a part or parts into a finished product. It can

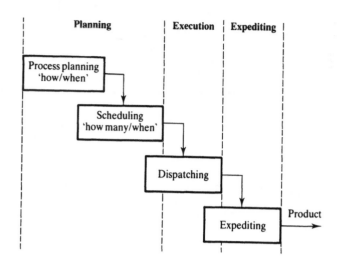

Figure 8.1
Production planning
and control.

be seen that the process planning activity has no time element associated with it. It is the manufacturing scheduling which has this time component. It involves the time ordered arrangement of a set of jobs (parts) to be processed on a set of processors (such as machines) to optimize some measure of performance. A job may consist of more than one operation. Scheduling could be deterministic or non-dynamic. Deterministic scheduling implies that all operational parameters are assumed to be known in advance. Figure 8.1 shows the relationship between these two planning systems in the production planning environment. The main difference between process planning and scheduling is that scheduling is time driven while process planning explains what will happen with the scheduled time zones and how it will happen. In this chapter, we will explore both process planning and scheduling systems in manufacturing.

8.2 Process Planning

Most of the process planning activities in today's industry still involve either manual preparation of the plans or semi-automated process planning. We will describe process activities in general; discuss the available works in process planning and then show what the desired automated process planning should embody.

Process planning involves the translation of design data to work instructions to produce a part of a product. The process planner normally uses the information presented on the engineering drawing and the bill of materials in the generation of an executable plan. Process planning is still prepared manually in the majority of industries. The plan can contain elaborate details or it can be simple. For instance, in a model shop where the machinists are highly skilled, the process plans are quite unique, and generally show workstation routes. The other aspects of the plan are left to the machinists to develop. In contrast, when a product is produced by an automated transfer line, the process plan would normally contain stage by stage details of activities. These two examples are extreme cases; the batch shop is the normal process planning environment and the following knowledge must be possessed by the typical process planner, also discussed in Chang and Wysk (1985):

- ability to interpret an engineering drawing,
- knowledge of manufacturing processes and practices,
- knowledge of tooling and fixtures,
- knowledge of shop's available resources,
- knowledge of how to use reference books, such as machinability data handbooks,
- knowledge of how to perform operation time and cost analyses,
- knowledge of raw materials,
- knowledge of the relative costs of processes, tooling and raw materials.

With these prerequisites, the process planner can use the steps shown in Figure 3.38 to evolve a suitable plan. These steps involve going from the study of the overall shape of the part to classify the part into a particular geometric family through to the preparation of the final process plan document.

8.3 Automated Process Planning

With the availability of the computer, attempts are made to automate most of the above steps. The automated process planner should have the following features, see also:

- It should operate as an integrated planning aid that obtains input data automatically from engineering and sales to generate a complete set of process plans to be used by production planning as well as manufacturing, material and quality control.
- It should provide basic data for work order routing, fabrication schedules, payroll accounting and material release.
- It should be generalized in nature in order to accommodate a variety of parts.
- It should possess a good interactive user interface in order to maximize fully the potential of the computer.
- It should be user friendly and provide operator guidance.
- It should be modular to allow for easy expansion, modification and maintenance.
- It should also be cost effective in implementation.

In summary, the key elements of process planning include material selection, operation selection, machine selection, operation sequencing, tool selection, and jig and fixture selection. Numerous factors affect the process planning activity: geometric shape, tolerance, surface finish, size, material type, quantity and the manufacturing system itself, as explained in Chang and Wysk (1985).

As mentioned earlier, many of the above activities are still performed manually. Automation requires computerization of these activities. When many of the functions above have been computerized, they require little or no time on the part of a human process planner to execute. It is the system in which these functions are automated that is of interest in this book. There are several advantages to computerized process planning:

- reduction in process planning time,
- reduction in the required skill of the process planner,
- reduction in costs due to efficient use of resources,
- increased productivity and process rationalization,
- production of accurate and consistent plans,
- maximization of just-in-time performance.

There are two types of process planning techniques: the **variant** and the **generative** techniques. The variant technique uses the classification and coding of parts to initiate the process planning activity. When a plan is to be generated for the production of a new product, a standard plan for a similar product is retrieved and modified for the new product. The plan may be a non-parameterized model of the part, and the user just enters the parameters of the part he needs to describe. This approach is generally useful in cases where there is a lot of similarities between products. This technique is illustrated by CAPP, see Figure 8.2. CAPP will be discussed in more detail in Section 8.8.7.

8.3.1 Generative technique

The generative approach on the other hand, does not use any stored standard plan. When a plan is generated, the system uses information about a part's geometry, machining, or assembly data, machines (including robots) and their parameters, as well as process planning rules. There is currently no truly generative process planner. Existing systems still require human interaction and only work for non-complex geometric shapes.

If a family of parts has many common form elements then it is often possible to describe planning macros for these elements. Therefore, a hybrid process planning method can be conceived using both variant and generative process planning principles. Figure 8.3 illustrates the generative process planning system.

This figure shows how the generative process plan works for machining operations. The design data is transmitted to the planners by a modeling system which captures the design features, functions and general designer intentions for the product. This information along with process knowledge and raw materials are used to perform the process selection. The rest of the processes are very close

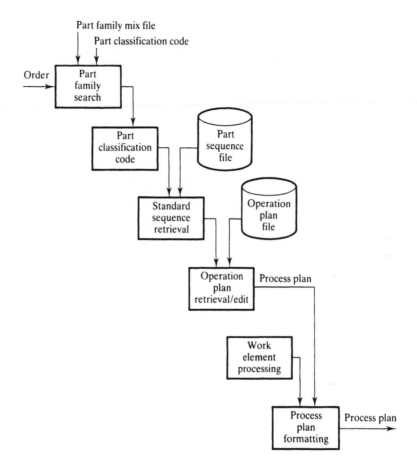

Figure 8.2
CAPP system
(courtesy of
Computer-Aided
Manufacturing
International Inc).

to the variant technique transactions. The major difference is that the CAD system data plays a major role in the generation of new plans and therefore the part description no longer needs to be done by code which accesses pre-stored routes and plans. These can now be generated to fit the part geometry and manufacturing context of the geometry.

8.4 The Operational Sheet

In order to understand the nature of the document normally produced by the conventional process planning system, a simplified manual operation sheet is presented in Figure 8.4. The sheet contains the following information: the heading identifies the planning sheet, its origin, important dates, the number of pieces to be manufactured, and any significant signatures to verify its contents. Next, there is workpiece-related information which identifies the part, the drawing, the

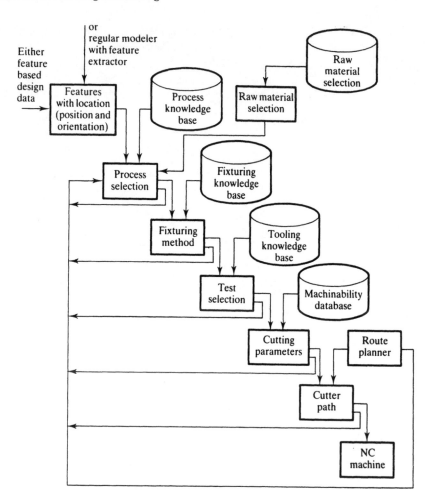

Figure 8.3

Automated process planning – generative technique.

Operation sheet No.:			Data:			
Part No.:		Part name:		Drawing No.:		
Orig:		Checked:		Changes:	Approved:	
Pieces:		Matl:		Weight:		
Op No.	Operation	Machine Tool	Tools	Fixtures	Set-up time (hrs)	Operation time (hrs)
5	Rough turning	Lathe 4	T5	Chuck	0.2	0.2
10	Fine turning	Lathe 2	T3	Chuck	0.1	0.2
15	Drilling	D Press 2	D2	Drill jig	0.15	0.1
20	Chamfer	D Press 2	Ch3	Drill jig	0.1	0.07
25	Counterboring	D Press 2	D1	Drill jig	0.1	0.09
30	Heat-treat	Furnace			0.15	0.09
35	Grinding	Grind S			0.15	0.06

Figure 8.4

A conventional manual operation sheet.

classification, the part family and its physical parameters. The third aspect of the information shows the operation number and description, machine tools to be used, tools, and set up and operation times.

This information is needed regardless of whether the process planning method is manual or automated; an automated process planner must be capable of producing such information.

8.5 Group Technology

When a part is designed for manufacturing, it may require several succeeding manufacturing operations. If there are many different parts requiring different operations, it will be necessary for workpieces to share common processing equipment. But only parts which by their attributes share common operations can benefit from the sharing of processing equipment. Therefore, recognition of similarities is essential for the classification of parts according to their geometric or fabrication method similarities.

Figures 8.5 to 8.8 show different similarity attributes, see also Rembold (1985). In Figure 8.5, the three parts belong to the same rotational part family and require similar turning operations. Figure 8.6 shows parts that do not appear to be similar but can in fact be made by the same multi-axis machining center which can employ the same tools. Figure 8.7 shows dissimilar parts which have one operation in common, namely the drilling of four holes. Finally, Figure 8.8 shows two parts completely identical in design, but made of different materials. Because of this, the plastic part will be injection molded while the steel part will be turned.

Group technology is an attempt to find common manufacturing processes and to schedule parts through a manufacturing facility to maximize utilization of available resources. The groups obtained normally carry sets of codes in which each element of the code describes some attribute associated with the part. These attributes can range from geometric information to manufacturing process information. The code is quite different from the part number; although some organizations use the group technology code as a part's number. The part numbers

Figure 8.5
Rotational part family requiring similar turning operations.

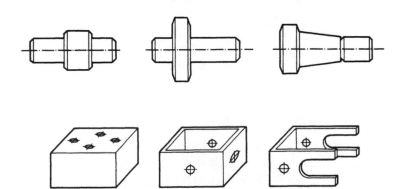

Figure 8.6
Similar cubic parts requiring similar milling operations.

Figure 8.7
Dissimilar parts
requiring similar
machining operations
(hole drilling, surface
milling).

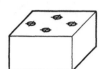

Figure 8.8
Identically designed
parts requiring
completely different
manufacturing
processes.

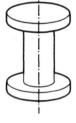

Material: Plastic   Material: Steel

are merely an identity for the individual part. A part may also change its number from one process to another.

The concept of group technology (GT) has wide applications. It is a concept with which humans solve a variety of problems. Our languages are meaningful because of group technology. Group technology is a realization that various objects and situations have similarities which make it possible to provide generic solutions or analyses to all the members of the given group, thereby saving tremendous resources, time and effort.

In manufacturing, group technology is used to classify parts, operations, and processes on some set of attributes for retrieval and analysis with applications in the following:

- marketing
- engineering
- manufacturing engineering
- manufacturing
- purchasing
- material control

The general principle is that when classification is completed in group technology, it is normally coded for efficient use. We will therefore discuss the coding system in the following section.

8.6 Coding Structure

A part coding system consists of a sequence of symbols that identify the part's design and/or manufacturing attributes as described in Groover and Zimmers (1984). The symbols are usually alphanumeric, although most systems use only numerals. The three basic coding structures are:

(1) chain-type structure (Figure 8.9),

(2) hierarchical structure (Figure 8.10),

(3) hybrid structure, a combination of hierarchical and chain-type structures (Figure 8.11).

In the chain-type structure, the interpretation of each symbol in the sequence is fixed and does not depend on the value of the preceding digits. Another name commonly given to this structure is polycode. This code can easily be read by humans, however it wastes computer memory.

In the hierarchical structure, the interpretation of each succeeding symbol depends on the value of the preceding symbols. Other names commonly used for this structure are monocode and tree structure. The hierarchical code provides a

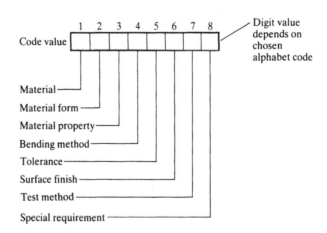

Figure 8.9
Example of chain-type code.

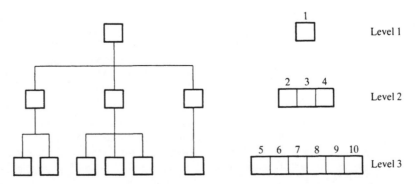

Figure 8.10
Hierarchical coding.

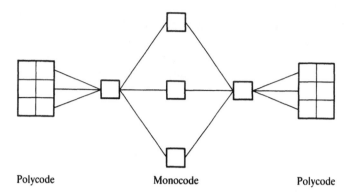

Figure 8.11
Hybrid coding
structure.

Polycode   Monocode   Polycode

relatively compact structure which conveys much information about the part in a limited number of digits. This code is well suited for processing by computer. However it is difficult for people to understand.

The problem associated with polycodes is that they tend to be relatively long. On the other hand, the use of a polycode allows for convenient visual identification of specific part attributes by the user. This can be helpful in recognizing parts with similar processing requirements.

To illustrate the difference between the hierarchical structure and the chain-type structure, consider a two-digit code such as 37 or 47. Assuming that the first digit represents the overall part shape, the symbol 3 may mean cylindrical part and 4 may mean polyhedral geometry. The symbol 7 is independent of the preceding digit and is interpreted the same way no matter what the value of the first digit is, for example, a 7 could indicate a length. In a hierarchical code system, the interpretation of the second digit would then depend on the value of the first digit. If preceded by 3, the 7 might indicate some length/diameter ratio, and if preceded by 4, the 7 might be interpreted to specify some overall length.

Most of the commercial parts coding systems used in industry are a combination of the two pure structures. The hybrid structure is an attempt to achieve the best features of monocodes and polycodes. Hybrid codes are typically constructed as a series of short polycodes. Within each of these shorter chains, the digits are independent, but one or more symbols in the complete code number are used to classify the part population into groups, as in the hierarchical structure. Thus hybrid coding seems to best serve the needs of both design and production. These number codes typically range from 6 to 30 digits. When implementing a parts classification and coding system, most companies elect to purchase a commercially available package rather than develop their own.

In general, the coding can be applied enterprise-wide. The applications (given in the prior checklist: marketing, engineering, manufacturing engineering, manufacturing, purchasing, and material control) can each be examined according to the type of data required and within a given specific scope of the application. For instance, manufacturing engineering can be coded for process data and within the scope of mechanical components. We will present a further breakdown of the application areas for the type of retrieval that it could be coded for.

Although the intention in the applications checklist in Table 8.1 is to be able to develop coding for each segment, it is possible to develop coding for a combination of segments. For instance, in manufacturing engineering, process plans retrieval could be combined with process plan standardization, manufacturing routing, tooling retrieval, and machinery planning to obtain a coding structure that is geared toward an automated process planning system.

There are usually four generic types of data:

(1) product data

(2) process data

(3) resource data

(4) measurement data

The product data can include part number, drawing number, shape, and tolerances (see Table 8.2). The process data includes process code, standard plan,

Table 8.1 The applications checklist for manufacturing coding

Application	Data requirements
Marketing	• design data retrieval • manufacturing data retrieval • cost data retrieval • product analysis
Engineering	• design analysis • design standardization • value engineering • value analysis • cost estimating • manufacturing cost/design guide
Manufacturing engineering	• process plan standardization and control • manufacturing routing and cost analysis • NC family of parts programming • tooling retrieval and standardization • machinery planning • flexible manufacturing system (FMS) design • design of small parts feeding (orientation)
Manufacturing	• work cells • flexible manufacturing systems
Purchasing	• family of item bought • vendor analysis • substitution
Materials control	• raw material standardization • in-process inventory reduction • salvage and effective re-utilization of inventory • quick response to service parts requirements • inventory of partially completed items to produce a family of parts

Table 8.2 Requirements by generic class

Generic data class	Requirements	
Product	part number	material
	drawing number	hardness
	description	tolerances
	revision level	surface finish
	drawing size	coating
	design location	color
	shape code	design standards
	function code	inch/metric
	dimensions	
Resource	work center number	tooling
	machine code	skill
	manufacturing section	primary vendor
	plant location	alternate vendor
	facility	
Measures	annual quantity	weight
	lot size	lead time
	material cost	set-up time
	manufacturing cost	run time
Process	process code	test data
	heat treat	process planner
	standard plan	engineered time standards
	alternate plan	effective date
	NC family program number	plan revision level
	make/buy identifier	process standard

alternative plan and so on. Resources data can include the work center number, tooling, skill and machine code. Finally, the measurement data includes the annual quantity, weight, lot size, and other measurables.

Table 8.2 shows a breakdown of the requirements in each of the generic data classes.

The scope for the coding system determines whether the application is in raw material or finished components and the type of finished components. The scope checklist includes the following:

- raw material
- finished components
  - mechanical
  - electronic
  - electrical

Each of these can be seen from the standpoint of coding intentions as follows:

- commercial items
- fabrication
- assembly
- tooling
- machines and equipment

In constructing the coding system for a components representation, the following factors are of interest:

(1) the type of components (sheet metal, rotation and prismatic parts, and so on),
(2) the scope of the code representation,
(3) the type of coding structure,
(4) the digital representation (binary, octal, decimal, hexadecimal, alphanumeric, and so on).

The fewer the number of digits which can completely represent a component the better. The choice of digital code representation is usually dependent on the ease of processing the code.

Coding systems are usually constructed in such a manner that the most important parameters are described in the first fields (for example, the length/diameter relationship of a workpiece) and the lesser important ones in the following fields. The process planning system must be able to select the manufacturing process and tools from the code.

8.7 Available Coding Systems

There are several commercial coding systems available, including the Opitz coding system, the CODE system, the KK-3 system, the MICLASS system, the DCLASS system, and COFORM. Some of them are design-oriented and some are manufacturing-oriented. We will discuss the Opitz, KK-3, and MICLASS systems in more detail.

8.7.1 The Opitz coding system

The Opitz coding system is perhaps the best known. It was developed by H. Opitz at the Technical University of Aachen, Germany. The system is made up of five main digits with four supplementary digits (Figure 8.12). The first five digits represent the geometric information describing rotational, flat, long, or cubic parts. Thus, these first five fields indicate the shape or form of a component. With this information, an automatic process planning system can select the machine tools and tools needed for processing the part. The next four depict attributes of particular

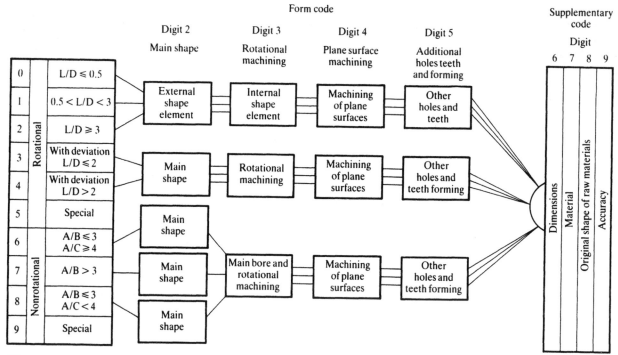

Figure 8.12

Basic structure of the Opitz system of parts classification and coding (Groover (1980)).

importance to manufacturing, such as dimensions, work material, starting raw workpiece and accuracy. It is possible to extend the Opitz code by four additional digits called secondary code. Usually, the secondary code is designed by the company using the Opitz code to suit its needs.

8.7.2 The KK-3 system

The KK-3 system was developed by the Japan Society for the Promotion of Machine Industry as a general purpose classification and coding system for machinable parts. The cutting system is used for metal cutting and grinding. There are 21 decimal digits in the KK-3 system. Because of the long code, it can carry more information than the Opitz coding system.

Figure 8.13 shows the concept behind the KK-3 system while Figure 8.14 shows an example of the application of the KK-3 system. A part name is described by two digits. The first digit classifies the general function such as gears, shafts, drive and moving parts, fixing parts and so on, as described in Chang and Wysk (1985). More detailed functions are presented in the second digit. Up to 100 functional names for rotational and non-rotational components can be classified.

Two digits are used to classify materials. The first shows material type and the

Digit	Items		(Rotational component)
1	Parts name		General classification
2			Detail classification
3	Materials		General classification
4			Detail classification
5	Major dimensions		Length
6			Diameter
7	Primary shapes and ratio of major dimensions		
8	Shape details and kinds of processes	External surface	External surface and outer primary shape
9			Concentric screw threaded parts
10			Functional cut-off parts
11			Extraordinary shaped parts
12			Forming
13			Cylindrical surface
14		Internal surface	Internal primary shape
15			Internal curved surface
16			Internal flat surface and cylindrical surface
17		End surface	
18		Nonconcentric holes	Regularly located holes
19			Special holes
20		Noncutting process	
21	Accuracy		

Figure 8.13
Structure of the KK-3
(rotational
components) (courtesy
of the Japan Society
for the Promotion of
Machine Industry).

second digit depicts the shape of the raw material. The KK-3 system also classifies
dimensions and dimension ratios. By studying Figure 8.13, the reader can find out
how the other fields are used for classification.

8.7.3 The MICLASS system

MICLASS, which stands for Metal Institute Classification System, was developed
in the Netherlands and is maintained in the USA by the Organization for Industrial
Research (OIR). MICLASS was designed to standardize many design, manufacturing,
and management functions. It carries information regarding main shape, shape
elements, position of the shape elements, main dimension, ratio of the dimensions,
auxiliary dimension, form tolerance, and machinability of material (Figure 8.15).
These are represented in the twelve main fields. There can be up to an additional
18 fields for user-defined functions which capture part functions, lot sizes, major
machining operations, and so on. MICLASS is used by many companies in Europe
and the United States.

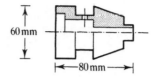

Code digit	Item	Component condition	Code
1	⎫ Name	Control valve	0
2	⎭	(others)	9
3	⎫ Material	Copper bar	7
4	⎭		
5	Dimension length	80 mm	2
6	Dimension diameter	60 mm	2
7	Primary shape and ratio of chief dimension	L/D 1.3	2
8	External surface	With functional tapered surface	3
9	Concentric screw	None	0
10	Functional cutoff	None	0
11	Extraordinary shaped	None	0
12	Forming	None	0
13	Cylindrical surface > 3	None	0
14	Internal primary	Piercing hole with diameter variation, NO cutoff	2
15	Internal curved surface	None	0
16	Internal flat surface	None	0
17	End surface	Flat	0
18	Regularly located hole	Holes located on circumferential line	3
19	Spatial hole	None	0
20	Noncutting process	None	0
21	Accuracy	Grinding process on external surface	4

Figure 8.14

Example of a KK-3 coding system (source: Chang and Wysk (1985)).

8.8 Design Data and Automated Process Planning

Most of the existing automated process planning systems are not of the generative type. Of the few generative varieties available, some use the CAD system as the basis for planning. Yet the big bottleneck in the automated generation of process plans is the issue of geometric recognition of shapes and mapping of the shapes into functions that can trigger machining operations. In the previous chapter we presented the CAD system. The CAD system of the future must be capable of capturing more than geometric information. It must capture the designer's view of constructing and manufacturing the part, which includes the preparation of process information. But understanding the geometry of parts to be fabricated or assembled is a very complex undertaking. It is obvious that parts features are the essential elements in the manufacturing of the parts.

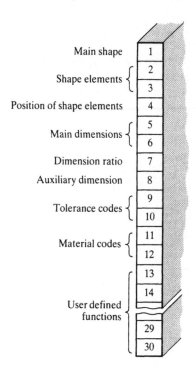

Main shape — 1
Shape elements { 2, 3
Position of shape elements — 4
Main dimensions { 5, 6
Dimension ratio — 7
Auxiliary dimension — 8
Tolerance codes { 9, 10
Material codes { 11, 12
User defined functions { 13, 14 ... 29, 30

Figure 8.15
MICLASS code
structure.

A feature can be defined as a set of surfaces together with specifications of the bounding relationships between them, which imply an engineering function on an object.

There are three ways of understanding what features exist in a part. The first one is by using manual methods to input the manufacturing information in a process planning system. The method is tedious and makes the system non-automated. The second approach is to freely design parts and then extract the features from the part using a combination of algorithms in computational geometry and artificial intelligence (see Section 3.5.4). This approach is very difficult to implement and solutions in feature extraction have only worked for limited domains of geometry, although more recent works report more robust feature extractors. The third approach requires that features be symbolically designed into a part so that the part 'remembers' that it carries the features. This approach is generally called **design-with-features** or **design-by-features**.

8.8.1 Feature extraction research

The conventional approach to feature-extraction is accomplished by the human planner examining the part and recognizing the features designed into the part.

This process only requires the planner to be familiar with the various features in the domain of interest.

With the use of a CAD system to generate the features, it will no longer be necessary to specify descriptions of a part in the form of group technology code. It will now be possible to capture information about tolerance, surface finish and so on, through the CAD system. Even assembly information, such as spatial relationships, can be captured using this design system.

Future CAD systems can help us work out, for example, that a certain type of feature is a hole and that a hole can be drilled, thereby requiring a drill press to use a drill tool. With this knowledge, we can discuss the invariant types of information, namely process selection, tool selection, and precedence creation.

What will largely be invariant in the future of process planning will be machinability data, tool information, machining force and process parameters.

This information can be stored in knowledge bases, with each knowledge base possessing rules which enable the appropriate decisions to be made about planning.

The contents of the database will be application-specific. For instance, if the application is assembly, then the crucial information will be a series of configuration states and the corresponding spatial relationships that exist among linking parts in the assembly sequence, as discussed in Nnaji (1990). Additional information may include grasp surfaces and manipulator orientations, and so on.

We will use the machining application to illustrate automated process planning data requirements for planning. We will discuss process selection, fixture selection, tool selection, cutting parameters, and tool path determination. Figure 8.3 shows the sequence of automated process planning.

8.8.2 Process selection

Process selection can be aided by information which is obtained from the design system. The number of design features and the process information enable the selection of the appropriate set of operations. Associated with this selection is the set of tools, auxiliary devices, and operations to be performed at a given cost.

While it would take little training for a worker to be able to associate a process with some geometric shape, it is, however, an arduous task for a machine reasoning system. Current automated process planning systems use codes to associate the part form with the appropriate process. The designed approach is, however, to use an automated system which can take the geometry and topology and map them into operations by first, symbolically, extracting features from geometry and topology. Figure 8.16 illustrates the two methods of obtaining features. One is through raw geometry of free designs and the other is by using a designer system such as ProMod, see Nnaji (1990), which produces the features automatically.

Design features can be obtained from a regular designer system which does not directly show which features are being used in a design. This type of designer system is still the most prevalent. The geometric data obtained from this system must undergo some reasoning process to obtain the designer's implied features – a process which can be likened to a 're-invention of the wheel'. This is because the

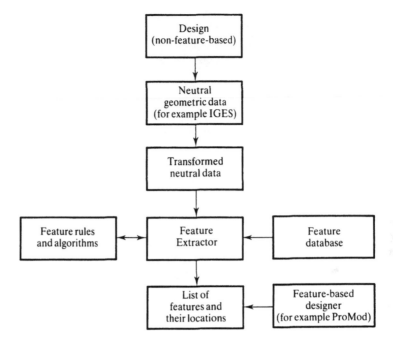

Figure 8.16

Feature extraction methods (source: Nnaji (1990) Automation and Robotics Laboratory, University of Massachusetts at Amherst).

designer may have initially conceived the features but has no way of capturing them in today's CAD system. A planner must therefore interpret the geometry to infer the implied features. In Figure 8.16, the design data from a regular designer system will be passed through a neutral geometric data conversion system such as the Initial Graphics Exchange Specifications (IGES), and a data restructuring system which produces geometric data in the form of computable vertices, edges, and faces. The transformation to neutral data is certainly optional, but it guarantees that the resulting data is CAD-system independent. The resulting set of computable vertices, edges, and faces is used in feature extraction where sets of topological and geometric patterns are matched against a database of features using some feature rules and algorithms. The result will be a set of features in part represented in a computable manner, sometimes in frames, semantic networks, graphs and so on.

The alternative approach is to obtain the design features directly at the time of design by using a design with a feature system such as ProMod, see Nnaji and Lui (1990). With this approach, a designer calls features symbolically (implied functional meaning); and the modeling system places the features at the desired locations on the design. The feature symbols implicitly activate the protocols which define the explicit specifications for the features. By using either of these approaches, a list of features and their locations can be obtained.

Table 8.3 is a simple example of how shapes can be associated with processes in the process selection. In the table, it can be seen that three different operations can be associated with a hole feature. How does one know which process to use for a given hole? The answer lies in the process boundaries. This is where the

Table 8.3   Some geometric forms and associated processes

Process	Geometric shape or feature
Boring	hole
Drilling	hole
Reaming	hole
Face milling	flat surface
Grinding	flat face, hole, outside cylindrical surface
Tapping	outside surface which can be generated by rotating a line or curve around an axis

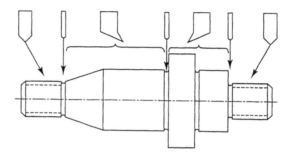

Figure 8.17

The selection of
cutting bits for turning.

limiting surface areas, tolerances, surface finish, and so on help to determine the process. A process boundary table can be developed for each process.

Boundaries can be set for the smallest tool sizes, largest tool sizes, positive and negative tolerances and so on. An equation for tolerance in the case of the hole-making process can be obtained as follows and is discussed in Chang and Wysk (1985).

$$Tol = A(D)^N + B$$

where:

A = coefficient of the process
N = exponent describing the process
B = constant describing the best tolerance attainable by the process
D = hole diameter

The value of A, B, and N can only be obtained by experimentation and no universal set of parameters exists to obtain these, making them system dependent. Figure 8.17 shows the selection of cutting bits for turning.

Selection of blanks or stock

This activity investigates the dimensions of the part, processing requirements (for example, heat treatment) and the number of parts to be made. The latter information has an influence on the type of the raw part to be used, for example, a low volume part may be machined from a rod and a high volume part from a forging. The rules can be categorized as follows:

- selection of material,
- determination of the type of blanks or stock,
- calculation of the machining allowance.

Machine tool selection

This is a more straightforward problem. However, there are also many parameters to be considered, including:

- size of part
- required surface finish
- machining sequences
- process variants
- required accuracy
- piece rate
- available machine tools

Process planning of bent sheet-metal parts as shown in Figure 8.18 creates different problems, see also Ehrismann and Reissner (1987). The rules determining the bending sequences must search for a collision-free space for the tools as well as for the selection of the sheet-metal parts which are being bent. The collision check is made for each two-dimensional sheet metal surface which is moved through three-dimensional space into its final position. Thereby, it may be necessary to try out a great number of sequences until one is found which permits complete bending of parts. With complex parts, as shown in Figure 8.18, the search may become a combinatorial problem resulting in long computer runs. For this reason, the search space should be pruned to a minimum.

8.8.3 Fixturing selection

Just as process selection is evolving into a CAD-based inference system, fixturing is going to be mainly done on the computer where the tool designer can select fixtures and perform fixturing of the raw material, and test the set up for structural integrity building without constructing a physical test model. We have presented some of the artificial intelligence based on approaches to fixturing in Chapter 3. Expert systems will probably be the most important tools for fixture solutions. This is because the fixturing experience can conveniently be represented by rule-based presentation of knowledge. Figure 8.19 show methods of presenting knowledge for fixturing, see Eversheim 1989.

In a study by Kanumury, Shah and Chang (1989), the functions of fixture

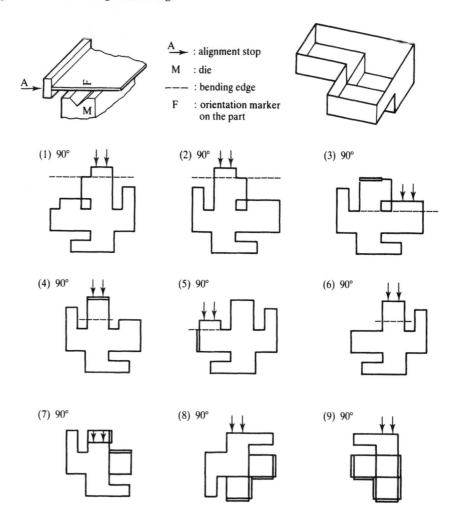

Figure 8.18
Bending sequences for
a three-dimensional
sheet metal part.

planning are as follows:

- the selection of approximate fixturing elements,
- maintain them in a database and handle the relevant queries,
- 'understand' the part globally and choose the appropriate fixturing scheme,
- make a detailed geometrical analysis for determining the location and orientation of the fixturing elements with respect to the part,
- determine the interferences with features for a particular set up,
- map the relative orientations of the fixturing elements from the part coordinates to the machine coordinates,
- conduct a force analysis based on the selected tools, and
- evaluate the resulting fixturing scheme.

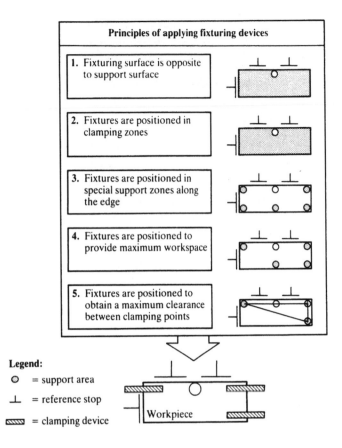

Figure 8.19
Various principles of
applying fixturing
devices to a cubic
part.

8.8.4 Tool selection

Tool selection involves not only the tool but also the machine on which the tool
is going to be used. Tools normally come in different shapes, diameters, lengths,
number of teeth, and alternative tool materials. The most popular tools are those
made of high-speed steel (HSS) and tungsten carbide. Usually, tungsten carbide
tools can be operated at higher cutting speed than HSS, but are more costly and are
sometimes difficult or impossible to regrind. Tungsten carbide is brittle and is
therefore not suitable for interrupted cuts.

The tool selection activity is related to the process sequence activity and,
therefore, sometimes difficult to separate from process sequencing. The key source
of data for tool selection is the machine tool master file which contains a description
of the plant's tools. In the machining section, the workpiece is taken to be composed
of unit machining surface. Each process or machine tool will possess a process
capability file containing parameters such as:

- types of surfaces and dimensions that can be machined,
- workpiece material and initial condition,

- types of machining operations and machinability data,
- types of tools and tool approach directions,
- obtainable tolerance and surface finish,
- set up and machining times and cost data,
- surface treatment and finish,
- necessary tool changes,
- required clamping space for fixtures.

Tool selection algorithms employ look-up tables. However, more contemporary tool selection systems use artificial intelligence tools to make decisions which are based on the data in the process capability file. Typical problems of tool selection are shown in Chapter 3.

8.8.5 Cutting parameters

Although machining forces and power are seldom classified as process boundaries, they are necessary for the process parameter selection including feed, speed, and depth of cut. Since force and power are functions of process parameters and since they are constrained by machine output, it is useful to know the power requirements as they relate to the process parameters. The cutting force can be expressed as a product of the specific cutting resistance and the cross sectional area of the undeformed chips. The cutting power can be calculated as the product of the cutting speed and the cutting force. Table 8.4 shows a summary of the equations for cutting operations. It also includes tool life calculations.

Machining optimization

Machining optimization or process optimization can be classified into two types: single pass and multipass models. The single pass approach assumes that only one pass is needed to generate the desired geometry. In this case, the depth of the cut will be fixed. In the multipass, there will be more than one pass and here the depth of cut becomes one of the decision variables.

Single pass model

The following notations are used in the modeling of the single pass problem:

t_m = machining time
l_h = material handling time
t_t = tool changing time
t = tool life

An expanded Taylor tool life equation can be represented as:

$$t = \frac{\lambda C}{V^{\alpha_T} f^{\beta_T} a_p^{\gamma_T}}$$

Table 8.4 Summary of equations for cutting operations

Operation	Machining time t_m	Tool life t	Cutting force F_c	Power P_m	Surface finish R_a
Turning Boring	$\dfrac{l_w}{f \cdot n_w}$	$K_T \cdot v^{\alpha T} \cdot f^{\beta T} \cdot a_p^{\gamma T}$	$K_F \cdot f^{\beta F} \cdot a_p^{\gamma F}$	$\dfrac{F_c \cdot v}{6120\eta_m}$	$\dfrac{32 \cdot f^2}{r_\epsilon}$
Facing	$\dfrac{D_m}{2 \cdot f \cdot n_w}$				
Parting	$\dfrac{b_w}{f \cdot n_r}$				
Shaping and planing	$\dfrac{l_w}{f \cdot n_t}$				
Drilling and reaming		$K_T \cdot v^{\alpha T} \cdot f^{\beta T} \cdot a_p^{\gamma T} \cdot D_t^{\delta T}$	$K_F \cdot f^{\beta F} \cdot a_p^{\gamma F} \cdot D_t^{\delta F}$	$\dfrac{M \cdot n_s}{9.74 \times 10^5 \eta_m}$	$\dfrac{64 \cdot 2 f^2}{D_t}$
Slab milling[a] Side and face milling	$\dfrac{l_w \sqrt[\kappa]{a(D_t - a)}}{f \cdot n_t}$	$K_T \cdot v^{\alpha T} \cdot a_f^{\beta T} \cdot a^{\gamma T} \cdot D_t^{\delta T} \cdot b_w^{\epsilon T} \cdot z^{\xi T} \cdot \lambda_\beta^{\eta T}$	$K_F \cdot f^{\beta F} \cdot a_e^{\gamma F} \cdot D_t^{\delta F} \cdot b_w \cdot z$	$\dfrac{F_c \cdot v}{6120\eta_m}$	$\dfrac{64\,2 f^2}{D_t + e}$
Face milling	$\dfrac{l_w + D_t}{f \bullet n_t}$		$K_F \cdot v^{\alpha F} \cdot a_f^{\beta F} \cdot a_p^{\gamma F} \cdot d_t^{\delta F} \cdot b_w^{\epsilon F} \cdot z^{\xi F}$		$K_R \cdot a_f^{1.4}$
Broaching	$\dfrac{l_t}{v}$	$K_T \cdot v^{\alpha T} \cdot a_f^{\beta T}$	$K_F \cdot a_f^{\beta F} \cdot D_m \cdot z_c$		

where:

$$\lambda, C = \text{constants for a specific tool/workpiece combination}$$
$$V = \text{cutting speed}$$
$$f = \text{feed rate}$$
$$a_p = \text{depth of cut}$$
$$\alpha_T = \text{speed exponent}$$
$$\beta_T = \text{feed rate exponent}$$
$$\gamma_T = \text{depth of cut exponent}$$

From the above, the single-pass machining for prismatic parts can be considered. The processing time t_{pr} can be expressed as the sum of machining time l_m material handling time t_h and tool handling time t_t.

$$\min t_{pr} = t_m + t_h + t_t\left(\frac{t_m}{t}\right)$$

Table 8.5   Variables from equations for cutting operations

Process	Geometric shape or feature
a, a_e, a_f, a_p	depth of cut
b_w	width of workpiece
D_m	diameter of the machined surface
D_t	diameter of the tool
f	feed (including revolution)
K_F, K_T, K_R	constant for cutting-force, tool-life, and surface-roughness empirical equations
l_t	length of tool
l_w	length of surface to be machined
n_r	frequency of reciprocation (strokes/min)
n_t	tool spindle speed (rpm)
n_w	rotational speed of the workpiece (rpm)
r_ε	tool nose radius
v	cutting speed (inches/minute)
z	number of teeth on the cutting tool
z_c	number of teeth cutting simultaneously in a tool
$\alpha_T, \beta_T, \gamma_T, \varepsilon_T, \xi_T, Y_T, n_T, \delta_T$	cutting speed, feed, depth of cut, tool diameter, machined surface width, number of teeth in the cutting tool, and tool cutting edge inclination ($L°$) exponents for cutting-force, surface-roughness, and tool-life equations, respectively
h_m	overall efficiency of the machine tool motor and drive systems
e	tool cutting edge inclination

where t/t_m represents the number of parts that can be produced before requiring changing or resharpening of a tool.

A cost model can also be formulated for the production cost per component C_{pr} as follows, see also Chang and Wysk (1985):

$$\min C_{pr} = \frac{C_b}{N_b} + C_m\left[t_m + t_h + \frac{t_m}{t}\left(t_t + \frac{C_r}{C_m}\right)\right]$$

where:

C_b = set up cost for a batch

C_m = total machine and operator cost (including overhead)

C_r = cost of regrinding (for tungsten carbide tool, it is the cost of one new insert cutting edge)

N_b = batch size

The following constraints apply to both the production time model and the cost model.

There is a minimum and a maximum allowable rotational speed. Thus the spindle speed constraint is:

$$n_{w\text{min}} < n_w < n_{w\text{max}} \quad \text{(for workpiece)}$$

$$n'_{t_{\min}} < n_t < n'_{t\text{max}} \quad \text{(for tool)}$$

Similarly there exist upper and lower bounds for the feed rate producing the following feed constraint:

$$f_{\min} < f < f_{\max}$$

There is a maximum cutting force allowable:

$$F_c < F_{c\text{max}}$$

Finally, there exists a constraint on the surface finish:

$$R_a < R_{a\text{max}}$$

Each of the variables n_w, n_t, F_c, C_m, and R_a can be found in Table 8.5.

A number of researchers have proposed solutions to the models. Chang and Wysk (1985) presented the cost most model. Berra and Barash (1968) and Wysk (1977) used iterative search procedures which approached optimum solutions. Groover (1987) used an 'evolutionary operation' procedure which is somewhat similar to a Hooke-Jeeves search procedure. Hati and Rao (1976) applied a sequential unconstrained minimization technique (SUMT), and Fiacco and McCormick (1968) in conjunction with the Davidson-Fletcher-Powell (D-F-P) found an algorithm for their solution of the problems. Other researchers have applied dynamic programming and other mathematical programming methods to solve the problem. None of these models can achieve universal solutions to the problem.

Multipass model

For the multipass model, a_p is a variable and n is the number of passes. The time required per part can be formulated in this system, see Chang and Wysk (1985) for details.

$$t_{pr} = t_h - \sum_{i=1}^{n_p} \left(t_m^i + \left(\frac{t_m^i}{t} \right) t_t \right)$$

where t_m^i is the time required for machining pass i. The cost per part model is:

$$C_{pr} = \frac{C_b}{N_b} + C_m t_h + \sum_{i-1}^{n_p} C_{pr}^i$$

and where

$$C_{pr}^i = C_m \left[t_m^i + \frac{t_m^i}{t} \left(t_t + \frac{C_r}{C_m} \right) \right]$$

The models for C_{pr} and C_{pr}^i are subject to the same constraints as those for a single pass model:

$$a_{p_{min}} < a_p^i < a_{p_{max}}$$

$$a_t = \sum_{i=1}^{n_p} a_p^i$$

The additional variable a_p^i makes the multipass problem more difficult than the single pass equivalent. In solving the problem, Challa and Berra (1976) used a modified Rosen's gradient search method. Subarrao and Jacobs (1978) used goal programming in solving the problem. Iwata, Murotsu, and Oka introduced in 1977 dynamic programming techniques to solve a multistage machining optimization problem. Davis, Hayes, and Wysk (1981) and Chang et al. (1982) mapped some variables of the problem into a discrete domain and then applied a dynamic programming technique in its solution. There is no general solution either which exists for solving this kind of problem.

8.8.6 Cutting path

The cutting path problem is determined by technological restrictions, in general. The algorithm for determining a cutting path can come from a decomposition of machinable volume and a sequencing of the machinable features resulting from this, see Kusiak (1990). If the modeling system in which the part is represented provides features symbolically, it will be possible to reason about the features based on their locations on the part and generate a sequence of cuts. For problems in which the order of cut is unimportant, the traveling salesman problem has been applied.

8.8.7 Examples of existing automated process planning systems

CAPP

Computer-automated process planning system (CAPP), as discussed in Link (1977) was developed under a CAM-I sponsored research project by McDonnell Douglas Automation Company (McAuto). It has emerged as perhaps one of the most widely used process planning systems. Since it is very difficult to design a general process planning system which can be used for any part, the CAPP system provides a framework for implementing custom designed process planners.

CAPP is written in ANSI Fortran and can be viewed as a database management system (Figure 8.2). The system provides a format for a database, retrieval logic, and interactive editing. The user has to enter his own classification scheme via a 36 character alphanumeric array. In addition, the part sequence file and operation plan file must be provided.

CAPP is a variant system which can handle rotational, prismatic, and sheet metal components. A five digit code is used to define standard plans and to retrieve them. The operation sequences are also represented by codes. Each operation sequence code in the sequence produces an operation plan defining the detailed operation steps, machine tools, fixtures, operation time and so on.

CAPP is easy to learn and allows the use of existing group technology systems for process plan searches. The range of components which can be planned by CAPP is dependent on the capabilities of the coding scheme entered by the user. The CAM-I library and the Society of Manufacturing Engineering division CASA have been responsible for providing the numerous CAPP system variants in use today.

AUTAP

Of all the existing process systems, AUTAP is probably the most complete. The system can perform material selection, process sequencing, machine tool selection, tool selection, lathe chuck selection (for turned parts), and part program generation, as discussed in Chang and Wysk (1985). A key aspect of AUTAP is its method of part representation. In AUTAP, parts are represented by a subset of the constructive solid geometry (CSG) using Boolean operations. A user must rigorously adhere to the tenets of the AUTAP representation scheme in order to generate an AUTAP model, see Figure 8.20.

With the aid of the selected language representation scheme, it is possible to capture in AUTAP many of the manufacturing process data such as geometric tolerances, feature designations, lot sizes and so on. The system is a generative planner with rotational and sheet metal parts capability.

AUTAP uses a decision table logic such as the one shown in Figure 8.21 to select the processes, the process sequences, and the machine tools. The contents of the decision table are supplied by the user.

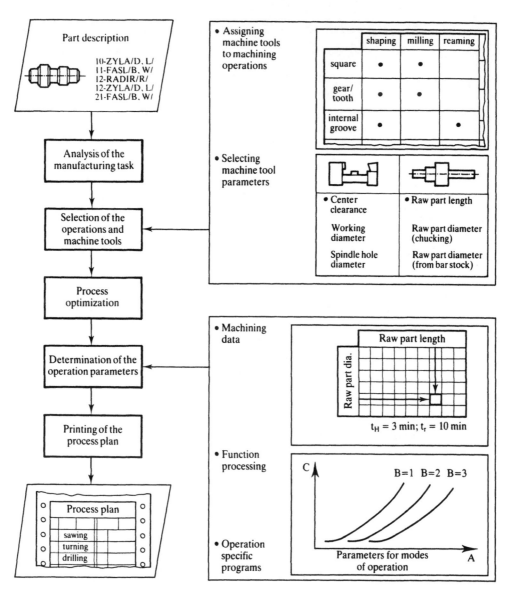

Figure 8.20

Automatic generation of a process plan by the AUTAP system (courtesy of Laboratory for Machine Tools and Production Engineering of the Technical University of Aachen).

The AUTAP-NC is an extended AUTAP to be used for NC tools, it performs additional functions, such as tool selection, fixture selection, and part program generation; it is designed for rotational parts. By coupling AUTAP with the EXAPT programming system, see Eversheim and Fuchs (1980), a final Cutter Location Data (CLDATA) output and a verification drawing can also be obtained.

AUTAP is one of the few systems integrated with a CAD system and is being widely used in Germany.

Table 8.6 shows other prominent available process planning systems, and their capabilities.

			Machine tool selection					
		No.	1	2	3	4	5
Condition	200 < length < 600	1		•	•		
	Dia < 300	2					
	Max speed < 300	3				•	
	Tolerance < 0.02	4	•		•		•
	Lot Size > 200	5	•	•			
	Fixture 225 exist	6		•	•		
	Fixture 345 exist	7	•	•			
Conclusion	Machine 2110	1	•		•		•
	Machine 2111	2		•			
	Machine 2113	3			•		

Figure 8.21
Decision table logic of the AUTAP process planning system.

8.9 Manufacturing Resource Planning

The issues in manufacturing resource planning include master production scheduling (MPS), capacity requirements planning, order release planning and operation sequencing.

8.9.1 Resource planning

Resource planning starts with a long-range plan where the production activities are projected into the future. Market planning provides research information on the market potential, customer demand, demographics, resources, processes, funds, new inventions, competition, etc. From these long-range plans, intermediate-range plans and short-range plans are extracted. Long-range plans may have a horizon of five years, intermediate-range plans a horizon of less than 12 months, and short-range plans of two weeks. Before the intermediate and/or short-term plans are activated, there should be an order placed and the design of the product should be known, as well as its manufacturing methods and sequences.

The master production schedule is produced by considering the customer orders and the forecasting based on market research or history of product performance in the market. The master production schedule consists of a list of end products to be manufactured, the quantities ordered and the due dates. The end product may be assemblies, sub-assemblies or individual parts. The plan contains a matrix listing the number of products to be completed in a given time scale. The planning horizon is normally divided into two: the firm and the tentative planning periods (Figure 8.22). The firm planning time horizon reflects the end products for which committed orders have been received. The master production schedule normally serves as input into the manufacturing resource planning which includes the manufacturing inventory planning, capacity planning and order release planning. Figure 8.23 shows how these plans evolve from the master production schedule.

Table 8.6 Available automated process planning systems and capabilities

Systems	Designer	Input				Parts			Automated functions			
		Variant	Generative	CAD	Others	Rotational	Prismatic	Sheet metal	Process seq.	Machine	Tool	Parameter
CAP	Lockheed	•			Part number			•	•			
AUTAP	Aachen Tech		•		•	•		•	•	•	•	•
AUTOPLAN	Metcut		•	•		•			•	•	•	•
APPAS	Wysk/Purdue		•		Coform		•				•	•
ACUDATA/UNIVATION	Allis Chalmers	•			Part number	•						
CADCAM	Chang/Va Tech		•	•			•		•			•
CIMSPRO	Iwata et al. Japan		•		•	•	•		•	•	•	
CAPP	McAuto/CAM-I	•				•	•		•	•	•	
COBAPP	Phillips/Purdue	•			Appocc	•	•	•	•			
COMCAPPV	MSDI	•			CODE		•					
CPPP	UTC Research				CODE	•	•	•	•			•
DCLASS	Allen/Utah		•		DCLASS		•			•	•	
EXAPT	EXAPT	NA	NA		APT	•	•			•	•	
GARI	Descott et al.	•	•				•					
GENPLAN	Lockheed	•			OPTIZ		•	•	•	•		
GENTURN	GE	NA	NA		•	•	•		•			•
MIAPP	OIR	•			MICLASS	•	•		•			
MIPLAN	GE/OIR	•			MICLASS	•	•		•			
MITURN	OIR	•			MICLASS	•	•		•			
SISPA	Siemens	NA	NA			•	•			•	•	
RPO	GE/MECUT	•			•	•			•			•

NA = not applicable

	Quarterly Period							
	Year 1				Year 2			
	1	2	3	4	5	6	7	8
Type of product	Fixed orders				Tentative orders			
A	5	10	15	6	7	3	5	8
B	20	15	30	25	10	8	11	3
C	2	2	2	3	1	4	3	1
D	50	60	44	30	50	20	15	15

Figure 8.22
A master production schedule.

Item number			Gross requirements for:	Week 1	Week 2	Week 3	Week 4	Week 5	Week 6	Week 7	Week 8
Lead time			Parent item No. 346								
On hand											
Safety stock											
Allocated for:			Service orders								
Parent	Date	Quantity	Total								
			Scheduled receipts								
			Available								
			Net requirements								
			Planned order receipts								
			Planned order releases								

Item number			Gross requirements for:								
Lead time			Parent item No. 350								
On hand											
Safety stock											
Allocated for:			Service orders								
Parent	Date	Quantity	Total								
			Scheduled receipts								
			Available								
			Net requirements								
			Planned order receipts								
			Planned order releases								

Item number			Gross requirements for:								
Lead time			Parent item No. 476								
On hand											
Safety stock											
Allocated for:			Service orders								
Parent	Date	Quantity	Total								
			Scheduled receipts								
			Available								
			Net requirements								
			Planned order receipts								
			Planned order releases								

Figure 8.23
Typical materials requirement transaction chart.

The manufacturing resource planning starts with the conception of the master production schedule. The horizon for scheduling produced here is usually one or several years. This schedule is used to make capacity decisions for the plant, new manufacturing processes, equipment and labor.

Capacity planning is concerned with providing the manufacturing resources when they are needed for production. The planning horizon is usually in months in time increments of weeks. When a typical capacity adjustment is necessary it may include production equipment, subcontracting of work, and planning of labor resources.

Order release planning provides details for inhouse production. When capacity cannot be scheduled economically or when problems occur with production equipment then subcontracting or overtime has to be made. The planning horizon here is in days for several weeks ahead.

Operation sequencing is concerned with queueing of operations at each work center. The operation sequence system attempts to maximize equipment utilization and to meet deadlines. The planning horizon here is the shift or the work day, and the increments are minutes or hours.

8.10 Material Inventory Systems

8.10.1 Introduction

The holding of inventory is very expensive because it ties up funds which could be used for other manufacturing resources. In modern manufacturing structures, an attempt is made to work without inventory or to reduce it to a minimum. Often the supplier is required to hold the inventory and to use the parts on a need basis. In the majority of functions we do need some inventory or safety stock.

Material Requirements Planning (MRP) is a type of inventory management system which provides insight into the manufacturing inventory. Figure 8.24 shows the general relationship of material requirements planning and other planning activities. Inventory control techniques use statistical approaches to manage independent demand items.

Traditional inventory control of stock replenishment aims at restoring inventory to its state of original capacity. But manufacturing inventories should not be restored to full capacity. The principle of stock replenishment requires that inventory items be available in stock at all times. This approach compensates for the inability to determine the precise quantity and time of need in the near future. Inventory control should be aimed at providing items at the time of need. It is therefore necessary to investigate the point of inventory level, at which a new supply should be ordered, to bring inventory to its desired level. This point is called the re-order point.

Re-order point techniques represent the implementation of the stock

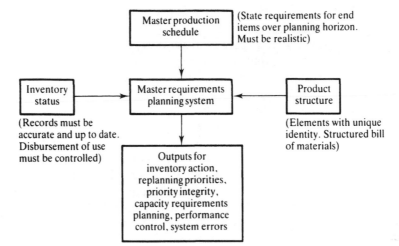

Master production schedule — (State requirements for end items over planning horizon. Must be realistic)

Inventory status — Master requirements planning system — Product structure

(Records must be accurate and up to date. Disbursement of use must be controlled)

(Elements with unique identity. Structured bill of materials)

Outputs for inventory action, replanning priorities, priority integrity, capacity requirements planning, performance control, system errors

Figure 8.24
The relationship of the master resource planning system to other planning systems.

replenishment notion; they typically forecast demand during the replenishment lead time. Almost all of them attempt to provide for some safety stock to compensate for fluctuations in demand. There are many problems associated with the re-order point techniques. They include:

- false assumptions about observed demand,
- lack of ability to determine specific timing of future demand.

As a result of the above two problems, there is unnecessary high overall inventory level, inventory imbalance and stockouts (shortages caused by the system itself).

In view of the discussions above, order point or stock replenishment is a set of procedures, decision rules, and records intended to ensure continuous physical availability of all items comprising an inventory in the face of uncertain demand. Whenever supply reaches a re-order point, then inventory is replenished. To ensure continuous availability of supply, the order point is some positive quantity and this value (stock) is carried in inventory to prevent lost sales.

8.10.2 Vilfredo Pareto's Law (1897) and the ABC Inventory

Some managers have looked upon inventory management as finding that technique which ensures that materials are available when needed and that excessive costs are not incurred in stocking items. One method, which managers have applied, and which is simple but possesses the capability of classifying inventory for better management, comes from the application of Vilfredo Pareto's Law (1897). In his study of wealth and income distribution in Italy, Vilfredo Pareto found that the national income was concentrated in a small percentage of the population. He believed that this represented a universal trend and thus formulated an axiom that: significant items in a given group normally constitute a small percentage of the total items in the group and the majority of the items in the total will, in aggregate, be of minor significance.

A rough pattern of Pareto's Law expressed in mathematical empirical relationship shows that 80% of the distribution is being accounted for by 20% of the group membership. The same pattern applies in inventory. Approximately 20% of the items account for 80% of total cost.

Pareto's Law gave birth, in inventory management, to what is called 'ABC Inventory'. In the ABC classification method, this 20% is designated as A-items, and represents 80% of cost. Thirty percent will be B-items and will represent 15% of cost while the remaining 50% of items will be classified as C-items and would represent only 5% of the total cost.

The idea behind ABC principle is to apply most planning and control resources to the A-items at the expense of the other items that have little effect on the total cost or moderately to control the A-items more 'tightly' than the other classes. This will result in tighter control and more frequent reviews. An example of the review pattern is as shown in Table 8.7.

It is important to mention that within each class of ABC classification, a computational method of controlling inventory can be employed.

8.10.3 Types of manufacturing inventory

In a manufacturing system, various types of materials, compounds, assemblies and sub-assemblies are used. Some objects are kept in stock and others are stored in in-process buffers. We can classify inventory as follows:

- raw materials in stock
- semi-finished component parts in stock
- finished component parts in stock
- sub-assemblies in stock
- component parts in stock
- sub-assemblies in process

Inventory policies for the various types of materials and compounds may be quite different. For example, raw materials can often be used for many components, whereas sub-assemblies may be specific to an order.

A general inventory management system must be concerned with the following four basic activities: planning, acquisition, stock-keeping and disposition.

Table 8.7 The ABC classification of stock control

Inventory class	Review frequency	Order quantity
A	Monthly	1 month's supply
B	Quarterly	3 month's supply
C	Annually	12 month's supply

Manufacturing inventory has its own distinct characteristics and has distinct features from non-manufacturing inventory in all four attributes:

(1) Planning: various types of planning and forecasting methods may be applied to secure an adequate stock of materials and parts. Objects of interest are:

    (a) inventory policy

    (b) inventory planning

    (c) forecasting

(2) Acquisition: this activity is concerned with the ordering of inventories, including:

    (a) positive order (place or increase)

    (b) negative order (decrease or cancel)

The material in manufacturing process is always being acquired and re-acquired as it progresses through the multiple production stages. An order for a manufactured item cannot be cancelled once started without penalty for scrap or re-work and also cannot be increased or decreased. Scrap and re-work are factors that go into determination of order quantity.

(3) Stock-keeping: this activity manages the inventory. Its components are:

    (a) receiving

    (b) physical inventory control

    (c) inventory accounting (bookkeeping)

The inventory accounting function may be integrated into or merged with the inventory planning function.

(4) Disposition: is concerned with the use and depletion of the inventory. Its activities are:

    (a) purging (scrap and write-off of obsolete items)

    (b) disbursement (delivery of source of demand)

Manufacturing inventory is typically for inhouse consumption. Demand is therefore represented by a production requirement or a production schedule. If it is shippable upon completion, then it becomes distribution inventory.

8.11 Material Requirements Planning

8.11.1 Introduction

Material requirements planning (MRP) is a viable method of assuring that items are available at their times of need; material requirements planning consists of a set of logically related procedures, decision rules, and records (records could be viewed as material requirements input) designed to translate the master production schedule into time-phased net requirements, for each inventory item needed to implement this schedule.

The general attributes of the material requirements planning system include:

(1) product structure or bill of materials
(2) bill of materials explosion process
(3) establishment of independent and dependent demand
(4) time phasing

They will be discussed in the following section.

8.11.2 The product structure

The product structure provides a hierarchical classification of the items which form a product. With the product structure, the understanding of the components which compose a product as well as their attributes, can be represented. This presentation is used to produce the production product structure of the process which contains several manufacturing levels of material, component parts and sub-assemblies. In determining material requirements, the following two cases must be considered:

- a given component can exist in its own right as a uniquely identified physical entity (for example, raw material, component part, sub-assembly),
- a compound can exist physically as an already assembled object of another inventory item (that is, it has lost its identity).

It is necessary to identify all the basic items in developing the material requirements planning system. This is done by using the bill of materials which reflects the product structure levels by listing components of each assembly and sub-assembly. A typical structure (material conversion stages) of a product is shown in Figure 8.25, see Section 7.13.

To define a product structure it is necessary that:

(1) Each component, part, sub-assembly, assembly and end item must have a unique part number.

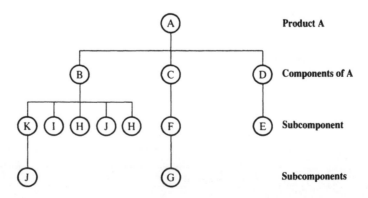

Figure 8.25
Material conversion
stage for a product A.

(2) The structural bill of materials (BOM) must be up-to-date and accurate including:

(a) usage,

(b) lead times (time of material receipt = time of decision to order + lead time),

(c) low level coding, which is the presentation of components which could have been made from a higher level to a lower level (as described in Chapter 7) to take advantage of capacity and lead time.

Using Figure 8.26 to illustrate the MRP transaction for item D in Figure 8.25, the gross requirement is the quantity of the items that will have to be disbursed (that is, issued to support a parent order or orders), rather than the total quantity that will be consumed by the end item. These two quantities may or may not be the same. For instance, in those cases where usage is not only by one type of parent item but by more than one parent item, the quantity will not be identical.

In evaluating the product structure in Figure 8.25, the following observations can be made:

- It proceeds level by level.
- It begins with the top level and proceeds to the bottom level. As a consequence, this procedure accounts for those products or components that exist as consumed items at higher levels, by bringing them down to the lowest level.
- It is assumed that no inventories existed from which demand could have been met.

Lead times

Lead time is the time between issuing an order and arrival of such issued order. Lead times are used to determine order release dates. The lead time for parts of the same order may differ considerably. Simple items usually have a short lead time and complex parts a long lead time. Planned or normal lead times are used by material requirements planning for purposes of planning and their accuracy is very crucial. The following are the elements that contribute to lead time in manufacturing:

(1) Queue time: waiting to be worked on;

(2) Process time: machining, fabrication, assembly and so on;

Figure 8.26
Inventory coverage using calculated EOQ for item D.

Period	1	2	3	4	5	6	7	8
Net requirements		1100	930	30	455			
Planned Order Release		1100	930	30	455			
EOQ Required	251	251	251	251	251	251	251	251
Inv. Strt. #753	753	1004	155	−528	−247	−451	−200	51

(3) Set-up time;

(4) Waiting time for transportation;

(5) Transportation time;

(6) Other elements; such as safety lead time.

Dependent vs independent demand

Demand for an item is independent when such demand is unrelated to demand for other items or when it is not a function of demand for some other inventory items. Independent demands must be forecast. However, demand for an item is dependent when directly related to, or derived from, the demand for another inventory item or product.

A dependency is 'vertical' when an item is needed to build a sub-assembly or product; it is 'horizontal' when it accompanies another object, as in the case of an attachment or owner's manual shipped with the product. In manufacturing systems, the bulk of the total inventory is in raw material, component parts, and sub-assemblies, all subject to dependent demand. Therefore, dependent demand need not, and should not, be separately forecast as it can be precisely determined from those items that are its sole cause.

In material requirements planning, the goal is usually to determine the net requirements for a product which can then be transformed into planned order receipts. The net requirements and the planned order receipts are usually the same quantity and fall into the same time slot. It is merely a matter of re-labelling the net requirements to reflect the intention which is order receipts at the time. The planned order release is the time phased depiction of the quantity in the planned order receipts. For example, if 20 items of item B are due to be received by week 4 and the lead time for making item B is two weeks, then the planned order release now will show 20 items of item B in week two.

Using Figure 8.26, the following observations can be made:

(1) Net requirements are developed by allocating quantities in inventory to quantities of gross requirements in a level-by-level process which proceeds downward from the parent item.

(2) At each level:

Net requirement = gross requirements

     − on hand

     − on order

     + allocated

     + safety stock

(3) The uncashed requisitions are allocated due to time lag between order release and the filling of the respective material requisition. They will be those items which are uncashed before the start of the planning scale.

Time phasing

Time phasing is used to determine when items expected for receipt at given times should be released for production or purchase.

This is the division of time's continuous flow into increments suitable for measuring its passage. In production planning, particularly material requirements planning, these increments provide time slots for which discrete planning is possible. The Gregorian calendar does not employ a decimal base. It has months of uneven number of days, and has an irregular pattern of holidays. As a result of these complicating factors, different calendars needed to be established to suit the manufacturing control systems.

Some systems use a 100 week year (each period has 28 days); others use a 1000 day year scheduling calendar (working days only). Whichever method is used, the result is to split the continuous flow of time into discrete time increments.

8.12 Lot Sizing Techniques

Suppose that after performing the material requirements planning transactions, the net requirements row is as shown in Figure 8.27 for item B of Figure 8.25.

Ordering the items required in each period when they are needed means that items must be ordered in each period. Therefore, advantage is not taken of quantity discounts, ordering costs and so on. Most companies usually use some ordering policy to maintain availability of inventory for the item.

The results of an MRP transaction show the quantities to be ordered and when. Lot sizing techniques are used to determine the efficient ordering of the quantities to minimize cost.

The ordering policy would establish the lot size to be ordered and how frequently such lot sizes will be ordered. There are many lot sizing techniques available. These techniques generally fall into two categories: static and dynamic lot sizing techniques. The static lot sizing technique assumes that an order quantity, once computed, remains unchanged in the planned order schedule. Examples include fixed period quantity (FPQ), where fixed orders are placed at fixed intervals, and economic order quantity (EOQ). The dynamic lot sizing technique allows for an order quantity to be subject to recomputation (such as those required by changes in the net requirements data). An example is the Wagner–Whitin algorithm.

We will present, in detail, two lot sizing techniques: the economic order quantity and the Wagner–Whitin algorithm. This illustrates both the static and dynamic

Figure 8.27
Net requirements for item B.

Periods	1	2	3	4	5	6	7	8	9	Total
Net requirements	25	30		40	35			25	10	145
Planned order release	25	30		40	35			25	10	145
Coverage										

---

lot sizing techniques. The economic order quantity has been very widely used in industry. The Wagner–Whitin algorithm assumes coverage of the items during the planning horizon and lends itself to computational optimal programming.

8.12.1   Economic order quantity (EOQ)

This type of policy aims at finding the economic lot size which will give the lowest yearly inventory cost. Several simplifying assumptions are involved:

- The order is placed, and it is received at the time stock is depleted: no emergency reserve.
- Inventory carrying costs is applied to the average inventory value: other costs which do not vary with inventory amount are assumed negligible.
- Unit cost will remain constant throughout the range of production under consideration.

Figure 8.28 shows the composition of the economic order quantity and the behavior of the components. The total yearly inventory cost (TYIC) is composed of three costs:

- cost of preparation (or set up) for production,
- cost of carrying the inventory,
- cost of products in inventory.

Cost of set up

From Figure 8.28, this cost function is a curve which is steepest when more lots per year (small size lots) are used and becomes more horizontal as the lot size increases:

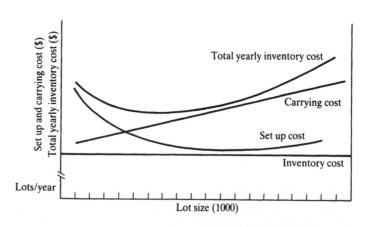

Figure 8.28
Economic order quantity technique.

Cost of set up for production

The cost of set up per year can be calculated as follows.

Let:

$$Y = \text{yearly demand in units/hour}$$
$$Q = \text{economic lot size in units/lot}$$
$$S = \text{preparation (set up) costs in dollars/lot}$$

Cost of set up,

$$T_s = \frac{\text{Yearly demand (units/year)} \times \text{Cost of order preparation ($/lot)}}{\text{Quantity (units/lot)}}$$

$$= \frac{YS}{Q} \text{ in $/year}$$

Cost of carrying inventory per year

This cost varies directly with the amount of inventory, hence it will increase as fewer lots are produced where more inventory will result:

Let:

$\frac{Q}{2}$ = Average inventory carried during the inventory period.

U = Unit cost consisting of the sum of the direct labour, direct material and overhead in $/unit.

I = Total of all carrying charges expressed as a percentage of the cost of the parts carried in stock for the year.

Then the cost of carrying inventory is:

$$T_H = \frac{QUI}{2}$$

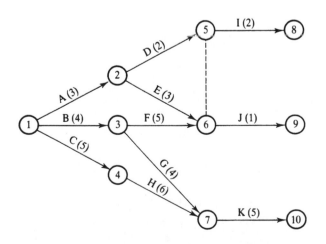

Figure 8.29
Activity on arcs
(AOA) graph.

Cost of inventory ($/yr)

This function arises from the cost of the inventoried items. Therefore the cost of the inventory (T_I) is obtained from the yearly demand ($\frac{\text{units}}{\text{year}}$) as follows:

$$T_I = \left(\tfrac{\text{unit}}{\text{year}}\right) \cdot \tfrac{\text{cost}}{\text{unit}} = YU$$

Then total yearly inventory cost (TYIC) is given by:

$$TYIC = \frac{YS}{Q} + \frac{QUI}{2} + YU$$

where Y might be available from inventory records and other information from cost records except for Q.

We need to know Q and we can use calculus. In particular we want to find which Q minimizes TYIC. Since the minimum position on a convex curve is the point where the slope goes to zero, we can differentiate with respect to Q and set this equation to zero. Figure 8.29 illustrates the EOQ technique.

$$\frac{d(TYIC)}{dQ} = \frac{YS}{Q^2} + \frac{UI}{2} + \frac{d(YU)}{dQ}$$

$$0 = -\frac{YS}{Q^2} + \frac{UI}{2} + 0 = -2YS + Q^2 UI$$

$$Q^2 = \frac{2YS}{UI}$$

$$Q^* = \sqrt{\frac{2YS}{UI}} \Leftarrow EOQ$$

Analysis of EOQ

The function $\dfrac{2YS}{UI}$ increases when S increases (that is, preparation cost goes up):

- penalty for ordering
- larger order quantity

When the product U*I decreases,

- it is cheaper to carry inventory
- more can be carried by way of larger EOQs.

Since the shape of TYIC is quite flat near the bottom (minimum), there really would be a considerable range of values for EOQ which nearly minimize the TYIC curve.

Example problem: Suppose that item D on the product structure (Figure 8.25) has the following data:

Table 8.8 Carrying cost and lot size relationship

Desired lot size reduction, %	Carrying cost	EOQ	EOQ²
	100	100	100
10	123	90	81
20	156	80	64
30	200	70	50

$$Y = \text{Yearly Demand}$$
$$= 1.5(1100 + 930 + 30 + 455)$$
$$= 3772.5 \text{ (number of ordering periods)}$$
$$Q = \text{Unknown}$$
$$S = \$100.00$$
$$U = \$50.00$$
$$I = 0.24/\text{annum}$$

Then $EOQ = \sqrt{\frac{2(3773)100}{.24*50}} = \sqrt{\frac{754\,500}{12}} = 251.$

The planned order releases (Figure 8.25) are as shown in Figure 8.29 from the EOQ solution. It can be seen that even with a starting on-hand inventory of 753 items the EOQ was unable to satisfy the demand from period 3. This is the typical outcome of using the EOQ in a widely varying demand schedule.

In manufacturing environments, the EOQ is not very useful because it is totally insensitive to timing of actual discrete demands (requirements) during the period the EOQ is intended to cover following the arrival of stock. The EOQ is derived on the basis of set-up cost, unit cost, carrying cost and annual usage. The EOQ formula rests solely on the assumption of uniform demand. The EOQ varies inversely with the square root of the carrying cost (see Table 8.8).

8.12.2 Wagner–Whitin algorithm

This technique is a dynamic programming approach to minimize the combined cost of set up and carrying charges. It optimizes the planning horizon (that is, implicitly assumes that requirements beyond the horizon are zero).

Stages are taken as periods on the horizon. Decisions are the quantity Q to produce during the period. Stage-by-stage costs contribute to the total cost over the horizon. Inventories from the previous period carry over to the current period and a carrying charge is made at that point.

Note: For the cost structure assumed, it can be shown that $Q_t = 0$ for $t = 1, 2, \ldots, N$. Otherwise, there will be no need to order. The requirements in a period are satisfied either entirely from procurement in the period or entirely from procurement in an earlier period.

Let periods $= 1, 2, \ldots, N$ and let demands $= D_1, D_2, \ldots, D_N$.

Assumptions

(1) Single lot may be procured in each period.

(2) No shortages are allowed.

(3) No inventory at the beginning of planning period.

Now, carriage cost, h_t, is incurred to carry a unit from period t to period $t + 1$.

Q_t = Lot size procured in period t
S_t = Set up cost for ordering in period t
U_t = Variable unit cost (this is the cost of direct labor, direct material and overhead, $\frac{\$}{\text{Unit}}$)
I_t = Inventory level at time t
D_t = Expected demand
h_t = Unit carrying cost
F_r = Minimum cost program for period $1, 2, \ldots, k$
C_{jk} = Total cost incurred in periods $j + 1$ through to k

Constraints

For the cost structure assumed, it can be shown that:

$$I_{t-1} Q_t = 0, \text{ for } t = 1, 2, \ldots, N$$

the requirements in a period are satisfied either entirely from procurement in that period or from procurement in a previous period.

Goal

To determine the lot sizes $Q_1, Q_2, \ldots, Q_N$, which mimimize the sum of procurement costs and inventory carrying costs over N periods. Then, we can investigate those programs where:

$$Q_t = 0 \text{ or}$$

$$Q_t = D_t + D_{t+1} + \ldots + D_k$$

for some $k = t, t + 1, \ldots, N$

To obtain a solution such as the above, we need to use the Wagner–Whitin algorithm.

Let F_k be the minimum cost program for periods $1, 2, \ldots, k$ when

$$I_k = 0 \text{ is required}$$

Let j be the last period prior to k for which the ending inventory was zero. Then,

$$I_j = 0 \text{ and } I_k = 0$$

$$I_t > 0 \text{ for } t = j + 1, j + 2, \ldots, k - 1$$

Then from the equations for Q_t we obtain

$$Q_{j+1} = D_{j+2} + \ldots + D_k$$

Define the total cost incurred in periods $j + 1$ through k as C_{jk}.

$$C_{jk} = S_{j+1} + U_{j+1}Q_{j+1} + \sum_{t=j+1}^{k-1} h_t I_t;$$

where

$$I_t = Q_{j+1} - \sum_{r=j+1}^{t} D_r = \sum_{r=t+1}^{k} D_r, \text{ for } j < t < k$$

Therefore,

$$C_{jk} = S_{j+1} + U_{j+1}Q_{j+1} + \sum_{t=j+1}^{k-1} h_t \cdot \sum_{r=t+1}^{k} D_r$$

We can now write the following recursive equation:

$$F_k = \text{Min}[F_j + C_{jk}](k = 1, 2, \ldots, N)$$
$$0 \leqslant j < k$$

where $F_0 \equiv 0$

Note:

$C_{jk} = 0$ if $D_{j+1} + D_{j+2} + \ldots + D_k = 0$ since $Q_{j+1} = 0$ in such case and no fixed cost is incurred.

Logic

For a k period horizon with zero initial and final inventories and no shortages allowed, there will be some period where the last procurement is made. Let this period be $j + 1$ and by the property of optimality, $I_j = 0$.

Now, suppose that the optimal policy is found at F_t with minimum cost ($I_t = 0$). Then, since $t < k$, F_j can be used to find C_{jk}. The minimum cost for a k-period horizon results from selecting the optimal period for the last procurement.

By trying all $j < k$ one can find the value of J, say J_k^*, which minimizes $F_j + C_{jk} = F_k$.

The last procurement period will be at $J_k^* + 1$.

- The procedure is then to determine in sequence the values of $F_1, F_2, \ldots, F_N$.
- When F_N is found, then we have found the minimum cost value for the N-period horizon and then J_N^* to trace backwards to extract the optimal lot sizes.

Consider the following example problem with associated data:

Month (+)	1	2	3	4	5	6
Set-up cost (S_t)	100	100	100	120	120	150
Unit variable cost (U_t)	2	3	4	2	5	5
Unit carriage cost (h_t)	2	1	1	2	2	1
Expected demand (D_t)	70	150	200	160	90	120

Now consider the stage process in problem solution here as stages in **dynamic programming**.

STAGE 1 – Period 1

Using equations for C_{ik} and F_0, we have:

$$F_1 = F_0 + C_{01}$$
$$= F_0 = S_1 + U_1 Q_1$$
$$= 0 + 100 + (2)(70)$$
$$= 240.$$

Notice that $Q_1 = D_1 = 70$

STAGE 2

Now consider the first two periods.

$$F_2 = \min
\begin{cases}
= F_0 + C_{02} \\[1em]
= F_1 + C_{12} \\
= F_0 + S_1 + U_1(D_1 + D_2) + h_1 D_2 \\
= 0 + 100 + 2(70 + 150) + (2)(150) = 840
\end{cases}$$

$$= \min
\begin{cases}
= F_1 + S_2 + U_2 D_2 \\
= 240 + 100 + 3(150) = 790
\end{cases}$$

We can see that the optimal $J_2^* = 1$ and so the optimal procurement period is $J_2^* + 1 = 2$. By following the technique above, the optimal procurement periods for stages 3 to 6 can be found. Table 8.9 shows the values of $F_j + C_{jk}$. Table 8.10 shows the resulting ordering policy.

8.13 Sequencing of Operations

The optimal planning of production systems has always posed a tremendous challenge to the production engineer. Today, with the increasing complexity of production systems, this problem is far more varied, complex and interactive than ever before. Sequencing problems can be found in all manufacturing operations.

Table 8.9 Values of $F_j + C_{jk}$

Period of last procurement	Last period with zero inv.	Planning horizon k					
$j + 1$	j	1	2	3	4	5	6
1	0	240	840	1840	2220	3520	4720
2	1		790	1590	2390	3020	4100
3	2			1690	2490	3120	4200
4	3				2030	2390	3110
5	4					2600	3440
6	5						3140
	F_k^*	240	790	1590	2030	2390	3110
	J_k^*	0	1	1	3	3	3

Table 8.10 Resulting ordering policy

Periods	1	2	3	4	5	6
Ordering quantity	70	350	0	250	0	120

For example, during fabrication families of parts must go through a sequence of machining operations to become completed. Yet each machining system may be used for several or all parts.

In some systems, sequencing may just mean the use of an internal timing device to determine when to initiate changes in output variables, see Groover (1987). For example, a washing machine has a sequencing device which is set for defined washing cycles. The operations in this case are fixed, and there is no competition for the machine by other jobs at any given time. Industrial tasks are however much more complex. It is usually the task of the planner to capture the complexity of the sequencing decision process.

There exist many planning and design decisions which have a significant effect on both the time and cost associated with manufacturing a product. Two of the most fundamental of these decision issues are: process sequence selection and machine requirements planning. We will present techniques developed by Nnaji and Davis, for modeling and analysis of process sequence selection decisions, and their effect on machine requirements in the context of manufacturing flow systems planning. It is intended that the algorithm developed here will be embedded in an expert system such as those explained in Kusiak (1988) and Groover and Zimmers (1984) which performs automatic manufacturing planning.

8.13.1 Background

Sequencing is the order in which activities (processes) are performed in a manufacturing system. Routes deal with the specific path taken to accomplish the various processes necessary for completion of a task.

The sequencing problem can be categorized into two broad classes: first, static or deterministic sequencing; second, dynamic or stochastic sequencing. We can further categorize the sequencing problem, as discussed in Day and Hottenstein (1970), by the nature of the production demands as arrivals to the manufacturing system. These arrivals are considered to be either deterministic or stochastic batches. Each type of arrival process is then classified by the number of machines required (single stage as opposed to multistage process).

Next comes the characteristic of the job route and finally, the nature of the set up and operation times (that is, dependent versus independent set-up and operation times). Of course, further refinement could be added to the classification scheme. For example, it is possible to have set up time dependence and operation time independence and vice versa. However, the above serves to capture the fundamental nature of sequencing problems.

In production planning, it is usual to account for the generation of defective parts at various stages in the manufacturing system. In some processes, the percentage of these defects can be very high. Of major importance to the manufacturing planner is ensuring that defective products will not be passed on to the consumer since the repercussions inherent in such a practice can be quite devastating. At the same time, the planner has to meet demands and due dates. This problem can be accommodated by having an in-process inventory. But the cost of in-process inventory has to be such that it will merit its existence. Another approach to this problem could be always to start production with more items than would be required by the demand.

As stated earlier, the problems of the planner are quite varied and complex. The manufacturing sequencing problem therefore consitutes a multi-faceted challenge to the planner in that it is directly related to both the batch size selected for production and the planned production capacity (the machine requirement).

8.13.2 Problem definition

The definition of manufacturing sequencing refers to the order in which operations are performed on a family of product (or products). The problem of job scheduling will be discussed in a later section of this chapter.

A definition of the sequencing problem is best established from the production system characteristics given below, and also discussed in Nnaji and Davis (1988):

(1) There are N products and M processes.
(2) Some of the products may go through fewer than the M processes.

(3) Each process required by a product may also be required by one of the other products.

(4) There is a certain percentage of defects generated by each of the processes performed on each of the products.

(5) The facility with which the sequencing problem is associated is an existing facility.

(6) There is a daily demand for each product.

(7) The processing time for each operation is a function of predecessor operations.

(8) Set-up time is a function of first, product predecessors, second, process predecessors.

(9) Travel time between processes is assumed to be negligible.

(10) Batch production is allowed for all products.

8.13.3 Assumptions

Certain basic assumptions apply to the problems addressed here:

(1) The technological order of all the products through all the processes is known in advance, see Goyal (1975).

(2) The technological order may be different for each product.

(3) The processing times and set-up times for each machine and for each product are deterministic and known, assuming knowledge of precedences.

(4) Preempting is not allowed. That is, no splitting of jobs. Once a job is started on a machine, it will be completed without interruption.

(5) The objective to be achieved can be either to minimize total elapsed time from the start of the first job to the completion of the last job, or to minimize the total cost of production.

(6) All jobs are of equal importance, see Gupta (1971).

(7) All jobs are processed as soon as possible.

(8) All N jobs are simultaneously available at the beginning of each planning period.

(9) All M machines are available at the beginning of the planning period and are ready to take on any of the N jobs.

(10) A machine can process only one batch at a time.

(11) In-process inventory is permitted.

(12) Excess time available on any machine can be used for some other useful and profitable work, see Gupta (1971).

The objective would be to develop a mathematical model of the operation sequencing decision process which will capture the problem characteristics, and also the assumptions as stated above.

To achieve this objective, certain control variables need to be stated. These control variables can be modified (or relaxed) to produce different models with possible different constraints. They also depict the nature of the product and its movement within the system. The amount of time spent in the system would be a function of these control variables, as would the production output of the system, and consequently, the overall result of accomplishing the objective of the system's operations. The specific control variables are identified as follows:

Sequence: The state of a product in the model will depend on the sequence of operations performed. This is generally true in sequencing problems; but, often, some other parameters are functions of the state of the product as well. These are as follows:

- Processing time: The processing time of a given machine or processor is not constant, rather, it varies depending on the manufacturing state of the product.
- Set-up time: The set-up time is also a function of the state of the product as that product comes to the machine in consideration.

Batch size: The quantity of products carried between stages will be affected by the state of the product. This means that replenishing of products between stages, to account for defects, will not be permitted.

Machines: The number of machines required at a particular processing station may depend on the state of the product as well as the batch size chosen for each product processed.

The determination of how these control variables are related in their effect on production planning for minimum cost (or time) constitutes the crux of the problem for investigation and analysis. The type of algorithm(s) employed in the solution of the model developed is portrayed by the nature and interrelationships of these control variables. A deterministic planning perspective is taken throughout this development.

It is not immediately obvious how one can isolate the sequencing problem, or determine the impact of sequencing decisions on other decisions, or vice versa, see Day and Hottenstein (1970). It is quite possible for a facility not to be aware of its sequencing problems because they are shielded, from showing a strong interdependence, by other components of the total cost of production (for example, accounting, inventory, capital budgeting, loading, engineering).

The choice of one criterion as a measure of performance in the system is really a drastic simplification. This automatically eliminates specific consideration of some of the problems in an operational shop. Some of the factors affecting (or affected by) the sequencing problem, each of which may either be maximized or minimized to achieve an optimum solution, are as follows:

- Relative to the facility: idle capacity, set-up cost, output, material handling cost, reserve capacity, utilization, operation cost and so on.
- Relative to products: in-process inventory, due dates, routes, technological requirements, raw material and finished product inventory, makespan, obsolescence and deterioration losses and so on.

For management, these problems are very pertinent but no one has really been able to model a production system in such a way that individual and collective optimality can be achieved for every possible objective. Doing this would involve solving all facility planning, scheduling, and control problems ever investigated.

Sequencing problems arise from factors relative to both a facility and its products. As a result, it is important to model the sequencing problem in such a way as to incorporate components of these two areas. Modeling sequencing problems using cost of production and minimum elapsed time as criteria enables this.

8.13.4 Operational restrictions

The manufacturing sequencing problem has a number of constraints which must be accomplished in order to resolve the sequencing problems. Some of these restrictions make the problem more difficult.

The system can be constrained as follows, see also Nnaji and Davis (1988):

(1) Feasible sequences exist, that is, there may be precedence restrictions on operations.
(2) Amount of products carried between stages are limited.
(3) Investment cost is limited.
(4) Demand is known.
(5) Time (in machine time) is known.
(6) Homogeneity of machines at a workstation is assumed.
(7) Each workstation must have at least one machine.

The precedence restrictions on the processes for each product determine many of the system's other variables. Variables such as processing time and set-up time depend on this sequential order for products and processors.

The number of processors in the system is important, since this will establish whether it is multistage or one stage. The case of parallel processors is a consequence of using the values of demand, sequence, and due date to replicate the processors, with the aim of meeting demand and due dates.

Specifications of product types help establish product routes, given the precedence restrictions on products and machines. The process and set-up times are direct results of the sequence restrictions. These times are important to establish optimization criteria in the problem being considered. Lot sizes carried between stages influence other variables in the system. They can also influence in-process inventory. Demand and due dates are required to define more realistic capacity requirements.

We present two alternative mathematical models: a cost objective function and a time objective function as developed by Nnaji and Davis (1988).

8.13.5 **Mathematical model objective**

The aim is either to minimize total elapsed time from the start of the first job to the completion of the last job or to minimize the total cost of production per period. Both models are presented.

Before proceeding further, certain nomenclature must be listed and defined, as shown in Table 8.11.

8.13.6 **Cost function**

Using Table 8.11, we obtain

Min: Processing Cost + Set-Up Cost + Fixed Cost + Material Handling Cost

$$Cv_{i,j} + C_{i,j} + F_{i,j} + T_{i,j}$$

Model components

(1)   Let $Cv_{i,j}$ be the cost of processing one item of product type i through operation j for one unit of process time $v_{i,j}$. Then the total cost must be $Cv_{i,j} * v_{i,j} \langle s_i, D_i \rangle$

Table 8.11   Notation for cost model

i = product
j = operation
$n_{i,j}$ = number of machine performing operation j for product i
D_i = demand for product i
$v_{i,j}$ = operation time for product i on operation j
$k_{i,j}$ = set-up time for product i, operation j
$U_{i,j}$ = quantity (unit load) moved from operation j to operation $j + 1$ considering product i (also run quantity/set up)
s_i = state of product in the system (i.e., which operations have been performed on the product)
$C_{i,j}$ = cost associated with set-up time for product i, operation j
$C_{v_{ij}}$ = cost associated with processing product i, operation j, for a unit of time v
$F_{i,j}$ = fixed cost incurred by using each n machine
$T_{i,j}$ = cost for transporting a given batch of product i after operation j
I = investment money available at the beginning of planning period
$W_{\max}$ = maximum weight which can be transported between workstations
H = amount of time available for meeting demand D
$f_{i,j}\langle s_i, D_i \rangle$ = frequency of moves with a given demand for product type i through work center j
$\langle s_i, D_i \rangle$ = state of product i and demand for product i
$\delta_{i,j}$ = percentage of defects at workstation j for product i
R = real numbers
Cn_j = cost of one machine which can perform operation type j
$f_{i,j}$ = frequency of moves of product i through operation type j
$f_{i,o}$ = frequency of moves through the last station for product i
$v_{i,j}$ = unit of process time

for this one item. The process time, $v_{i,j}$, is a function of the state of the product i in the system, given demand D_i: hence, $v_{i,j}\langle s_i, D_i \rangle$. Therefore, for demand D_i the cost will be:

$$D_i C_{vi,j} v_{i,j} \langle s_i, D_i \rangle$$

(2) Set up for product i on machine $n_{i,j}$ is on a per batch basis. If $C_{i,j}$ is the cost associated with setting up machine $n_{i,j}$ once for product i, then the cost of setting up for $n_{i,j}$ machines will be:

$$D_i C_{i,j} n_{i,j} f_{i,j} \langle s_i, D_i \rangle$$

Notice that frequency of moves is also a function of the state of product i for a given demand D for product i.

(3) There is always a fixed cost due to using machine $n_{i,j}$ at the workstation j. This cost is:

$$F_{i,j} n_{i,j}$$

(4) In order to move product i between workstations, a certain cost is incurred due to material handling. This cost $T_{i,j}$, is cost per move. For $f_{i,j}$ moves, the cost will then be:

$$T_{i,j} f_{i,j} \langle s_i, D_i \rangle$$

The above four components of the objective function are summed over all products and for all operations performed on a given product.

For brevity in the formulation, we have the following:

$$f_{i,j} = f_{i,j} \langle s_i, D_i \rangle$$
$$v_{i,j} = v_{i,j} \langle s_i, D_i \rangle$$

Cost objective function:

$$\Phi = \text{Min} \sum\sum [D_i C_{vi,j} v_{i,j} \langle s_i, D_i \rangle + C_{i,j} n_{i,j} f_{i,j} \langle s_i, D_i \rangle$$
$$+ F_{i,j} n_{i,j} + T_{i,j} f_{i,j} \langle s_i, D_i \rangle]$$

Constraints

(1) The batch size at the last workstation must be less than or equal to demand:

$$U_{i,j} \leq D_i$$

(2) For each product i, the quantity carried through operation j for all the processes at demand D_i must equal the quantity produced:

$$U_{i,j} f_{i,j} = U_{i,o} f_{i,o}$$

(3) At a workstation, there must be at least one machine:

$$\sum_i n_{i,j} \geqslant 1$$

(4) The frequency of moves through the last workstation must be an integer for product i:

$$f_{i,o} = Z_i$$

where Z_i is an integer.

(5) The total number of products for a given production period must be equal to total demand D_i:

$$U_{i,o} f_{i,o} = D_i \qquad U_{i,j} \in R$$

(6) The weight of each batch moved between workstations must be less than or equal to a given maximum weight for every operation j:

$$w_i D_i \leqslant W \max f_{i,j} \qquad \forall_j$$

(7) The frequency of moves at a workstation is dependent on the percentage of defects at that workstation for product i:

$$f_{i,j} = f_{i,0}\left(\prod_{l=1}^{j-1} (1 - \delta_{i,l})\right)$$

(8) Cost of purchasing all machines for the entire system must be less than or equal to the available investment capital:

$$\sum_j C_{nj} \sum_i n_{i,j} \leqslant I \qquad \forall_j$$

(9) The total time utilized in setting up and processing product i at workstation j must be less than or equal to the machine time specified for processing demand D without violating the due date:

$$\frac{v_{i,j} D_i}{\prod_{i=1}^{j-1}(1 - \delta_{i,l})} + f_{i,j} k_{i,j} n_{i,j} \leqslant H n_{i,j} \qquad \forall_j$$

(10) The frequency of moves, the machine available at workstation j, and the batch moved between workstations must be greater than or equal to one:

$$f_{i,j}, n_{i,j}, U_{i,j} \geqslant 1$$

Solutions for this system can be found in Nnaji and Davis (1988).

Summary

Thus far, the following procedures have been presented:

(1) Calculate the lower bounds on frequency of moves using the material handling constraint and substituting into the relation for $f_{i,o}$ and $f_{i,j}$.

(2) Find the machine requirements by substituting given values along with the lower bound values for $f_{i,o}$ into the machine-time constraint.

(3) Solve the fixed-sequence problem by checking the capital investment constraint and the other constraints for non-violation.

(4) Calculate the objective function.

Variable sequence problems

The sequence selection problem can range from fixed sequence, fixed product to variable sequence, variable product. Certain features are common to all major classifications of the sequence selection problem: the procedure for calculating the lower bounded frequency of moves between workstations, and the calculation of the machine requirements at the same workstation. These common features make the solution of these problems more tractable. The solution of multiple product-variable sequence problems will be combinatorial in nature; but the calculations are similar to those of the fixed-sequence problem.

The main difficulty of the sequence selection problem is examined here. As mentioned earlier, the variable sequence problem has many characteristics which make it complex and, at the same time, interesting. The combinatorial nature of this problem is first encountered at the capital investment constraint calculation:

$$\sum_j C_{nj} \sum_i n_{ij} \leqslant I \qquad \forall_j$$

It is easy to see that one must choose a sequence for each product in order to sum all the $n_{i,j}$ over all products. Each product may have more than one sequence and this introduces a combinatorial problem. Searching for the optimal sequence is necessary if one must consider feasibility requirements and the possible numerous objective values that go with these combinations.

The number of combinations build rapidly with greatest dependence on how many products one is considering. In fact, the number of possible sequence combinations is related to the product and number of sequences/product as follows: Combinations $= S^N$ where S is the number of sequences/product and N is the number of products. One must realize that this combinatorial problem is significant because there is a need to choose sequences for each product which will combine with other sequences to produce an optimal collective result.

We now present an enumeration procedure for the selection of an optimal sequence combination.

Enumeration procedure

(1) Enumerate all possible sequence combinations for the product.

(2) For each product, calculate the frequency of moves at each workstation.

(3) For each sequence of operations for each product, calculate the machines required to process that product at the various workstations.

(4) If a product line is under consideration, take integer values of the machine requirements for each product at each workstation (that is, $n_{ij} \in Z_{ij}$ is an integer). Otherwise, sum all the required machines for all products at workstation $j \left(\sum_i n_{i,j} \in Z_j \right)$ where Z_j is also an integer.

(5) For each sequence combination, calculate for each product type and sequence of operations the process cost, set-up cost, material handling cost and the fixed cost at workstation j. Sum these costs as the cost of production at station j. Sum for all j. This is the total production cost for sequence combination q.

(6) Also for each sequence combination, calculate for each product type and sequence, the investment cost. Sum these costs as the investment cost for each station j. Sum for all j. This is the investment cost of sequence combination q.

Search procedure

When all these calculations are performed then the search proceeds as follows:

(1) Let $CMIN$ be the minimum total cost set initially at a very large value.

(2) Then for all feasible sequence combinations $(q = 1, 2, \ldots, S)$,

(a) If $CMIN$ is greater than the total cost of sequence q then check for feasibility by comparing the total investment cost associated with sequence q against the given capital investment potential.

(i) If this sequence is feasible then the current total cost $CMIN$ is set equal to the cost of sequence q.

(ii) If all sequences are considered then go to step 4; otherwise, let $q = q + 1$ and return to (2(a)).

(b) If $CMIN$ is less than the total cost for sequence q, then go to the next sequence combination.

(3) If all sequences are unfeasible, then go to step 6.

(4) Publish results showing the attributes of the optimal sequence q.

(5) Stop.

(6) Problem unfeasible.

Determination of integer machines

The way in which integer values of machines are found from the required machines at each workstation influences the objective value and more especially, the investment cost. In many cases where the constraints are tight, the feasibility of the problem may depend totally on how these integer machines are obtained.

When the system is set up as a product line, it will be necessary to take integer values on the number of machines at each work station ($n_{i,j} = Z_{i,j}$ where $Z_{i,j}$ is an integer). Then, integer values are summed over all products and the sum is used for the investment cost calculation. However, if a homogeneous workstation is needed (that is, a system where all products can share machines – a general flowline), then the machines required to process each product can be summed over all products and the resulting integer value used to calculate the cost. Extensive computational experience shows that the product-line case is more expensive and leads to unfeasibility of the problem in some cases while the flowline reduces costs including the investment cost, which determines whether a system can be implemented or not.

Problems which produce results for the computational experience presented below can be found in Nnaji (1982). A typical example is presented in Table 8.12 for a three-product case with four operations. Each product has two possible sequences so one obtains a total of eight possible sequence combinations from the formula: combinations $= S^N$.

From these results one can see that the optimal sequence out of the eight sequence combinations was sequence four. The results indicate that the problem was feasible.

To gain appreciation for the combinatorial difficulties associated with obtaining solutions to this general class of problem, the following scenerios were tested. The calculations were for fixed product cases. The execution times for these cases are shown in Tables 8.13 and 8.14, and refer to an IBM 3032 processor.

One can see that the execution time increases almost as fast as the sequence combinations, but is most sensitive to the number of products involved. The time required to perform these calculations is also relatively small. This means that

Table 8.12   Summary of a three product example problem

Seq. no.	1	2	3	4	Obj. cost ($)	Invst. cost ($)
1	8,	13,	14,	10	19631.83	2250.00
2	8,	14,	12,	11	19061.26	2250.00
3	8,	13,	13,	11	18812.98	2250.00
4	8,	14,	11,	12	18242.40	2250.00
5	8,	15,	14,	10	20567.50	2350.00
6	8,	15,	12,	11	19996.93	2300.00
7	8,	15,	13,	12	19748.65	2400.00
8	8,	15,	11,	13	19178.07	2350.00

Table 8.13   Multiple fixed products, variable sequence (number of products = 3)

Number of sequences	Seq. combinations	Time
1	1	0.10
2	8	0.30
3	27	0.80
4	64	1.79
5	125	2.6

Table 8.14   Fixed number of sequences, variable number of products (number of sequences = 2)

Number of sequences	Seq. combinations	Time
1	2	0.08
2	4	0.12
3	8	0.31
4	16	0.61
5	32	1.32
6	64	2.93

enumeration of all feasible sequences is the only cumbersome aspect of the solution of the sequence selection problem. This procedure assumed that all sequences had been enumerated in advance as part of a manufacturer's process planning activity.

8.14 Scheduling Systems

There are three major types of manufacturing scheduling systems including project scheduling, job shop scheduling, and assembly line balancing. There is a variety of methods for analyzing the scheduling problem and they include: graphical techniques (for example, Gantt charts); project networks (for example, critical path methods (CPM) and project evaluation and review techniques (PERT)); job shop models; and dispatching rules.

Scheduling normally deals with intermittent systems of production as opposed to high-volume manufacture in production line. Thus scheduling helps to determine when to start each job and when to finish. The job shop is the general scheduling system and can be described in terms of:

- n jobs arrive simultaneously in an idle shop (static job shop) or jobs arrive intermittently (dynamic job shop),
- number of machines – m machines,
- flow process – if all jobs follow the same route (flow shop) or not,
- criterion for evaluating performance.

Performance measures can include:

(1) Minimize makespan – total amount of time required to completely process all jobs.

(2) Minimize average lateness/job

   d_i = due date of job i

   C_i = Completion time of job i

   $L_i = C_i - d_i$ = Measure of lateness

(3) Find average tardiness/job

$$T_i = \max(0, L_i)$$

(4) Average time in the system/job (average flowtime).

(5) Machine utilization/job

(6) Average waiting time/job

(7) Labor utilization

(8) Shortest processing time (SPT) and longest processing time (LPT)

We can characterize the general scheduling system as follows A/B/C/D, where:

A = number of jobs

B = number of machines

C = type of flow

D = criterion for performance measure

Using the above letter notations, the system can be expressed. For example, $n/m/G/F_{max}$ or $n/m/F/F_{max}$, where F_{max} = maximum flow time (min F_{max}); where G and F refer to general job shop and flow shop, respectively.

If there are n jobs to be processed on one machine, then we could minimize processing by sequencing them in increasing order (SPT method), also discussed in Conway *et al.* (1967):

$$P_{[1]} \leqslant P_{[2]} \leqslant P_{[3]} \leqslant \ldots \leqslant P_{[n]}$$

where $P_{[k]}$ denotes the job that is the k position in the sequence. This can easily be proven as follows:

Let,

$$F_{[k]} = \sum_{i=1}^{k} P_{[i]} = \text{Flowtime of job in the } i\text{th position}$$

The, mean flowtime of n jobs in the sequence can be calculated as:

$$\bar{F} = \frac{1}{n} \sum_{k=1}^{n} F_{[k]}$$

$$= \frac{1}{n} \sum_{k=1}^{n} \sum_{i=1}^{k} P_{[i]}$$

$$= \frac{\sum_{i=1}^{n} (n - i + 1) P_{[i]}}{n}$$

Axiom:

Sum of pairwise products of two sequences of numbers can be minimized by arranging one sequence in non-increasing order and the other sequence in non-decreasing order. Since $(n - i + 1)$ is in non-increasing order then arranging processing times in non-decreasing order will minimize mean flowtime.

SPT minimizes the following measures of performance:

- mean waiting time
- mean lateness
- mean weighted flowtime

For instance, let, w_i = importance weight for job i. Then,

$$\bar{F}_w = \frac{\sum_{i=1}^{n} w_i F_i}{n}$$

SPT will yield:

$$\frac{P_{[1]}}{w_{[1]}} \leqslant \frac{P_{[2]}}{w_{[2]}} \leqslant \ldots \leqslant \frac{P_{[n]}}{w_{[n]}}$$

Graphical methods

There are many graphical techniques, but the most prominent is the Gantt chart. Developed by Henry L. Gantt, this graphical technique is equivalent to a jigsaw puzzle in which one arranges a set of g_i blocks in which each block carries information about the operation number or name; number of the operation j $(j = 1, 2, \ldots, g_i)$; and the number of the machine k required to perform the operation (Figure 8.30).

The following constitutes the type of scheduling jobs for the general job shop:

Scheduling problems

(1) $n/1$- n jobs, 1 machine
(2) $n/2$- n jobs, 2 machines
(3) $n/3$- n jobs, 3 machines
(4) $1/m$-   1 job, m machines
(5) $2/m$ 2 jobs, m machines
(6) $3/m$ 3 jobs, m machines
(7) n/m n jobs, m machines

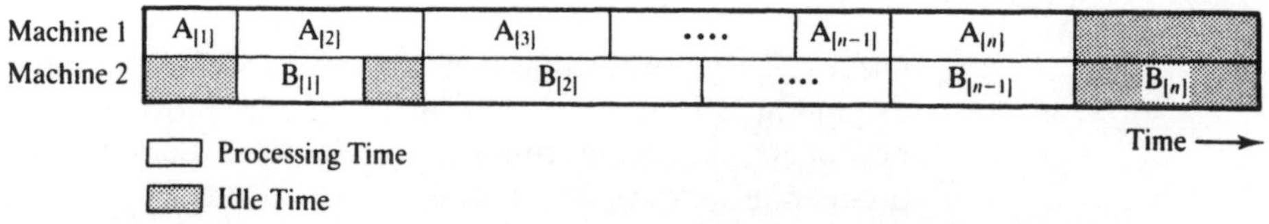

Figure 8.30
Gantt chart for $n/2$ problem.

□ Processing Time
▨ Idle Time

Time ⟶

8.14.1 The *n*/1 problem

The *n*/1 problem can easily be solved using the shortest processing time (SPT) method provided that the machines perform one operation:

$$P_{[1]} \leqslant P_{[2]} \leqslant \ldots \leqslant P_{[n]}$$

This sequence gives minimum flowtime for all jobs. So the jobs could be scheduled according to that sequence.

8.14.2 The *n*/2 flow shop

The *n*/2 problem can be solved using Johnson's rule with the following assumptions (see Conway *et al.* (1967)):

(1) All jobs are simultaneously available.

(2) Each machine can process only one job at a time.

(3) Each job can be in process on only one machine at a time for each job *i*, A_i must be completed before B_i can begin.

(4) No preemption, no insertion allowed.

Let

$A_i = P_{i,1}$ the processing time (including set up, if any) of the first operation of the *i*th job.

$B_i = P_{i,2}$ the processing time (including set up, if any) of the second operation of the *i*th job.

F_i the time at which the *i*th job is completed.

Johnson's Procedure

When one is scheduling an $n/m/F/F_{\max}$ problem with all jobs simultaneously available, to minimize the regular measure of performance, one needs to consider only schedules in which the same job order is prescribed on machines 1 and 2 and on machines $m - 1$ and m.

The positions of these two jobs can be reversed on machine *m* without increasing the maximum flowtime of these particular jobs and without changing the flowtime of any other job.

With this in mind, Johnson's procedure can be applied as follows:

● Each job consists of a pair (A_i, B_i)

● This ordering is the same for all machines.

● A_i or B_i could be zero if some job has single operation.

Given the 2*n* values, $A_1, A_2, \ldots, A_n, B_1, B_2, \ldots, B_n$, the aim is to find the ordering of these jobs on each of the two machines so that precedence (routing) and the occupancy constraints are not violated and so that the maximum of the F_i is made as small as possible.

From Figure 8.30, it is clear that the last job cannot be completed earlier than the time required to process each job on machine 1 plus the time required to perform the second operation of the last job, since $B_{[n]}$ cannot overlap $A_{[n]}$. Thus:

$$F_{max} \geqslant \sum_{i=1}^{n} A_{[i]} + B_{[n]}$$

Similarly, the last job cannot be completed in less time than it takes to process each job on machine 2 plus the time caused by the delay before machine 2 can begin, since $B_{[1]}$ cannot overlap $A_{[1]}$. Thus:

$$F_{max} \geqslant A_{[1]} + \sum_{i=1}^{n} B_{[i]}$$

Note that the sum of the As and the sum of the Bs are direct consequences of the given information and entirely unaffected by the ordering of the jobs.

To reduce these bounds, one can only influence $B_{[n]}$ and $A_{[1]}$ by the choice of the sequence. The following procedure can therefore evolve:

- Choose the smallest of the $2n$ values of the As and Bs.
- If this value happened to be A_i, put that job first in the sequence so as to make $A_{[1]}$ as small as possible.
- If it happened to be B_i, put that job last in the sequence so as to make $B_{[n]}$ as small as possible.
- With the position of that job determined, repeat the same argument for the set of $n - 1$ jobs left.

8.14.3 Calculation of idle time in $n/2$ flow shop

Due to the nature of the flow shop, one can restrict attention to schedules in which work on machine 1 is compact to the left; that is, there is no idle time on machine 1 until $A_{[n]}$ has been completed.

The assignments of the $A_{[i]}$ values of any arbitrary schedule can be shifted to the left to achieve this compactness without increasing the maximum flowtime, and one can make the following definition. Let $X_{[i]}$ be the idle time on machine 2 immediately preceding $B_{[i]}$. The values of $X_{[i]}$ can be expressed in terms of As and Bs so that:

$$X_{[1]} = A_{[1]}$$

$$X_{[2]} = \max(A_{[1]} + A_{[2]} - B_{[1]} - X_{[1]}, 0)$$

$$X_{[3]} = \max(A_{[1]} + A_{[2]} + A_{[3]} - B_{[1]} - B_{[2]} - X_{[1]} - X_{[2]}, 0)$$

In general,

$$X_{[J]} = \max\left(\sum_{i=1}^{j} A_{[i]} - \sum_{i=1}^{j-1} B_{[i]} - \sum_{i=1}^{j=1} X_{[i]}, 0 \right).$$

Partial sums of Xs can be obtained from:

$$X_{[1]} = A_{[1]}$$

$$X_{[1]} + X_{[2]} = \max(A_{[1]} + A_{[2]} - B_{[1]}, A_{[1]})$$

$$X_{[1]} + X_{[2]} + X_{[3]} = \max(A_{[1]} + A_{[2]} + A_{[3]} - B_{[1]} - B_{[2]}, X_{[1]} + X_{[2]})$$

$$= \max(\textstyle\sum_{i=1}^{3} A_{[i]} - \sum_{i=1}^{2} B_{[i]}, \sum_{i=1}^{2} A_{[i]} - B_{[1]}, A_{[1]})$$

In general,

$$\textstyle\sum_{i=1}^{j} X_{[i]} = \max(\sum_{i=1}^{j} A_{[i]} - \sum_{i=1}^{j-1} B_{[i]}, \sum_{i=1}^{j-1} A_{[i]} - \sum_{i=1}^{j-2} B_{[i]}, \ldots,$$

$$\textstyle\sum_{i=1}^{2} A_{[i]} - B_{[1]}, A_{[1]})$$

Let $Y_J = \sum_{i=1}^{j} A_{[i]} - \sum_{i=1}^{j-1} B_{[i]}$. Then,

$$\sum_{i=1}^{j} X_{[i]} = \max(Y_1, Y_2, \ldots, Y_J).$$

Let $F_{\max}(S)$ be the maximum flowtime of a particular schedule S, then:

$$F_{\max}(S) = \textstyle\sum_{i=1}^{n} B_{[i]} + \sum_{i=1}^{n} X_{[i]}$$

$$= \textstyle\sum_{i=1}^{n} B_{[i]} + \max(Y_1, Y_2, \ldots, Y_n)$$

Since the sum of B_i's is independent of sequence, the maximum flowtime depends entirely on the sum of the intervals of idle time on the second machine and this is equivalent to the maximum of the Ys. We must then find a schedule S^* such that $F_{\max}(S^*) \leqslant F_{\max}(S)$ for any S.

This is simply finding the maximum of the n values of Y_i's and an ordering that minimizes that maximum. We can now obtain total idle time and minimum (maximum flowtime).

8.14.4 *n/2* job shop

Johnson's algorithm can be generalized for the job shop with the constraint that each job can have at most, two operations. This modification was due to Jackson's theory, see Jackson (1956).

Jackson's Procedure

(1) Partition the n jobs into four sets as follows:
   (a) (A) = Set of jobs that have only one operation and which is performed on machine 1.
   (b) (B) = Set of jobs that have only one operation, which is to be performed on machine 2.
   (c) (AB) = Set of jobs to be processed on machine 1 then machine 2 and so have two operations.

(d) (BA)=Set of jobs to be processed on machine 2 then machine 1 and so have two operations.

(2) Initially sequence AB by Johnson's procedure as if they were the only work to be done.

(3) Similarly determine BA by Johnson's procedure.

(4) Since the ordering of jobs within A and B has no effect on the maximum flowtime, select an arbitrary ordering of these goods.

(5) Combine the sets to obtain an optimal schedule without changing the order within the sets you have arranged by Johnson's rule:

(a) On Machine 1: Jobs in AB before jobs in A before jobs in BA.

(b) On Machine 2: Jobs in BA before jobs in B before jobs in AB.

8.14.5 *n*/3 job shop

The *n*/3 job shop can be solved using Johnson's algorithm if:

either $\min P_{i1} \geqslant \max P_{i2}$

or $\min P_{i3} \geqslant \max P_{i2}$

That is, machine 2 is completely dominated by either machine 1 or machine 3. If the above condition is obtained, then follow this procedure:

(1) Define dummy machines 1′ and 2′ with the following processing times.

$$P_{i1'} = P_{i1} + P_{i2}$$

$$P_{i2'} = P_{i2} + P_{i3}$$

(2) Apply Johnson's two machine procedures.

8.14.6 Two jobs, *m* machines (2/*m* job shop)

While it was possible to schedule *n*/2 problem, there is no optimal method of scheduling the 2/*m* job shop problem. A method that works graphically but is difficult computationally is the one suggested by Akers and Friedman (1955) and stated more completely by Hardgrave and Nemhauser (1963). For example,

PROBLEM MACHINE

Job	1	2	3	4	5
1	3	4	4	3	3
2	2	5	3	5	4

Technological order for job 1 = (2,1,3,4,5)
Technological order for job 2 = (3,1,5,2,4)

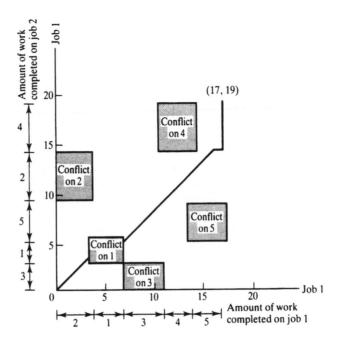

Figure 8.31

Example solution to 2/m job shop.

We find that $\sum_{j=1}^{m} P_{1j} = 17$, and that $\sum_{j=1}^{m} P_{2j} = 19$. A solution is any line drawn from (0,0) to the point $(\sum_{j}^{m} P_{1j}, \sum_{j}^{m} P_{2j})$ which does not pass through a boxed region. Vertical segments represent work on job 2 only. Horizontal lines imply work on job 1 only. 45° implies simultaneous processing.

A minimum makespan schedule is one that maximizes the amount of simultaneous processing. Scheduling must be determined by trial and error (see Figure 8.31).

8.14.7 *n/m* Job shop

Scheduling problems for 3 jobs/*m* machines up to *n* jobs/*m* machines have proven very difficult to solve to completion. Heuristics have been developed to deal with these kinds of problems. Unfortunately, these are the predominant problems in industry and therefore computationally efficient techniques are very important for their solution. Some of the problems are made even more difficult when the rate of demand for products varies. This dynamic element violates some of the convenient assumptions about approaches used in static cases.

The problems usually have some or all of the following characteristics:

● Multiple products competing for limited resources such as machine time, labor, working capital and so on.

● Multiple resources, or processes by which a product can be produced. Examples include different possible product routings within a plant, different plants, make versus buy alternatives, overtime versus regular time and so on.

- Multiple facilities arranged in a multistage structure in such a way that the output from one stage becomes the input to a subsequent stage.

As stated, the general n jobs/m machines job shop scheduling problem has proved to be very difficult. The difficulty is not in modeling but in computation since there is a total of $(n!)^m$ possible schedules, and each one must be examined to select the one that gives the minimum makespan or optimizes whatever measure of effectiveness is chosen. There are no known efficient, exact solution procedures. Some authors have formulated integer programming models but the computational results (computational complexity) have not been encouraging.

For small problems, the branch-and-bound technique has been shown to be very good and better than integer programming formulations, but this has been computationally prohibitive for large problems.

Heuristic methods have shown significant promise in solving the general job shop scheduling problem. The most widely used heuristics belong to the class of procedures called dispatching rules. These are simply logical decision rules that enable a decision maker to select the next job for processing at a machine when that machine becomes available, see Johnson and Montgomery (1974). In this way, decisions are made about the schedule sequentially over time instead of all at once. The concept of job priority is always included in the dispatching algorithm. A priority is merely a numerical attribute for a job which is defined in such a way that the highest numerical priority is scheduled next for processing.

In the dynamic job shop case, priority queuing models have been developed. However, only limited success has been achieved using queuing models. The extent of success in the area has been with single-channel, single server queues which correspond to single-facility job shops, and multi-channel queues that correspond to queues of several facilities in parallel. There are some limited approaches to queues in tandem (series), and sufficient conditions have been shown under which a general network of queues, which corresponds to a general job shop, may be decomposed and treated as independent queues.

An alternative to the queuing model is an experimentation with various scheduling procedures, such as priority dispatching in an actual job shop. This approach has problems in that it is usually not clear whether a particular result is because of the scheduling procedure being tested, or due to some undetected condition in the job shop. This solution can lead to loss of generality.

Simulation methods can be used to overcome this problem. The simulation procedure essentially amounts to experimentation with different priority dispatching rules using the computer. The general procedure involves developing a computer program which simulates job arrivals and controls the flow of these jobs through various processing. Such programs can assign due dates or other job attributes and contain lists and files to record the state of the job shop along with a computation of various measures of effectiveness. It has been shown that the shortest processing time (SPT) priority rule provides a superior performance over many other measures of effectiveness. SPT has already shown optimal results for mean flowtime in static machine problems.

8.15 Project Scheduling

Project scheduling is typically for jobs that have 'one of a kind' (one-time effort) characteristics, for example, buildings, a complex software system or a factory. Methods of analysis include CPM, PERT, and others. These methods are based on activity networks in which the work is broken down into activities and the activities are depicted in a network.

CPM – Critical Path Method

The critical path method (CPM) was developed by E.I. duPont de Nemours in 1957 for the construction and maintenance of chemical plants. Since then, CPM has been applied to numerous other areas with many articles and computer programs appearing on its use.

CPM starts with the construction of a network diagram which shows the precedence among activities in the project. The construction of the network is the most difficult aspect of the CPM since it is the actual planning function and requires detailed analysis and thought. This planning activity is the best part of CPM since it shows a clear and logical sequence for the activities.

A project would normally consist of activities, time duration, and precedence relationships as shown in Table 8.15. Figure 8.32 represents the network of the activities shown in Table 8.15. In this figure, arcs represent the activities while the nodes represent events. An event is the point which marks the end of an activity and the commencement of a new group of activities. In Figure 8.32, the letter and number along the arcs represent the activity labels and the activity duration respectively. A network system represented this way is called an activity-on-arc (AOA) network. Another way to represent the same information is by using nodes to represent the activities and the arcs (usually arrows) represent only the precedence relationships between activities. This kind of network is called an activity-on-nodes (AON) network. This AON network is not widely used and therefore does not

Table 8.15 Information for AOA and AON networks

Activity	Depends on	Duration
A	–	3
B	–	4
C	–	5
D	A	2
E	A	3
F	B	5
G	B	4
H	C	6
I	D	2
J	D, E, F	1
K	G, H	5

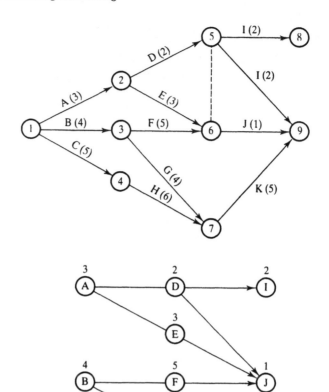

Figure 8.32
Activity on arcs
(AOA) graph.

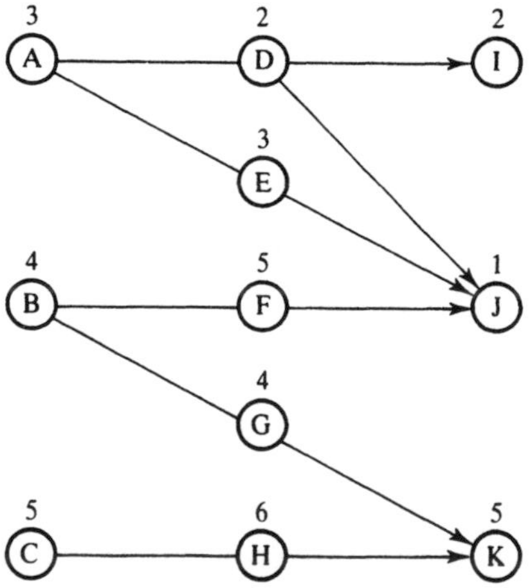

Figure 8.33
Equivalent node
representation (AON).

have many CPM computational programs for solving problems as represented. Figure 8.32, as mentioned earlier, is the AOA network for the activities in Table 8.15. Figure 8.33 is the AON representation of the same activities.

The basic rules for representing any of the networks are as follows:

(1) Each activity is represented by one, and only one, arrow in the network.
(2) No two activities can be identified by the same start and tail events.
(3) To ensure correct precedence relationships, the following questions must be checked as every activity is added to the network:

    (a) What activities must be completed before this activity can start?

    (b) What activities must follow this activity?

    (c) What activities must occur concurrently with this activity?

In Figure 8.32, the dashed line for arc (5,6) represents a dummy activity and has a zero duration. This dummy activity is introduced to represent the dependence of activity J on activity D. Sometimes, the dummy activity is used to represent partial dependence.

To illustrate how this scheduling system works, let the events on AOA be indexed by i $(i = 1, 2, \ldots, n)$. Then an activity can be represented by (i,j) where $i < j$.

Let,

t_{ij} = duration for activity (i,j).
E_i = earliest occurrence time for event i.
L_i = latest occurrence time for event i.
ES_{ij} = earliest start time for activity (i,j).
EF_{ij} = earliest finish time for activity (i,j).
LS_{ij} = latest allowable start time for (i,j).
S_{ij} = total slack or float for activity (i,j).
FS_{ij} = free slack or float for activity (i,j).
LF_{ij} = latest finish time for activity (i, j).
S_i = slack time for event i.

CPM computation

We will now show how the CPM method can be used to find the critical path through a network and to calculate the latest finish time, latest start time, total slack time, and the free slack. This can be done by using a forward pass and a backward pass.

(1) Forward pass

   (a) Set the start time of the initial event to zero.

   (b) Start each activity as soon as its predecessor event occurs.

   (c) Calculate early event time as the maximum of the earliest finish time of activities terminating at the event.

   Mathematically, the procedure is as follows:

   (a) $E_1 = 0$

   (b) $E_j = \max\{E_{i1} + t_{i1,j}, E_{i2} + t_{i2,j}, \ldots, E_{ik} + t_{ik,j}\}$

   $j > 1$. $i_1, i_2, \ldots, i_k$ show the preceding events of the k activities, that terminate at event j.

   $ES_{ij} = E_i$

   $EF_{ij} = E_i + t_{ij}$

(2) Backward pass

   (a) Set the latest time for the terminal event equal to the earliest time for that event.

   (b) Start each activity at the latest time of its successor event less the duration of the activity.

   (c) Determine event time as the minimum of the latest start times of all activities emanating from the event.

Mathematically:

(a) $L_n = E_n$

(b) $L_i = \min[L_{j1} - t_{i,j1}, L_{j2} - t_{i,j2}, \ldots, L_{jm} - t_{i,jm}]$

$1 < n. j_1, j_2, \ldots, j_m$ show the successor events of the m activities that emanate from event i. One can also obtain:

Latest finish time:

$$LF_{ij} = L_j$$

Latest start time:

$$LS_{ij} = L_j - t_{ij} = LF_{ij} - t_{ij}$$

(3) Compute slack time as the difference between its latest and earliest occurrence times.

$$S_i = L_i - E_i$$

Total slack time is the time by which that activity could be extended without changing the total duration.

$$S_{ij} = L_j - E_i - t_{ij}$$

that is, the difference between the latest occurrence time of event j and the earliest finish time for (i,j).

Free slack for any activity is the amount of time that activity completion can be delayed without affecting the earliest start time for any other activity in the network. Mathematically:

$$FS_{ij} = E_j - E_i - t_{ij}$$

(4) Determine critical path, which consists of all activities with zero slack ($S_{ij} = 0$).

This path is a path of activities upon which any delay will result in the delay of the completion time of the project.

PERT-Project Evaluation and Review Technique

The difference between CPM and PERT is that PERT takes account of the uncertainty in the activity duration. A probability distribution is typically given by three estimates for a given activity (i,j) as shown in Figure 8.34:

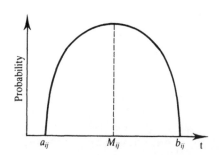

Figure 8.34

Beta distribution for activity duration.

(1) the most likely duration: M_{ij}

(2) the optimistic duration: a_{ij}

(3) the pessimistic duration: b_{ij}

In the absence of historical data, one can use the beta distribution. Using these three estimates, one can derive the mean and variance:

Mean: $$\bar{t}_{ij} = \frac{a_{ij} + 4M_{ij} + b_{ij}}{6}$$

Variance: $$\sigma_{ij}^2 = \frac{(b_{ij} - a_{ij})^2}{36}$$

The PERT network is the same as in the CPM. Computation of earliest event time (E_i) and computation of latest event time (L_i), which are random variables, yields mean event times of $\bar{E}_i$ and $\bar{L}_i$.

Assumption

Assume that for a given critical path on the CPM network, the mean and variance of the entire project could be the sum of individual means and variances, where p indicates the project and U indicates the set of activities on the critical path.

$$\bar{t}_p = \sum_{(i,j) \in u} \bar{t}_{ij}$$

$$\sigma_p^2 = \sum_{(i,j) \in u} \sigma_{ij}^2$$

This assumption requires that the activity durations be independent random variables.

8.16 Assembly Line Balancing

The assembly line can be composed of many workstations. The workstations can be connected by conveyors or by other types of material transfer devices or human movers. Each workstation's task can range from simple to complicated. The workstations may also perform their tasks at different rates because of their production capabilities. The goal is to ensure that there is maximum utilization of the workstations and that the production rate of the stages are about equal to ensure that some stations do not slow others down; and that others do not get starved of work. Assigning tasks to a work place ensures that appropriate measures of performance are optimized. If a line is balanced perfectly, then all stations will have an equal amount of work and smooth product flow with no delay. The assembly line is normally characterized by movement of workpieces from one station to the next. The workpieces remain at a certain station for a certain duration of time called cycle time. The overall design of the production line includes:

(1) proper sequencing of operations planning,
(2) setting of a production rate for the line,
(3) balancing the load on individual workstations.

Suppose there is a set of n tasks to be performed on a workpiece. Then let:

P_i = Performance time (processing time) of job i (or task i) on the workpiece.

k = Number of stations $k \geqslant n$.

$\sum_i^n P_i$ = Total work content of the product.

I = Total idle time of the product on all machines.

c = Cycle time.

d = Balance delay.

P_{max} = Longest performance of all sets of tasks.

A number of objectives can be modeled for this problem, including:

- Minimize I
- Minimize d
- Minimize k and I
- Minimize k and c, equivalent to minimize I.

We can define idle time as follows:

$$I = kc - \sum_{i=1}^{n} P_i$$

Similarly, we can define balance delay (equivalent to average idle time) as follows:

$$d = \frac{kc - \sum_{i=1}^{n} P_i}{kc}$$

For a perfect balance, I and d could be zero (that is $I = d = 0$). When this happens, then:

$$kc = \sum_{i=1}^{n} P_i$$

and the number of workstations k can be found as:

$$k = \frac{\sum_{i=1}^{n} P_i}{c}$$

The goal is, generally, therefore to find the number of workstations for which $\sum_{i=1}^{n} P_i/c$ is an integer and $I = 0$. An important constraint in the assembly line balancing problem is: the longest performance time of the set of tasks (P_{max}^*) must be less than or equal to the chosen cycle time which must be less than or equal to the total cycle time.

$$P^*_{max} \leqslant c \leqslant \sum_{i=1}^{n} P_i$$

These upper and lower bound considerations are based on the practical considerations that, usually, the expected demand for the product implies a cycle time, since the production rate (in units of product/unit time) is just $\frac{1}{c}$, see Johnson and Montgomery (1974). When this implied cycle time is smaller than the cycle time which minimizes idle time, then, in the long run, a second shift, duplicating bottleneck workstations, or a second production line may be required to meet demand. In the case where the implied cycle time is larger than the cycle time which minimizes idle time, then the production line need not be operated continuously or the resources of such a line can be shared with other products.

There are mathematical programming formulations by Held *et al.* (dynamic programming), by Thangavelu and Shetty (integer programming), and by Elmaghraby. There are heuristics available. These do not really always show optimal solutions but are very good to use. In fact, engineers prefer to use them.

Consider the problem shown in Figure 8.35 for a six-station assembly line. The sum of the processing times can be obtained as follows:

$$\sum_{i=1}^{n} P_i = 3 + 2 + 3 + 1 + 3 + 4 = 16$$

The following flow sequences through the six stations are allowable:

123546

132456

Clearly, six stations will be feasible for this problem but will not balance the line since work content for the various stations will be unequal. Five stations will have the same problem. Suppose that we choose four stations, especially since the sum of the work content is an integer multiple of the number of possible stations of four. But the sequences will not work:

$$\frac{\sum_{i=1}^{n} P_i}{k} = c = \frac{16}{4} = 4 \text{ time units}$$

Figure 8.35

A six-station assembly line with precedence.

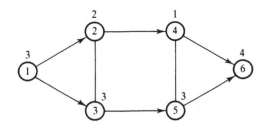

We must now try three stations:

$$c = \frac{16}{3} = 5.333$$

$$P_{max} \leqslant c \leqslant \sum_{i=1}^{n} nP_i$$

The sequences will not allow for work balance either in the three station case. Then we try two stations, for which:

$$c = \frac{16}{2} = 8$$

This will be found to work for the two sequences. The above manual method is quite cumbersome. The Kilbridge and Wester (1961) algorithm is a more systematic approach to the balancing problem. In this system the network is divided into segments that obey the sequence of jobs. Figure 8.36 shows a modified Figure 8.35 with four regions of the network. We again notice that $\sum_{i=1}^{n} P_i = 16$.

We can write this work content as a product of prime numbers:

$$16 = 2 * 2 * 2 * 2$$

Since $P_{max} \leqslant c \leqslant \sum_{i=1}^{n} P_i$ is required then $\sum_{i=1}^{n} P_i/c$ is an integer.

$$4 \leqslant c \leqslant 16$$

Using the prime numbers, we find the following station cycle times, c_i:

$$c_1 = 2 * 2 * 2 * 2 = 16$$
$$c_2 = 2 * 2 * 2 \quad = 8$$
$$c_3 = 2 * 2 \qquad = 4$$

We now apply each of the above cycle times to obtain the stations as follows:

$$k_1 = \frac{\sum_{i=1}^{n} P_i}{C_1} = \frac{16}{16} = 1 \text{ station (trivial)}$$

$$k_2 = \frac{\sum_{i=1}^{n} P_i}{C_2} = \frac{16}{8} = 2 \text{ stations}$$

$$k_3 = \frac{\sum_{i=1}^{n} P_i}{C_3} = \frac{16}{4} = 4 \text{ stations}$$

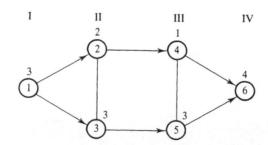

Figure 8.36

Kilbridge and Wester regionalization.

Table 8.16   Task assignments in Kilbridge and Wester method

Column no. (Figure 8.36)	Task	P_i	$\Sigma_{i=1}^{n} P_i$	Cumulation of $\Sigma_{i=1}^{n} P_i$
I	1	3	3	3
II	2, 3	2, 3	5	8
III	4, 5	1, 3	4	12
IV	6	4	4	16

Procedure:

Pick $C = 8$, for example

(1) Scan the cumulative sum, find the smallest sum greater than or equal to cycle time 8. This sum is in column II of Figure 8.36 (Table 8.16). Assign tasks in I to station I. Obtain $C - 3 = 5$ time units left. Find the combination of tasks in II that totals five time units. Assign all two tasks in II to station I. Obtain $c = \sum_{k=1}^{3} P_k = 8$.

(2) Find the smallest cumulative sum greater than or equal to $2 \times 8 = 16$. This is the cumulative of column IV of Figure 8.26. Assign all remaining jobs to station 2.

In general, the Kilbridge and Wester algorithms are very good for large cycle times when one station crosses columns. In cases of low cycle times, where one row may require two or more stations, much manipulation of the tasks will be necessary resulting in heavy computational burden, and yet good results cannot be guaranteed.

8.17 Summary

In this chapter, we have presented the components of process planning and manufacturing scheduling issues. The subtopics were presented to show an adequate understanding of their relationship in the overall process planning and manufacturing scheduling domain. The computational and algorithmic aspects of the subjects were vigorously covered where necessary.

8.18 Problems

(1) What is the main difference between process planning and scheduling?

(2) Describe the traditional process planning system.

(3) How does the generative process planning work? What is the major problem in this approach and how can it be overcome?

(4) Develop a coding system for the process planning of gear assembly. Do the same for the planning of injection molded products.

(5) What are the major disadvantages of the traditional inventory control? What computational problems are inherent in this approach?

(6) Develop the architecture of a program for the material requirement planning of microcomputer assembly. You may decide what level components, detailing the computer you want to have.

(7) What is the relationship between Pareto's laws and the ABC inventory? How can you apply the ABC inventory in the electronics industry?

(8) Provide the product structure for the major components of an automobile. Describe the architecture of an MRP program for this product.

(9) Why do you need lot sizing techniques for MRP?

(10) Give the advantage of the Wagner–Whitin algorithm.

(11) Use the Wagner–Whitin algorithm to provide the ordering schedule for computer chassis given the following cost and MRP data:

Month	1	2	3	4
Set-up cost	50	50	100	100
Unit variable cost	2	3	4	5
Unit carrying cost	4	3	3	4
Net requirements	300	400	200	150

(12) Describe the characteristics of a sequencing problem.

(13) Suppose that the following processing times were provided for a robot assembling electronic chips onto PC boards: 3, 4, 1, 5, 6, 7, 8, 2 minutes for a seven operation assembly. Prove that ordering them in order of SPT will minimize the mean flow time.

(14) Describe how you would apply Jackson's rule in a small machine shop. Can you extrapolate this to a large machine shop?

(15) How does Johnson's rule help in solving CIM scheduling problems?

(16) What are the limitations in finding optimal solutions for n/m job shop?

(17) How can you use project scheduling techniques (CPM and PERT) to solve automated manufacturing scheduling problems?

(18) How can you apply the Kilbridge and Wester algorithm in the electronics manufacturing industry?

9

Robotics

CHAPTER CONTENTS

9.1	Introduction	449
9.2	Industrial Robots	450
9.3	Classification of Robots	450
9.4	Major Components and Functions of Robots	455
9.5	End-Effector	461
9.6	Robot Coordinate Systems	464
9.7	Manipulator Kinematics	476
9.8	Dynamics of Kinematic Chains	482
9.9	Robot Programming	492
9.10	Task-level Programming	499
9.11	Applications	515
9.12	Problems	520

9.1 Introduction

This chapter provides very detailed descriptions of the derivation of various aspects of robot programming. This level of detail is provided because of the complexity of the robot compiler. People who are interested in developing robot programs must be conversant with kinematics, dynamics and path planning. In addition, the ongoing developments in the generation of 3-D drawings of robot action plans cannot be done without a good knowledge of the necessary mathematical algorithms. In multi-robot arm operation, for instance, where the integration of sensors and vision is necessary, the knowledge of this level of description is essential.

9.2 Industrial Robots

A manipulator is a mechanical emulation of the human arm or hand. It is an anthropomorphic device. Early robots evolved from the concept of modeling the human arm. The evolution started from teleoperation, a handling approach in which a manipulator is tightly coupled to its human operator's arm enabling the operator to perform tasks in environments not normally suitable for man. A teleoperator may typically be used in a radioactive environment.

With modern robots, the human operator is replaced by a programmable device, such as a computer, which controls the motions of the mechanical equivalent of the manipulator in the teleoperation. Robots normally perform tasks in various environments but in this chapter, we will be concerned with those that are applied to industrial tasks.

This chapter introduces the various aspects of robotics but will concentrate on the subject of programming, which is the computational link to the mechanical system.

An industrial robot is a general purpose programmable, multi-functional manipulator designed to move material, parts, or tools through specialized variable programmed motions for the performance of a variety of tasks. The robot, therefore, represents flexible automation as opposed to hard, or dedicated automation and so it fits well in the framework of CIM.

9.3 Classification of Robots

There are two major methods of classifying robots. One is in terms of their physical or geometric attributes. The second approach is in terms of the way they are controlled.

9.3.1 Geometric classification

Industry uses various robot designs which have their advantages and disadvantages. In principle, a robot must have 3 degrees of freedom to reach any point in space. However, it must have an additional 3 degrees of freedom to handle an object in space. There are five major geometric classes of robots. These are:

(1) Cartesian (rectangular) geometry (x, y, z)
(2) Cylindrical (post type) geometry (r, θ, z)
(3) Spherical (polar type) geometry (r, θ, ϕ)
(4) Jointed (anthropomorphic or articulated) geometry $(\theta_1, \theta_2, \theta_3)$
(5) SCARA (selective compliance assembly robot arm) $(\theta_1, \theta_2, \theta_3)$

Each of the above classes is described according to the first three joints excluding the wrist degrees of freedom. Each can be mathematically described in its own coordinates which can be mapped into the Cartesian coordinate system. The following is the mathematical characterization of these robot configurations. (Note that the translational and rotational movements of the arm are restricted by the robot design.)

(1) Cartesian geometry: the Cartesian robot (Figure 9.1), sometimes also called the gantry manipulator, has joints that move in rectangular orthogonal directions. It is the easiest to model and handle mathematically since the coordinates correspond to the Cartesian coordinate system, as shown in Figure 9.1. This robot is used where a very high positioning accuracy is required. However, it has a limited work space.

(2) Cylindrical geometry: this robot has one rotational and two translational joints. The first three joints of the cylindrical robot corresponds to the three principal variables of the cylindrical coordinate system as shown in Figure 9.2.

$$\text{rotation} = \theta$$
$$\text{height} = h$$
$$\text{reach} = r$$

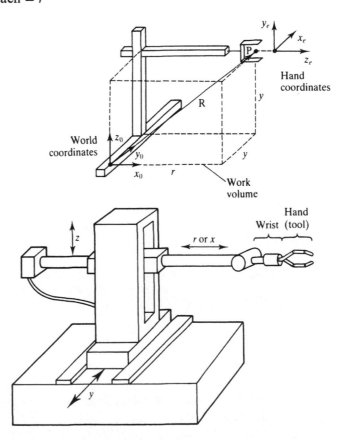

Figure 9.1

Rectangular (Cartesian) robot: three axes.

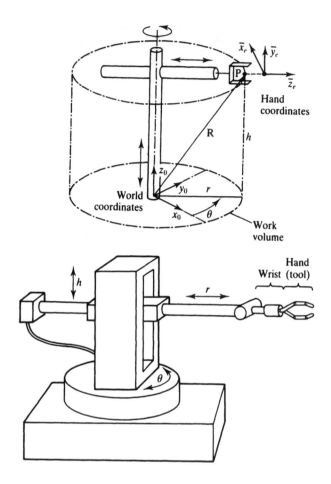

Figure 9.2
Cylindrical (post-type)
robot: two linear and
one rotary axes.

If the position of the hand reference point is specified in (θ, h, r), then the position will be easily known at all times, but if the position was specified in, say (x, y, z), as is the usual case, then some transformation is needed to relate the two coordinates. Using a non-rigorous approach and solving for the first three joints, the z-axis will be seen to be the same as the vertical axis of the robot:

$$z = h$$

The xy-plane is perpendicular to the z-axis and is, therefore, parallel to the plane in which θ rotates (about the z axis) and r extends. We now perform the following kinematic transforms:

$$x = r \cos \theta \qquad (9.1)$$

and

$$y = r \sin \theta \qquad (9.2)$$

This robot is more universal than the Cartesian robot. However, it has a lower accuracy.

(3) Spherical geometry: this robot has one translational and two rotational joints. The first three joints of the spherical geometric robot correspond directly to the three principal variables of a spherical coordinate system (Figure 9.3):

rotation $= \phi_1$

rotation perpendicular to the plane of ϕ (rotation about z-axis) $= \theta$

reach or radius $= r$

In order to know the position of the hand, it must be maintained in (ϕ, θ, r). However, if we have the hand specified in (x, y, z), then we must perform some transformation – the kinematic transform:

$$x = r' \cos \theta \qquad\qquad (9.3)$$

$$y = r' \sin \theta \qquad\qquad (9.4)$$

$$z = r \sin \phi \qquad\qquad (9.5)$$

$$r' = r \cos \phi \qquad\qquad (9.6)$$

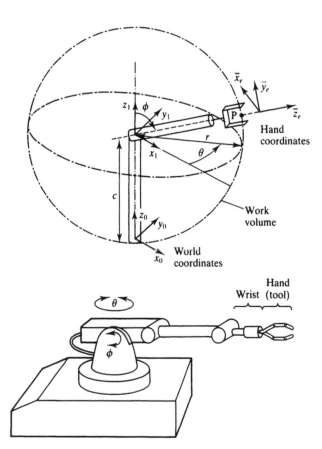

Figure 9.3

Spherical (polar) robot: three rotary axes.

In a non-rigorous way similar to the cylindrical case, we can determine what the joint coordinates (r, θ, ϕ) must be to place the hand at a particular Cartesian point.

With equations 9.3 to 9.6, we obtain

$$\theta = \tan^{-1}\left(\frac{y}{x}\right), \; r' = \frac{x}{\cos\theta} = \frac{y}{\sin\theta}$$

$$\phi = \tan^{-1}\left(\frac{z}{r'}\right), \text{ and } r = \frac{r'}{\cos\phi}$$

The reader should note that the inverse tangent does not exist in all sectors of the 360° range. A solution is found by using the ATAN2 function which properly apportions the tangent value to the proper sector, see Paul (1981).

The addition of a second rotational joint increases the versatility of the robot. However, it decreases the accuracy.

(4) Jointed or articulated geometry: This robot has three rotational joints. The transformation of the jointed geometry, see Figure 9.4, into the Cartesian coordinate system is a much more complex task than in the previous cases. The robot is very versatile. This type of geometry is very good for cases where the robot must reach over obstacles. However, its accuracy is lower than that of all other designs. We will discuss a rigorous method of handling the mathematics of geometries like this, later in this chapter.

(5) SCARA – selective compliance assembly robot arm: the SCARA is the newest geometry introduced for modeling robots. It was first developed in 1979 at the Yamanashi University in Japan for assembly operations. The motions of the major joints of the robot are planar in nature. The robot has shoulder and elbow joints which rotate about vertical axes. The configuration provides substantial rigidity for the robot in the vertical direction while allowing for compliance in the horizontal plane, see Figure 9.5. The motions of all the

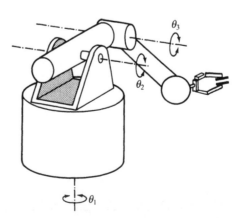

Figure 9.4

Jointed
(anthropomorphic or
articulated): three axes.

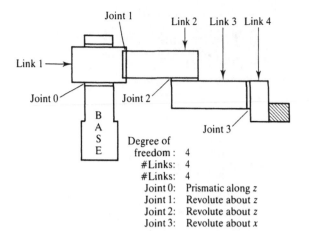

Figure 9.5
Selective compliance
assembly robot arm
(SCARA).

major joints are rotary and hence the mathematics are not trivial. We will
discuss kinematics that relate to the general manipulator later in this chapter.
The robot is very good for highly accurate assembly tasks.

9.3.2 Control classification

There are two methods for controlling a robot. They include non-servo and servo
control techniques. The non-servo control uses mechanical stops to provide the
extreme ranges of motion and when motion command is used, the joint is driven
until the mechanical stop is reached. This technique is now obsolete. The servo
control technique involves the use of a feedback architecture to respond to
locational changes in the robot joints.

There are two types of servo control techniques: point-to-point and continuous
path. The point-to-point involves the specification of the starting point and end
point (and often intermediate points) of the robot motion requiring a control
system which renders some feedback at those points. This technique is used for
spot welding, pick-and-place operations and so on. The continuous path control
requires the robot end effector to follow a stated path from the starting point to
the end point. This technique is required in many applications that require the
actual tracing of a contour, for instance, in arc welding or spray painting.

The continuous path robots usually follow a series of closely spaced points on
a path and these points are defined by the control unit rather than the programmer.
In many cases, the paths between points are straight lines.

9.4 Major Components and Functions of Robots

A robot is a very complex automaton which may contain mechanical, electrical,
pneumatic, hydraulic and microelectronic components. The following are the

major components of the robot:

(1) drive (actuators) system

(2) control system

(3) measuring system

(4) sensors

9.4.1 Drive system

The drive system is responsible for conversion and transmission of the required power to all axes of movement. There are three major types of drive systems used in robots. They include:

(1) hydraulic drive

(2) pneumatic drive

(3) electric drive

The choice of one drive over another in design and configuration of a robotic system is largely dependent on the type of tasks and the environment in which the robot works. Each device has characteristics which make it either attractive or non-attractive to an application. For instance, the hydraulic drive is excellent for heavy loads but leaks early and therefore cannot be used in environments where cleanliness is an absolute necessity. Similarly, the pneumatic drive is clean and can be used for heavy loads but is highly unreliable because of air leaks from the drive. The electric drive is excellent for digital and analog control but may not be suitable if an environment is explosive. A compendium of the drive sub-classification characteristics can be found in Nnaji and Davis (1985).

Kinematics

Associated with driving and controlling the robot is the issue of kinematics. Kinematics deals with the geometry and time-dependent aspects of motion without considering the forces involved with the motion. The parameters of kinematics include: position, displacement, velocity, acceleration, and time. These parameters are used to obtain the relationship between a workpiece and production equipment. The relationship can be between links and joints or between the position of the hand and the joint angles.

9.4.2 Control system

The control system provides a logical sequence for the operating program. It provides the theoretical values required for each program step, it continuously measures the actual position during motion, and it processes the theoretical versus actual difference, see Snyder (1985). As mentioned earlier, there are two types of control systems: the point-to-point system and the continuous path system.

The controller

The controller is usually a microprocessor or a computer and has the following functions:

(1) **Servo:** Given the current position and/or velocity of an actuator, determine the appropriate drive signal to move that actuator towards its desired position.

(2) **Kinematics:** Given the current state of the actuators (position and velocity), determine the current state of the gripper. Conversely, given the current state of the hand, determine the desired state of each actuator.

(3) **Dynamics:** Given knowledge of the loads on the arm (inertia, friction, gravity, acceleration), use this information to adjust the servo operation to achieve better performance.

(4) **Workplace sensor analysis:** Given the knowledge of the tasks to be performed; for example, threading a nut, determine the appropriate robot motion commands. This may be accomplished through vision or tactile sensing techniques, or by measuring and compensating for forces applied at the end-effector, and so on.

9.4.3 Sensors

Sensors permit robots to interact with their environments in an adaptive and intelligent manner. There are numerous transducers available which measure the most important physical variables of interest to manufacturing. The sensors include tactile, proximity, ultrasonic ranging, vision sensors, and other transducers.

Tactile sensors

Usually, if objects that are to be manipulated are delicate or possess positional problems which cannot be addressed using vision techniques, then a sense of touch must be used to understand the attributes of the part. There are two main deficiencies in the vision systems which make tactile sensing attractive in some applications. These deficiencies are: low accuracy and inability to see hidden objects. Because of inaccuracy, parts may be positioned outside an acceptable range of tolerance. When this happens, the precise position must be sensed using some adaptive approach and this can be accomplished by a sense of touch. In a similar way, the fact that vision systems are unable to see the rear of the objects, allowing for at most $2\frac{1}{2}$-D vision makes it impossible to know how to manipulate in the rear of the objects. Therefore, the use of tactile sensors becomes crucial here.

Sophisticated sensors try to emulate the human sense of touch. The human physiological sense of touch has two distinct aspects: the cutaneous sense, which denotes the ability to perceive textual patterns encountered by the skin surface; and the kinesthetic sense, which denotes the human ability to detect forces and moments, as discussed in Snyder (1985).

The tactile sensor typically possesses the capability to detect the following:

(1) presence of objects,
(2) part shape, location, orientation,
(3) contact area, pressure and pressure distribution,
(4) force magnitude, location, and direction,
(5) moments magnitude, plane, and direction.

A tactile sensor may work according to different principles. Typical sensors have the following structure:

(1) A touch surface consisting of a matrix of pressure pins, variable resistors, capacitors or optical devices.
(2) A transduction medium, which converts local forces or moments to electrical impulses.
(3) A multiplexer which connects the elements of the tone matrix with a measuring and amplification circuit.
(4) A control/interface to bring the measured data to a computer for evaluation.

There are essentially two types of tactile sensors: touch and force sensors. The touch sensors are generally used to detect the presence or absence of an object without regard to the contacting force. Included in this category are limit switches, microswitches and so on. The simpler devices are frequently used in the development of a robot's interlock devices. The force sensors measure local touch forces. In case of a matrix sensor, the computer evaluates the touch pattern with the help of algorithms similar to those used for machine vision.

Force sensors

These sensors are normally used to measure the robotic system's forces as it performs various operations. Forces which the robot uses in manipulation and assembly are usually of great importance. Force sensing can be achieved using a force-sensing wrist, as discussed in Groover *et al.* (1986), which consists of a special load cell mounted between the gripper and the wrist. Another method is to measure the torque exerted at each joint and this can be done by measuring motor current at each of the joint motors.

Proximity sensors

Proximity sensors are used for detecting the properties of the surface of an object without touching it. The typical strategy is to detect the presence or absence of a surface. Proximity sensors are used in collision detection situations and in detecting approach situations for the arm. There are various types of proximity sensors. They include:

(1) Optical proximity detectors, which project light beams and use reflectivity of any encountered surfaces to detect the presence of an object.

(2) An optical range sensor, which projects a calibrated light source onto object surfaces and then measures the reflected intensity.

(3) Triangulation proximity sensors, which project light at an angle to the object and then, by using the triangle created by the light projection, the object and the detectors, detect the distance of the object.

Ultrasonic ranging

The ultrasonic transducer is another sensor which is used to sense the distance of an object. The transducer emits a pulse of high-frequency sound and then listens to the echo. Since the speed of sound is known, the time between emission and hearing the echo can be measured to provide the distance.

Infrared

An infrared sensor can be used to measure the presence or absence of an object between the fingers of the gripper. The arrangement can be as shown in Figure 9.6. When an object is located between the sensors, the infrared light emitted by the transmitter will be reflected to the receivers of the transmitting arrangement. In the case where the object surface is parallel to the transmitting arrangement, both receivers will detect an equal amount of reflected light. If the object deviates from parallel alignment, with respect to the gripper, surface signals of different intensity will be received. It will become possible for the robot to re-position its hand to accommodate this skewness. The infrared sensor is also used to sense the proximity of an object (workpiece) by evaluating the intensity of the two reflected signals.

Figure 9.6
(a) Arrangement for infrared sensor.
(b) Detecting the presence of an object with infrared sensor.

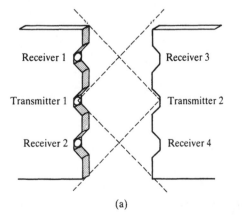

Receiver 1

Transmitter 1

Receiver 2

Receiver 3

Transmitter 2

Receiver 4

(a)

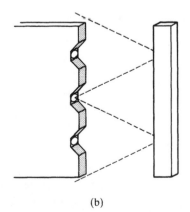

(b)

Vision

The vision system is usually used for viewing the work area of the robot to provide information about the environment, in real time. It can also be used for augmenting work area information to make precise decisions for the robotic system. Finally, it can be used to confirm the existence of selected entities in the work space. This last application usually amounts to pattern matching of new vision information against existing information.

Modern vision systems possess mechanisms which enable them to acquire models of objects from the user. These models can be CAD-based representations of the objects to provide a richer understanding or they can be as simple as stored pictures of parts which are then extracted to be matched against the new information.

Machine vision is concerned with the sensing of vision data and its consequent interpretation using the computer. A typical vision system consists of the camera and digitizing hardware, a digital computer, and hardware and software necessary to process the vision information as well as to link the various devices. Vision is normally achieved by sensing through pictures and digitizing the image data. The image is then processed and analyzed to refine it and infer the geometric entities involved, which are then matched against existing duplication data.

The processing and analysis of images are intended to reduce the data for interpretation. Clearly, imprecise data can lead to high inaccuracies in data which in turn leads to inaccuracies in object position and orientation. An image is normally thresholded to generate a binary image. The binary image will then undergo various feature measurements to further reduce the data representation and produce more precise data. The data reduction can, in fact, result in data which is of the order of one thousandth the size of the original data.

Image characteristics can be obtained as features of objects. Features can be inferred from the geometric elements which are obtained from approximation of the original image in a pattern such as Figure 9.7(a) to the segmented image of Figure 9.7(b). Feature descriptors are normally programmed as software and a variety of geometric entities could combine to yield a class of features.

In 2-D image processing, features can include gray scale (maximum, average, or minimum), area, perimeter, length, diameter, minimum enclosing rectangle and so on. Other indirect features include center of gravity, which can be calculated as follows. Let x and y designate a pixel location of the xy-plane. The center of gravity can be calculated for n pixels as:

$$CG_x = \frac{1}{n} \sum_x x \tag{9.7}$$

$$CG_y = \frac{1}{n} \sum_y y \tag{9.8}$$

Other basic indirect features include: eccentricity which is a measure of 'elongation'; thinness – a measure of how thin an object is; number of holes,

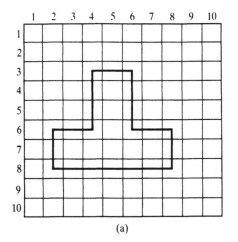 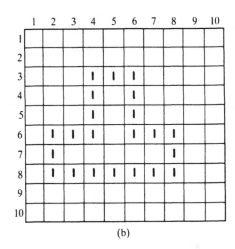

Figure 9.7

(a) Image segmentation – image pattern entered on grid. (b) Segmented image after processing.

(a)

(b)

moments; aspect ratio – the ratio of length to width of an enclosing rectangle and so on.

In 3-D feature recognition for vision, CAD models of the objects are used. Stereo vision is an attempt to obtain a 3-D image of an object. 3-D images can be obtained by using two cameras or by using structured lighting and optical triangulation techniques to obtain a stereoscopic view of the scene. Such images normally apply the types of algorithms used in feature extraction for CAD data in understanding the 3-D data.

In image processing and analysis, the vision system is normally trained by having it store a series of images which can then be recalled for pattern matching when the system is in real operation.

The ability of a vision system to recognize objects with precision is dependent on a number of factors. The main ones include:

(1) the power of the camera, such as field of view, focal length, resolution, geometric distortion and so on;

(2) the robustness of the associated software;

(3) the power of the associated computational device.

9.5 End-Effector

The robot end-effector is that part of the robot which makes the connection to the workpiece. There are two types of end-effectors: first, those intended to grip objects in some manner (grippers) and second, those which are special purpose tools attached to the wrist of the robots. The grippers do machine loading and unloading; picking objects from one location and moving them to another location and putting them on pallets. The special-purpose tool end-effectors are those which directly perform tasks. These are typically found in process applications such as spot welding, arc welding, spray coating, machining applications and so on.

9.5.1 Different types of gripper systems

Grippers can be purely mechanical grasping devices or devices which hold objects by attaching to them using magnets, suction cups, adhesives, hooks, scoops and so on. Some grippers grasp objects by clasping to the outside surfaces of the object and others clasp at the inside surfaces of the object. These are sometimes called external and internal grippers, respectively.

A gripper normally has an actuation mechanism that is not considered to be part of the robot since it does not change the position and orientation of the robotic system. Grippers usually operate in jaw type fashion by having fingers which either attach to the gripper, or are part of the construction, open and close. The attached fingers can be replaced with new or different fingers, allowing for flexibility, see Figure 9.8. Grippers can operate with two fingers or more. The more fingers there are, the more difficult it is to control the fingers and usually the more dexterous the grippers will be.

9.5.2 Gripper actuators

The gripper mechanism translates some form of power into grasping or holding action of the fingers on the part. The gripper drives can be hydraulic, pneumatic, or electric. The choice of one method over the other usually depends on the attributes of the robotic system. If a robot possesses the faculties to transmit air pressure, then a pneumatic actuator may be ideal for the gripper. In general, the control of the signals to regulate the end-effector is usually provided by controlling the transmission of the actuating power. Feedback is usually needed from sensors to indicate the amount of force being applied to the object held by the gripper or some other type of necessary feedback information. Pneumatic power usually uses shop air to deliver the air through a piston arrangement. Pneumatic power is, however, limited to open and close operations but is quite common as a source of gripper power.

Hydraulic power is less frequently used because of its inherent leakage problems. The hydraulic drive has, however, the potential of delivering higher holding force. It is also limited to opening and closing actions.

The electric drive at the gripper enables the gripper to operate many jointed fingers at controlled forces and exercise feedback control. This is very useful in manipulating nonrigid objects which would require careful grasping.

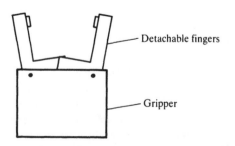

Figure 9.8

9.5.3 Gripper designs

There are many approaches to gripper designs. Usually, the designs are based on the kinematic structure of the intended device. The prominent designs include: linkage gear-and-rack; cam; screw; and rope-and-pulley actuators. Generally, each design approach would accomplish its gripping actuation by pivoting or prismatic motion. Figure 9.9 is an example of prismatic gripper design while Figure 9.10 shows the various linkages which result in pivoting action for gripping. In each figure, F_a represents the applied force which gets transformed into gripping force F_g applied at the fingers.

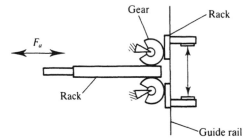

Figure 9.9

Gear and rack gripper actuation method.

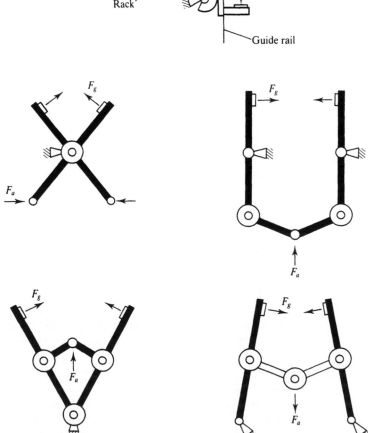

Figure 9.10

Some possible gripper linkages.

In addition to the gear-and-rack and linkage methods, cam, screw, rope-and-pulley actuation methods are also used. Of these methods, the rope-and-pulley method has been applied to multi-jointed, multi-degree of freedom grippers. Some of the advanced robot hands which have been demonstrated, but do not have commercial applications yet, include the Stanford/JPL three-fingered anthropomorphic hand, the Karlsruhe three finger hand and the MIT/Utah hands which are the most dexterous to date.

9.6 Robot Coordinate Systems

The robot's coordinate system representation of position and velocity of the effector can be in Cartesian space, which is the normal measurement space in which the robot sits, or in joint space, which in the case of a robot with six degrees of freedom is an ordered six-tuple describing the angular position of each joint. It is usual to describe tasks to the robot in Cartesian space. For instance, the position of a part on the conveyor can be expressed by the coordinate location of the part, the parts relationship to the conveyor, and the conveyor's relationship to the world. In order for the robot to relate the Cartesian location to its own location, it must find its appropriate joint configurations which will yield the Cartesian location. In other words, some transformation must be performed to map the Cartesian space into joint space.

9.6.1 Position and orientation

In order to perform a full range of manipulation tasks, a robot must be able to reach any point in its work envelope. For this, the robot needs 3 degrees of freedom; for reaching any point at any arbitrary orientation the robot hand needs an additional 3 degrees of freedom. The various positions and orientations within this envelope can be infinite. For any point, however, there must be a method of calculating the exact position and orientation. The issue of position and orientation can be captured in relative frames.

Relative frames

A frame is a representation for a coordinate system so that the representation includes the possibility that the coordinate system may be displaced (translated) and/or rotated with respect to another coordinate system, as discussed in Snyder (1985). The concept of frame is not meaningful in itself, but only as it may be related to another coordinate (frame). It is a frame of reference.

Consider the problem of a robot grasping a part from a table and then holding the part to have a hole drilled in it on an NC machine and then putting the part on a conveyor.

(1) The robot first grasps the part in a specified way and inserts it into the machine for drilling.

(2) After drilling, the robot grasps the part in a certain way and puts it onto a moving conveyor.

The problem is how to exactly specify those grasping positions? Some possible specifications can be to 'grasp the part at a defined contact point in its middle for the first operation and then near the top perimeter for putting it on the conveyor'. That is, we specify the grasp position with a coordinate system defined on the part.

The grasp points are specified relative to a coordinate system defined on the part. But the location of the part itself (its coordinate system) is a variable, known at any instant in time relative to some other coordinate system. If we know these, we know the robot configuration (the 3-D position and orientation of a robot component). We must learn how to find this relative position and orientation.

Consider another problem as shown in Figure 9.11: Given a location **a**, which is known in a coordinate system **B**, and given that the location of the coordinate system, **B**, is known in another coordinate system **C**; an objective may be to find the location of the point **a** as measured in coordinate system **C**. Coordinate system **B** is displaced and rotated from system **C**. **a** is known in **B**. Where is **a** in **C**? This is a problem of rotation and displacement.

Transformations

We first deal with the issue of displacement where no rotations are allowed as shown in Figure 9.12.

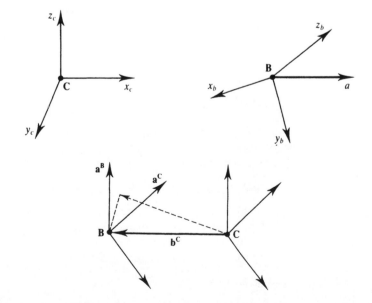

Figure 9.11
Coordinate system B
is displaced and
rotated from system
C, a as known in B.

Figure 9.12
Transformations
showing a as known
in C.

If the coordinate axes of **B** are parallel to those of **C**, then $\mathbf{a^B}$ may be described by a vector:

$$\mathbf{a^B} = (b_{x_a}, b_{y_a}, b_{z_a})^T \tag{9.9}$$

The superscript denotes that the variable **a** is measured with respect to the coordinate system named **B**. Similarly, the vector $\mathbf{b^C}$ which is the origin of the coordinates **B** as known in **C**:

$$\mathbf{b^B} = (c_{x_b}, c_{y_b}, c_{z_b})^T \tag{9.10}$$

The transpose is used for typographical convenience since the vectors are actually column vectors. Finally, **a** relative to **c** is:

$$\mathbf{a^C} = \mathbf{b^C} + \mathbf{a^B} \tag{9.11}$$
$$= (c_{x_b} + b_{x_a}, c_{y_b} + b_{y_a}, c_{z_b} + b_{z_a})^T$$

Using the relative frame concept above, suppose that a point $[p_x, p_y, p_z]^T$ is known relative to a coordinate system named OXYZ which is currently aligned with fixed coordinate system OABC (Figure 9.13). We can find a new point $[p_a, p_b, p_c]^T$ that results from rotating the point about the x-axis of OXYZ through angle θ. We use a rotation matrix to multiply and obtain:

$$\begin{pmatrix} p_a \\ p_b \\ p_c \end{pmatrix} = \begin{pmatrix} 1 & 0 & 0 \\ 0 & \cos\theta & -\sin\theta \\ 0 & \sin\theta & \cos\theta \end{pmatrix} \cdot \begin{pmatrix} p_x \\ p_y \\ p_z \end{pmatrix} \tag{9.12}$$

Later, a more general case of the rotation will be shown.

9.6.2 Derivation of rotation and translation matrices

Rotation

Let i_x, j_y, k_z and i_a, j_b, k_c be the unit vectors along the coordinate systems OXYZ and OABC, respectively (Figure 9.14). These two coordinates are initially superimposed onto each other, achieving alignment. A point **p** can be referenced to any of the frames. Assume **p** is at rest and fixed with respect to OABC and OXYZ:

$$\mathbf{p}_{abc} = (p_a, p_b, p_c)^T \quad \text{and} \quad \mathbf{p}_{xyz} = (p_x, p_y, p_z)^T \tag{9.13}$$

$\mathbf{p}_{abc}$ and $\mathbf{p}_{xyz}$ represent the same point in space with reference to different coordinates. We design a 3×3 transformation matrix R which transforms the coordinates of $\mathbf{p}_{abc}$ to the coordinates expressed with respect to OXYZ coordinate system after the OABC coordinate system has been rotated:

$$\mathbf{p}_{xyz} = R\mathbf{p}_{abc} \tag{9.14}$$

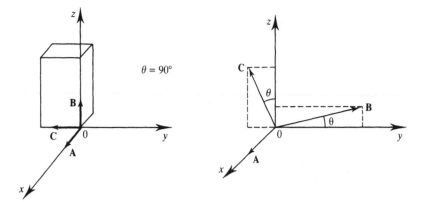

Figure 9.13
Coordinate system
OABC is rotated from
system OXYZ about
the x-axis through
angle θ.

Note that the coordinate system OABC and OXYZ are identical before being rotated (Figure 9.13).

Physically, the point $\mathbf{p}_{abc}$ has been rotated together with the OABC coordinate system.

$$\mathbf{p}_{abc} = p_a i_a + p_b j_b + p_c k_c \tag{9.15}$$

p_x, p_y, p_z represent the components of $\mathbf{p}$ along OX, OY, OZ axis respectively, or the projections of $\mathbf{p}$ onto the respective axes. Using the definition of a scalar product and the preceeding equation, we obtain:

$$
\begin{aligned}
p_x &= i_x^T \cdot \mathbf{p} = i_x^T \cdot i_a p_a + i_x^T \cdot j_b p_b + i_x^T \cdot k_c p_c \\
p_y &= j_y^T \cdot \mathbf{p} = j_y^T \cdot i_a p_a + j_y^T \cdot j_b p_b + j_y^T \cdot k_c p_c \\
p_z &= k_z^T \cdot \mathbf{p} = k_z^T \cdot i_a p_a + k_z^T \cdot j_b p_b + k_z^T \cdot k_c p_c
\end{aligned}
\tag{9.16}
$$

In matrix form, this can be expressed as:

$$
\begin{pmatrix} p_x \\ p_y \\ p_z \end{pmatrix} =
\begin{pmatrix}
i_x^T \cdot i_a & i_x^T \cdot j_b & i_x^T \cdot k_c \\
j_y^T \cdot i_a & j_y^T \cdot j_b & j_y^T \cdot k_c \\
k_z^T \cdot i_a & k_z^T \cdot j_b & k_z^T \cdot k_c
\end{pmatrix} \cdot
\begin{pmatrix} p_a \\ p_b \\ p_c \end{pmatrix}
\tag{9.17}
$$

The rotation matrix is therefore:

$$
R =
\begin{pmatrix}
i_x^T \cdot i_a & i_x^T \cdot j_b & i_x^T \cdot k_c \\
j_y^T \cdot i_a & j_y^T \cdot j_b & j_y^T \cdot k_c \\
k_z^T \cdot i_a & k_z^T \cdot j_b & k_z^T \cdot k_c
\end{pmatrix}
\tag{9.18}
$$

Similarly, one can obtain $\mathbf{p}_{abc}$ from the coordinate $\mathbf{p}_{xyz}$:

$$\mathbf{p}_{abc} = S\mathbf{p}_{xyz} \tag{9.19}$$

$$
\begin{pmatrix} p_a \\ p_b \\ p_c \end{pmatrix} =
\begin{pmatrix}
i_a^T \cdot i_x & i_a^T \cdot j_y & i_a^T \cdot k_z \\
j_b^T \cdot i_x & j_b^T \cdot j_y & j_b^T \cdot k_z \\
k_c^T \cdot i_x & k_c^T \cdot j_y & k_c^T \cdot k_z
\end{pmatrix}
\begin{pmatrix} p_x \\ p_y \\ p_z \end{pmatrix}
\tag{9.20}
$$

Because of the commutativity of the dot products, it can be seen that $R^T = S$; and because of the orthogonality of the transformation matrices:

$$S = R^{-1} \tag{9.21}$$

$$SR = R^T R = R^{-1} R = I_3 \tag{9.22}$$

where I_3 is a 3×3 identity matrix.

EXAMPLE 1

Consider a case where OABC coordinate system is rotated θ angle about OX axis to arrive at a new location in space: $\mathbf{p}_{abc}$ with coordinates $(p_a, p_b, p_c)^T$ with respect to OABC will have different coordinates $(p_x, p_y, p_z)^T$ with respect to OXYZ. We can find the necessary transformation for this $R(x, \theta)$:

$$\mathbf{p}_{xyz} = R(x, \theta)\mathbf{p}_{abc} \tag{9.23}$$

Since the point is rotated about the x-axis of OXYZ then the following is true:

$$i_a = i_x$$

$$j_b = j_y \cos \theta + k_z \sin \theta$$

$$k_c = -j_y \sin \theta + k_z \cos \theta$$

$$R(x, \theta) = \begin{pmatrix} i_x^T \cdot i_a & i_x^T \cdot j_b & i_x^T \cdot k_c \\ j_y^T \cdot i_a & j_y^T \cdot j_b & j_y^T \cdot k_c \\ k_z^T \cdot i_a & k_z^T \cdot j_b & k_z^T \cdot k_c \end{pmatrix} \tag{9.24}$$

$$\begin{pmatrix} 1 & 0 & 0 \\ 0 & \cos\theta & -\sin\theta \\ 0 & \sin\theta & \cos\theta \end{pmatrix}$$

We can find other rotational matrices in like manner: $R(y, \theta)$ and $R(z, \theta)$.

Special case of example 1 with $\theta = 90°$:

Given point $(7, 2, 5)^T$ in frame OABC, what are its coordinates in frame OXYZ if frame OABC is obtained from OXYZ by a rotation about the x-axis through an angle $\theta = 90°$?

Solution:

$$\begin{pmatrix} p_x \\ p_y \\ p_z \end{pmatrix} = \begin{pmatrix} 7 \\ -5 \\ 2 \end{pmatrix} = \begin{pmatrix} 1 & 0 & 0 \\ 0 & 0 & -1 \\ 0 & 1 & 0 \end{pmatrix} \begin{pmatrix} 7 \\ 2 \\ 5 \end{pmatrix} \tag{9.25}$$

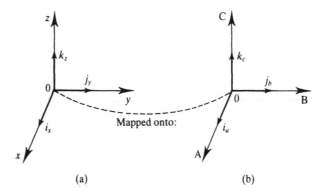

Figure 9.14
Unit vectors in
(a) OXYZ and
(b) OABC.

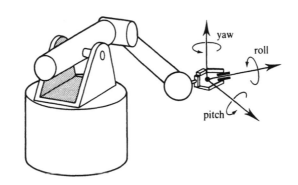

Figure 9.15
Roll, pitch and yaw
by the robot hand.

Translation

Using similar reasoning as that for rotations, the translation of a point vector can be obtained as,

$$\mathbf{p}_{xyz} = \mathbf{T}(x, d_x)\mathbf{p}_{abc} \tag{9.26}$$

where

$$\mathbf{T}(x, d_x) = (d_x, 0, 0)^T \tag{9.27}$$

General form of the rotation matrix

One of the generalized methods for specifying rotation and orientation is by rotating about some known axis. One of these approaches is the **roll, pitch, yaw** (as shown in Figures 9.15 and 9.16).

Roll (ϕ_z) = rotate ϕ_z about z.

Pitch (ϕ_y) = rotate ϕ_y about y.

Yaw (ϕ_x) = rotate ϕ_x about x.

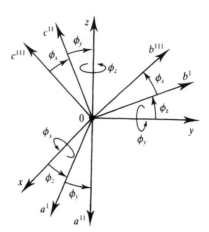

Figure 9.16
Roll, pitch and yaw
representation in
kinematics.

We can combine these rotations as follows:

$$\mathbf{RPY}(\phi_z, \phi_y, \phi_x) = Rot(z, \phi_z)Rot(y, \phi_y)Rot(x, \phi_x) \qquad (9.28)$$

Besides specifying roll, pitch and yaw, there are other conventions for specifying arbitrary rotations by the angles of three successive rotations about fixed axes. The most common method uses Euler angles, which specify a rotation about the z-axis through an angle ϕ, followed by a rotation about the new y-axis through an angle θ, followed by a rotation about the new z-axis through an angle ψ. Both of these descriptions, RPY and Euler, suffer from the same shortcomings. Most important is the computational burden due to the multiplications. Matrix representation of rotations simplifies many operations. Moreover, nine elements are required to completely describe the orientation of a rotating rigid body using matrices, and the representation does not lead directly to a complete set of generalized coordinates which can be used to describe the orientation of a rotating rigid body with respect to a reference coordinate frame.

Homogeneous transforms

The goal, however, in robot motion is often to describe the effect of combined motions resulting from both translation and rotation. We desire a generalized homogeneous transform that combines the effect of rotation and translation by a single matrix. This transform maps a position vector expressed in homogeneous coordinates from one coordinate system to another coordinate system. It consists of four submatrices:

$$\begin{bmatrix} \text{rotation matrix} & \text{position vector} \\ \text{perspective transformation} & \text{scaling} \end{bmatrix}$$

The upper left submatrix is a 3×3 rotation matrix, the upper right submatrix is the position vector which is a 3×1 matrix. The lower left submatrix contains the perspective vector which is not used in robot kinematics. It is usually employed when dealing with vision to establish a perspective. The lower right submatrix is the scaling factor which is a 1×1 submatrix. For the case of translation and rotation, the perspective transformation is a null vector. The scaling is always 1. Hence, we obtain the following 4×4 homogeneous matrices for **pure rotations** about an axis of a Cartesian coordinate system:

For the x-axis:

$$R(x, \theta) = \begin{pmatrix} 1 & 0 & 0 & 0 \\ 0 & \cos\theta & -\sin\theta & 0 \\ 0 & \sin\theta & \cos\theta & 0 \\ 0 & 0 & 0 & 1 \end{pmatrix} \tag{9.29}$$

For the y-axis:

$$R(y, \theta) = \begin{pmatrix} \cos\theta & 0 & \sin\theta & 0 \\ 0 & 1 & 0 & 0 \\ -\sin\theta & 0 & \cos\theta & 0 \\ 0 & 0 & 0 & 1 \end{pmatrix} \tag{9.30}$$

For the z-axis:

$$R(z, \theta) = \begin{pmatrix} \cos\theta & -\sin\theta & 0 & 0 \\ \sin\theta & \cos\theta & 0 & 0 \\ 0 & 0 & 1 & 0 \\ 0 & 0 & 0 & 1 \end{pmatrix} \tag{9.31}$$

For a **pure translation**, we obtain the following matrix:

$$Trans(d_x, d_y, d_z) = \begin{pmatrix} 1 & 0 & 0 & d_x \\ 0 & 1 & 0 & d_y \\ 0 & 0 & 1 & d_z \\ 0 & 0 & 0 & 1 \end{pmatrix} \tag{9.32}$$

With the help of the four basic matrices, it is possible to describe any motion of a robot joint. Thus, we can say that homogeneous position coordinates $\bar{d}(d_x, d_y, d_z, 1)^T$ can be expressed as follows:

$$\bar{d}_{x,y,z} = T\bar{d}_{abc},$$

where

$$T = \begin{pmatrix} n_x & s_x & a_x & d_x \\ n_y & s_y & a_y & d_y \\ n_z & s_z & a_z & d_z \\ 0 & 0 & 0 & 1 \end{pmatrix} = \begin{pmatrix} \mathbf{n} & \mathbf{s} & \mathbf{a} & \mathbf{d} \\ 0 & 0 & 0 & 1 \end{pmatrix} \qquad \textbf{(9.33)}$$

See Figure 9.17, for a geometric description of these vectors. Descriptions of the vectors as described by Fu *et al.* (1987) are:

n = Normal vector of the hand. Assuming a parallel jaw hand, it is orthogonal to the fingers of the robot arm.

s = Sliding vector of the hand. It is pointing in the direction of the finger motion as the gripper opens and closes. This depends on the type of gripper and can vary depending on the gripper construction.

a = Approach vector of the hand. It is pointing in the direction normal to the palm of the hand (that is, normal to the tool mounting plate of the arm).

d = Position vector of the hand. It points from the origin of the base coordinate system to the origin of the hand coordinate system, which is usually located at the center of the fully closed fingers.

Problem: Find the effect of acting on the point $(p_a, p_b, p_c)^T$ by the combined motions:

(1) a rotation of 90° about z-axis followed by
(2) a rotation of 90° about x-axis followed by
(3) a translation of (d_1, d_2, d_3)

Solution: The matrix multiplications are ordered by first rotating the point $(p_a, p_b, p_c)^T$ about the z-axis. Then this is followed by rotating the new point about

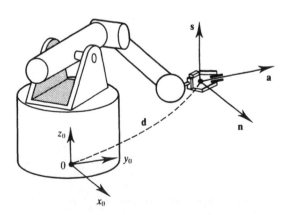

Figure 9.17
Hand coordinate
system and [**n**, **s**, **a**].

the x-axis and finally translating the resulting point along distances, d_1, d_2 and d_3 in the x, y and z directions.

$$\text{Trans}(d_1, d_2, d_3) \times Rot(x, 90) \times Rot(z, 90) \cdot \begin{pmatrix} p_a \\ p_b \\ p_c \\ 1 \end{pmatrix}$$

$$\begin{pmatrix} 1 & 0 & 0 & d_1 \\ 0 & 1 & 0 & d_2 \\ 0 & 0 & 1 & d_3 \\ 0 & 0 & 0 & 1 \end{pmatrix} \cdot \begin{pmatrix} 1 & 0 & 0 & 0 \\ 0 & 0 & -1 & 0 \\ 0 & 1 & 0 & 0 \\ 0 & 0 & 0 & 1 \end{pmatrix} \cdot \begin{pmatrix} 0 & -1 & 0 & 0 \\ 1 & 0 & 0 & 0 \\ 0 & 0 & 1 & 0 \\ 0 & 0 & 0 & 1 \end{pmatrix} \begin{pmatrix} p_a \\ p_b \\ p_c \\ 1 \end{pmatrix}$$

$$= \begin{pmatrix} -p_b + d_1 \\ -p_c + d_2 \\ p_a + d_3 \\ 1 \end{pmatrix}$$

This process can be accomplished with a single homogeneous matrix as follows, in general:

$$\text{Trans}(d_1, d_2, d_3) \times Rot(x, 90) \times Rot(z, 90):$$

$$\begin{pmatrix} 1 & 0 & 0 & d_1 \\ 0 & 1 & 0 & d_2 \\ 0 & 0 & 1 & d_3 \\ 0 & 0 & 0 & 1 \end{pmatrix} \begin{pmatrix} 1 & 0 & 0 & 0 \\ 0 & 0 & -1 & 0 \\ 0 & 1 & 0 & 0 \\ 0 & 0 & 0 & 1 \end{pmatrix} \begin{pmatrix} 0 & -1 & 0 & 0 \\ 1 & 0 & 0 & 0 \\ 0 & 0 & 1 & 0 \\ 0 & 0 & 0 & 1 \end{pmatrix}$$

$$= \begin{pmatrix} 0 & -1 & 0 & d_1 \\ 0 & 0 & -1 & d_2 \\ 1 & 0 & 0 & d_3 \\ 0 & 0 & 0 & 1 \end{pmatrix}$$

Notice the right-to-left nature of the operation. Matrix multiplication is not commutative and neither are rotations:

$$Rot(x, \theta)Rot(y, \phi) \neq Rot(y, \phi)Rot(x, \theta), \text{ in general.}$$

Inverse of a homogeneous matrix

The following can be said about homogeneous transformation matrices. The inverse of a rotation submatrix is equivalent to its transpose. Therefore, the row vectors represent the principal axis of the reference coordinate system with respect to the rotated coordinate system OABC. But the inverse of a homogeneous transpose matrix is not equivalent to its transpose. In general, the inverse of a homogeneous transform matrix can be found as follows:

Given a homogeneous transform matrix T:

$$T = \begin{pmatrix} n_x & s_x & a_x & p_x \\ n_y & s_y & a_y & p_y \\ n_z & s_z & a_z & p_z \\ 0 & 0 & 0 & 1 \end{pmatrix}$$

(9.34)

The inverse of the matrix T is:

$$T^{-1} = \begin{pmatrix} n_x & n_y & n_z & -\mathbf{n}^T\mathbf{p} \\ s_x & s_y & s_z & -\mathbf{s}^T\mathbf{p} \\ a_x & a_y & a_z & -\mathbf{a}^T\mathbf{p} \\ 0 & 0 & 0 & 1 \end{pmatrix}$$

(9.35)

$$= \begin{pmatrix} & & & -\mathbf{n}^T\mathbf{p} \\ & R^T_{3\times 3} & & -\mathbf{s}^T\mathbf{p} \\ & & & -\mathbf{a}^T\mathbf{p} \\ 0 & 0 & 0 & 1 \end{pmatrix}$$

Problem: Find the homogeneous transformation matrix T that represents a rotation of angle α about OY axis, followed by a translation of d_1 units along the OX axis, followed by a translation of d_2 units along the OA axis, followed by a rotation of θ radians about the OZ axis.

Solution:

$$T = T_{z,\theta}((T_{x,d_1} \cdot T_{y,\alpha}) \cdot T_{x,d_2})$$

Notice that the rotation or translation matrix is pre-multiplied against the existing matrix when the operation is rotated about a fixed frame. It is a post-multiplication when the operation is about the moving frame in OABC.

We have identified two coordinate systems: the fixed coordinate system OXYZ and the moving (rotation and translation) coordinate system OABC. We used a 4 × 4 homogeneous transformation matrix to describe their spatial displacement. When operating on position vectors expressed in the coordinate system, the homogeneous matrix has the combined effect of rotation, translation, perspective and global scaling.

When the two coordinates are assigned to each of the adjacent links of a robot arm, link $i - 1$ and link i, then:

link $i - 1 \equiv$ reference system

link $i \equiv$ moving coordinate system when joint i is activated

Using the T matrix, we specify a point p_i at rest in link i and expressed in the link i (OABC) coordinate system in terms of the link $i-1$ (or OXYZ) coordinate system:

$$\bar{\mathbf{p}}_{i-1} = T\bar{\mathbf{p}}_i$$

T = 4 × 4 homogeneous transformation matrix relating the two coordinates

$\bar{p}_i$ = 4 × 1 augmented position vector $(x_i, y_i, z_i, 1)^T$ representing a point in the link i coordinate system expressed in homogeneous coordinates

$\bar{p}_{i-1}$ = 4 × 1 augmented position vector $(x_{i-1}, y_{i-1}, z_{i-1}, 1)^T$ representing the same point p_i in terms of the link $i-1$ coordinate system.

9.6.3 Relating the robot to its world

Having derived the homogeneous transformation matrices which relate the end-effector to the base of the robot, we can now pursue the task of relating the robot, through its end-effector, to the world. This way we can describe a task and the world containing the task to the robot for execution.

To explain further the nature of this problem, consider the following problem. Imagine that a robot's end-effector can be represented by a homogeneous transformation matrix, $^{R}T_H$, with respect to some base frame. The base frame is related to the world by $^{W}T_R$ as shown in Figure 9.18. The center of the top surface of the box is b_1. The reference point of a cylinder is located at some point P known in the world frame. The top and bottom of the cylinder are C_1 and C_2, respectively located at the middle point of the faces.

Find $^{R}T_H$ when the bottom of the cylinder is placed at the center of the box's top with the robot's hand being aligned to the principal axis of the cylinder.

The following notation is used for the transformations in the example:

$^{R}T_H \Rightarrow$ relates the hand to the robot base.

$^{W}T_R \Rightarrow$ relates the robot base to the world frame, W.

$^{W}T_B \Rightarrow$ relates the box frame to the world frame.

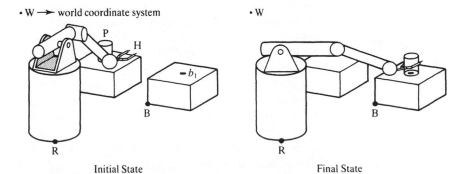

Figure 9.18
Relating the robot to the world coordinate W.

Initial State   Final State

$^{B}T_{b_1} \Rightarrow$ relates the point b_1 on the box to the box frame.

$^{W}T_{P} \Rightarrow$ relates the point P to the world frame.

$^{H}T_{c_1} \Rightarrow$ relates the top center of the cylinder to the robot hand frame.

$^{C_2}T_{C_1} \Rightarrow$ relates the top center of cylinder to its bottom (cylinder frame).

$^{P}T_{C_2} \Rightarrow$ relates the cylinder frame to the point p.

The symbols $^{W}T_{H}$, $^{W}T_{C_1}$, $^{W}T_{C_2}$, $^{W}T_{b_1}$ and $^{H}T_{C_2}$ are similarly defined.

At point of pick up, the hand position is the same as the top of the cylinder:

$$^{H}T_{C_1} = I_4 \tag{9.36}$$

$$\textit{also } ^{W}T_{H} = {}^{W}T_{C_1} \tag{9.37}$$

$$^{W}T_{R} \cdot {}^{R}T_{H} \cdot {}^{H}T_{C_1} = {}^{W}T_{C_1} \tag{9.38}$$

$$^{W}T_{C_1} = {}^{W}T_{P} \cdot {}^{P}T_{C_2} \cdot {}^{C_2}T_{C_1} \tag{9.39}$$

At point of deposit of the cylinder on top of the box:

$$^{W}T_{C_2} = {}^{W}T_{b_1} \tag{9.40}$$

$$^{W}T_{R} \cdot {}^{R}T_{H} \cdot {}^{H}T_{C_2} \cdot {}^{C_2}T_{C_1} = {}^{W}T_{B} \cdot {}^{B}T_{b_1} \tag{9.41}$$

One can solve for $^{R}T_{H}$ in either case above. It is clear that although relationships may exist among objects in the world, not all the relationships are useful in computation of various locations. They, however, provide redundant constraints.

9.7 Manipulator Kinematics

9.7.1 Parameters of links and joints

The robot manipulator is composed of a sequence of rigid bodies which are connected by either a revolute or prismatic joint. There are many other types of joints (Figure 9.19) but the most commonly used are the revolute and prismatic types.

Each joint-link pair constitutes one degree of freedom of the manipulator. Hence, for a manipulator with n joint-link pairs, there are n degrees of freedom. In this arrangement, the first link (not considered part of the manipulator) is attached to the supporting base which serves as the inertial coordinate frame.

The last link is attached with a tool. The joints and links are numbered outwardly from the base: joint 1 is the point of connection between link and base. In general, each link is connected by a lower pair joint which has two surfaces sliding over one another while remaining in contact.

Revolute
Planar
Cylindrical
Prismatic
Spherical
Screw

Figure 9.19
The lower pair of joints.

Joint $i + 1$
θ_{i+1}
Joint i
Link $i - 1$
Link i
Link $i + 1$
α_i
z_i
a_i
x_i
d_i
z_{i+1}
x_{i+1}

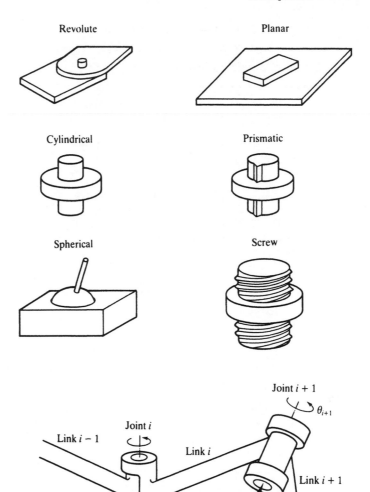

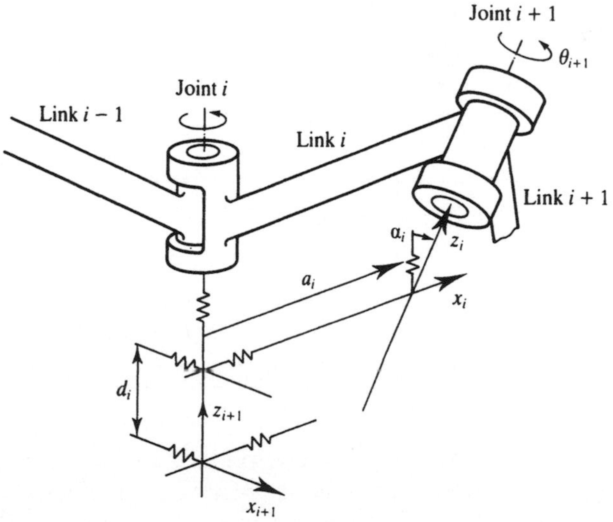

Figure 9.20
Link coordinate system and its parameters.

In general, the joint axis for joint i is established as follows (Figure 9.20):

- The joint axis will have two normals connected to it, one for each of the links, x_{i-1} and x_i.
- The relative distance of two such connected links (link $i - 1$ and i) is given by d_i which is the distance measured along the joint axis between the normals.

- The joint angle θ_i is measured in a plane normal to the joint axis.
- d_i and θ_i are distance and angle between the adjacent links respectively.
- They determine the relative position of neighboring links.
- The significance of links from a kinematic perspective is that they maintain a fixed configuration between their joints which can be characterized by two parameters: a_i and α_i.
- a_i is the shortest distance measured along the common normals between the joint axes (x_{i-1} and x_i).
- α_i is the twist angle of link i from $i - 1$.
- a_i and α_i are length and twist angle pair, and help determine structure of link i.

In summary, the set d_i, θ_i, a_i, and α_i completely determine the kinematic configuration of each link of the robot arm.

9.7.2 Denavit–Hartenberg representation

In 1955, Denavit and Hartenberg described translational and rotational relationships between adjacent links using a matrix method of systematically establishing a coordinate system (body-attached frame) to each link of an articulated chain:

- The nth coordinate frame moves with the hand (link n).
- The base coordinates are defined as the 0th coordinate frame (X_0, Y_0, Z_0) which is also the inertial coordinate frame of the robot.
- So for the PUMA, see Figure 9.21, with six degrees of freedom we have seven coordinate frames:

$$(X_0, Y_0, Z_0) \dots (X_6, Y_6, Z_6)$$

Every coordinate system is established on the basis of three rules as discussed in Fu *et al.* (1987):

(1) The Z_{i-1} axis lies along the axis of motion of the ith joint.
(2) The X_i axis is normal to the Z_{i-1} axis, and pointing away from it.
(3) The Y_i axis completes the (right-hand screw rule) coordinate system as required.

Note that by these rules, the location of the coordinate frame can be chosen anywhere in the supporting base, provided that Z_0 lies along the axis of motion of the first joint.

- The last coordinate can be placed anywhere on the hand provided that X_n axis is normal to Z_{n-1} axis.

Thus, for the **rotary joint**, the following are the variables and fixed parameters:

$$d_i, a_i, \alpha_i = \text{joint parameters; remain fixed for a robot,}$$

$$\theta_i = \text{joint variable; changes as robot moves.}$$

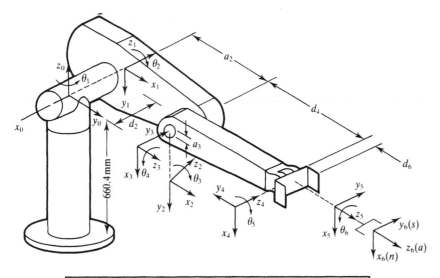

PUMA robot arm link coordinate parameters					
Joint i	θ_i	α_i	a_i	d_i	Joint range
1	90	-90	0	0	-160 to $+160$
2	0	0	431.8 mm	149.09 mm	-225 to 45
3	90	90	-20.32 mm	0	-45 to 225
4	0	-90	0	433.07 mm	-110 to 170
5	0	90	0	0	-100 to 100
6	0	0	0	56.25 mm	-266 to 266

Figure 9.21

Establishing link coordinate systems for a PUMA robot (source: Fu *et al.* (1987)).

For the **prismatic joint**,

$\theta_i, a_i, \alpha_i = $ joint parameters; remain fixed for a robot,

$d_i = $ joint variable; changes as robot moves.

An algorithm which, given n degrees of freedom of a robot arm, assigns an orthonormal coordinate system to each link of the robot arm according to configurations similar to a human arm geometry can be found in Fu *et al.* (1987).

The result of an algorithm established by Denavit and Hartenberg can be expressed by a basic homogeneous matrix of four homogeneous matrices to yield a composite homogeneous transform matrix $^{i-1}A_i$ known as the D–H transformation matrix for adjacent coordinate frames, i and $i-1$:

$$^{i-1}A_i = T_{z,d_i} T_{z,\theta_i} T_{x,\alpha_i} T_{x,\alpha_i}$$

$$= \begin{pmatrix} \cos\theta_i & -\cos\alpha_i \sin\theta_i & \sin\alpha_i \sin\theta_i & a_i\cos\theta_i \\ \sin\theta_i & \cos\alpha_i \cos\theta_i & -\sin\alpha_i \cos\theta_i & a_i\sin\theta_i \\ 0 & \sin\alpha_i & \cos\alpha_i & d_i \\ 0 & 0 & 0 & 1 \end{pmatrix} \quad (9.42)$$

where α_i, a_i, d_i are constants while θ_i is the joint variable for a revolute joint. For a prismatic joint, d_i = joint variable and α_i, a_i and θ_i are constants:

$$^{i-1}A_i = T_{z,\theta_i} T_{z,d_i} T_{x,\alpha_i}$$

Using the $^{i-1}A_i$ matrix, one can relate a point $\mathbf{p}_i$ at rest in link i, and expressed in homogeneous coordination with respect to coordinate system i, to the coordinate system $i-1$ established in link $i-1$ by

$$\bar{\mathbf{p}}_{i-1} = {}^{i-1}\mathbf{A}_i \bar{\mathbf{p}}_i \tag{9.43}$$

$$\bar{\mathbf{p}}_{i-1} = (x_{i-1}, y_{i-1}, z_{i-1}, 1)^T \quad \text{and} \quad \mathbf{p}_i = (x_i, y_i, z_i, 1)^T \tag{9.44}$$

The six $^{i-1}A_i$ transformation matrices for the six PUMA robot arm (Figure 9.21) can be found as:

$$^0A_1 = \begin{pmatrix} c_1 & 0 & -s_1 & 0 \\ s_1 & 0 & c_1 & 0 \\ 0 & -1 & 0 & 0 \\ 0 & 0 & 0 & 1 \end{pmatrix} \quad ^1A_2 = \begin{pmatrix} c_2 & -s_2 & 0 & a_2 c_2 \\ s_2 & c_2 & 0 & a_2 s_2 \\ 0 & 0 & 1 & d_2 \\ 0 & 0 & 0 & 1 \end{pmatrix}$$

$$^2A_3 = \begin{pmatrix} c_3 & 0 & s_3 & a_3 c_3 \\ s_3 & 0 & -c_3 & a_3 s_3 \\ 0 & 1 & 0 & 0 \\ 0 & 0 & 0 & 1 \end{pmatrix} \quad ^3A_4 = \begin{pmatrix} c_4 & 0 & -s_4 & 0 \\ s_4 & 0 & c_4 & 0 \\ 0 & -1 & 0 & d_4 \\ 0 & 0 & 0 & 1 \end{pmatrix}$$

$$^4A_5 = \begin{pmatrix} c_5 & 0 & s_5 & 0 \\ s_5 & 0 & -c_5 & 0 \\ 0 & 1 & 0 & 0 \\ 0 & 0 & 0 & 1 \end{pmatrix} \quad ^5A_6 = \begin{pmatrix} c_6 & -s_6 & 0 & 0 \\ s_6 & c_6 & 0 & 0 \\ 0 & 0 & 1 & d_6 \\ 0 & 0 & 0 & 1 \end{pmatrix}$$

In these matrices, $\sin \theta_i$ and $\cos \theta_i$ are abbreviated to s_i and c_i, respectively.

9.7.3 Kinematic chains

From the $^{i-1}A_i$ matrices, the homogeneous matrix ^0T_i which specifies the relationship of the ith coordinate frame to the base frame can be obtained as a chain product of successive coordinate transformation matrices of $^{i-1}A_i$ as follows:

$$^0T_i = {}^0A_1 {}^1A_2 {}^2A_3 \ldots {}^{i-1}A_i = \prod_{j=1}^{i} {}^{j-1}A_j, \text{ for } i = 1, 2, \ldots, n \tag{9.45}$$

From this matrix, a five degree of freedom manipulator can yield a transformation matrix $T = {}^0A_5$ that specifies the position and orientation of the end point of the manipulator relative to the base coordinate frame.

Method of multiplying the n *matrices*

(1) Hand multiplication,
(2) Showing the symbolic multiplication of all matrices manually and then using the computer to evaluate the elements,
(3) Using the computer all the way,
(4) $T_1 T_2$ where $T_1 = {}^0 A_3 = {}^0 A_1 {}^1 A_2 {}^2 A_3$ and $T_2 = {}^3 A_6 = {}^3 A_4 {}^4 A_5 {}^5 A_6$ (in the case of $n = 6$)

From an analysis done by Fu *et al.* (1987), there are 12 transcendental function calls, 40 multiplications and 20 additions if we only compute the upper 3×3 submatrix of T and the normal vector **n** is found from the cross product of $\mathbf{n} = \mathbf{s} \times \mathbf{a}$. Also, if we combine d_6 with the tool length of the terminal device, then $d_6 = 0$ and the new tool length will be increased by d_6. The computation is then 12 transcendental functions, 35 multiplications, and 16 additions. The arm matrix, T, for the PUMA robot arm is, according to Fu *et al.* (1987):

$$\mathbf{T} = \mathbf{T_1 T_2} = {}^0\mathbf{A_1}{}^1\mathbf{A_2}{}^2\mathbf{A_3}{}^3\mathbf{A_4}{}^4\mathbf{A_5}{}^5\mathbf{A_6} = \begin{pmatrix} n_x & s_x & a_x & d_x \\ n_y & s_y & a_y & d_y \\ n_z & s_z & a_z & d_z \\ 0 & 0 & 0 & 1 \end{pmatrix}$$

where:

$$[n_x] = c_1[c_{23}(c_4 c_5 c_6 - s_4 s_6) - s_{23} s_5 c_6] + s_1(s_4 c_5 c_6 + c_4 s_6)$$
$$[n_y] = s_1[c_{23}(c_4 c_5 c_6 - s_4 s_6) - s_{23} s_5 c_6] - c_1(s_4 c_5 c_6 + c_4 s_6)$$
$$[n_z] = -s_{23}[c_4 c_5 c_6 - s_4 s_6] - c_{23} s_5 c_6$$
$$[s_x] = c_1[-c_{23}(c_4 c_5 s_6 + s_4 c_6) + s_{23} s_5 s_6] - s_1(-s_4 c_5 s_6 + c_4 c_6)$$
$$[s_y] = s_1[-c_{23}(c_4 c_5 s_6 + s_4 c_6) + s_{23} s_5 s_6] - c_1(-s_4 c_5 s_6 + c_4 c_6)$$
$$[s_z] = s_{23}(c_4 c_5 s_6 + s_4 c_6) + c_{23} s_5 s_6$$
$$[a_x] = -c_1(c_{23} c_4 s_5 + s_{23} c_5) - s_1 s_4 s_5$$
$$[a_y] = -s_1(c_{23} c_4 s_5 + s_{23} c_5) + c_1 s_4 s_5$$
$$[a_z] = s_{23} c_4 s_5 - c^2{}_3 c_5$$
$$[d_x] = c_1[-s_{23} c_5) + s_{23} d_4 + a_3 c_{23} + a_2 c_2] - s_1 d_3$$
$$[d_y] = s_1[-s_{23} d_4) + s_{23} d_4 + a_3 c_{23} + a_2 c_2] + c_1 d_3$$
$$[d_z] = -c_{23} d_4 - a_3 s_{23} - a_2 s_2$$

where c_{ij} and s_{ij} are $\cos(\theta_i + \theta_j)$ and $\sin(\theta_i + \theta_j)$ respectively.
We conclude our discussion on kinematics dealing with geometry and

time-independent aspects of motion without considering the forces causing the motion. We will now investigate the aspects of kinematics which take into account the forces that cause motion.

9.8 Dynamics of Kinematic Chains

Robot dynamics constitutes the mathematical formulations of the equations of robot arm motion. It is a set of mathematical equations describing the dynamic behavior of the arm. These equations can be used for:

(1) computer simulation of the robot arm motion,

(2) design of suitable control equations for the arm,

(3) evaluation of kinematic design and structure of the arm.

Dynamic models are usually obtained from known physical laws: Newtonian mechanics and Lagrangian mechanics. From this, it is possible to develop the dynamic equations of the various articulated joints, in terms of specified geometry and inertial parameters of the links. Conventional approaches include Lagrange–Euler (L–E) and Newton–Euler (N–E). A variety of robot arm motion equations describing rigid body arm dynamics can be obtained from these two formulations.

9.8.1 Attributes of the Lagrange–Euler formulation

The Lagrange–Euler formulation is a simple and systematic one. Assuming rigid body motion, we can obtain a set of second-order coupled non-linear differential equations (excluding the dynamics of electronic control devices, backlash, and gear friction). These can be used to solve the **forward kinematic** problem. Given the desired torques/forces, the dynamic equations are used to solve for joint accelerations which are then integrated to solve for the generalized coordinates and their velocities.

It enables the solution of the inverse kinematic problem. Given the desired generalized coordinates and their first and second derivatives, we can find the forces/torques. In each case, the dynamic coefficients may have to be computed. These coefficients require a fair amount of arithmetic operations. In L–E, the computations are very tedious and thus cannot be used in real time unless they are simplified.

9.8.2 Attributes of the Newton–Euler formulation

The Newton–Euler formulation is simple but awkward. It involves vector product terms. The resulting dynamic equations (excluding the dynamics of control devices, backlash, and gear friction) are a set of forward and backward recursive equations.

This set can be applied sequentially to the robot links. The **forward recursion** propagates kinematic information such as: linear velocities, angular velocities, angular accelerations, and linear accelerations at the center of mass of each link, from the inertial coordinate frame to the hand coordinate frame. The **backward recursion** propagates the forces and moments exerted on each link from the end-effector of the manipulator to the base reference frame. The computation time here is found to be linearly proportional to the number of joints of an arm and independent of the arm configuration. The Newton–Euler approach makes it possible to implement realtime control of the manipulator.

9.8.3 Attributes of the generalized d'Alembert approach

Another approach is the generalized d'Alembert principle. It is a faster computation than L–E. Dynamics are expressed in vector-matrix form suitable for control analysis. It explicitly identifies the contributions of the translational and rotational effects of the links. It is useful in designing a controller in state space. Computational efficiency is achieved by using Euler transformation matrices (or rotation matrices) and relative position vectors between joints.

Computational complexity of the systems are of the order:

(1) d'Alembert $\Rightarrow$ Order O (n^3)
(2) L–E $\Rightarrow$ Order O (n^4) (or order $O(n^3)$ if optimized).
(3) N–E $\Rightarrow$ order $O(n)$.

Where $n = \#$ of degrees of freedom.

9.8.4 Lagrange–Euler formulation

The Lagrange dynamics can be combined with the D–H representation of spatial relationships between adjacent link pairs to obtain the Lagrange–Euler formulation for a manipulator. It employs Lagrangian dynamic technique to derive the dynamic equations of the manipulator. Dynamic equations for an n degree of freedom manipulator are based on the following, see also Fu *et al.* (1987):

(1) 4×4 homog. matrix, $^{i-1}A_i$ from D–H.
(2) The Lagrangian–Euler equation:

$$\frac{d}{dt}\left[\frac{\partial L}{\partial \dot{q}}\right] - \frac{\partial L}{\partial q} = \tau_i \quad i = 1, 2, \ldots, n \tag{9.46}$$

where

L = Lagrangian function = $K - P$
K = Total kinetic energy of the arm
P = Total potential energy of the arm
q_i = Generalized coordinates of the arm

$\dot{q}_i$ = First derivative of the generalized coordinate q_i

τ_i = Generalized force (or torque) applied to the system at joint i to drive link i

- The generalized coordinates are normally chosen to completely describe the location and position of the system with respect to a reference coordinate system. See Figure 9.22 for a point in a robot link i.

- Angular positions of the joints are readily available since they can be measured by sensing devices (potentiometers, encoders and so on), they provide natural correspondence to the generalized coordinate systems.

- It corresponds to the generalized coordinates of the joint variable defined in each of the 4×4 link coordinate transformation matrices.

- For a rotary joint, $q_i = \theta_i$ is the angle this joint spans.

- For a prismatic joint, $q_i = d_i$ is the distance traveled by the joint.

9.8.5 Trajectory planning and control

Trajectory planning converts a description of the desired motion, such as **move** P_1, see Figure 9.23, to a trajectory defining the time sequence of intermediate

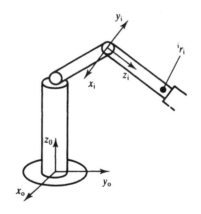

Figure 9.22

A point in link i.

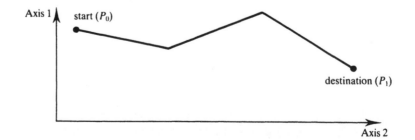

Figure 9.23

A motion in the configuration space for a two-link manipulation.

configurations of the manipulator between the start point P_0 and the destination point P_1.

When the trajectory is executed, the reference point of the end-effector traces a curve in space and sometimes the end-effector changes its orientation. The space curve traced by the end-effector is the path of the trajectory. The result of the motion would be a new position and orientation, which we call location.

The trajectory planning system will produce a sequence of joint configurations for the manipulator $\{\theta_i\}$, $1 \leqslant i \leqslant I$. Also the first and second derivatives of the sequence $\{\theta_i\}$ will produce velocities and acceleration parameters necessary to control the servo mechanisms that perform the desired motion.

It has been seen that manipulator configuration can be expressed either in joint space or Cartesian space. It is easier to express the arm configuration in joint space to enable control and proper formulation of kinematics and dynamics. But when the arm is expressed in joint vectors, humans find great difficulty in relating them to locations in Cartesian space, which is suitably described in orthogonal, cylindrical, or spherical coordinate frames, as discussed in Brady et al. (1983). By using kinematics it is, however, relatively easy to compute the Cartesian space configuration from the joint vector. The inverse problem of relating the Cartesian location of the end-effector in space to joint configuration of the arm is called the inverse kinematic problem, which is not only difficult but sometimes intractable. Sometimes, multiple solutions are obtained for the same Cartesian location. This can, at times, result in disastrous consequences for the task being performed. The careful design of the manipulator is one way of limiting some of these problems or being able to predict what the behavior of such a kinematic system will be.

Because of the problem of relating the Cartesian configuration to joint vectors, it is easier to perform trajectory calculations using joint space. Two popular methods for computing trajectories involve first, using a suitably parameterized class of functions as, for example, polynomials (as discussed in Brady et al. (1983)) to provide the class of trajectories. The parameter values for each move are chosen to satisfy programmer-defined constraints on position, velocity, and acceleration of the manipulator during the move process. In the second method, a trajectory is obtained which constrains the manipulator end-effector to traverse a straight line path in Cartesian space between the points P_0 and P_1. We now examine both the Cartesian and joint control approaches.

Cartesian space control

We have stated that it is more convenient to compute the manipulator trajectory in joint space. It is, however, useful to discuss the Cartesian motion, at least to expose the problems that can be inherent.

In Cartesian control, the configurations of the arm are specified in Cartesian coordinate frames at each point along the path. Complex transformations are usually necessary to achieve Cartesian control. It is possible to compute the path parameters offline. This is accomplished by choosing the path in advance, and at intervals of time computing the Cartesian or joint configurations.

These computations can be in terms of hand location in space or the joint angles.

The hand configuration can be $(x, y, z, \theta_x, \theta_y, \theta_z)$ where $\theta_x, \theta_y, \theta_z$ correspond to roll, pitch, yaw angles of the wrist. The joint angles are in terms of $(\theta_1, \theta_2, \ldots, \theta_n)$ each corresponding to the joint angle value in the computed configuration. But computations done this way invariably do not take accelerations into account, and this can be disastrous, especially when the system requires the robot to achieve accelerations beyond its limit.

Polynomial functions can be used to accomplish a smooth motion in Cartesian space continuous in position, velocity and acceleration. The path of each variable (x, y, z) can be defined to be a polynomial function of time with coefficients which we can find.

Imagine the simple case where the manipulator must move along the x-axis (Figure 9.24). Let $x(t)$ be the position of the hand in x-direction at time t. The arm can be accelerated from rest until time τ when it uses the achieved velocity. At time τ, the end-effector will be at point $x = K$ and is moving at a constant velocity $\dot{x} = K/\tau$. For mathematical expediency, let us assume that the motion commences at $t = -\tau$. We desire to specify six parameters (position, velocity, and acceleration) at both ends of the motion in some n-order polynomial. Because of the symmetry of the problem, we choose a fourth order polynomial expression with unknown coefficients:

$$x(t) = a_4 t^4 + a_3 t^3 + a_2 t^2 + a_1 t + a_0 \tag{9.47}$$

We can differentiate equation 9.1 with respect to time to obtain $\dot{x}$ and $\ddot{x}$. We can specify boundary conditions as follows:

$$x(-\tau) = 0 \quad \dot{x}(-\tau) = 0 \quad \ddot{x}(-\tau) = 0 \tag{9.48}$$

$$x(\tau) = K \quad \dot{x}(\tau) = \frac{K}{\tau} \quad \ddot{x}(\tau) = 0 \tag{9.49}$$

By using the sets of equations obtained from x, $\dot{x}$ and $\ddot{x}$ with the boundary conditions, the coefficients of the polynomial can be obtained:

$$a_0 = \frac{3K}{16}; \ a_1 = \frac{K}{2\tau}; \ a_2 = \frac{3K}{8\tau^2} \tag{9.50}$$

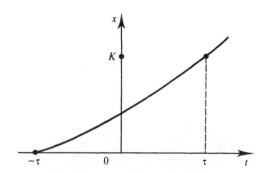

Figure 9.24

Trajectory for $x(t)$.

$$a_3 = 0; \quad a_4 = \frac{-K}{16\tau^4} \tag{9.51}$$

In realtime computation of the path, hand position relative to the robot base will need to be computed at each interval and the joint configurations obtained to compute the polynomial path. The chained expression which may relate the current location P_0 to the new location P_1 will be found.

$$R_{T_H} = R_{T_{P_0}} \cdot X_{1 T P_0} \ldots X_{n-1_{T_{X_n}}} \ldots X_{n_{T P_1}} \tag{9.52}$$

where the point P_1 and the hand position H are the same.

Joint space control

Of course, the Cartesian space control approaches described above are not efficient and do not always guarantee the desired result. Computing the trajectory in joint space is therefore more convenient and accurate.

Consider the points P_0 and P_1 being the configuration of the two-link manipulator above. These points allow for positional constraints which can be satisfied by a trajectory of the nature:

$$\hat{\theta}(t) = g(t)\hat{\theta}_1 + (1 - g(t))\hat{\theta}_0 \tag{9.53}$$

where $g: [0, 1] \to [0, 1]$ is any continuous function satisfying $g(0) = 0$, $g(1) = 1$. It is clear that g converts the path traced by the hand into a trajectory. The simplest function which satisfies the above requirement is $g(t) = t$, which produces the linear combination:

$$\hat{\theta}(t) = t\hat{\theta}_1 + (1 - t)\hat{\theta}_0 \tag{9.54}$$

The problem with this function even with its advantage of simplicity is that joint space angular velocity is a constant: $\hat{\omega} = \hat{\theta}_1 - \hat{\theta}_0$. Therefore, the velocity cannot be specified independently at both ends of the trajectory and this requires 'infinite' acceleration between movements. In addition to this, the joint solution $\hat{\theta}(t)$ cannot always be guaranteed to lie in the workspace.

It is, however, possible to constrain a trajectory by specifying its initial and final velocities such that $\hat{\theta}(0) = \omega_0$ and $\hat{\theta}(1) = \omega_1$. The four constraints will not be met by either a linear or a quadratic function of time, but by a cubic function, which has this symmetric form:

$$\hat{\theta}(t) = (1 - t)^2\{\hat{\theta}_0 + (2\hat{\theta}_0 + \hat{\omega}_0)t\} + t^2\{\hat{\theta}_1 + (2\hat{\theta}_1 - \hat{\omega}_1)(1 - t)\} \tag{9.55}$$

But the cubic function is still an oversimplification when one considers that there is a maximum attainable velocity. In this approach, the acceleration, and therefore the torque, cannot be independently specified at both ends of the trajectory since it grows linearly, as discussed in Brady *et al.* (1983). When the accelerations α_i are required to be specified at both ends of the trajectory along with the velocities and accelerations, there are a total of six constraints. Then this fifth order polynomial suffices:

$$\hat{\theta}(t) = (1 - t)^3 \left\{ \hat{\theta}_0 + (3\theta_0 + \hat{\omega}_0)t + (\hat{\alpha}_0 + 6\omega_0 + 12\hat{\theta}_0)^z \frac{t}{2} \right\}$$

$$+ t^3 \left\{ \hat{\theta}_1 + (3\hat{\theta}_1 - \hat{\omega}_1)(1 - t) + (\hat{\alpha}_1 - 6\hat{\omega}_1 + 12\hat{\theta}_1) \frac{(1 - t)^2}{2} \right\} \quad \textbf{(9.56)}$$

The problem with this quintic is that practical difficulties exist in general with trajectories planned by constraint satisfaction and with higher order polynomials in joint space. Even though the constraint satisfaction will work from simple descriptions of positions, velocities and accelerations, the main problem lies in the fact that it is weakly constrained. Specifically, the path and the rotation curve followed by the end-effector are not explicitly specified. Avoiding overshoot of the arm cannot be guaranteed because there is no constraint to prevent this.

The basic approach in trajectory planning is to treat the planner as a black box which accepts input of location, velocity and acceleration expressed either in joint or Cartesian coordinates from starting location to destination location. Two approaches from the planners perspective are, first, explicitly stating a set of constraints, for example, continuity and smoothness, on position, velocity, and acceleration of the manipulator's generalized coordinates at selected locations called **knot points** or **interpolation points** along the trajectory. The planner can then select a parameterized trajectory from a class of functions (polynomial functions) within the interval $[0, \tau]$ that interpolates and satisfies the constraints at the interpolation points. The second approach requires the user to explicitly specify the path which the end-effector must traverse by an analytical function and then the planner will determine a trajectory either in joint or Cartesian coordinates that approximates the desired path. As stated earlier, the first approach has problems because the end-effector is not constrained on the path, but in the second approach path constraints are specified in Cartesian coordinates, and the joint actuators are servoed in joint coordinates. This is why the problem is one of finding a parameterized trajectory that satisfies the joint path constraints.

Many joint interpolation algorithms have been proposed to minimize the problems discussed above. One such method is to use splines of low order polynomials (usually cubics) rather than higher order polynomials.

Joint interpolated control

The joint interpolation constitutes a compromise between point to point and continuous Cartesian path control. The goal is to provide a smooth and accurate path without incurring a heavy computational burden. There are certain desirable characteristics in developing the strategy for path control:

(1) precise control of motion at or close to the workpiece,
(2) continuity of position, velocity, and acceleration,

(3) mid-range motion which is predictable and only stops when necessary.

    (a) The first characteristic can only be achieved by using Cartesian control or joint control as discussed.

    (b) Characteristics 2 and 3 can be achieved by the following technique, as discussed in Snyder (1985):

        (i) Divide the path into a number of segments k.

        (ii) For each segment, transform the Cartesian configuration at each end, and determine starting and ending angular positions for each joint.

        (iii) Determine the time τ_i required to traverse each segment where ω_i is the known (*a priori*) maximum velocity for joint i.

$$\tau_1 = \max \frac{\theta_{1i} - \theta_{2i}}{\omega_i^{\max}} \tag{9.57}$$

        where $1 \leqslant i \leqslant n$

        (iv) Divide τ_1 into m equal intervals, $\Delta\tau$, where:

$$m = \tau f$$

$$f = \text{sampling frequency}$$

$$= 50 \text{ to } 60 \text{ Hz} \tag{9.58}$$

$$\Rightarrow \Delta\tau = \frac{1}{f}$$

        (v) For each joint, determine the angular distance to be traveled during $\Delta\tau$:

$$\Delta\theta_i = \frac{\theta_{1i} - \theta_{2i}}{m} \tag{9.59}$$

        (vi) When motion begins, at the nth sampling time, joint servo i receives set point $n\theta_i$.

        —Using this algorithm, all joints begin and end their motion simultaneously.

        —Path is smooth and predictable, though not necessarily along a straight line.

Another approach which is more commonly used is the one described by Paul (1981), also found in Fu *et al.* (1987). In this scheme, a path is divided into segments in which certain chosen polynomial functions are applied to satisfy the position, velocity and acceleration at the knot points of these segments. Using Figure 9.25, the knot points are chosen with the following considerations:

(1) When picking up an object, the motion of the hand must be directed away from an object; otherwise, the hand may crash into the supporting surface of the object.

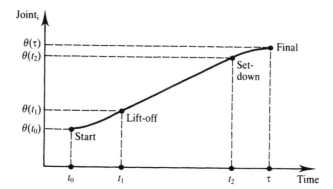

Figure 9.25

Position conditions
for a joint trajectory.

(2) If we specify a departure position (lift-off point) along the normal vector to the surface, away from the initial position, and if we require the hand (that is, the origin of the hand coordinate frame) to pass through this point, we then have an admissible departure motion. If we further specify the time required to reach this position, we could then control the speed at which the object is to be lifted.

(3) The same set of lift-off requirements for the arm motion is also true for the set-down point of the final position motion (that is, we must move to a normal point out from the surface and then slow down to the final position) so that the correct approach direction can be obtained and controlled.

(4) From the above, we have four positions for each arm motion: initial, lift-off, set-down, and final.

(5) Position constraints:

    (a) initial position: velocity and acceleration are given (normally zero),

    (b) lift-off position: continuous motion for intermediate points,

    (c) set-down position: same as lift-off position,

    (d) final position: velocity and acceleration are given (normally zero).

(6) In addition to these constraints, the extrema of all the joint trajectories must be within the physical and geometric limits of each joint.

(7) Time considerations:

    (a) Initial and final trajectory segments: time is based on the rate of approach of the hand to and from the surface and is some fixed constant based on the characteristics of the joint motors.

    (b) Intermediate points or mid-trajectory segment: time is based on maximum velocity and acceleration of the joints, and the maximum of these times is used (that is, the maximum time of the slowest joint is used for normalization).

The following constraints apply to the planning of joint-interpolated trajectory, see Fu *et al.* (1987).

Initial position:

(1) position (given)

(2) velocity (given, normally zero)

(3) acceleration (given, normally zero)

Intermediate positions:

(1) lift-off position (given)

(2) lift-off position (continuous with previous trajectory segment)

(3) velocity (continuous with previous trajectory segment)

(4) acceleration (continuous with previous trajectory segment)

(5) set-down position (given)

(6) set-down position (continuous with next trajectory segment)

(7) velocity (continuous with next trajectory segment)

(8) acceleration (continuous with next trajectory segment)

Final position:

(1) position (given)

(2) velocity (given, normally zero)

(3) acceleration (given, normally zero)

   With the above constraints, some polynomial function of degree n or less can be chosen that satisfies the position, velocity and acceleration at the knot points (initial, lift-off, set-down, and final positions). An approach is to specify a seventh degree polynomial for each joint i:

$$q_i(t) = a_7 t^7 + a_6 t^6 + a_5 t^5 + a_4 t^4 + a_3 t^3 + a_2 t^2 + a_1 t + a_0 \qquad \textbf{(9.60)}$$

where the unknown coefficients a_j can be determined from the known positions and continuity conditions. But, the use of such a high degree polynomial to interpolate the given knot points can be problematic. It is difficult to find the extrema of this polynomial and it tends to have extraneous motion. Another approach is to split the entire joint trajectory into several trajectory segments so that different interpolating polynomials of a lower degree can be used to interpolate in each trajectory segment. There are many splitting approaches and each method has its attributes. The most common methods are the following, also discussed in Fu *et al.* (1987).

4–3–4 trajectory:

Each joint has the following three trajectory segments: the first segment is a fourth-degree polynomial specifying the trajectory from the start position to the lift-off position. The second trajectory segment (or mid-trajectory segment) is a third-degree polynomial specifying the trajectory from the lift-off position to the set-down position. The last trajectory segment is a fourth-degree polynomial specifying the trajectory from the set-down position to the final position.

3–5–3 trajectory:

Same as 4–3–4 trajectory, but uses polynomials of different degrees for each segment: a third-degree polynomial for the first segment, a fifth-degree polynomial for the second segment, and a third-degree polynomial for the last segment.

Cubic trajectory:

Cubic spline functions of third-degree polynomials for five trajectory segments are employed here.

In summary, trajectory planning can be accomplished either in Cartesian space or in joint space. However, since manipulators are served in joint space and task specifications are accomplished in Cartesian space, the appropriate method to specify constraints depends on the trajectory to be specified. These constraints relate to the position, velocity, and acceleration. Using higher-order polynomials to solve the trajectory problem can be problematic due to the extraneous motion and finding the extrema. A more manageable approach is to split the trajectory into segments and apply lower degree polynomials.

9.9 Robot Programming

Robot programming is concerned with describing the sequencing of robot motions and activities in order to accomplish a desired set of tasks and to teach the robot those motions or activities. The activities can include interpretation of sensor data, sending and receiving of signals and data from other devices or actuation of the joints and end-effector. The teaching of the motion paths and activities can be accomplished in two major ways: first, lead-through techniques (or teaching by showing) or second, by language-based techniques, or by a combination of both techniques.

The lead-through methods are accomplished in powered or in non-powered modes of the live robot. In non-powered mode, the robot is manually moved to the desired location. The powered lead-through method essentially allows for the robot to be guided to its destination using a teach pendant to control the various joint actuators.

Language-based programming techniques are quite similar to regular computer programming. Programs are normally specified in some type of language which can be low-level or high-level. There are two levels of language-based techniques: first, explicit robot programming languages and second, task-level programming.

The explicit robot programming languages are robot level languages and require the robot programmer to be experienced in computer programming and in the design of sensor-based motion strategies, which is a skill not usually possessed by workers on the manufacturing floor. The lead-through method is a type of explicit programming technique but we will restrict ourselves to a language-based approach and the graphical approach in our description of explicit robot programming.

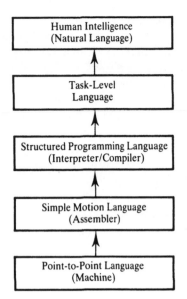

Figure 9.26
Levels of robot programming languages.

Task-level programming, on the other hand, allows the user to specify the effects of the actions on objects rather than the sequence of manipulator motions needed to achieve the task as is the case in explicit robot programming languages. Since task-level programming is object-oriented, and not robot-oriented, it deals with the effects on objects, and not on the motions of the robot. It is certainly an easier programming method. Figure 9.26 shows various levels of programming languages available.

9.9.1 Explicit robot programming

Explicit robot programming means that the programmer has to provide commands on the robot level, that is, he has to specify particular robot motions, for example, MOVE gripper TO POSITION (x, y, z). This may be performed through special textual robot languages or through graphical-interactive techniques using a graphical simulation system.

Language-based techniques

This explicit robot programming is the act of specifying robot executable instructions in some textual language which can be translated by a system which is either an interpreter or a compiler. This programming scheme requires the instructions to be specified as a series of motion instructions which will accomplish the desired task. When the translation system is an interpreter, a program can be decoded instruction by instruction, and the robot control will also execute it in stepwise manner. In cases when a compiler is the translation system, an intermediate code is generated which is executed by an interpreter program.

Although the compiler method is more powerful, it often requires more powerful hardware and software. Therefore, the more prevalent explicit robot programming method is the interpreter method.

With the explicit programming method, it is possible to program robots both offline and online. The offline programming is accomplished by specification of trajectories and sensor calls which must be made in textual mode. Conditional statements are possible in this programming scheme.

There are two methods of specifying the trajectories as points of movement. In the first approach, the trajectory coordinates are defined explicitly in the program as numerical variables or constants. These values can be either measured from CAD drawings or read from an existing database into the program. Since this method can be very cumbersome, the second approach appears to be more promising.

In the second method, symbols are used to specify the trajectory. For instance, the symbolic variables can be for the start, intermediate, and end points, in addition to orientation. The actual values can be obtained through teaching and are assigned to the variables. A typical robot program can be supplemented by non-motion commands, including computation of spatial relationships of objects in the world, commands for sensors and other peripherals, elements for interrupt handling and synchronization of processes and so on.

As can be seen from Figure 9.26, the versatility of a robot language is dependent on the complexity of the programming scheme. The most comfortable approach is task level but it is the most difficult to develop, as will be seen later. The most verbose is machine language.

There are currently over 100 explicit robot programming languages, almost as many as there are varieties of robots. Some of the languages include VALII, or extended Pascal or Basic. The development of robot languages started with the extension of the standard computer languages to include some robot control commands. Table 9.1 contains some of the more prominent robot languages which have been implemented. Most of the extended languages are not used by industry because they are too complex to apply.

We will now illustrate the structure of explicit robot programming using VALII. The VAL language was developed in the USA for programming of the PUMA series robots in Unimation. An improved language, VALII, was released in 1984. It can also handle sensor information, even in difficult situations. The following include some of the important constructs of the language:

(1) Locations definition. Several methods can be used to define locations and for determining the current location of the robot. For instance:

HERE P1

defines the variable P1 to be the current location of the robot. Similarly, the command

WHERE

Table 9.1 Prominent robot languages

Developer	Language	Robot implementation
Cincinnati Milacron	T3	T3
Unimation	RPL, VALII	PUMA
Adept	VALII	Adept1
Sheinmann	AL, PAL	Stanford arm
IBM	AML, Funky, Emily, Maple Autopass	IBM arm
Bendix	RCL	PACS arm
General Electric	Help	Allegro
Anorad	Anorad	Anomatic
Olivetti	Sigla	Sigma
Stanford	WAVE	Stanford arm, Robovision
Automatix	RAIL	Autovision, Cybervision

requests the robot system to display the current location of the robot. The command TEACH is used to lead the robot through and record, for the robot, a series of locations using the teach pendant. A third method for defining locations for the robot is by using the POINT command. For instance:

POINT PX = P1

sets the value of PX, which is a location variable, to be equal to the value of P1.

(2) Editing and control of programs. To initiate the creation or editing of a program, the following commands may be issued:

```
EDIT  PICKUP1
    :
E
```

The EDIT command opens a program named PICKUP1 and the E command, which stands for EXIT, enables switching from the programming mode to the monitor mode.

Before the execution of a program, the robot speed is normally given as a percentage of a specified speed, and this can be accomplished using the SPEED command. For instance:

SPEED 70

specifies that the speed of the manipulator is 70 units on a scale which may range from 0.39 (very slow) to 1.2800 (very fast), with 1.00 as 'normal' speed. With the value of 70, one would be setting the robot speed below normal speed. However, the actual speed that is used for programs is obtained from the product of the speed specified in the monitor mode (such as 70 above) and the speed specified within the text of the program. For instance, if 50 was set as the speed within the program, it means that the actual speed will be (0.7 × 50 = 35) 35 units for the 'normal' speed. Execution of a specified program can be accomplished using the command:

EXECUTE PICKUP1

In cases where the program needs to be executed several iterations, the EXECUTE command may read as follows:

EXECUTE PICKUP1, n

where n is the number of iterations of execution.

There are many additional monitor commands which can be used in VALII. They include STORE, COPY, LOAD, FLIST, RENAME, DELETE and so on.

(3) Motion commands. VALII possesses commands which are used for translational movements by the robot. The motion instruction,

MOVE PX1

enables the robot to move by joint interpolation to the point PX1. Joint interpolation allows the robot to start its motions at the various joints and synchronize the completion of the motions even when their distances are not the same.

Another command related to this is:

MOVES PX1

which causes the robot to move along a straight line from its current position to the point PX1.

Similarly, the commands APPRO and DEPART are used to move the tool towards and away from some specified position and orientation, respectively. For instance:

APPRO PX1 70
MOVE PX1
DEPART 70

means that the tool should move to the location (position and orientation) specified by PX1, but offset from the point along the tool z-axis by a distance of 70 mm (distances can also be measured in inches or other units). The DEPART command returns the tool to its original position by moving it away from the new point PX1 70 mm along the tool z-axis. Although both the APPRO and DEPART commands are carried out in joint-interpolated mode, it is possible to carry them out in straight line motions by using APPROS and DEPARTS. Other commands that apply to motion include SPEED, DRIVE, and ALIGN. The DRIVE command can be used to change a single joint from its current location to another. For instance, if a joint is rotary, a command might be issued as follows:

DRIVE 5, 55, 60

This implies that the angle of joint 5 should be changed by driving the joint 55° in the positive direction at a speed of 60% of the monitor speed.

(4) Control of the hand. The gripper of the hand can be controlled using gripper control commands such as OPEN, CLOSE, GRASP. The commands can be specified as follows:

> OPEN 60
> :
> CLOSE 60

The open statement causes the robot gripper to open to 60 mm and the CLOSE command causes the robot to close to 60 mm. GRASP is another command that causes the robot to close immediately. A GRASP command

> GRASP 10.5, 100

causes the gripper to close immediately while checking to ensure that the final opening is less than the 10.5 mm specified. After closing, the program should go to statement 100 in the program.

(5) Configuration control and interlock commands. There are configuration commands that ensure that the manipulator is spatially configured to approach its task like the human arm of shoulder, elbow, wrist and so on. The command of RIGHTY or LEFTY will imply that the elbow is pointed to the right-hand side or left-hand side, (with respect to the upper arm) respectively. Other such commands include, ABOVE, BELOW and so on.

Interlock commands are used to communicate about input or output signals. The RESET command turns off all external output signals and is typically used for initialization. The SIGNAL command is used to turn output signals on or off. A positive signal number is used to turn the corresponding signal on and a negative number turns it off. For instance:

> SIGNAL 2, −5

turns output signal 2 on and output signal 5 off. There is also the REACT command which is used to interrupt the program due to external sensor signals. For example:

> REACT VAR2, SUBR5

would cause the continuous monitoring of the external binary signal specified as variable VAR2. When the expected signal occurs, depending on the signal sign, control is transferred to subroutine SUBR5.

There are many other VALII commands which this book does not have space to discuss. There are INPUT/OUTPUT control commands for terminals such as PROMPT and TYPE; program control instruction sets such as IF ... THEN ... ELSE ... END, WHILE ... DO, DO ... UNTIL and so on.

This section is intended to show the structure of a commercially available explicit programming language. A complete set of VALII commands can be obtained from Unimation, see references for details.

Graphics-based techniques and simulation

There is a saying that a picture is worth more than a thousand words. The importance of visual representation of task programming in graphics embodies this saying. The graphical programming system uses models of the real objects and the robot itself in the world model to illustrate the programming process and execution. The attributes of the world model in task-level programming apply here except that some of the subsystems that need not be described in normal task-level programming must now be modeled. For instance, sensors and controllers must be modeled with their characteristics to depict the processes within what could have been merely described as 'black boxes.' With this approach, it becomes possible, for example, to predict program execution times or cycle times for programmed tasks.

Many of the graphical approaches use wireframe models to provide the animation of the program execution. Some, however, use hidden line removal graphics for the animation. The trade-off is in the speed versus the visual realism. The wireframe provides complex environments, difficult visual images of the environment, with high speed, while the hidden line removal system provides more realistic images but at reduced speed due to the increased computations necessary to perform the hidden line removals.

There are two methods of actual programming using graphical methods. The first one involves use of devices such as dials to move the robot through the needed motions. The second approach requires the use of spatial (geometrical) relations between the robot and its world. In this case, the relationships are expressed in terms of configuration states which can be shown by the joint configurations for those states. Additional information will normally be included as textual information to supplement the program. For instance, motion parameters can be specified this way. Some systems enable easier text inclusion by providing a general purpose robot programming language which can be mapped into a specific robot programming language, using some driver (post-processor). Others have specialized programming environments.

In general, the difference between the idealized robot environment and that of the real world of the robot necessitate changes in the resulting robot programs.

The type of graphical programming described above tends to be cumbersome and non-generalized. An ideal simulator is one which uses the task-level programming system as described in Section 9.10 to generate a program which can be simulated to test the validity of the program so generated. This type of simulator incorporates robot motions and sensory information into a dynamic world model which provides the planner with feedback about events in the world. A program can then be changed according to the results of the program. A detailed description of this type of simulator can be found in Rist (1990).

9.10 Task-level Programming

Although task-level programming is a type of robot programming scheme and therefore should be treated under robot programming, it has become the most important method of designing robot programs. For this reason, we have provided a separate section for task-level programming.

Developing a task-level language is very difficult. Many researchers have worked on aspects of the language but few have been able to produce reasonable language programs. Even those that have been developed operate within very restricted domains. The more prominent ones include: AUTOPASS, see Lieberman and Wesley (1977); RAPT, see Ambler and Popplestone (1975) and Popplestone *et al.* (1978) and LAMA, see Lozano-Perez and Winston (1977). The elements of task-planning include, world modeling, task specification, and manipulator program synthesis as discussed in Lozano-Perez (1983) and Brady *et al.* (1983).

World modeling tasks must include the following information, discussed in Nnaji (1988).

(1) geometric description of all objects and manipulators in the task environment;
(2) physical description of all objects, for example mass and inertia;
(3) kinematic description of all linkages;
(4) description of manipulator characteristics, for example joint limits, acceleration bounds, and sensor characteristics, see also Lozano-Perez (1983).

In addition to the configurations of all objects in the manipulator world, the models should include uncertainty associated with each configuration. Geometric description of the objects in the robot world is the major element of a world model. The geometric data can be acquired from vision or from the CAD system. Since parts to be manufactured are usually modeled in the CAD system, it seems appropriate that the information should come from the CAD. In addition, the data represented is more accurate and easier to reason with than the data which can be acquired by today's vision systems. The impasse in complete development and implementation of a good automated process planner is due to the range of knowledge needed by the robot to interact with its world, the problems associated with acquiring the world knowledge, and the difficulty in developing a robust reasoning paradigm. Research in the 1980s, including Ambler and Popplestone (1975); Brady *et al.* (1983); Lozano-Perez (1983) and Nnaji and Chu (1990), has shown that it is best to represent world knowledge using the CAD system and to deal with the dynamic world of the robot using sensors such as the vision system.

Task specification deals with the specification of the sequence of configuration states from the initial state of manufacturing parts to the final state of the product. The specification can be accomplished in three ways:

(1) using a CAD system to model and locate a sequence of object configurations, see Brady *et al.* 1983; Frommherz and Werling (1989) and Nnaji and Chu (1990),

(2) specifying the manipulator configurations and locating the features of the objects using the manipulator itself, see Grossman and Taylor (1978),

(3) using symbolic spatial relationships among object features to constrain the configurations of objects, see Popplestone *et al.* (1978).

The first and second methods are prone to inaccuracy since the light-pen or mouse is often used in the first case and in the second case the robot has limited accuracy. The third method holds out more promise. It involves the description of the robot configurations by a set of symbolic spatial relationships that are required to hold between objects in that configuration. Some advantages can be seen from this approach. For instance, the method does not depend on the accuracy of the light-pen or the manipulator. Also, the families of configurations such as those on a surface or along an edge can be expressed. Although using symbolic spatial relationships is a more elegant way of specifying tasks, the fact that they do not specify tasks directly is a drawback. They require conversion into numbers and equations before use.

In this book, we present a method which exploits the powers of the CAD specification approach and the symbolic spatial relationship to provide the designer with a robust task specification system which can be used to capture the designer's intent and to pass on this intent to the process planner. A product modeling system, described by Nnaji (1992), is needed which can allow for the use of a mouse or light-pen to point to features which link up and which can provide the designer with the syntax and protocol to specify orientation by using symbolic spatial relationships. It is also possible to derive operations by examining binary sets of parts relationships on the bill of materials tree. The product modeler can build the bill of materials automatically by providing the designer with data regarding the object-feature interconnections.

In summary, task specification can be achieved by using a sequence of operations and/or symbolic spatial relationships to specify the sequence of states of the objects in the realization of the task. But rather than have a sequence of intermediate model states on a CAD system, an initial state on the CAD system coupled with the symbolic spatial relationships can completely specify the task. It makes it possible to use, for instance, a disassembly method to obtain the individual model states, approach directions and linking features.

The typical components of manipulator program synthesis are grasp planning, motion planning and error detection. The output is normally a sequence of grasp commands, motion specifications and error tests. It may be instructive to note that when the product modeler is used, it produces the relationships, including functions of features of parts, which are essential in program synthesis. For instance, grasp planning is well constrained if spatial relationships have been specified which must exist between features of parts in an assembly. The search for grasp surfaces will therefore be constrained by such relationships.

The description here is concerned with robot task-level programming where the world of the robot can be specified in a CAD environment. Since robot programming can be for a range of applications, we show how automatic robot programming works by restricting ourselves to the assembly domain, specifically where the robot is the executor of the assembly task.

9.10.1 World modeling

Various publications, including Nnaji (1992); Hoermann (1989); Lozano-Perez (1983) and Lozano-Perez and Winston (1977), have demonstrated a robot assembly task planner which processes the knowledge of the working environment and generates a sequence of general or robot independent commands when the robot world is static. By means of acquiring the world knowledge from the CAD system and generating these commands, a robot is able to automatically perform the task. The major drawback in the system so far developed (such as RAPT discussed in Popplestone *et al.* (1978) and AUTOPASS discussed in Lieberman and Wesley (1977)), is that the system must be capable of interpreting geometry to recognize geometric features and inferring functions which the features are intended to serve. Even for the most simple geometries – the polyhedral objects – this has proved to be very cumbersome. The problem of inference of function from form (geometry) is particularly difficult because it involves the regeneration of the designer's intent for the part or product. It invariably requires an experienced process planner to do this and amounts to a 're-invention of the wheel'.

Representing information to the robot (task specification) using the CAD system is in itself fraught with problems. This is largely due to the limited capacity of today's CAD systems. This problem is not limited to assembly but rather spans all domains of applications of automatic process planning concepts. Geometric data currently obtained from CAD systems are CAD system dependent. In addition, current CAD systems are incapable of carrying manufacturing process information other than the product modeling system – a CAD system which is capable of capturing the designer's intent for the part or product and can transfer the product definition information to a process planner. This new CAD system should ideally be capable of translating the data, so acquired at design stage, into a neutral graphics medium in order to eliminate CAD system dependence of the geometric data. Function can be captured by the designer and transferred to the process planner, thus eliminating the huge stumbling block of today's automatic process planner – the inference of function from form. We will not describe the requirements for world modeling for the generalized task-level programming system as described by Nnaji and Chu (1990).

Desired CAD data

We have already mentioned that the future CAD system must produce both geometric and non-geometric data which are relevant to the process planning function. To ensure that future planning systems deal with data that is easy to manipulate, we have studied the type of data appropriate for reasoning about geometry, as discussed in Nnaji (1988). The following observations were made.

Since the essential geometric information in object reasoning is the boundary information, a good modeler must be able to present boundary information in the form of vertices, edges, and faces, from which features may be extracted. Also, a good modeler must be capable of translating its 3-D geometric data into a neutral format such as IGES (initial graphics exchange specification), see references for details. In addition, it seems a waste of very valuable time to completely design parts using the current approach of free design where models of parts are created and the process planner must 'tease out' embedded features on the part using some type of feature recognition system. It seems inevitable that at times a design will necessarily be reduced to 'raw geometry', or that elaborations of features will take them out of classes whose manufacturing properties are explicitly present in a database. The capability of recognizing instances of feature classes in an object description, which does not have them explicitly present, will be a continuing need. However, a more reliable and efficient technique will be to employ the concepts of design with features, discussed in Libardi and Dixon (1986) and Nnaji and Liu (1993). This means the existence of a particular form feature or 'module' has been directly expressed by the user of the system, or inferred as being part of a module so expressed. In contrast, the concept of free design merely requires the designer to describe with explicit geometry the form of the features in a part without symbolic reference to the feature.

Although there are many other available modeling schemes which can be used for geometric model representation on the CAD systems (for example, sweep representation, primitive instances, spatial occupancy enumeration, cell decomposition), we are concerned only with the boundary representation and constructive solid geometry (CSG) methods because they possess the best attributes of all the representation schemes, see also Voelker et al. (1978). The former is important because it is close to computer graphics, is unambiguous and is available to computing algorithms. Disadvantages include: verbosity, complicated validity checking and difficulty in creating objects. CSG is preferred because it has features such as simple validity checking, unambiguity, conciseness, ease of object creation and high description power. Its disadvantages include: inefficiency in producing line drawings, and inefficiency in graphic interaction operations. In addition, at least 95% of manufacturing parts do not require multi-curved surfaces found on ships' hulls, car bodies, and aircraft, and hence can be represented in CSG. These schemes essentially fit well in design with features, reasoning about geometry, transmutation of form into function, and the consequent product modeler, and result in a synergistic effect on the automatic process planning system.

It is well known that just as robot programming is currently manipulator dependent, modeling is still CAD system dependent. In other words, there is no standard internal data representation across CAD system manufacturers. However, the advent of IGES, which is a neutral graphics data representation medium, is the first step in this regard. For a CAD-driven programming system to be truly beneficial to the manufacturing world, there must also be a standard CAD representation. IGES is the most popular neutral graphics system but is severely limited in the types of data which it can communicate. It can only manage with geometric information. Even this geometric information is quite restricted. It has been shown by Yeh (1990) how CAD data translated into a neutral form such as IGES can be restructured into a form understandable by a reasoning system. This is useful in using such data for automatic process planning.

9.10.2 Task specification

The product modeler as described in Chapter 7 provides the vehicle for capturing and passing on the designer's intentions for a part. In the case of assembly, a modeling system such as PAM, see Nnaji and Liu (1990), enables the user to assign spatial relationships and obtain other attributes such as linking functions, assembly tolerances, required torque for screwing and the contact force exerted on each part that signals the robot to stop. Also, since in this book we advocate task specification by using symbolic spatial relationships, the product modeling system is best suited for this approach. Since the task specification has as its kernel the spatial relationship specification, we now present the spatial relationship approaches that can be used.

Spatial relationships

The linking relationships of parts represented in current CAD systems are based on the relationship between the coordinate frames attached to each part. This relationship is specified using a transformation matrix. This requires that the designer knows the details associated with the dimensions of each part. This information would be difficult to extract from 3-D drawings unless dimensions are labelled. This becomes cumbersome as the number of parts increases.

The new method consists of finding a linking frame for the assembly parts through spatial relationships. The assembly process can be defined using the constraints imposed on the degrees of freedom of motion between relative linking parts. When the 'against' relationship is assigned between two faces of different parts, the relative degrees of freedom of motion are one rotation and two translations, or 'against' between a cylindrical face of a part and a planar face of the other results in relative degrees of freedom of motion of two rotations and two translations. After another 'against' relationship is assigned in the first case, and the first 'against' face is not parallel to the second one, the degrees of freedom of motion will only have one translation left. Next, by adding non-parallel 'against' relations, there will be no degree of freedom of motion left, then the assembly parts are fixed. Through the combination of the spatial relationships between assembly parts which eliminate some or all of the degrees of freedom of motion the fixed linking frames for each part can be assigned as shown in Figure 9.27.

The spatial relationships describe how bodies are positioned relative to one another. According to the conventional manufacturing processes, all the positional dimensions are based on a planar face or a cylindrical centerline. For this reason, the features which hold the spatial relationships are:

(1) planar face,

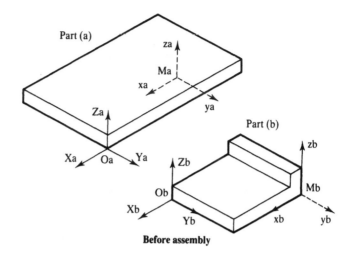

Figure 9.27

Assembly with linking frames (source: Automation and Robotics Lab., Univ. Massachusetts).

(2) centerline for cylindrical feature, and

(3) point for spherical features.

The relationship types shown in Figure 9.28 are classified as follows:

(1) **Against:** 'against relationship' means that the faces touch at some points. For planar linking features, the two planar faces touch over an area; a cylindrical feature touches a planar face along a line; a spherical feature touches another feature only at a point. The two planar faces can be against one another only if the direction normals are in opposite directions.

(2) **Parallel–offset** (general form of 'coplanar' in Figure 9.28): the parallel relation holds between planar faces, cylindrical and spherical features. This is similar to 'against' except that the outward normals of planar faces point in the same direction, and there exists an offset distance between selected features.

(3) **Parax–offset**: this is similar to 'parallel', but the outward normals or axes are anti-parallel.

(4) **Incline–offset:** the inclination relation holds between two planar faces. This relationship expresses the offset and the inclination angle between two features.

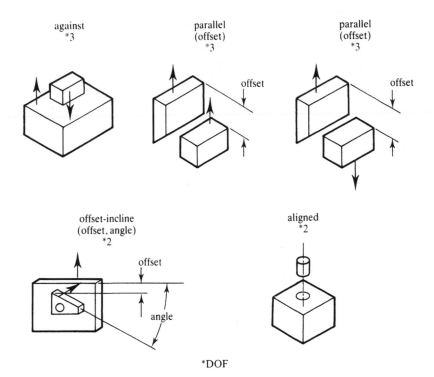

against
*3

parallel
(offset)
*3

parallel
(offset)
*3

offset

offset

offset-incline
(offset, angle)
*2

aligned
*2

offset

angle

*DOF

Figure 9.28
Spatial relationships
(source: Automation
and Robotics Lab.,
Univ. Massachusetts).

(5) **Aligned** is a type of relationship between cylindrical and spherical features corresponding to 'fits' in Figure 9.28. Two features are aligned if their axes are colinear.

The 'spatial relationships engine' is developed according to a product rule-based system. By inferring the symbolic spatial relationships the geometric calculations are performed to generate the linking frames.

Assembly operation attributes
In order to represent the assembly specifications, some important assembly operation attributes have to be assigned by the designer. These operation attributes contain the information about the nature of assembly (such as glue, weld, fit and so on) and the tolerances associated with positioning the parts. Through assignment of the spatial relationships these attributes are inferred. For example, the positional tolerance is acquired when the offset distance is requested. The assembly operation attributes within spatial relationships are described as follows:

- **Against:** contact force
- **Parallel:** offset distance and tolerance
- **Parax:** offset distance and tolerance
- **Incline:** offset distance and tolerance, and incline angle

Some operations with special attributes are assigned from the functions with a

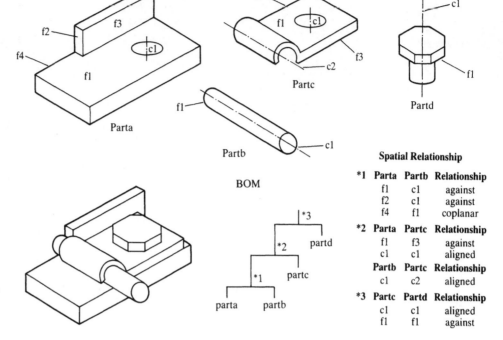

Figure 9.29
The BOM and spatial relationships output of the product assembly model.

Spatial Relationship		
*1 Parta	**Partb**	**Relationship**
f1	c1	against
f2	c1	against
f4	f1	coplanar
*2 Parta	**Partc**	**Relationship**
f1	f3	against
c1	c1	aligned
Partb	**Partc**	**Relationship**
c1	c2	aligned
*3 Partc	**Partd**	**Relationship**
c1	c1	aligned
f1	f1	against

combination of spatial relationships such as:

- **Fit:** this includes aligned and against or aligned and parallel relations. The acquired attribute is the applied force in cases dealing with aligned and against relationships. In situations that have to do with aligned and parallel relations, the offset distance, tolerance, and applied force are needed.

- **Screw:** this is similar to fit except the applied force is replaced with the torque.

- **Glue:** the same as against except by replacing contact force with the one based on glue specifications.

- **Weld:** this function will be used when the linking position is decided. The intersection of the two selected faces indicates the weld seam. The weld specifications are then requested.

9.10.3 Robot program synthesis

The synthesis of a robot program consists of the major steps: assembly sequence planning, motion planning, and plan checking. Output of the synthesis phase is a program in a robot-level language or a control structure such as, for example, a Petri net, that can be interpreted by a suitable robot controller, see Frommherz and Werling (1989).

The representation of plans is a central point in every plan generation and execution system. In some applications the use of a plan-representation with a non-linear structure, like precedence graphs, has proved to be very powerful and several AI planning systems have been built that pay attention to that item. Non-linear plans contain some degree of freedom concerning the execution sequence. For example, they allow for some parts of the plan to be executed in parallel, or that the order of the single actions can be determined at plan execution time.

However, precedence graphs do not allow representation of mutually exclusive alternatives of subplans, a problem which is very frequent, say, in assembly tasks. As a consequence, in some cases the resulting assembly plan will be under or overrestricted. One single precedence graph will not be sufficient, in general, for representing assembly plans.

An alternative representation is given by 'AND/OR-Graphs' as discussed in Homem de Mello and Sanderson (1986) (see also Section 3.5.5). These graphs theoretically allow one to represent all alternative assembly sequences. However, they are very complex even for simple assembly tasks and therefore cannot be used for practical applications.

A condensed representation of all possible assembly sequences is given by the diamond-diagram. This is an undirected graph, where each node represents a certain assembly state. If one node can be reached from another node by executing one assembly operation, they are linked by an edge, see Whitney *et al.* (1989). The process of generating all sequences is separated from evaluating and selecting good sequences. This ensures that all possible solutions are considered. However, the diagram is rather complex and it depends on the user's skill to select a good assembly sequence.

The approach described in this book uses precedence graphs for the representation of assembly plans.

Assembly sequence planning

There are several methods which have been shown to work when the medium for capturing the designer's intentions concerning the operation sequence have not been captured as part of CAD data. One of these methods is the precedence graphs approach.

In this method, each node of the graph represents one assembly operation and the arcs represent the precedence relations between two given operations. This kind of graph does not, however, allow for complete representation of all types of assembly interaction. A modification, whereby the ordering constraints are obtained from a geometric description of the parts and spatial relationships that exist between the parts when assembly is completed, will be more accurate. It can be seen that a CAD system which provides spatial relationship information, such as ProMod, see Nnaji and Liu (1993) or GEMOS, see Frommherz and Werling (1988), will satisfy the needed requirement for precedence graph generation. With this information, a disassembly of the final product is performed. This disassembly is a step by step process, whereby at each step, the possible departure direction, which will subsequently become the assembly approach direction, is determined using simple heuristics that are sometimes based on contact surfaces involved, as discussed in Frommherz and Werling (1989). Each departure direction is investigated for possible collision with another part. If collision is found, then some precedence constraint will be established accordingly. If no collision is discovered, then the part along with other parts, which can be simultaneously disassembled with it, will be removed. Then each part is again tested to check whether it can be removed. This process continues until all parts are disassembled.

Additional rules are normally applied to the precedence graphs. One rule investigates whether a subassembly or remaining subassembly will be unstable during or after the given disassembly process. If such a situation will occur, additional precedence constraints will be imposed.

The second type of rule ensures that there is no unallowable interaction ('side effects') between assembled parts. For instance, in Figure 9.29, the part B must be assembled with C and D or onto A before C and D. If C or D begin to be assembled onto A, instability will occur and at the same time, unallowable interaction will obtain.

Additional conditions regarding the assembly are imposed by several planning phases that are part of the overall system. All these conditions can lead to the generation of sets of precedence constraints. By observing the precedence constraints during assembly, these conditions are satisfied. Since many assembly tasks can be performed in several different ways, for example, a set of components may be assembled in different orientations on the worktable, the conditions can result in several alternative sets of constraints. These alternative sets are combined in so-called compound sets, that is, sets whose elements are sets of precedence constraints that are alternatively valid. To generate a precedence graph from a given set of compound sets, we have to solve the problem of finding one set of precedence constraints which has the characteristic requirements of a precedence graph. These requirements are as follows:

(1) It should contain at least one element of each compound set to guarantee that each of the mentioned rules is observed.

(2) It must not contain any cycles, since cyclic graphs cannot be executed.

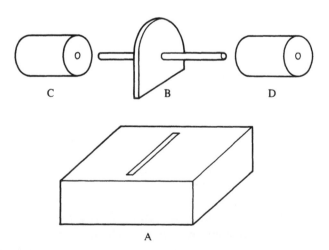

Figure 9.30
Assembly precedence constraint problem.

(3) It should be as minimal as possible to maintain a high degree of freedom concerning the order of execution that is useful for optimization.

The task of finding such a graph is accomplished using a heuristic search method. The reader is referred to Frommherz (1989) and Nnaji and Chu (1990) for a complete treatment of precedence graph generation.

The assembly precedence constraint can also be acquired from the bill of materials and the spatial relationships between parts. The bill of materials is represented in a binary tree with parts or subassemblies as leaves which are sequentially generated when the spatial relationships of parts are assigned (Figure 9.29). In the bill of material tree, if the parts have the spatial relationships with the same ancestor(s) and do not block the assembly of its offspring(s), then these parts have the same precedence. For example, in surface mounting, chips are mounted on a PCB, the chips will always have the spatial relationships with PCB only. Consequently, the assembly precedence is the same for these chips. After all, the precedence of assembly operations is built for determining operations sequence, as described in the next section.

Motion planning

Having dealt with determining the assembly sequence, we now discuss how a robot actually implements particular operations. There are three issues to be considered:

(1) **Part-linking planning** (also called fine motion planning): involves the determination of fine motions which enable the parts to connect or disconnect.

(2) **Grasp planning:** involves finding the appropriate fine motion that will result in a stable grasp or release of a given object.

(3) **Gross motion planning:** involves finding a trajectory or path for the robot arm with its gripper (or end-effector) and payload that may be at the end-effector.

Figure 9.31 shows how these motions can combine into a homogeneous complex motion which results in the accomplishment of a desired task. The following section discusses these critical issues in more detail.

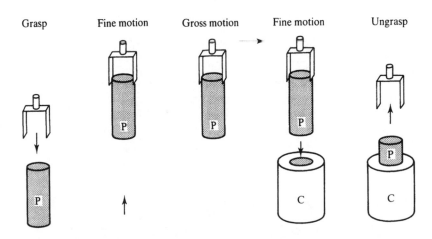

Figure 9.31
Motion classification using pick-and-place operation.

9.10.4 Collision-free motion planning

The most common manipulator transfer movements are of the 'pick-and-place' operation, which consists of first, moving the manipulator from its current configuration state to a grasp configuration state on some object P; second, grasping P; and third moving P to some specified configuration. The pick-and-place synthesis problem is one of deriving the manipulator motions that will carry out a pick-and-place transfer movement, given as input the following data, see also Brady *et al.* (1983).

(1) A geometric description of the manipulator and the objects in the work space.
(2) The current configurations of the manipulator and the objects in the work space.
(3) The desired final configuration of object P.
(4) The (optional) grasp configuration on P.

The major focus is on the geometric aspects of the pick-and-place synthesis problem. For instance, when the grasp configuration is known, the pick-and-place synthesis problem is equivalent to finding collision-free paths for the manipulator and P between the configurations in items 2, 3, and 4 above. On the other hand, when the grasp configuration is unknown, there is the additional task of choosing a configuration such that:

(1) the manipulator's fingers are in contact with P,
(2) the manipulator does not collide with nearby objects,
(3) the configuration is reachable,
(4) the object is stable in the manipulator's hand.

The first three conditions reflect geometric constraints on the manipulator configuration relative to P and to other objects in the work space, see Brady *et al.* (1983) for details. The stability condition reflects issues of grasping beyond the geometric, but when P is small, relative to the manipulator hand, and when part linking effects are ignored, then stability considerations can typically be reduced to geometric heuristics. Using a product modeling system, the geometric issues can easily be captured for planning.

One good formulation of the geometric aspects of pick-and-place in terms of two fundamental spatial planning problems, Findspace and Findpath, which occur in many applications, is discussed in Lozano-Perez (1983). The definition of these basic problems are presented for the case of polyhedral objects.

Let W be a convex polyhedron that bounds the work space and which contains R_Q other, possibly overlapping, convex polyhedra Q_j designated as obstacles. Let P, the object being moved, be the union of k_p convex polyhedra P_i, that is, $P = \bigcup_{i=1}^{k_p} P_i$.

(1) **Findspace.** Find a configuration for P, inside W, such that for all i and for all j; $P_i \cap Q_j = \emptyset$. This is called a safe configuration.

(2) **Findpath.** Find a path for P from configuration C_1 to configuration C_2 such that P is always in W and all configurations of P on the path are safe. This is called safe path.

The pick-and-place with a known grasp configuration can be viewed as a sequence of two Findpath problems. Also, the configurations that are legal candidates for grasping can be derived from solutions to the Findspace problem.

One problem is that the reduction of the pick-and-place problem to these more fundamental geometric problems assumes that the locations of all objects are known to high accuracy and that the path of the manipulator can be controlled to the same precision. In reality, uncertainty always exists in the positions of objects and error in the control of the manipulator. Discussion about the effects of uncertainty can be found in Brady *et al.* (1983).

The position and orientation of a rigid solid can be specified by a single six-dimensional vector, called its configuration, see Lozano-Perez (1983). The six-dimensional space of configurations for a solid, P, is called its **configuration space** and denoted *C-space*$_P$. For example, a configuration may have one coordinate value for each of the x, y, z coordinates of a selected point on the object and one coordinate value for each of the object's Euler angles. In general, an n-dimensional configuration space can be used to model any system for which the position of every point on the object(s) can be specified with n parameters: An example is the configuration of an industrial robot with n joints, where n is typically 5 or 6. In *C-space*$_P$, the set of configurations of P where P overlaps Q, that is, $A \cap Q \neq \varnothing$, will be denoted $CO_P(Q)$, the *C-space*$_P$ obstacle due to Q. Similarly, those configurations of P where P is completely inside Q, i.e. $P \subseteq Q$, will be denoted $CI_P(Q)$, the *C-space*$_P$ interior of Q.

Gross motion planning

Much of robot motion is associated with gross motion where the arm requires little or no sensor interaction. In pick-and-place operations, it will be that aspect of the trajectory between the lifting of the object and the preparation for the placing of the object. Since little or no sensor interaction is involved, most of the task of the robot planner is finding collision free motion(s) among the obstacles in the work environment.

The set of parameters which specify completely the locations of all points of an object is known as the configuration of the object. These parameters provide the constraints which form the **configuration space** (**C-space**) (Figure 9.32). Any point within the C-space defines a configuration of the object. An example of a usual C-space is the joint space of a robot.

When an object exists in the work space of the manipulator, that object constitutes an obstacle. Such an obstacle is a C-space obstacle within the configuration space. Any point within a C-space obstacle represents a configuration of the manipulator which could cause a collision with the object.

In gross motion planning, the problem involves transforming the obstacles into

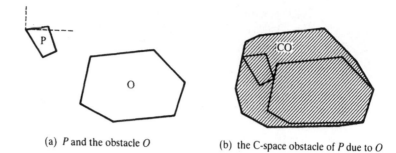

Figure 9.32
The configuration of
space of obstacles.

(a) *P* and the obstacle *O*   (b) the C-space obstacle of *P* due to *O*

C-space obstacles and then finding a free path through the free space which connects the starting point to the destination point.

Several approaches have been used for gross motion planning. The three prominent ones include using a free space search in the configuration space of some of the links of the robot. This approach suffers from poor interface capability with respect to task planning to other geometry-based planners such as grasp and fine motion planning. Work that discusses this approach can be found in Brady *et al.* (1983). Another approach is by using freeways for the hand and payload, a C-space representation for joints 1 and 2 and constraint propagation to communicate between these two representations. This approach solves the problem for four joints of a PUMA-class robot but it suffers from its limitations in the representation of the objects.

The third approach, which is more fruitful in task-level programming, is using the Cartesian free space representative for the first three joints of the robot. In this method, the manipulator links are decoupled, preventing the normal dependence of each link's shape on the configuration of the preceding links. With this, one can compute the free space obstacle for each of the links separately.

The work at the University of Karlsruhe, see Hoermann (1989) for further details, illustrates this Cartesian free space representation very well. In this work, the Cartesian free space is constructed with respect to some reference point in the wrist of the robot. However, the Cartesian location of such a reference point does not unambiguously specify the configuration of the robot. This problem can be solved by using **kinematic states** of the robot and solving for free space obstacles for each of the states of the manipulator. The result is that when searching for a path for the reference point, only those configurations which match the kinematic states are of interest.

The approach enables the approximate computation of boundaries of a free space obstacle for faster performance. The procedure involves making a space grid of cubes of uniform dimensions and using this to represent the Cartesian configuration space. Then, free space obstacles are encapsulated in this cube space and fill the interior with the material of the hull. Any cube which is either partially or completely occupied by the free space obstacles is then marked. The resulting free space outside all the free space obstacles is known as **free configuration space** (FCS), while the free space outside of all the real obstacles is called **free space** (FS).

A configuration of the manipulator is considered safe if the manipulator reference point is placed within an FCS cube, as discussed in Hoermann and Rembold (1988). By using the orthogonal space grid defined by the centers of the FCS cubes, the **cube space skeleton** (CSS) can be obtained, with connecting vertices and edges as CSS edges and vertices respectively. The position of the reference point of the arm is the point where all three rotation axes of the hand intersect.

The range of orientation can be subdivided into discrete orientations. By placing the reference point at a CSS vertex, it is possible to map the hand and payload into the cube space. The result is a pattern for each discrete orientation which corresponds to the cube space occupancy of the end-effector for the given orientation. In the same way, a pattern for each rotational transition between two adjacent orientations can be computed. The computational procedure is followed for the whole set of discrete orientations and transitions.

Using the above, a free configuration for the arm and end-effector can be represented by a CSS vertex, and a free pattern for the end-effector configuration at that vertex. A safe translation between two CSS vertices is one in which two orientation patterns are safe. A state-space graph with nodes representing free configurations and arcs representing free translations and rotations can be constructed. This Karlsruhe approach uses A* algorithm to search the graph with the heuristic function as the sum of the shortest orthogonal distance and the number of rotations between the start and destination orientation.

Grasp planning

One major problem in task-level programming is finding feasible grasp configurations for the robot. Most of the known approaches to find these grasp configurations include two parallel jaws. We will briefly discuss the new approach, which is based on the geometry of the object to be grasped and uses information from the product modeling system related to spatial relationships between the object and the gripper.

The two main known approaches for the two jaw gripper are as follows:

(1) Reduction of the task to a 2-dimensional problem. In this case, grasping feasibility is investigated with two parallel jaws by first evaluating pairs of parallel faces for suitable gripping. Using each pair of faces, the obstacle created by both the object and other items in the work space are projected on an intermediate plane (gripping plane) between the faces. Within the gripping plane, a path is sought for the projection of the jaws and the palm from the initial configuration to the intersection of the two faces.

(2) The second approach involves the computation of an explicit free space specification in the configuration space of the gripper. Using this specification, a free space graph can be constructed for searching for a free path for the reference coordinate frame of the gripper.

In either approach, additional knowledge of the environment will be needed to reduce the possible grasping configurations, and to assume collision-free motion for the gripper.

A grasp configuration is defined as a position and orientation of the gripper with respect to the object. To achieve good grasp, some constraints must be taken into account.

(1) The gripper must not collide with other objects in the environment or with the object itself.

(2) The grasp must be stable, which means there should be no change in spatial relationship between the object and the gripper during motion.

(3) The feasible grasp must accommodate the goal spatial relationship between the gripper and the object configuration and the environment.

(4) All the motions for grasping and releasing the object must be collision-free for the gripper, payload, and for the manipulator.

The range of objects which the grasp planner discussed in this book can handle are the CSG modelable objects. The actual geometric data processed, however, is in boundary representation. This planner assumes a manipulator that can use a jaw-type gripper, fingered gripper or even suction gripper. For each gripper, a geometric model will normally be used in the grasping and associated motion planning.

To assume collision-free motion for the gripper, the gripper (and subsequently the gripper and payload) will be encapsulated in an appropriate polyhedral envelope. A collision-free path will then be planned for this polyhedron.

Potential grasping configurations are sets of object surfaces for which the following conditions must hold:

(1) The surfaces are reachable.

(2) The distance between the surfaces is between the maximum and minimum opening width of the fingers. In other words, the object is within the convex hull formed by the inside surfaces of the finger and the palm.

(3) In the case of the jaw-type gripper, the overlap between the two faces is large enough. This overlap is referred to as the **gripping face**. The plane parallel to these faces and in the middle between them is called the **gripping plane**.

(4) The manipulator's fingers must be in contact with the object.

(5) The object must be stable in the manipulator's hand; that is, the spatial relationships are manipulated with no slippage during motion.

Fine motion planning

The fine motion is often a straight line motion near the task to be performed. For instance, the final deliberate motion to pick up or place down an object by a manipulator is a fine motion. Associated with fine motion is a guarded motion which is planned to establish an initial contact between the object and the end-effector. In cases of grasping, especially in assembly, compliant checks are built into the system. If a system such as ProMod is used to provide planning information, then the assembly parts would have been designed for assembly, thus making compliance easier.

9.11 Applications

We have described assembly of robots, which is one of the areas of application, in some detail. There are however many other application areas which are equally important. These include material transfer and machine loading/unloading, arc welding, spray coating, space, assembly and inspection, underwater, construction and so on.

9.11.1 Material transfer and machine loading

The role of the robot in material movement is so important that it is incorporated in the definition of robot: '... multi-functional manipulator designed to move materials, parts, or tools ...'. The most basic type of this role for the robot is in the pick-and-place operations of moving parts from one location to another. The robot can also be used to load and unload machines.

Material transfer operations can include pick-and-place operations and palletizing and related operations. The work cell can be arranged so that parts are presented by some mechanical feeding devices or conveyor in specified location. The pick-and-place can range from the case where the location (position and orientation) is fixed, to the complex situation where the robot must track a moving part on a conveyor. In each case, the operation involves the robot grasping or connecting to the part using some end-effector and transferring the part to another location.

In palletizing operations, the robot operation includes picking up containers and stacking them onto the pallet in some pattern. The loads could be a single container on the pallet or many containers. There are many variations in this type of work. The robot can load parts or objects into a container and a robot can load the container onto a pallet. Depalletizing involves the reverse process and robots are suited for these kinds of activities, especially since they involve no skill on the part of humans to perform, and are therefore considered quite boring.

Machine loading and unloading refers to the category of material transfer where the robot transfers parts to and/or from a production machine. For instance, in machining, a robot may load the raw material into the machine and remove it from the machine when the machinery operation is completed. In other cases, the robot is responsible for either loading or unloading operations. In this case, some other method is used to perform the operation which the robot is not responsible for. Some examples of where robots have been successfully used for machine loading and/or unloading include: plastic molding, die-casting, forging, stamping press operations, and machinery operations. In each of these cases, robots have been applied to save the human from tasks that are either repetitive, dirty, unpleasant, or hazardous.

9.11.2 Processing operations

The processing operations are those which require the robot to actually perform work on the part. Usually, the end-effector of the robot is not a conventional gripper but the tool that performs the task. Some of these tasks include spot welding, continuous arc welding and spray coating.

Spot welding

Spot welding has over the years been one of the most important applications for robots because of their use for this process in the automobile industry. This is a process where two sheets of metal parts are fused together at localized points by passing a high electric current through the components where they are joined by the weld. The fusion is done at low voltage levels by using two copper (or copper alloy) electrodes to squeeze the parts together at contact points and apply the current to the weld area, see Groover *et al.* (1986). The electric current produces sufficient heat to fuse the two metal parts. The actual weld takes only a fraction of a second. The major time required in welding is actually spent in positioning the two parts which are to be welded and the two electrodes (which are shaped like pincers) to the weld area.

The welding equipment can be bulky and heavy (sometimes over 45 kg). Even with suspension of the apparatus from an overhead hoist, it is still difficult to manipulate. Because of these problems associated with welding equipment, especially in automobile industries, the robot can be fitted with a spot welding gun to perform the task. Essentially, the welding gun is attached as an end-effector to the robot's wrist and the robot is programmed to perform a sequence of welds on the product.

Continuous arc welding

In contrast to the spot welding where points on work parts are locally welded, arc welding is a continuous welding process used to make long welded joints where a solid connection between the metal parts is required. High electric current is supplied to an electrode which has the form of a rod or metal wire. Currents range from 100 A to 300 A at voltages of 10 V to 30 V. Typically the arc between the welding rod or wire and the metal parts to be joined produces sufficiently high temperature to fuse the parts together using a formation of molten metal. Often, the material of the electrode is the source of metal for the molten pool.

Arc welding is hazardous to the human worker. The arc from the weld usually produces ultraviolet radiation, which is harmful to the human eye. This is usually prevented by making the human workers wear a head mask with a dark window. This window filters out the radiation. But because of the mask, the work environment is dark to the worker until the arc is ignited during the welding process. The robot is therefore quite suitable for this type of operation when it can follow a defined welding trajectory. One of the problems a robot may have is that the welding task is often performed in confined places which might be difficult for the robot to reach. There are also problems of variations in the components to be welded, namely dimensional variation, and variations in the shape of the edges and surfaces of the part to be joined. The dimensional variation implies that the arc-welding path to be followed will change slightly from part to part and that the robot cannot compensate for it. Problems with edges or surfaces may cause the gap between the parts to vary so much that the pieces cannot be linked together by the welding operation. These are difficult problems for the robot; but advances have been made to compensate for them with the help of sensors.

For arc welding, contact sensors make use of mechanical probes to touch the sides of the groove ahead of the welding torch and to feed back location data to the controller, which makes corrections to the robotic system.

Spray coating

Metallic products usually require some type of painting finish before delivery to the customer. There are two types of approaches used: immersion and flow coating methods, and spray-coating methods. For both methods robots are used.

Immersion methods involve the dipping of the part by the robot into a tank of liquid paint. After dipping the part is removed from the container and the access paint is allowed to drain back into the container. Closely related to the immersion process is the flow coating where the part is positioned above a container and a stream of paint is made to flow over the part.

Both methods are quite inefficient when a high finish coating is needed, also paint is wasted and the method of the delivery of the part may require considerable mechanization.

Spray coating involves the use of spray guns by the robot to apply the paint or other coating to the object. There are various types, ranging from air spray to airless spray to electrostatic spray. The conventional air spray combines air and the paint to atomize the spray into a high velocity stream with the help of a nozzle to direct the stream at the object to be painted. The airless spray on the other hand uses liquid paint which is caused to flow under high fluid pressure through the nozzle. The sudden drop in pressure as the fluid exits the nozzle causes the liquid to break up into fine droplets. The electrostatic spray method uses either the air or airless approach except that the object to be sprayed is electrically grounded while the spray droplets are given a negative charge, causing the paint to adhere better to the object.

Just as in welding, spray coating operations are hazardous to human health. The fumes and constant mist in the air, noise from the nozzle, fire hazards and potential cancer hazards are some of the dangers of this type of task to the human operator.

Robots can be programmed to accomplish spray painting efficiently and better control can be achieved. This type of process would require continuous path control, hydraulic drive (since electric drive would cause fire hazards), and a good programming system.

There are numerous other processing operations for which robots have been used. These include machining operations such as drilling, grinding, polishing, deburring, wirebrushing; riveting operations; water jet cutting operations; and laser drilling and cutting operations.

9.11.3 Assembly application

It is axiomatic to note that assembly is normally accomplished when two or more parts are fitted together. There are basically two types of assembly operations: parts linking and parts joining operations. Parts are said to be assembled when two or more parts are brought into contact with each other, resulting in the reduction of degrees of freedom for each part yet maintaining stable contact. The parts linking operation includes peg-in-hole (Figure 9.33(a)), hole-in-peg (Figure 9.33(b)), and multiple peg-in-hole (Figure 9.33(c)) operations. On the other hand, parts are said to be joined when two or more parts are linked as described above and additional steps are taken to ensure that they maintain their spatial relationships. These operations can only be done properly by a robot if the gripper has sufficient compliance or if three-dimensional force-torque sensors are used in connection with feedback control to the hand (Figure 9.34).

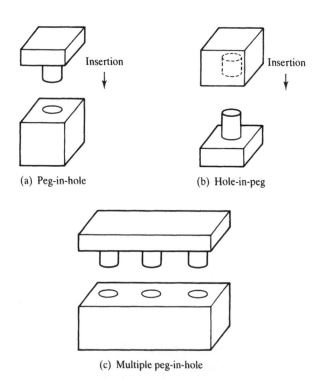

(a) Peg-in-hole   (b) Hole-in-peg

(c) Multiple peg-in-hole

Figure 9.33
Types of part linking operations.

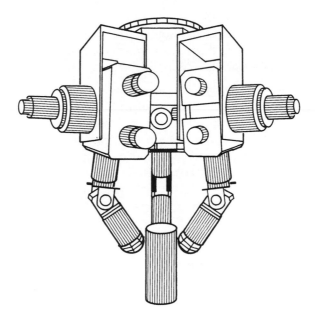

Figure 9.34
The Karlsruhe three-finger hand.

There are various types of parts joining operations:

- Fastening screws: the robot can perform the screw-fastening operation by either driving the screw by advancing and simultaneously rotating the wrist or by using a special end-effector fitted with a power screwdriver. The latter approach is often better because of the difficulty in simultaneously rotating and advancing into the hole. With screwing operations the robot must have torque control to observe the fastening process.

- Retainers: pins are inserted through the joining parts to maintain their spatial relationships or a ring clamps onto one part to establish its relationship with the other part. C-rings and snap rings are examples. Here the robot must be equipped with custom-designed tools.

- Press fits: these are linking operations similar in nature to the peg-in-hole operations where the peg is slightly larger than the hole. By pressing the peg into the hole strong assembly can be obtained. Because much force may be required to press the parts together, and since the robot invariably may not possess the power for such force, the robot may merely have to load the parts into a power press and the actual operation is performed by a plunger.

- Snap fits: these possess the features of both the retainer and the press-fit operations. In this operation, the two parts possess some temporary interference which exists only during assembly operation. During this operation, one (or both) of the parts elastically deforms to accommodate the interference, as described in Groover *et al.* (1986), and catches into the linking elements of the other parts.

- Welding processes: discussed above, are other methods of joining.

- With crimping a portion of one part (usually a sheet metal part) is changed to fasten it to another part. For instance, the squeezing (crimping) of electrical connector onto a wire is a crimping operation. The robot normally requires special tooling or a pressing device which is attached to the end-effector.

Other types of joining operations include the use of adhesives to join the linking parts, and sewing used to join soft parts such as leather.

Future robots can be expected to perform more complex tasks, possessing intelligence and sensors to deal with them. In some instances, they will be expected to possess mobility and have the ability to understand at least a limited type of natural language to make it possible for the average factory worker to program them.

9.12 Problems

(1) In modern robots, what part does the computer play that was formerly taken by human teleoperation?

(2) Suppose that the robot hand coordinate is given by (2, 4, 5). What would be the transformation in spherical coordinate system?

(3) What kind of control system would be used for a spot welding robot?

(4) Describe how you would arrange a sensor fusion architecture for machine operation by a robot.

(5) A football must be sewn together using a robot. Describe the end-effector design you would use for this kind of task.

(6) Why is the concept of frames useless when making reference to another frame?

(7) Derive the 3×3 rotation matrix about the y-axis. How can you augment this to obtain a homogeneous matrix for a rotation about the y-axis, $60°$ and a translation of (2, 3, 4)?

(8) Develop a chain of matrix formulation to relate a robot to a conveyor, to a vision system, and to parts on a conveyor.

(9) Solve the inverse kinematics for the formulation developed above if the vision system must see the part on the conveyor for the robot to pick it up. Note that the conveyor is moving!

(10) Develop the coordinates and kinematics for a SCARA robot. Program this to obtain the robot positions as desired.

(11) Provide a Lagrange dynamic model for a robot which has a rotation at its base and a prismatic motion, and no other joints. This type of robot is called the O-r manipulator.

(12) Suppose that a simple manipulator must travel vertically in the z-axis. Let its hand position be $Z(t)$ at time T. The arm can achieve a cruise velocity after time T and reached a distance $z = P$. Find a fourth order polynominal

expression for the trajectory if its acceleration at the end point is zero, and the position and velocity at the beginning are zero.

(13) Suppose that a robot with six joints must start and finish its task in five seconds, describe a joint interpolation algorithm (in general) that you can use to achieve this. Assume a sampling frequency of 60 Hz.

(14) What are the major disadvantages of language-based programming. How does task-level programming solve these?

(15) Describe how to integrate world modeling and task specification using a product modeler.

(16) Use spatial relationships to constrain a gearbox assembly so that a robot can automatically perform the assembly.

(17) How can you automatically generate the precedence relationship for the gearbox above?

(18) Show an algorithm based in geometry for performing collision detection in robot programming.

(19) How can grasp planning for a jaw type gripper be generated in 3-D space?

10 Material handling

Material handling

CHAPTER CONTENTS

10.1 Preview and Summary		522
10.2 Modern Material Handling Concepts		523
10.3 Controlling the Material Flow		535
10.4 Hardware for Material Handling		558
10.5 Warehousing		565
10.6 Problems		570

10.1 Preview and Summary

The material handling system has a key role in flexible manufacturing; it interconnects the various manufacturing processes, machine tools, robots and storages to a functional unit. The material and parts must be brought at the right time to the right workstation to achieve an optimal utilization of the production resources. With the present trend towards customized products and small batch production, the proliferation of models and parts requires new material distribution strategies and methods. Typically, within the last ten years, the number of parts which are handled and processed by the automotive industry has doubled. A´ particular problem is created by the various options of many product models and leads to a situation where a proportionally large number of part types is used in only a small number of cars. This frequently leads to a situation where a single part has to be brought to a workstation to be ready when a car arrives that has been configured to a customer specification.

A manufacturing process has an input for parts and materials and an output for the finished products. During their time in the factory, materials and parts are processed in sequential and parallel operations and compete for manufacturing resources. The processing times for the various operations may differ considerably. In addition, some product models may have to go through all manufacturing operations, whereas others only go through some of them. This usually requires complex scheduling and the set up of an overhead inventory. In general, a complete utilization of all manufacturing resources is not possible. With flexible manufacturing, an attempt is made to view the factory as a virtual fabrication facility, where the manufacturing resources are configured for making a product under the following constraints: flowline production, no inventory, high flexibility, adherence to delivery dates and maximum utilization of equipment. Such requirements make necessary a comprehensive solution for the entire factory, which may be easy to do for a new plant, but difficult to retrofit to an old plant.

The computer and a flexible material handling system are the principal means of configuration.

The material handling system comprises warehouses, in-process buffers, conveyors, transportation vehicles, part sorters and feeders, and manipulators. For automation, it must be possible to identify materials and parts with the help of the computer and to guide and trace them through the factory. The computer produces the operating parameters for the material flow control and therefore it must be part of the comprehensive planning and control system of the factory. The algorithms producing the control signals will be set up according to the constraints mentioned before. A major contribution to an integrated material flow system can be expected by the use of software, the development of dynamic planning modules and simulation systems. In this chapter we will be discussing the most important components of a modern computer-controlled material handling system. We will convey only a basic understanding of the topic since a more detailed discussion would fill several volumes.

10.2 Modern Material Handling Concepts

10.2.1 A simple material handling problem

A well-designed material handling system is crucial for the operation of a computer-controlled factory. Figure 1.15 shows that, of the total time a workpiece spends in a factory, 85% of it is in waiting. In other words, a considerable amount of capital is tied up in inventory doing nothing. With modern material handling systems an attempt is made to reduce the residence time of a workpiece considerably and to streamline the flow of material through a plant, thereby minimizing inventory in storage and online. This goal can only be reached by a systematic integration of the manufacturing process, manufacturing methods, material handling equipment and the information flow.

Figure 10.1 shows a typical sequence of material handling operations needed for bringing raw material from storage to a processing station and returning the product to storage.

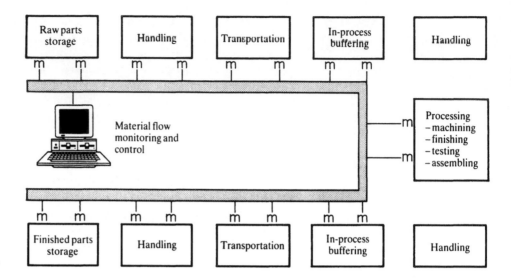

Figure 10.1

Typical sequence of material handling operations in a factory.

Figure 10.2 contains a schema of the software needs for this operation. First, the raw part has to be located in storage, recognized and prepared for picking. This requires the reading of the part number, the location of the part on the shelf, and the determination of how the part is to be gripped. Second, a handling device, like a robot, has to be made ready for handling, and a transportation vehicle has to be driven to a loading position. Now the robot has to recognize the part, pick it up and position it in a container or on a pallet. It is assumed that all manipulators have a vision system. Third, the vehicle has to be steered to the processing area and brought into a docking position. Fourth, a second handling device has to recognize the part, pick it up and load it in a machine tool for processing. Often, direct loading is not done, however, and the part is placed in an in-process buffer. The loading of the part in the machine tool may also require vision. Fifth, when the processing is finished the part is handled again and placed either in an in-process buffer or on a transportation vehicle. Sixth, the vehicle is steered to the finished part storage and brought into a docking position. Seventh, a manipulator is actuated and the part is placed in storage. In reality this chain of handling operations is more complex than described. Numerous sensors have to be actuated and synchronization has to take place. To keep track of the part, a computer system is needed to observe the operation of each component of the material handling system and to verify that the part has properly entered and left each station.

The reader can anticipate all the details that go into the layout and design of the simple system just described. In the factory, most material handling and transportation tasks are very involved and must be well tuned with the manufacturing processes and the information flow. In recent years, the term 'logistics' was coined for planning and operating a complex manufacturing system.

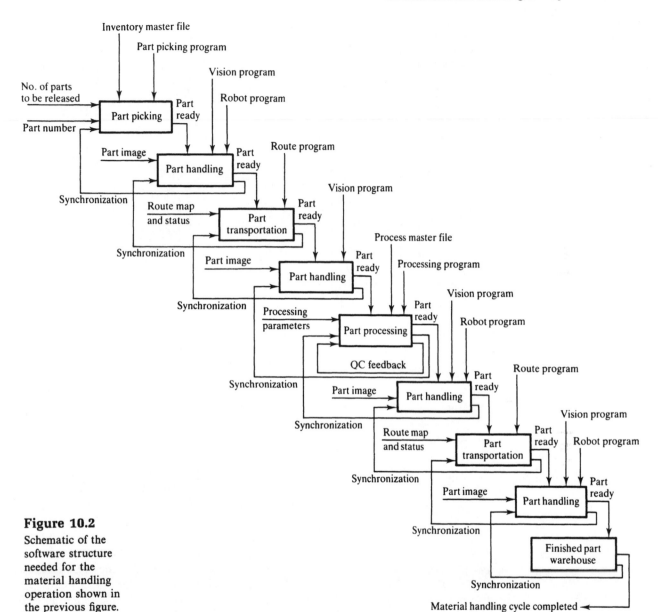

Figure 10.2
Schematic of the software structure needed for the material handling operation shown in the previous figure.

10.2.2 Logistics of a material handling system

The term 'logistics' can be interpreted as supervising the flow of material, energy, information and labor of an enterprise. The task of the logistics is supplying the right amount of resources to the various entities of an enterprise at the right time, required quality and reasonable cost in order to achieve the objectives set by the policy of a manufacturing operation.

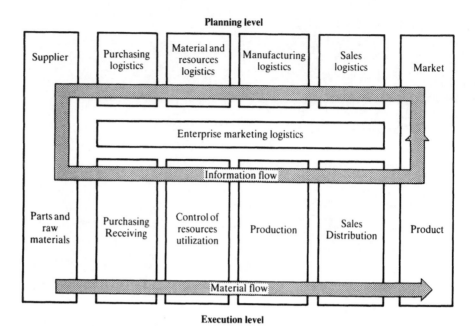

Figure 10.3
Enterprise marketing logistics.

Figure 10.3 shows the information and material flow of an enterprise and the various logistics functions, see Jünemann (1989). The global logistics are set by the marketing strategy of the firm. The customer is the center of the market. He sets the standards for the functions, features, quality and price of the product. The market is influenced by various factors such as availability of resources, purchasing power, competition, demography, saturation etc. In order to operate the individual functions of the enterprise, logistics must be applied to every one of them. In Figure 10.3 there are four logistic subfunctions, namely for sales, manufacturing, material and resource management and purchasing. These sub-functions are highly dependent on one another and are coordinated by global marketing logistics. Of particular interest in the context of this chapter is the material and resource logistics, which is responsible for providing the planning and control mechanism of the flow of material, parts and products through the plant. The planner must be careful that he not only concentrates on the optimization of the material handling system but also on the factory world he tries to control. For an efficient material flow, many factors have to be considered, such as the factory layout, piece rate, lot size, part family, product mix, arrangement of production resources and many others. Several of these factors will be discussed below.

(1) Factory layout: Figure 10.4 shows a conventional factory in the top of the picture. Here the layout is departmentalized according to various machining operations. The material flow follows a very awkward route. Fork lift trucks or manually operated carts may be used for parts distribution. If one would want to optimize the material supply route, for example, with a traveling salesman algorithm, only a limited improvement may be achieved. In the lower

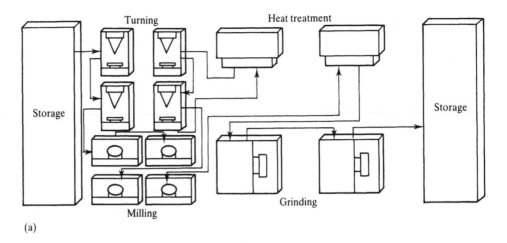

(a)

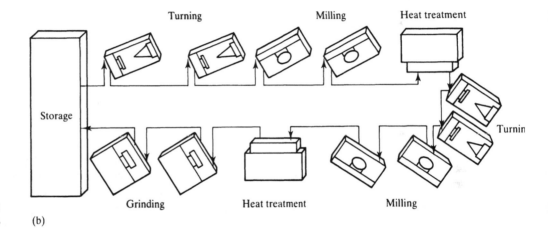

Figure 10.4
Factory layout (a)
batch-type operation
(b) flowline principle.

(b)

part of the picture, the manufacturing machines are rearranged by following
a part family production concept. Here a flow type operation is obtained. The
set up of the machine tools can be interconnected by an automatically guided
material transportion system. In this case, the vehicles can also be equipped
with automatic handling devices for loading and unloading of parts.

(2) Factory storage facilities: One of the major problems in material handling
is the large number of storage facilities and in-process buffers being used in
a factory. Figure 10.5 shows a plant layout having two storages and two
buffers, see Stübing (1988). Incoming parts are kept in storage 1 and brought
to the fabrication facility 1, where for example, a pre-assembly is done. The
assemblies are stored in storage 2 and brought to the fabrication facility 2. In
this facility a second set of parts is brought via storage 1 and buffer 1. The
parts are assembled and placed in buffer 2. From here they are taken and
assembled in the fabrication facility 3. In addition to the problem of having

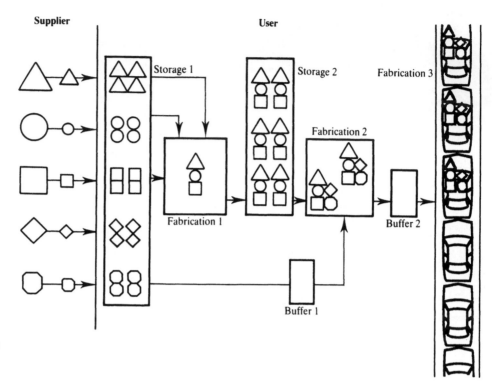

Figure 10.5
A manufacturing
facility with numerous
storages and
in-process buffers.

too many storage operations, the parts are handled too often. Figure 10.6 shows the same factory with only one storage, as discussed in Stübing (1988). Here, several storage facilities are eliminated by an improved material scheduling and handling strategy. A further improvement would be feasible by subcontracting parts of the assemblies or sub-assemblies to a supplier. Of course, such a set up is only possible if a 100% quality can be assured by the supplier.

(3) Transportation chain: Figure 10.7 shows a transportation chain between a supplier and user of parts. The parts are stored and buffered five times until they are finally assembled. In this case, the parts are also handled nine times. This material handling situation can be drastically improved by integrating the supplier with the user, thus, several of the storage facilities and handling operations are eliminated.

The reader can visualize from these three examples that there is a great potential for automation, if it is done from a logistic point of view. The logistics can only be applied successfully if there is an efficient production planning and control system available which furnishes the process plans, master production schedule, material requirement plans and manufacturing resource plans. Planning usually is done over a long, medium and short time period. An additional function is the load balancing of equipment and assembly lines (see Sections 8.14, 8.15 and 8.16).

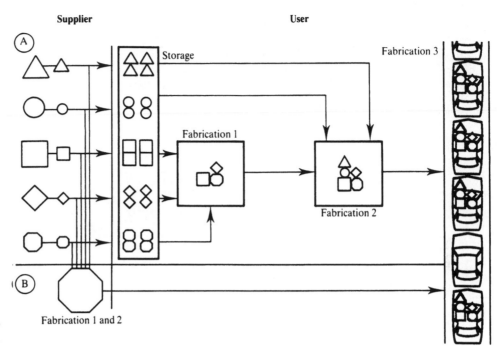

Figure 10.6
Manufacturing facility
with one storage area.

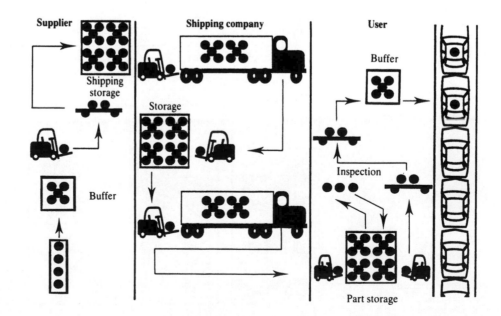

Figure 10.7
An inefficient material
and parts supply chain.

The proper operation of the manufacturing resources, the progress of a manufacturing run and the flow of material through a plant is monitored by a factory data acquisition system. The latter is an important part of an efficient material handling system and must be firmly integrated into the control of the plant. Figure 10.8 shows an integrated logistics system, and production and control system. The system combines the administrative and technical dataflow as discussed

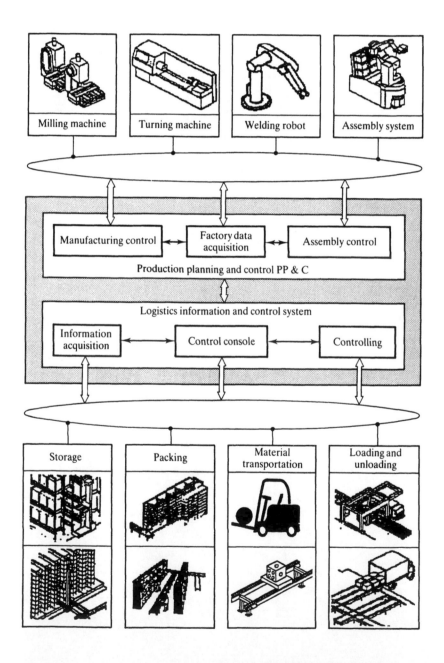

Figure 10.8

An integrated logistics and production and control system.

in Jünemann (1988). Its goals are:

- The interconnection of the flow of information with the flow of material.
- The automatic manufacturing of a variable product family under control of the computer.
- The provision of all parts and information at the exact time when they are needed.
- Online optimization of the production under the prevailing manufacturing orders.

The various manufacturing strategies which may be pursued to reach these goals will be discussed in the following section.

10.2.3 Material handling strategies

The flow of material through a manufacturing system can be planned and controlled according to various strategies. The principal factors affecting the strategy are the manufacturing concept and management objectives.

Figure 10.9 shows the flexibility of the most important manufacturing concepts plotted against the productivity, see also Rupper (1987). Each concept employs a specific inventory and material handling policy. For example with a conventional machine shop operation, common sheet and rod stock may be placed in inventory, and only the expensive components which are part of a product may be purchased with the receipt of an order. This, however, many require very long lead times for ordering components. On the other hand, in mass production using transfer lines, frequently excess inventory is built up to hedge against material shortage to fully utilize the expensive production equipment. In the case of flexible production systems, an attempt is often made to have zero inventory to reduce time in the factory of the material and parts.

The second influential factor includes management objectives, such as maximum utilization of manufacturing resources, maximum output, strict adherence to order

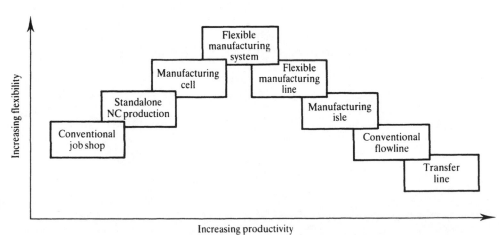

Figure 10.9

Flexibility of a manufacturing system as a function of productivity.

delivery dates, batch sizes, product mix and so on. Objectives are often set with no economic justification and may result in inefficient material control systems and manufacturing runs. The more raw material and parts that are used in a facility, the more complex the material handling problem becomes. From the entry of the raw material to the leaving of the product much time elapses, and the parts have to travel through a maze of machine tool and processing arrangements. Unfortunately, the machining and processing operations are of different length and for this reason it is extremely difficult to obtain a flow line operation, and inventory becomes a part of every fabrication facility. The objectives of almost all practiced material handling strategies is to operate with a minimum amount of inventory (see Section 8.10). The most important planning methods of material handling are scheduling and control with a material requirements planning system (MRP), and scheduling and control using the just-in-time (JIT) philosophy.

Material requirements planning (MRP)

MRP has many variants; the planning and control of the material supply is initiated with the processing of the master production schedule and the bill of materials. All part requirements are determined and the material is pushed through the production system. Every workstation is viewed as an independent agent to be utilized in an optimal manner. Usually, the jobs with the shortest delivery dates are processed first and the rest of the capacity is scheduled for less important orders. When a task of a job has been completed by one workstation, the job is routed to the next station. The completion of a job is reported to the central manufacturing control and a new job can be started.

The principle of a push production system is shown in Figure 10.10. The flow of material and flow of information go in the same direction. The MRP system has major problems. First, it is initiated by the master production schedule which actually contains a forecast production schedule. Thus, the system produces according to a schedule which does not present the actual market requirements. Second, with the push principle a large amount of material is moved through the plant regardless of whether or not succeeding processes are ready for production. Thus, a large amount of in-process inventory may build up. Third, the forecast production schedule requires the purchase and manufacture of material and parts for inventory and a special inventory control method must be used. Fourth, an attempt is usually made to minimize the set up cost for the production equipment

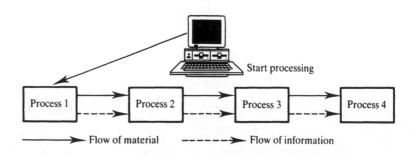

Figure 10.10
The push production principle.

and to run orders with large lot sizes through the plant. As a result the push principle cannot respond quickly to a dynamic market. The major inventory policy used is purchasing of materials and parts in economic order quantities, as shown in Figure 10.11 and discussed in Börnecke (1989). This policy takes into consideration the number of Y parts sold annually, the ordering and machine set up costs S, the unit cost of a part U and the carrying cost of inventory I. With this information the economic order quantity (EQC) can be calculated:

$$EOQ = Q^* = \sqrt{\frac{2YS}{UI}}$$

This formula represents only a simplified ordering strategy. First, the lot size is not always continuous over a year; and second, it is difficult to determine the exact carrying cost of inventory. For this reason, in practice, more detailed lot sizing techniques are used (see Section 8.12).

Just-in-time (JIT) planning

This method can be considered as a philosophy to improve the material flow and distribution in a manufacturing system. There are no ready made software systems which can be implemented and which can handle everyone's problem. A well-designed JIT planning and control system actually comprises all manufacturing functions, including design. For example, the designer can improve the production system by using as many common parts as possible. The overall objectives of JIT are the reduction of inventory carrying costs and the avoidance of production interruption costs. The method addresses the importance of the timing of the delivery of raw material, the quality of inspecting parts and products and the movement and distribution of material.

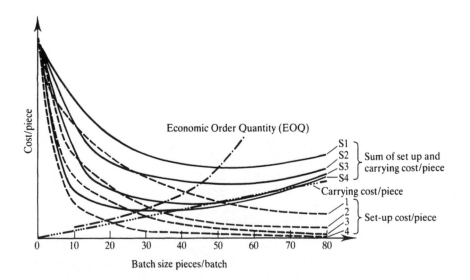

Figure 10.11
Economic order quantity.

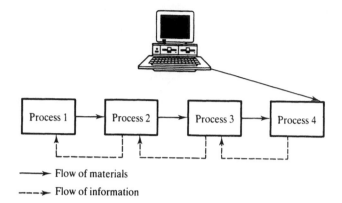

Figure 10.12
The pull production
principle.

The JIT method uses the pull production principle whereby the last processing operation of a job controls the manufacturing line as in Figure 10.12. All preceding processes orientate themselves on the last process. When a succeeding process needs material or parts it requests it from the preceding process. With this method each process only produces the exact amount of parts needed by a succeeding process. For this reason, inventory can be minimized, because at any time there is only as much material in the manufacturing operation as needed. The reduction of inventory presupposes that the production is done with zero defects and no delay occurs between processes. This, of course, is not possible and enough buffering of parts must be done to protect against production stops. JIT is the ideal method for controlling the material flow of flexible manufacturing systems. A strong effort must be placed on the reduction of lot sizes to be responsive to a dynamic market situation. This assumes that material and parts are always available when needed. For this reason, the supplier of parts will be integrated with the main user and he must assume a timely delivery of parts of a 100% quality.

Inventory can effectively be reduced by using automated warehouses and material handling systems. In addition, computer-controlled identification systems will greatly reduce the number of errors prevalent in manual operations. To reduce lot sizes, the set-up times for machine tools and processes must be considerably shortened. There are various ways of reducing the set-up times. First, rapid tool setting methods may be used which employ modular set-up methods. Second, complex and redundant set up procedures are eliminated. Third, more common parts are used. Another method of increasing the material flow through a plant is the implementation of redundant equipment and parallel processes, and the design of the material transportation system to allow bypassing of processing stations and damaged equipment.

When a JIT method is implemented, the critical lead times of all parts must be known so that delivery dates can be met. The lead time depends on the production volume and inventory cost, as shown in Figure 10.13 and discussed in Tompkins and Cramer (1987).

A well-designed JIT material handling system will render the following benefits:

● reduction of the in-factory time of material and parts

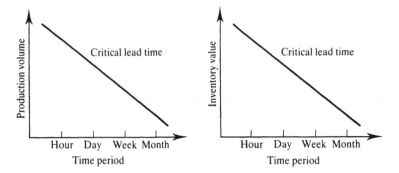

Figure 10.13
Dependence of critical
lead time on inventory
value and production
volume.

- low in-process inventory
- low perpetual inventory
- flow production
- high degree of serviceability

The Kanban system, introduced by the Japanese, means a visual record and designates a control method for manufacturing. With this method, the raw material and parts are requested from the previous process by cards (Kanbans). The use of cards is not mandatory, colored tokens, electronic signals and light may be used as Kanban signals. The Kanban system operates in conjunction with the pull production method and that is how it differs from a conventional move card principle.

10.3 Controlling the Material Flow

10.3.1 Computer architecture for material handling

The controller of the material handling operation is a submodule of the hierarchical computer system of the enterprise. Figure 1.1 shows a simple manufacturing operation in which the material flow is controlled by a small computer of a hierarchical computer system. Usually, in larger manufacturing plants the material handling is very complex and a distributed control system is needed for controlling that function. Material handling operations may be initiated by incoming material from a vendor, the release of an order to the factory to build a product or a manufacturing process reporting the readiness for a new part. Since a workpiece will be progressing through numerous operations, synchronization is necessary between manufacturing processes and the material handling equipment, as shown in Figure 10.12. Thus the material control has to have numerous links with the control computers of the machine tools, robots and sensors. An idealized control computer network for guiding material through a plant is shown in Figure 10.14. This network has two tiers. They are the planning and synchronization tier and

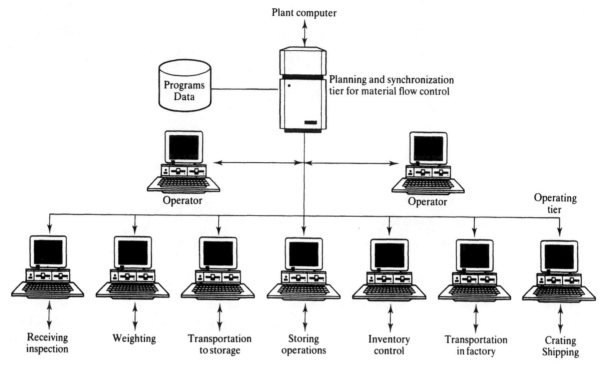

Plant computer

Programs
Data

Planning and synchronization
tier for material flow control

Operator

Operator

Operating
tier

Receiving
inspection

Weighting

Transportation
to storage

Storing
operations

Inventory
control

Transportation
in factory

Crating
Shipping

Figure 10.14

An idealized control
computer network for
a material handling
system.

the operating tier. An order for a material flow operation is sent to the planning
tier from the plant computer or from the factory floor. When an order has been
given, planning and dispatching algorithms will be called up to prepare the material
movement. For this purpose the planning computer must have access to a database
containing:

- a description of the configuration of the material flow system,
- a description of the material flow equipment, controllers and sensors,
- a route planner and route optimizer,
- dispatching and synchronization programs,
- the online status of all material handling equipment and processes,
- error recognition and recovery routines,
- operating programs for the material handling equipment (for down loading),
- statistical programs for evaluating manufacturing data,
- perpetual inventory,
- in-process inventory.

With this information, it must be capable of planning, for every part to be
manufactured, all required material handling operations. This requires that the
system is reconfigurable and open ended. The control components of the operating
tier may have various structures. With a small system, interfaces will be provided

for direct connection with the material handling equipment. With larger installations programmable controllers or process computers are used for every major subfunction.

For flexible manufacturing, it is necessary that the control programs and parameters can be changed online depending on the production run. The control programs are sent from the upper computer to the operating computers. In this case, each operating computer must be capable of controlling and supervising, independently, its material handling equipment. Operating data from the process is concentrated and sent to the planning computer. The operating computers typically contain the following programs and algorithms:

- control algorithms
- data collection programs
- sensor operating programs
- drivers for peripherals
- operator I/O programs
- error recognition and recovery procedures.

For the synchronization of the control computers of the operating tier, status data is communicated online to the planning computer. It accesses algorithms for sequencing the material flow in precise order. Material control is a key function in any manufacturing operation. Failure to supply parts to a workstation may have a detrimental effect on the rest of the production, and an entire plant can come to a standstill if an important component is missing. For this reason, it is essential that the control system has sufficient protection from computer failure. This makes it necessary that every computer can operate its manufacturing equipment independently upon breakdown of an upper tier. In this respect, the synchronization with other computers on the same tier is very critical, for example, to control a flowline operation. This synchronization is usually done by an upper tier. For important operations it may be essential to provide a backup computer on this tier with automatic switchover when a failure occurs.

10.3.2 Input/output devices

For the initiation and execution of a material movement or the dispatching of a job, various computer peripherals, sensors and coding devices are used. The selection of these devices depends on the application and the plant environment. One of the basic problems in material handling is the recognition and tracking of material which is carried through the plant by hand, conveyors and vehicles. There is administrative and physical data to be entered, processed and broadcast. Administrative data describes the customer order and is concerned with identification, time and piece rate information. Physical data is used for recognizing and tracking a workpiece. Of particular interest to flexible manufacturing are automatic data input and processing devices to make the material movement independent of human errors. However, in most cases automatic data entry is not possible and a human interface is required.

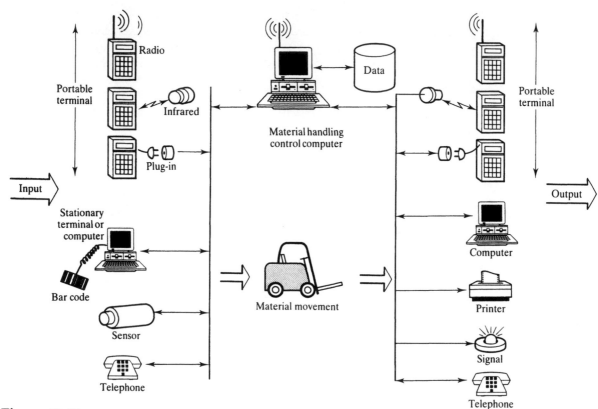

Figure 10.15

Various input/output peripherals for information processing in material handling.

Figure 10.15 shows various computer peripherals used in material handling applications. On the left side of the figure are the data entry peripherals and on the right side the output peripherals. A combination of any input device with any output device is feasible. With a conventional data processing system, an order entry is done by hand via a computer terminal and the order release is issued by a printout. The order entry can also be simplified with the help of move tickets, which may contain punched hole patterns, bar code or magnetic codes, as shown in Figure 10.16. Frequently, these data media, which accompany the workpieces through a plant, are also used for identification at the various workstations. Thus it is possible to keep track of the workpiece and its progress of manufacture. Portable terminals are a valuable addition to material handling. They can be placed in the vicinity of a manufacturing or material handling operation and data is entered by a keyboard or a code reader. For wireless transmission of data, radio or infrared communication is used. The output devices may be identical or similar to the input peripherals. However, in most operations printers are necessary to issue written or coded move tickets. The data terminals employed on the production floor must be rugged, insensitive to dust and abuse, and should be easy to operate. Since these terminals are used in a very busy manufacturing environment the code and data to be entered should be as simple as possible to avoid errors. If more

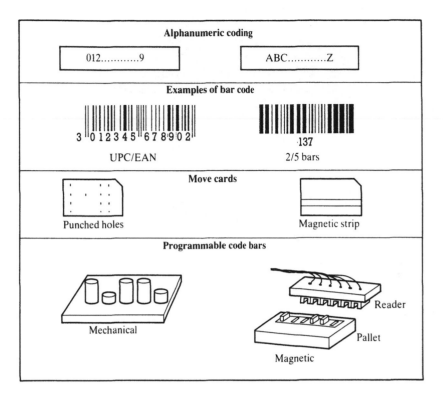

Figure 10.16
Various coding
methods.

complex input/output operations are to be done, operator guidance and error search routines are important features to simplify data processing.

In most manufacturing operations, the tracking of material and dispatching of material carriers is done in conjunction with a factory data acquisition system employed for production control. Thus, the computer has the dual function of production and material control; both of these functions are related to one another and are needed for monitoring and controlling the progression of an order. During the layout of such a data acquisition system it must be made certain that both of these control functions are well coordinated and supplied with their relevant information.

10.3.3 Data media for automatic material flow control

A material flow control system must be capable of identifying and directing the flow of parts and materials. For this purpose it is necessary to provide a part with information about part number, order number, manufacturing operations and sequences, due dates and so on. In a computer-controlled environment, it usually suffices to identify the part at strategic manufacturing states with the help of a reading station, for example, when a part leaves one machining operation and enters another one. By identifying the manufacturing operation, the part and the time, the computer can enter its database and obtain a status on the fabrication of the part. With this information a transportation order can be issued or the adherence to a due date can be verified. A particular problem which the part number faces is that different numbers may be assigned to a part when it progresses from one manufacturing operation to another. In principle, the identification of a part can be done by vision. However, machine vision systems are still very primitive. The recognition of even a simple part may be quite complex in a production environment where an object can assume any position and where it changes its shape during manufacturing. In addition, cleanliness of the surface and lighting conditions may affect the identification. For this reason, the use of machine vision is rather an exception than a rule in current manufacturing operations. There are various methods of identifying parts in a manufacturing system and of providing operational information about them. A part can be identified directly or its carrier can be marked. Typical representatives of these methods will be discussed.

Direct identification of a part

In many cases it is possible to attach the identifier directly to the workpiece. For the identification of the workpiece four basic methods are used:

(1) **Mass flow tracking**. With this method the product is identified by a simple light beam arrangement or a camera system. Often only one parameter, for example height, is measured to distinguish one product from another. This method gives information about the product type, but not about a specific part, as shown in Figure 10.17, see also Rembold *et al.* (1985).

(2) **Alphanumeric characters.** The information is either printed on the product or attached with labels to it, as shown in Figure 10.17. The print has to be held in a well-defined position to enable reading of it and to obtain a high recognition rate. Positioning is often a problem in production and for this reason the method is not used very often. However, it has the advantage that the print

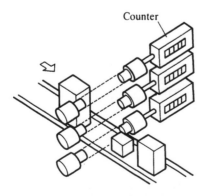
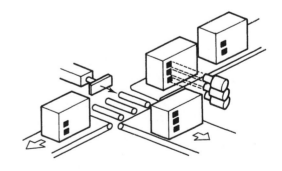

Figure 10.17
Various simple part recognizing methods.

Recognition of the type of parts with the help of light beams

Recognition of the identity of parts with the help of binary coding can be easily identified by production personnel. If information is directly printed on the part or attached by labels, there may be no heat treatment or washing of the part, otherwise the print may come off.

(3) **Bar code.** For this very popular method the product information is converted to a bar code such as the 2/5 or EAN-Code, as shown in Figure 10.16. The labelling of the product can be done directly with the printer, or a special label can be attached to the product by adhesive. The print is machine readable and does not have to be positioned very accurately, particularly when a laser scanner is used, as shown in Figure 10.18. Figure 10.19 and Rembold *et al.*

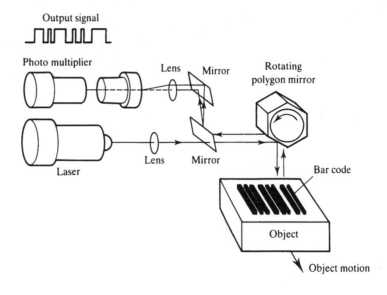

Figure 10.18
Laser scanner for reading bar codes.

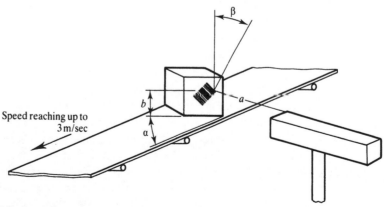

Figure 10.19
Tolerable misalignments of a bar code read by a laser device.

Misalignment

positional	$a = \pm 15\,\text{cm max}$
positional	$b = \pm 15\,\text{cm max}$
angular	$\alpha = \pm 45°\,\text{max}$
angular	$\alpha = \pm 25°\,\text{max}$

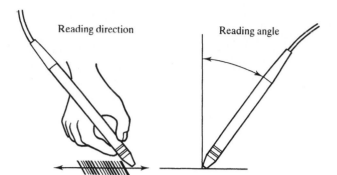

Figure 10.20
Identifying code with
a reading pen.

(1985) show typical positional misalignments which can be tolerated. The code can also be read by a hand held pen, as depicted in Figure 10.20.

(4) **Special coding.** Codes can also be directly applied to the product by cast-in dimples, punched hole patterns or other markings. This method has the advantage that subsequent heat treatment and washing operations can be applied to the part without danger of losing the code.

Identification of part on a carrier

Frequently, it is impractical, or not possible, to apply information directly to the product. In this case the part carrier may hold the code. When the part is placed on the carrier a code is applied to the carrier, uniquely identifying the part. Typical part carriers are pallets, hangers and baskets.

(1) **Move card.** A move card holding all part information is placed on the carrier and accompanies the part through the manufacturing facility, as shown in Figure 10.17. At a processing station the move cards will be placed in a reader to identify the part and report the work performed with it. Also hand held reading pens are used for decoding, as shown in Figure 10.20. Typical move cards contain punched holes, magnetic codes and even solid state memory. With the latter device, part information may be directly retrieved from the move card and also updated.

(2) **Code buffer.** Mechanical and magnetic methods are mainly used for code buffering, as shown in Figure 10.17. In a mechanical buffer the information can be stored by positioning pins or short metal flags which are set in an on or off position. The code is identified by a feeler or light barrier in the reading station. With magnetic code, little magnets are set in an on or off position, and the code is identified in the reading station by induction sensors. For example, a magnet which is close to a reader indicates a one and a magnet which is further away, a zero.

(3) **Programmable identification devices.** These devices contain a small microprocessor together with a memory, transmitter and receiver. They are attached to a pallet or in some cases to the product itself. At a reading station, informa-

tion transfer is done via ultrasound, infrared light or radio waves. The selected communication principle depends on the application and the factory environment. The memory may contain a full set of product information which can be interrogated by a reading station and updated with new manufacturing data. With a large memory it is also possible to carry process and quality control data, which is used for controlling a local manufacturing machine or a quality control test station.

10.3.4 Tracking of objects

Tracking of objects is an important task of a material flow system. For example, an object, or a carrier, has to be brought from a start position to various processing stations and to an end position. During its travel the object may be identified by a camera, light barrier or code attached to it. Various data media for object tracing were discussed in the previous section. For flow control the object is first localized and identified. Thereafter, a decision is made to control its flow, for example, by energizing a switch, as shown in Figure 10.21. When the object enters a workstation it may also be necessary to exchange processing or quality control information, which can readily be done with programmable identification devices.

In principle, two tracking methods are distinguished: direct and indirect tracking, see Figure 10.22 and Rembold *et al.* (1985).

With direct tracking, the workpiece identifier includes optical features, attached code, color or weight. The destination address of the workpiece can be either directly read from the identifier or indirectly obtained by reading the code and interrogating the computer about the routing.

With indirect tracking, the identifier is attached to the workpiece carrier, such as a hanger of a chain conveyor, a pallet, or a roller conveyor. During operation,

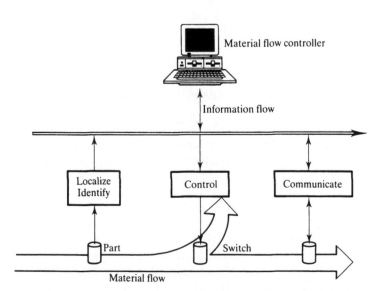

Figure 10.21

Basic information exchange between the material flow controller and the material flow system.

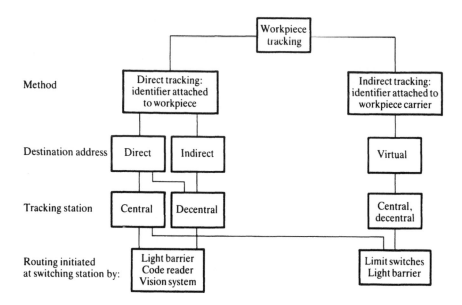

Figure 10.22
Various methods for
tracking workpieces.

the workpiece carrier is identified by a reader and its destination address is obtained
from the computer, which has a model of the transportation system and knows
the exact whereabouts of all parts on the carriers.

Problems may be encountered when the workpiece carrier oscillates; for example,
when a hanger swings back and forth. This may result in several repeated readings
for the same workpiece.

The simplest method of tracking the workpiece carrier uses optical tracking
stations along the conveyor system. Whenever a workpiece passes by a tracking
station, it is recorded. Since the computer knows the number of workpiece carriers
between two adjacent tracking stations, it will keep count of the number of carriers
passing each station. From the count it can determine when a carrier that has
previously left one station has arrived at the next station. In other words, the
computer has in its memory a model of the actual conveyor system. In the computer
the model has a ring buffer which contains the description about the part every
hanger carries; this information is stored in a contiguous manner. The movement
of the conveyor system is triggered by interrupts generated by hangers passing
through a light barrier. This interrupt either rotates a software pointer, or the ring
buffer synchronously with the conveyor system. The model is constantly updated
when a carrier passes a reading station. This method may have a problem with
swinging carriers, which interrupt the light beam several times at a station. The
computer may lose synchronization. With another method, the conveyor system
is divided into contiguous identification sections. There are two tracking guards
(for example, optical sensors) in each section, one at its entrance and one at its
exit. Each section can only accommodate one carrier. When the carrier enters a
section the entry guard records the move to the computer. The computer now
know the whereabouts of the part. Upon departure from this identification section

the computer obtains a signal from the exit guard and immediately records this move. Following the departure notification there must be a signal issued by the entry guard of the next identification section. This method resolves the synchronization problems the method discussed previously may have. However, many more readers are required and the computer is quite busy answering interrupts.

A more complex method is to use an analog system. Here a small analog (physical model) of the actual conveyor system tracks the movement of workpiece carriers. This analog is mechanically connected to the drive of the conveyor system to assure positive synchronization. Whenever a workpiece carrier enters the system, it is identified to the analog by means of a code carrier attached to it. From here the computer tracks this code carrier and is able to follow the workpiece carrier through the plant by tracking the analog.

Example of a component tracking and quality control monitoring system

In Figure 10.23 we show a material flow control system of a paint facility for car bodies using infrared communication (Siemens 1). The two functions performed by this system are product tracking and quality control monitoring. The car bodies are routed through various painting, touchup and, if necessary, repair stations. At the end of the facility the car bodies are stored. When an order is released, a car body is taken from storage and brought to the painting facility. The orders are grouped together into batches of the same color to avoid set up time for lengthy switchover operations. Upon entry of a car body in the paint facility, an infrared responder, having a unique code, is attached to it. An operator enters the order number and order data into the control computer of the paint facility via the data entry station A. The code is associated in the computer with the order number of the car body. Thus, the car body can be uniquely identified at every processing station. At an identification station, infrared light is cast on the responder by a reader located in a defined reading position. The light is reflected back to the reader, in the form of a six digit code, identifying the responder. Thus the computer can determine the location of the order in any section of the facility and check the progress of manufacture. At the end of the painting operation the responder is removed and the car body is stored.

The data entry stations B and C are used for quality control. When a car passes a quality control station it is identified by the infrared reader, and a signal is sent by an infrared relay station to a portable terminal of a quality control operator. Each terminal has a unique address. The operator checks the car body and enters any defects into the portable terminal through a keyboard. This data is signalled back to a relay station and transferred to the computer. The computer evaluates the data and sends a message to the portable terminal that the information has been processed. Thereafter, the portable terminal is released for a new quality control operation.

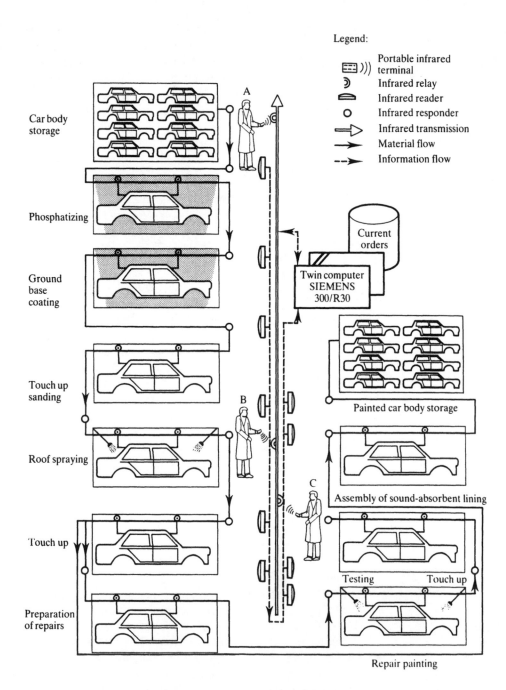

Legend:

⊟)))	Portable infrared terminal
⊃	Infrared relay
⊂⊃	Infrared reader
○	Infrared responder
⇨	Infrared transmission
→	Material flow
--→	Information flow

Car body storage

Phosphatizing

Ground base coating

Touch up sanding

Roof spraying

Touch up

Preparation of repairs

Current orders

Twin computer
SIEMENS
300/R30

Painted car body storage

Assembly of sound-absorbent lining

Testing   Touch up

Repair painting

Figure 10.23
Product tracking in a paint facility.

10.3.5 Control of material transportation vehicles

The flexibility of a material handling system can be greatly enhanced by online dispatch of the vehicles with the aid of a computer-operated control system. If only few vehicles are employed, centralized control is commonly used. For highly automated material distribution with numerous vehicles and a complex factory topology, decentralized control is advantageous. However, in this case the co-operation and synchronization of vehicles must be carefully planned to ensure an uninterrupted operation of vehicles and to avoid collision. The following three advanced concepts of material transportation vehicles are of interest in conjunction with a programmable factory:

- automatically guided vehicles (AGV)
- automatic dispatching of driver operated vehicles
- autonomous vehicles

The most important features of these concepts will be discussed next.

Automatically guided vehicles

An AGV is a battery powered platform having three or four wheels which move along the factory floor on a guided path. The vehicle may be used as a tractor, material carrier or a mobile base for handling devices. The path may be either a passive fluorescent or magnetic line painted on the floor or an active guide wire embedded in the floor.

(1) **Passive guide line.** This method generally uses paint lines on a floor which the vehicle follows. The controller, located in the vehicle, constantly observes the path on which the vehicle travels with sensors. If the vehicle veers off the path, a navigator calculates a corrective motion and sends a feedback control signal to the steering wheels to change the course. One of the basic problems with this control is the determination of the location of the vehicle. Various principles may be applied.

    (a) **Control markers or beacons on the floor or on the ceiling.** Figure 10.23 shows an infrared tracking system for tracking the flow of material. The principle of this control can also be applied to navigate a vehicle. During navigation, the vehicle stays in constant connection with the infrared relays mounted in its vicinity and calculates the vehicle location from the emitted code and its intensity. This, however, requires a complex triangulation procedure to locate the relay. A less accurate control mechanism is to contact the relay station only at strategic points which is called 'logic blocking.' In this case, the exact location of the vehicle between the strategic points has to be calculated by the vehicle controller monitoring the rotation of the wheels. The travel route of the vehicle is divided into contiguous identification sections where infrared relays serve as tracking markers to register the entry and departure of a vehicle in or from an identification section (logic block). Infrared communication allows data transmission speeds of 20 000 baud.

(b) **Tracking can also be done by cameras, radio transmitters, markers or laser beams** identifying logic blocks with the help of strategic points. With the latter it is even possible to guide the vehicle by a moving laser beam.

The passive control techniques have the advantage that they can be easily installed and extended. However, they require that manufacturing equipment is laid out in a well-structured arrangement and that the navigation system is free of noise.

With an advanced control system, the movement of all vehicles is monitored by the computer, which also observes the traffic flow and controls the right of way of vehicles. For this purpose all logic blocks are constantly updated, showing whether they are occupied or empty. The right of way of a vehicle is determined from traffic rules or the priority of an order. The system must also be capable of detecting any blockage of material flow and performing corrective measures. For operator supervision the traffic scenario can be displayed on a graphic terminal.

(2) **Active guide wire.** In this case, the wire embedded in the floor emits a radio-frequency signal. The signal may also be used for communication between the AGV and the computer. Each travel route of a factory broadcasts a defined frequency. The receiver of the automatically guided vehicle can be adjusted to these frequencies. The vehicle follows the guide wire, which is sending the frequency adjusted by the controller. If a vehicle has to follow another travel route, the controller must be adjusted to the corresponding frequency. At intersections and switches, the direction that the vehicle has to follow is determined by the guide wire which is sending this frequency.

Figure 10.24 shows the navigation hardware needed by an automatically guided vehicle, see References for information about 'Demag.' It consists of a system director which dispatches the vehicles and controls the navigation. The system director obtains its orders from a central plant computer. Communication with the vehicle is via the guide wire embedded in the floor. The system director is connected with all vehicles at all times. Each vehicle has its own guide frequency and follows the guide wire with the help of a sensor. A higher communication frequency is used for data transfer between the system director and the on-board computers. Thus, the system director is informed at all times about the location and loading status of all vehicles. The location of the vehicles can be displayed on a terminal. Material located on the vehicles is identified by reading bar code labels, and the information is transferred via the data channel to the system director. A vehicle traveling through a plant is identified at strategic points with the help of a responder in the floor and a receiver in the vehicle. At an identification point the vehicle get instructions to follow a given route. The essential functions of the system director are:

(a) selection of vehicles and administration of empty vehicles,

(b) control of the dispatching sequence,

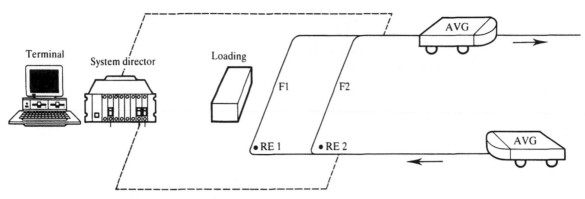

(a) Layout of an automatically guided vehicle system

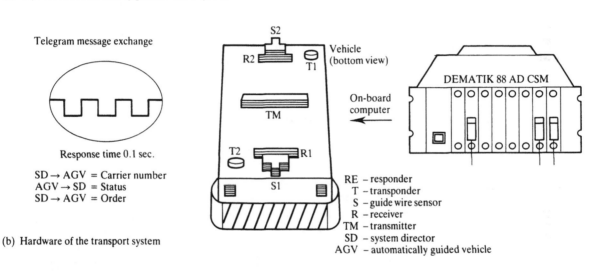

Telegram message exchange

Response time 0.1 sec.

SD → AGV = Carrier number
AGV → SD = Status
SD → AGV = Order

(b) Hardware of the transport system

Vehicle
(bottom view)

On-board
computer

DEMATIK 88 AD CSM

RE – responder
T – transponder
S – guide wire sensor
R – receiver
TM – transmitter
SD – system director
AGV – automatically guided vehicle

Figure 10.24
Navigation hardware
for automatically
guided vehicles
(courtesy of
Mannesmann Demag
Fördertechnik, Wetter,
Germany).

(c) tracking of the carriers,

(d) control of the right of way.

The on-board computer is responsible for the control of the steering, vehicle
drive system and the loading and unloading mechanism. In case a carrier
leaves the guide wire track for loading or in a curve, the on-board computer
takes over the temporary navigation. An additional function of the on-board
computer is the communication with the system director. This active control
technique is highly reliable and is being used in many factory installations.
Its basic disadvantage is that a change or extension of a travel route requires
expensive installation of new guide wires. It also has the problem with a low
communication rate of about 300 baud. This is partially due to the fact that
extensive safety protocols have to be provided.

Automatic dispatch of driver operated vehicles

Manufacturing operations requiring a highly maneuverable material transportation system often employ driver operated vehicles, such as fork lift trucks. These vehicles can be fully integrated into flexible manufacturing with the help of portable terminals, which are connected by radio or infrared links to the computer. The two methods of controlling the dispatch operation are use of voice operated receiver/ transmitters or portable computer terminals. The communication channels and methods will be similar to those described for AGVs in the preceding section. The general dispatching strategy will be done on the control computer which issues a list of materials to be brought from one station to another, and which may also contain optimization algorithms to minimize the distance to be traveled by a vehicle. There are also algorithms available to dispatch and control a multi-vehicle operation to fully utilize the material transportation equipment. The central computer may be in contact with the vehicle at all times, or periodically, depending on the dispatching strategy employed. It is also possible to track the vehicle and monitor its operation.

Autonomous material movement vehicle

Recently, an effort has been started to develop autonomous vehicles for material movement, see Rembold (1991). With such vehicles, it is possible to build manufacturing plants of great flexibility. Any combination of machine tools may be interconnected according to a virtual manufacturing concept. An increase in flexibility can be obtained by the use of knowledge-based planning, execution and supervision, which are sensor supported.

An autonomous system is capable of planning and executing a task according to a given assignment. When a work plan is available, its execution can start. A complex sensor system is activated which leads and supervises the travel of the vehicle. It is necessary to recognize and solve conflicts with the help of a knowledge processing module. The basic components of an autonomous intelligent vehicle are shown in Figure 10.25. We will discuss the sensor and control system for an autonomous vehicle. The vehicle must have the following capabilities:

- autonomous planning and preparation of a course of action according to a given task, such as transport of a workpiece from storage to a machine tool;
- independent planning execution and supervision of the course of actions;
- understanding of the environment with the help of sensors and interpretation of the results: the vehicle must carry a map of the plant floor it navigates on and must recognize all objects it encounters during the travel;
- independent reaction to unforeseen events: if an unforeseen object enters its path the vehicle must be able to recognize and bypass it;
- passive and active learning: the vehicle must be capable of learning from the work it performs to improve its operating capabilities.

```
┌─────────────────────────────────────┐
│   1. Mechanics and drive system      │
│                                      │
│   2. Sensor system                   │
│        – internal sensors            │
│        – external sensors            │
│                                      │
│   3. Planner and navigator           │
│        – planner                     │
│        – navigator                   │
│        – expert system               │
│        – knowledge base              │
│        – meta knowledge              │
│                                      │
│   4. World model                     │
│        – static component            │
│        – dynamic component           │
│                                      │
│   5. Knowledge acquisition and       │
│      world modeling                  │
│                                      │
│   6. The computer system             │
└─────────────────────────────────────┘
```

Figure 10.25
Components of an
autonomous vehicle.

The sensor system.
A sensor system of an autonomous vehicle for material movement consists of various sensors which are interconnected by hierarchical control. They furnish the planning and supervision modules with information about the status of the factory world. For each of the three major tasks of the vehicle, the navigation, docking and loading, a sensor system must be provided.

The sensoric has the following assignments:

- supervising the loading and unloading of workpieces
- supervising the vehicle navigation
- controlling the docking and undocking maneuvers

For the navigation of an autonomous vehicle, a multisensor system is necessary. A distinction is made between vehicle-based internal and external sensors and world-based sensors. Internal sensors are increment decoders in the drive wheels, the compass, inclinometer and so on. External sensors are TV cameras, range finders, approach and contact sensors. World-based sensor systems use sonic, infrared, laser or radiotelemetry principles. For navigation various principles may be used:

- dead reckoning
- navigation under the direction of a compass
- the use of world-based sensor systems
- driving under the guidance of floor markers and vehicle-based external sensors
- navigation by vehicle-based external sensors, such as a camera or a laser range finder
- the use of a combination of various navigation principles

A vehicle driving in an obstacle-free environment may use any of the first four principles. In case obstacles are entering or leaving the vehicle's path or when the vehicle veers off the course, vehicle-based external sensors must be used. For example, the vehicle must constantly monitor the path with a camera system. Most advanced autonomous vehicles use a combination of navigation systems to react to unforeseeable events. Recognition is by extracting, from the picture of the scenario, specific features and comparing these with a sensor hypothesis obtained from a world model. For scenes with many complex objects, the support of an expert system is needed for the picture evaluation.

A docking maneuver can be supported by optical, magnetic or mechanical proximity sensors. For coarse positioning a vision system may be used; and for fine positioning a mechanical feeler. Recognition of randomly oriented parts for loading and unloading can be done by vision systems, sonic sensors, touch sensors, or a combination of them. However, in most cases an attempt will be made to align the parts with a known position and orientation, thereby simplifying the sensor requirement.

A multisensor system may be designed according to the following concepts:

- a combination of various types of sensors
- the use of the same type of sensors at various locations
- the use of one sensor for the acquisition of various parameters
- the use of one sensor for interpreting moving scenes

The first two sensor principles are task dependent and must be carefully designed for the specific application. The last two sensor principles are difficult to implement. In all cases the capacity of the sensor channel and the picture evaluation algorithms must be carefully designed. For example, if a sonic sensor is used in conjunction with a laser scanner, the signals of both sensors have to be combined to one channel parameter.

Global planning and supervision

When parts are to be moved through a factory a transportation schedule is issued, as shown in Figure 10.26. A global action planner investigates the schedule and starts planning at the strategic level. The determination is made, how many parts have to be retrieved, where the parts are located and how they should be handled. Thereafter, the order of moving the parts is placed in a waiting queue, and the order will be processed when competing jobs have been completed. The strategy of processing an order in the queue may depend on its deadline, importance, processing time or other factors. Moving a part involves docking and undocking, navigation, and loading and unloading. For each of these functions a separate control module is provided, as shown in Figure 11.26. The modules have the following functions:

- **Navigation module.** It performs the route planning, fine planning, execution and supervision of the navigation.
- **Docking module.** For the docking maneuver the rough and the fine planning is performed and the operations are supervised.

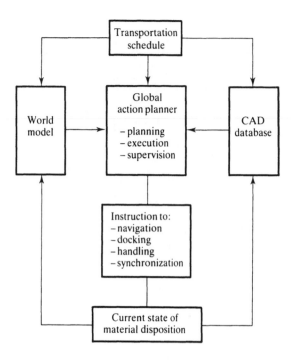

Figure 10.26
Global action
planning for an
autonomous vehicle.

- **Loading module.** Loading and unloading involves rough and fine planning, execution and supervision.

Normally, the three subtasks are executed sequentially. However, often the tasks have to be solved in quasi-parallel operations and must be prepared for parallel execution by the planner. The planner is also the executive who initiates the processing of the tasks and who controls asynchronous events. The planner may employ algorithms for path time optimization under the following boundary conditions:

- the precedance requirements of the task have to be obeyed
- due dates and priority assignments must be met
- the load capacity of the vehicle must not be exceeded

The planning module performs the planning under global considerations with the help of a very abstract model. At periodic intervals a data acquisition system reports the actual state of the world, for example the traffic density, jams, road blocks, station times, queues and so on. With the help of online information, the planner selects the proper vehicles and routes in case there are several alternatives. The execution module reports to the scheduler periodically, on the operation status of a task or possible problems which have arisen. If there is a problem, it may be necessary to do the task with another vehicle or to abort the whole endeavor and to remove the task from the queue. The supervision module keeps vigilance on the proper execution of all functions of a task. The three submodules for

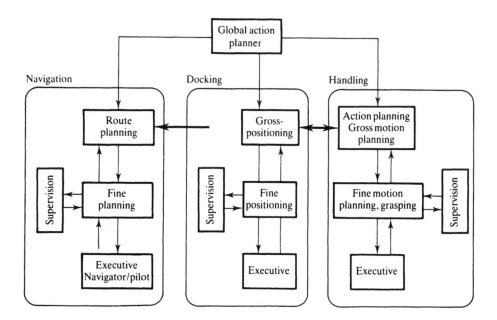

Figure 10.27
Planning and control
modules of a material
transportation vehicle.

navigation, docking and handling are responsible for the more detailed planning and controlling as shown in Figure 10.27. Each module has its specific planning, execution and supervision functions. The planning consists of coarse and fine planning. For example, for docking the vehicle it is brought to a rough position and then moved slowly to a fine position. The output of these modules are instructions for equipment controllers and sensor systems.

If a task is not solvable, the system has to be given information about the cause of the problem; that is, if a sensor has failed and a corrective action has to be initiated. For this purpose, it is required to store knowledge for the recognition and solution of problems. Thus, the system must have an expert system which is able to respond to any unusual situation.

Path planning and navigation system

With the help of the path planning and navigation module, the vehicle is capable of determining its path along the plant floor. It will attempt to plan an optimal route between the start and goal positions, and it will try to stay on the path, avoiding obstacles and possible collisions. The navigation system has to be tailored to the application. Often, a simple task in a material storage system can be solved by dead reckoning the vehicle from the start to the goal position. For this purpose, vehicle located external sensors may be used. However, in case unexpected obstacles are on the path, the navigation becomes very complex. In such a case, the world model has to be monitored constantly and the vehicle must be capable of recognizing dangerous situations and performing corrective actions. For the solution of conflicts, knowledge about new strategies is needed. The planner must have access to a map to know the action space of the vehicle and to interpret the map. For the interpretation of online sensor data, it is necessary to store in a world model sensor hypotheses for every situation.

Path planning is done according to a hierarchical scheme, whereby different tasks are solved at different levels, see Figure 10.28.

(1) At the planning level, the planner draws up a plausible and collision-free path by using expert knowledge. However, at first the vehicle has to determine its start position. This can be done by starting at a known position or by sensors. The planner may try to search for an optimal plan in case there are alternative solutions. Optimization criteria may be the shortest path or the route with the least number of obstacles to be encountered.

(2) On the navigation level there is a local planning module which has knowledge about the path, its objects and possible obstacles. Fine planning of the navigation is done on the basis of the plan which was drafted on the previous level, thereby all local situations must be considered. It is important to recognize unknown obstacles by the use of sensors and to report any abnormal situation to the planning system. All problems must be considered for further planning.

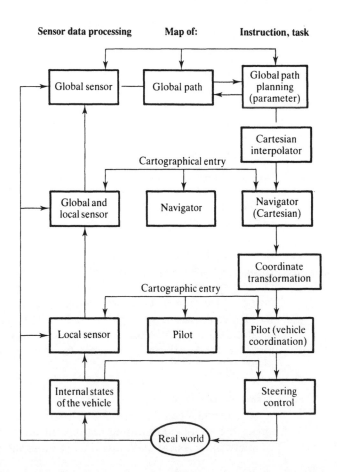

Figure 10.28
A hierarchical
navigation system.

(3) On the pilot level are performed the elementary control functions of the vehicle. For trajectory planning, a map of the world is used showing the path to be traveled in detail. For simple operations only a two-dimensional map of the routes, branches and crossings is needed. With the help of the map, trajectory segments, like straight lines or curves, are calculated. The passage of doorways or underpaths must be investigated with a three-dimensional map. The navigator corrects the map with the aid of realtime sensor data. A wrong way or obstacles entering or leaving the path must be recognized by the sensors. The information is processed quantitatively and qualitatively to plan the necessary corrective actions. Possible alternative moves are checked with the planner at the higher level. If the action is approved, detailed move instructions are sent to the pilot level for execution.

Planning and supervision of the docking

As soon as the autonomous vehicle has reached its goal, it must lock itself into a stable work position to be able to retrieve or store parts. In principle there are two docking maneuvers possible:

(1) The vehicle moves into a rough position and tries to perform fine positioning with the aid of sensors. Thereby, the vehicle aligns its own coordinate system with the reference coordinate system of the workbench.

(2) The robot locks itself into a rough position and measures, by activating sensors, its exact location with regard to the reference position of the work area. A coordinate transformation matrix is computed and with its help the position of a workpiece in the work area can be described in the coordinate system of the mobile robot. Now the robot is able to perform the given task.

Planning and supervision of loading and unloading

For loading and unloading, the vehicle must be placed in a fixed position. The parts may be handled directly by a robot, special manipulator, or they may be located in boxes or pallets. If the parts or their container are located and oriented randomly, a very complex sensor system must be provided to determine the gripping position for the handling device. If the parts are in a fixed and known position, simple sensors may suffice, for example, to determine the presence of a workpiece. In both situations the actions of the handling device are planned and controlled in their proper sequence. In case of a robotic device, the program to handle the part is entered into the robot controller and the sensors are actuated. When the sensors have recognized the part, a sensor hypothesis is consulted to identify the location and orientation of the workpiece. It may be necessary to constantly monitor the handling to observe for part slippage and possible collision with obstacles.

The world model

A world model can be considered in a broad sense as the computer's knowledge about the factory containing equipment and workpieces which are described by geometrical, physical and other data. A material handling vehicle needs information from the world model for the navigation, docking and part handling. Thus, three different types of world model are needed in which the following information is stored:

- geometry of the objects and their geometry relations;
- functions, operation parameters and capability of the objects (for example, of the vehicle, robots and auxiliary equipment);
- interpolation routines for the vehicle motions and the robot trajectories;
- motion parameters for all moving parts (speed, acceleration and deceleration);
- sensor hypotheses for planning and supervision of the action sequences;
- preferred routes and standard motions;
- behavior rules for the avoidance and removal of errors and collisions.

Since the world may change during the execution of a task, part of the world model has to be conceived as a dynamic world model. For this reason, the world is constantly monitored by sensors. For example, if a part falls on the floor, the event must be reported by updating the world model. Thus, information for the following tasks have to be supplied to the dynamic world model:

- evaluation results of the online sensor system,
- reports about changes in the world,
- updates about the world model.

Knowledge acquisition and world modeling

To plan, execute and supervise its actions, an autonomous vehicle must be able to make decisions on several levels. Thus, it must be equipped with a hierarchy of knowledge-based modules and submodules. A special problem is presented by the acquisition and storage of expert knowledge. For some functions, knowledge may be stored in independent databanks, in other cases knowledge has to be shared via a blackboard architecture.

An autonomous vehicle for material transportation will have the following knowledge-based modules:

- the global action planner and supervisor on the level of the factory or manufacturing cell,
- the route planner and supervisor for the navigation,
- the action planner and supervisor for the parts handling,
- the sensor data processor for the navigation,
- the sensor data processor for the assembly,
- the error recognition and recovery module for the navigation,
- the error recognition and recovery module for the parts handling.

The computer system

With computers currently available, two different computer systems are necessary for the planning and control of the autonomous vehicle. Planning is offline on a powerful scientific computer and the execution and supervision is online by using a vehicle-based realtime computer. The main computer contains the global planner and various expert systems for the planning and supervision of the subfunctions of the vehicle. For example, the main computer gives the action program to the vehicle computer for execution. The vehicle computer interprets the instructions stepwise and executes them. In addition, expert knowledge is given to the vehicle computer to process sensor information and to solve conflicts which may arise during the navigation, docking or parts handling. Since the size of the vehicle computer is restricted, it can only solve simple problems. In serious situations the main computer will be notified and it in turn will try to find a solution. It will also prepare and issue a situation report for the operator.

10.4 Hardware for Material Handling

The equipment used for material handling is very numerous and will not be covered in detail in this book, the concern of which is the use of computers in manufacturing. An attempt will be made in this section to discuss only typical representatives of material handling equipment to give the reader a glimpse of the most important classes of conveyors and transportation vehicles. Part feeders and robots are very important components for interconnecting machine tools and manufacturing processes. Chapters 5 and 6 discussed flexible manufacturing equipment in which the materials handling systems are important tools for configuring production tools to produce products. In order to use material handling equipment in conjunction with the computer, sensors and controllers must be provided which can be electrically connected to communication channels.

10.4.1 Ground conveyors

The ground conveyor is a basic material transportation device found in almost every discrete manufacturing plant. It interconnects various workstations to obtain a flowline principle. Typical ground conveyors are shown in Figure 10.29; they are the belt conveyor, roller conveyor, chain link conveyor and numerous others. A part can be placed directly on the conveyor or on a pallet. In the first case, the part to be manufactured has to be taken from the conveyor and placed back on it upon completion. This requires that the part is identified and handled by man or a sensor-equipped manipulator. For identification, the part can be directly recognized by a camera or with the help of a coding label attached to it. In the second case the part may be fixed to the pallet, which can also serve as a machining table, and a pallet handler takes the part from the conveyor and inserts it in a machine tool for processing. After completion, the part is removed and carried on the same pallet to the next operation. Again, the pallet serves as a machining table.

Conveyor systems are shown in Figure 10.29; they can be interconnected by turntables and curve sections. Figure 10.30 shows various conveyor system configurations as they may be used for assembly stations.

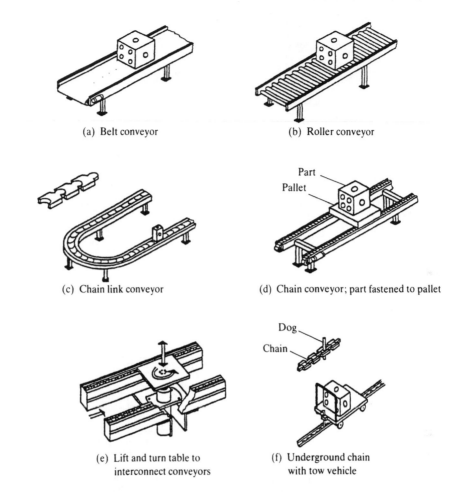

(a) Belt conveyor

(b) Roller conveyor

(c) Chain link conveyor

(d) Chain conveyor; part fastened to pallet

(e) Lift and turn table to interconnect conveyors

(f) Underground chain with tow vehicle

Figure 10.29
Various ground conveyor systems and tow vehicle.

One of the basic problems of a ground conveyor system is the synchronization of the product flow. If, for example, parts have to be brought together to an assembly station it is difficult to track the parts and to synchronize their arrival. Cameras, bar code readers or light barriers are used for synchronization, see Section 10.3.3.

10.4.2 Overhead material transportation systems

The next most important material transportation devices are the overhead transportation systems. There are basically three principles used, as shown in Figure 10.31.

Overhead conveyors with fixed hangers are low cost systems and, for that reason, are very commonly used by industry. A major problem with this equipment is that the hangers are fixed and cannot be routed to other conveyor tracks. Synchronization of material delivery is possible, however, irregular operation of the

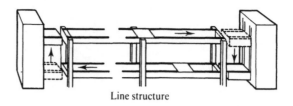

Line structure

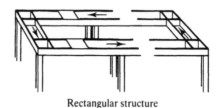

Rectangular structure

Figure 10.30
Various conveyor
configurations
(courtesy of
Mannesmann Demag
Fördertechnik, 6050
Offenburg 16,
Germany).

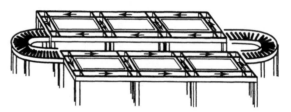

Structure containing rectangular and curved sections

conveyor may lead to problems for example, by swinging hangers, if only simple light barriers are used. The chain is moved by a central drive system.

The power and free conveyor is a more versatile device. Hangers can be engaged or disengaged at special switching stations to interconnect other conveyor tracks. A typical layout of a power and free conveyor system is shown in Figure 10.32. In a loading and unloading area, material is placed on or removed from the hangers. The hangers bring the material to the work area. The destination of the parts can be either read from code attached to the part or from code attached to a hanger. From the work area the parts are returned to the loading and unloading station. There are storage facilities provided for loaded hangers and empty hangers. The chain for pulling the hanger is also operated by a central drive system.

With the monorail system the trollies are operated by their own motors. The power is brought to the trolley via sliding contacts. The individual drives make the monorail system very versatile and adaptable to production variations. The routing of trollies is done in a similar manner to the routing of hangers with the power and free system. The trollies can be equipped with hoists, thus giving the system an additional degree of freedom in the vertical direction, as shown in Figure 10.33, see also the reference to 'Demag'.

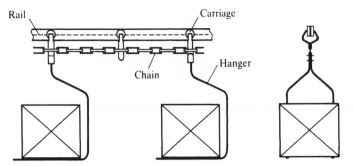

(a) Overhead chain conveyor with fixed hangers

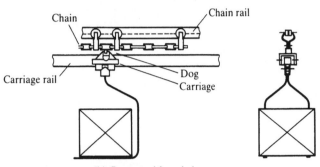

(b) Power and free chain conveyor

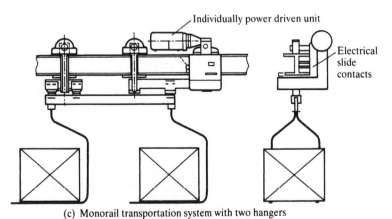

Figure 10.31
Various overhead
material
transportation systems.

(c) Monorail transportation system with two hangers

10.4.3 Material transportation vehicles

For flexible manufacturing systems, there are numerous designs of material transportation vehicles being used. The product, the design of the plant, the piece rate and numerous other factors determine the selection of the material transportation vehicles.

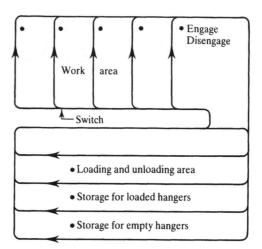

Figure 10.32

Layout of power and free tracks.

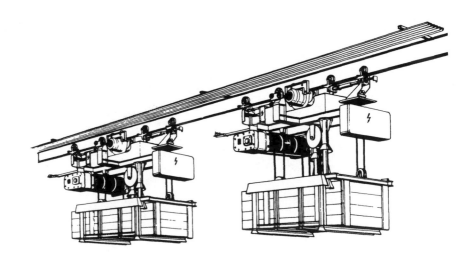

Figure 10.33

A monorail material transportation system (courtesy of Mannesmann Demag Fördertechnik, 5802 Wetter 1, Germany).

- **Chain pulled vehicles.** With the simplest method, the vehicle is pulled along the production facility by an under floor chain, as shown in Figure 10.29(f). In this case the vehicle follows the chain and is engaged or disengaged at a workstation or loading station. This transportation method offers only a limited flexibility.

- **Automatically guided vehicles.** More flexibility can be obtained with automatically guided vehicles which follow a guide line. With advanced systems, the vehicle can leave the guide line at a loading station or workstation and can assume a work position by docking in a defined location. The vehicle may even serve as a workstation where assembly operations are performed, see Figure 10.34 and Eisenmann. There are various types of superstructures available, such as a lift platform, horizontal telescope loading fork, roller conveyor etc, see Figure 10.35. The superstructures are often custom designed. Figure 10.36 depicts a work cell where an AGV has a specially designed transport mechanism for carrying

Figure 10.34
An AGV carrying a chassis; the platform may be used as a workstation (courtesy of Eisenmann Fördertechnik KG, 7038 Holzgerlingen, Germany).

workpieces clamped to pallets. The track patterns of an AGV are similar to the one shown in Figure 10.32.

- **Driver operated vehicles.** These transportation vehicles are highly flexible, however, they do have the disadvantage of needing a driver. Most of these vehicles are fork lift trucks, but there are also many special designs available which are adapted to the work environment.

- **Autonomous mobile vehicles.** With these devices an attempt is being made to completely automate the transportation task, and often the entire work assignment. Because an autonomous vehicle has to have an elaborate sensor system and knowledge about the work environment, it needs a very complex sensor and control system. Until now, such systems have not been available. Figure 10.37 shows the Karlsruhe autonomous mobile assembly robot KAMRO which is being developed by the University of Karlsruhe. A typical assignment for this robot is to transport and assemble a product. With the help of expert knowledge and a sophisticated sensor arrangement the system will plan a navigation route to a storage, drive there, pick up parts and bring them to an assembly station. At this station the parts will be viewed by a camera and their location and orientation identified. The robot will then draft an assembly plan and actuate two arms and various sensors. The assembly will be sensor-conducted and supervised. Upon completion of the work a visual test will be done.

For autonomous mobile robots, drive systems with three degrees of mobility

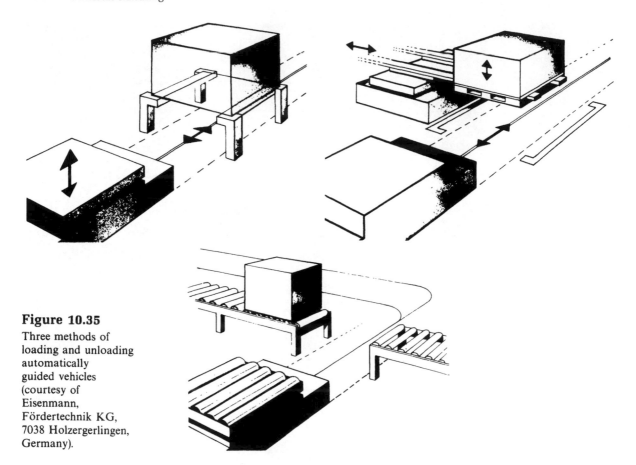

Figure 10.35
Three methods of loading and unloading automatically guided vehicles (courtesy of Eisenmann, Fördertechnik KG, 7038 Holzergerlingen, Germany).

are of interest. Figure 10.38 shows such drive systems. Concept A has three active wheels with passive rollers fastened along the circumference of these wheels. The axes of the passive rollers are perpendicular to that of an active wheel. Thus, a wheel is capable of driving in one direction and coasting orthogonally in another direction. If the vehicle is driven forward, both rear wheels are active and move the vehicle via the rollers of the idle front wheel into the desired direction. With a proper control strategy, the vehicle is capable of moving in any direction. Drive B has four active wheels. Along the circumference of each wheel, passive rollers are fastened in a fixed angular orientation with regard to the main axis. The rollers of the front wheels are arranged in a positive herringbone pattern, and those of the real wheels in a negative herringbone pattern. If the vehicle drives forward, all active wheels are rotated in the same direction. For a sideward motion the front wheels are rotated in the opposite direction to the rear wheels. Both sets of wheels rotate at the same speed. Thus the vector of the drive force acts perpendicular to the center-line of the vehicle and pushes the vehicle sideward via the passive rollers. This vehicle can also be moved in any direction or is able to rotate on the spot under the guidance of a proper control circuit. Drive C has independently powered and controlled wheels. Each wheel can swivel about its vertical axis with

Figure 10.36
A work cell tended by
an AGV; the
workpiece is fastened
to a pallet (courtesy of
Mannesmann Demag
Fördertechnik, 6050
Offenbach 16,
Germany).

the help of its own steering motor. For this drive system there are two basic
concepts. First, all four wheels are powered and with the other, two wheels are
powered and two are idle.

10.5 Warehousing

In most manufacturing systems, material and parts must be kept in stock because
ideal just-in-time operations are not feasible. Storage places must be provided for
raw material and parts, semi-finished goods and the completed product. A
warehouse operation may be manual, semi-automatic or fully automatic depending
on the affordable degree of automation. The design of the warehouse, storage
racks, material transportation vehicles, and storage and picking devices depends
on the product, the piece rate, management philosophy, safety requirement and
many other factors. Normally, the material transportation system used for
production is directly interfaced with the warehouse. For this reason the same
material transportation devices are used in the warehouse as in the factory.

A schema of a warehouse computer control system is shown in Figure 10.39,
see also Siemens (2). The incoming material from a vendor is identified by the
receiving station. Move and release papers are prepared which accompany the

Figure 10.37
The Karlsruhe autonomous mobile assembly robot, KAMRO.

Figure 10.38
Various means of propulsion for autonomous vehicles with three degrees of mobility. (a) Three independently powered active drive wheels with passive rollers, (b) four independently powered active drive wheels with passive rollers, (c) four independently powered and steered swivel wheels.

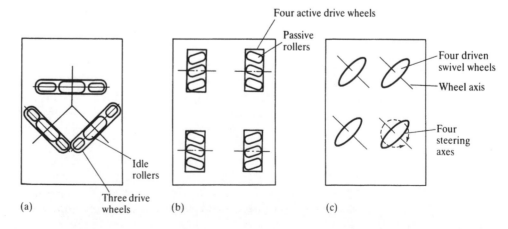

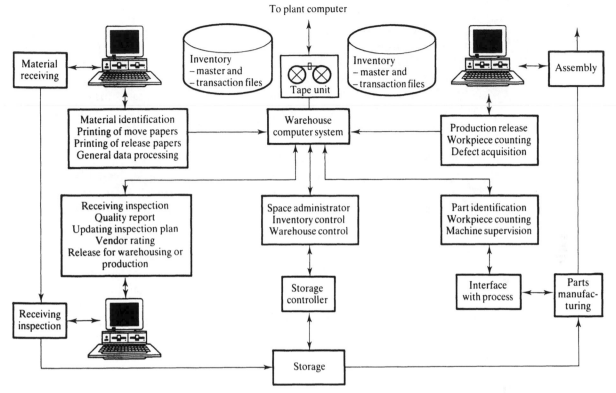

Figure 10.39

Warehouse computer
control system and its
interface with
production.

shipment. The material proceeds to receive inspection and is tested in accordance
with a test plan issued by the plant computer. A quality report is made and, if
necessary, material is back-ordered. The test results may necessitate updating of
the inspection plan, for example, the request for 100% testing of future shipments.
Thereafter, the vendor is rated and the material is released for warehousing or
production. The material destined for warehousing is stored. Storage may be done
according to various strategies:

- Material frequently used may be stored close to the loading and unloading
  system.

- Strategic material may be stored in parallel bays to assure their availability in
  case of a failure of a storage mechanism.

- Material subject to aging will be marked in the computer and retrieved in
  proper order to ensure that it is not spoiled when needed.

- Algorithms may be invoked to distribute the material strategically so that, upon
  retrieval, parts which belong to the same order are located close to each other.

- Hazardous material may be stored in special racks.

The master inventory and transaction files are kept in duplicate on two disk
drives to ensure that when an equipment fails data does not get lost. In addition,

periodically, the inventory data is stored on tape. The warehouse computer controls the inventory and also has a connection to the local computers which, for example, control the motions of the stacker cranes. Thus direct orders to store and retrieve parts can be given to the stacker cranes. Upon release of an order, parts are retrieved and brought to the material handling system. The parts are identified, counted and dispatched to the processes. Semi-finished parts may have to be restored and later released for assembly. The computer keeps a record of all transactions and updates its inventory master file. Also defects are reported to release back-orders to vendors. Figure 10.40 shows a photograph of a typical control system for material handling.

Figure 10.41 depicts an automatic warehouse for storing palletized parts. A typical storage cycle may be as follows. Pallets are delivered by truck to the unloading area (8). A buffer is provided for material that cannot be handled immediately (9). The pallets are moved along a roller conveyor and belt conveyor to a pallet checking station (2) to control proper outer dimensions. Thereafter, the pallets are identified by station (3) and moved to the mobile stacker facility (5). The stacker crane (6) picks up the pallet and the stacker is moved to and positioned in the selected bay. The stacker crane then travels along the bay and stores the parts in a designated storage place. The entire storage cycle is supervised by a control computer system. The pallets may be retrieved in a similar manner.

In Figure 10.42 a manual part picking operation is shown. A material request card is sent to the picking facility. An operator takes the card and picks the parts defined by the order. The parts are placed in a box and put on a conveyor system

Figure 10.40
Control center for material handling (courtesy of PSB GmbH Förderanlagen + Lagertechnik, 6780 Pirmasens, Germany).

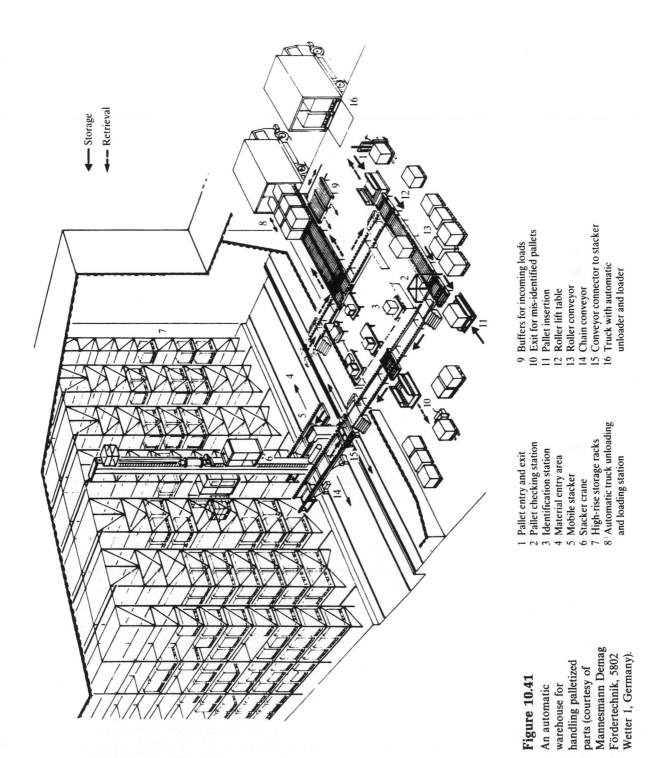

Storage

Retrieval

Figure 10.41
An automatic
warehouse for
handling palletized
parts (courtesy of
Mannesmann Demag
Fördertechnik, 5802
Wetter 1, Germany).

1 Pallet entry and exit
2 Pallet checking station
3 Identification station
4 Material entry area
5 Mobile stacker
6 Stacker crane
7 High-rise storage racks
8 Automatic truck unloading
  and loading station
9 Buffers for incoming loads
10 Exit for mis-identified pallets
11 Pallet insertion
12 Roller lift table
13 Roller conveyor
14 Chain conveyor
15 Conveyor connector to stacker
16 Truck with automatic
   unloader and loader

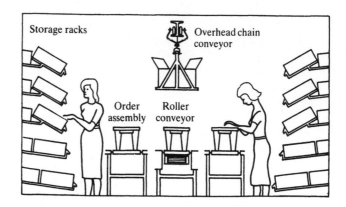

Figure 10.42
Simple part storage operation (courtesy of Krauskopf Verlag, Mainz, Germany).

together with move cards. The box travels under computer control to the specified workstation. At the workstation the box and the move card are taken from the conveyor and manufacture is started.

10.6 Problems

(1) The material handling devices are the interconnecting link needed to set up a specific manufacturing operation. How can a virtual manufacturing concept be conceived with a flexible material flow system?

(2) Why is the use of computers so important for a modern material handling system?

(3) Explain in detail the various software modules needed for the material flow shown in Figure 10.1. How would you design the interface between the software modules and how can the activities of the various modules be synchronized?

(4) What do you understand by the logistics of a material handling system? What activities and components of a factory are part of the logistics?

(5) Why is the factory layout so important for a well-functioning material handling system?

(6) There are various types of material handling strategies which can be followed. Explain them in detail and discuss their advantages and disadvantages.

(7) Which activities of a manufacturing system are concerned with 'just-in-time' planning?

(8) What kind of input/output devices used for material handling do you know? What are their features?

(9) How can the identifiction of a workpiece be done? Please explain basic principles and current coding systems.

(10) Explain direct and indirect tracking.

(11) Design a small factory for the assembly of pocket calculators. Set up a tracking system for components and assemblies.

(12) How can an automatically guided vehicle find its way through the factory? Please explain the basic principles.

(13) Which are the basic components of an autonomous vehicle? What are typical tasks of these components? What is a world model and how would you design one for an autonomous vehicle? What is the difference between a static and dynamic world model?

(14) Which different types of overhead conveyors do you know and what are their features? How can a computer track parts on these conveyor systems and how can the arrival of hangers be synchronized with the operation of machine tools?

(15) Which different types of material transportation vehicles do you know and how are they directed through the plant?

(16) What is the task of a pallet in material handling? What is a typical path of a pallet from a truck unloading area to a storage area? How can a computer track the pallets and direct their way through a plant?

(17) Explain a warehouse computer control system for a factory making plastic parts. Materials to be handled are raw material, parts and finished components. Should a computer hierarchy be used for this plant?

11

Quality assurance

CHAPTER CONTENTS

11.1	Introduction	572
11.2	An Integrated Quality Assurance Concept	573
11.3	Tasks of Quality Assurance	580
11.4	Performing the Quality Control Operations	585
11.5	Components of a Computer-Controlled Test System	590
11.6	Hierarchical Computer System for Quality Assurance	601
11.7	Programming of Test Systems	605
11.8	Problems	612

11.1 Introduction

Quality assurance is an essential function of an ongoing manufacturing operation. Its responsibility is to make sure that high quality is designed into the product and that the manufacturing operation produces the desired quality. The standards for the quality are set by the customer and the competitive market environment. The effectiveness of quality assurance depends on how well it is integrated into the entire process of designing, creating and making the product. Quality assurance must have a close contact to customer service and sales to monitor the acceptance and performance of the product and to take immediate corrective actions, if necessary.

It was recognized very early that the computer is an effective tool for aiding quality operations. It can control test stations, monitor the performance of processes, collect and evaluate data, and issue quality reports. One of the greatest values of the computer is its ability to integrate the quality assurance activities of the entire manufacturing system. For this purpose, a system approach must be taken to quality planning to understand the complex interaction between the basic manufacturing functions and the market, and to design a comprehensive quality control operation.

Usually, a quality control system is of hierarchical design and implemented on various tiers of a distributed computer system. Typically, on the lowest level the test stations are operated and controlled. On the second level the quality produced by a manufacturing cell is supervised. On the third level, quality data of the entire plant is collected and compared with quality standards and quality data from field operations. If quality deviations are observed corrective action is taken, which may involve a design change, selection of more durable material, the improvement of a process, retraining of operators or a change of a test procedure.

Due to its inherent speed and memory capacity, the computer has numerous capabilities normally not found in conventional measuring and test systems. For this reason quality assurance systems, designed with the computer in mind, will differ considerably from those used to the present day. Typical advantages are:

- performance of 100% testing, rather than testing on a sample basis;
- automatically operating test stands at different load conditions;
- testing of product variants with one test set up, just by switching test programs;
- observation of online quality control trends;
- quick performance of complicated online calculations;
- storing of a large amount of test data over a long period;
- learning of test limits;
- integration of quality data from various test stations and other sources.

An engineer designing a quality control system should have in-depth knowledge of the computer to utilize it to its full extent.

An important feature of the computer is its programmability which is needed when many product variants have to be tested or when the life cycle of a product is short. For this reason a quality assurance system should be projected into the future to be more adaptive to a rapidly changing market.

Quality control planning, measurement and data evaluation requires the design and use of control algorithms based on probability theory and other statistical tools. Often, these algorithms are computing intensive and their execution can substantially speed up with the help of the computer. Probability and quality control tools are covered in the literature and will not be discussed in this chapter; however, the reader concerned with CAQ must be familiar with them.

11.2 An Integrated Quality Assurance Concept

11.2.1 A comprehensive quality assurance system

The performance of a product and its quality is described in the requirements specification sheet and is a contractual document between the manufacturer and the customer. It is the responsibility of all functions of the manufacturing operation to adhere to this specification and to work together to deliver a product of the desired quality. The quality control loop starts with engineering, leads to manufacturing and ends with the field operation of the product by the customer. The feedback information from the customer is used to correct any problem with the product.

A quality assurance operation must be viewed as a complex control system which supervises the entire manufacturing process. It comprises quality planning and quality testing, and is performed through the entire process of creating and making the product. The operation of a quality assurance control loop is shown schematically in Figure 11.1. The reference input to the control loop comes from the engineering and manufacturing documents and specifies the desired quality. An error signal is calculated by subtracting the feedback value (returned from the plant) from the reference value, and it enters the controller which produces the manipulated variable. This variable enters the actuator, consisting of planning and scheduling; it produces the control information to drive the plant. The output of the plant is the controlled variable which is measured and the information is returned to the input of the control loop as the feedback signal. An important characteristic of this control loop is the reaction time, which is the elapsed time between the detection and correction of an error. It is the task of the computer to optimize the reaction of the manufacturing system to disturbances and to automate the function of the control loop.

Quality assurance must comprise all activities needed to conduct a product through its life cycle. Every design, planning and manufacturing activity of a product contributes to the quality. For this reason it needs thorough knowledge of design principles and manufacturing methods to build a high-quality product. A quality problem entered early into the product life cycle usually has a detrimental effect on the field performance of the product, and is very difficult and expensive to eliminate.

In addition, quality problems often accumulate throughout the various early stages of a product life cycle and are extremely difficult to trace back to their source. To be effective, quality assurance must be designed concurrently with the creation of the product and the development of the manufacturing processes. The goal of a well-conceived quality assurance system must be prevention rather than discovery of defects. In other words, it is more important to supervise the key points of the process parameters rather than the quality features of the product.

Figure 11.2 shows an integrated quality assurance concept, comprising development, manufacture and field operation of the product, see also Rembold *et al.* (1989). Each phase of the design, manufacture and use of the product has its own quality control loop, consisting of requirements analysis, planning, methods selection, execution, collection and evaluation of results, and comparison of the designed with the achieved quality. In a well-designed manufacturing system the functions of the control loops of each phase must be properly tuned and optimized to ensure that each activity maximizes its contribution to the quality of the product. The phase control loops are usually well structured in manufacturing and assembly. They can be found in almost any production facility. In engineering and process

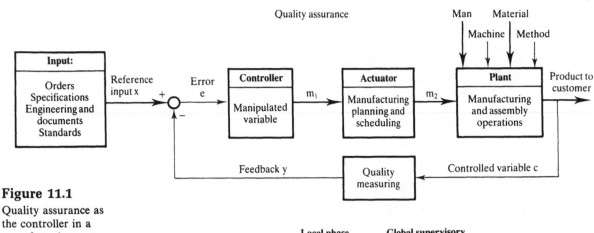

Figure 11.1

Quality assurance as the controller in a manufacturing system.

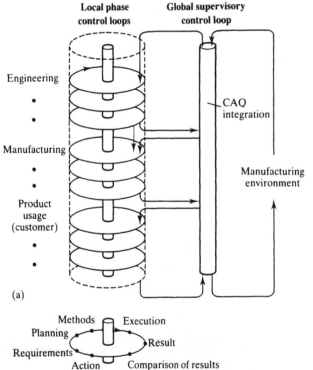

Figure 11.2

Concept of an integrated quality assurance system: (a) the total system, (b) the function of one phase.

planning they exist in a rather implicit form, where an attempt is made to design the quality into the product by providing a well thought out design. Since almost any quality control measure taken will influence the entire production process, the design of a control loop must be coordinated with various other activities.

The operation of a quality control system must be an independent activity which cannot be tampered with by manufacturing personnel. It is important, however, that all quality standards are agreed upon during production set up. This necessitates consent on the types of tests to be used and the amount of data to be processed. Usually, there is a lot of data accumulated from a quality control test. For the operation of a plant, the average performance data and the data on deviations from the test limits are of most interest. For this reason, this type of data must be contained in the quality control output reports. It will be the input to all phase control loops. The exchange of this data is the responsibility of the global supervising control loop which provides the test limits for every phase control loop, processes its final results and coordinates the operation of the entire quality assurance system.

Quality assurance measures taken in engineering and manufacturing differ considerably, and they will be discussed separately.

11.2.2 Quality assurance in engineering

These activities comprise product development and planning. For every product to be developed a requirements specification is drafted, containing information about the product features, product performance and product quality. The product quality is not very easy to predict. To design a specified quality into a product, considerable experience with similar products is needed and much testing must be done. The designer actually relies on the results of two phase control loops. One is the engineering performance test system and the other is the field test, done by the customer who uses the product. The results of the field tests from the customer are often returned to engineering one year later. For this reason the engineering test must be as thorough as possible by subjecting the product to normal and adverse operating conditions. Products which are designed for an extended service life must be subjected to long-term tests, which can be very expensive, when an average life expectancy has to be predicted from many test runs. It may be necessary to design test procedures which allow the prediction of the life expectancy of a product from short-term tests. It is not possible to engineer a perfect product, even with the best test methods. For this reason, engineering must observe the performance of the product when it is used by the customer.

The planning process can also considerably influence the quality of a product. For example, the best quality of a part may be obtained with a set of preferred machine tools and processing sequences. It requires much manufacturing experience to know the best choice of machines and processes. This is an organizational problem, because many parts compete for the same machines. A selection algorithm is needed which knows how to produce the best quality with the available machines and how to select manufacturing alternatives when the machines of choice are occupied.

11.2.3 Quality assurance in manufacturing

Basically, there are three types of phase control loops involved in manufacturing: receiving inspection for the material purchased, the direct control of the manufacturing process, and the acceptance testing of the product.

A prerequisite for the manufacturing process is a supply of good materials and parts. The quality for the material is specified by engineering. Purchasing will, however, make the decision where to buy the specified material and if necessary provide substitutes. For this reason, purchasing must be done in close contact with engineering to agree on quality standards and test procedures for inspection. A particular problem is the inspection of incoming material when the manufacturing operation uses Kanban, just-in-time or other methods, employing the flowline principle. More inspection is done by the vendor, and he must use the test methods specified by the purchaser of the parts. It is important that the supplier is absolutely reliable, otherwise the production system may come to a standstill when parts are defective. Inspection of incoming parts may be done on the basis of sampling or 100% testing. Often, with the computer, 100% testing is possible, thereby ensuring that every part meets specification.

When a machine is selected, its process capabilities must be known. Most manufacturing machines are provided with controllers to supervise their work. The mode of control can be open loop, feedforward, feedback or a combination of these principles. When a very high accuracy is required or for highly automated equipment, adaptive control is often used. With many manufacturing methods it is difficult to perform good or automatic measurements and to bring the measured quality back into the feedback loop. For example, if a laser cut contour is off by several millimeters the workpiece may have to be scrapped and correction can only be made for the succeeding part. A particular problem involves quality parameters which can not be measured well, for example, orange peel appearance or scratches on painted surface. In this case a human operator is a substitute for the sensor and controller. Often, with this type of operation, excess scrap is produced because of the long time it takes to correct a defect.

When a product is made in a sequence of operations, several successive tests have to be performed. If a problem occurs at the beginning of the process, and remains unnoticed, it will be carried through the entire operation and the defects may accumulate; this results in wasted time and unnecessary use of manufacturing equipment. This problem may be very difficult to control if the measurements are done offline and corrections are entered by hand, as shown in Figure 11.3. Such a manufacturing process can be substantially improved with online computer control working in a hierarchical mode. Here, every process is controlled by its own computer and a supervisor computer coordinates the cooperation of the individual test systems. Of course, this type of operation presupposes that the measurements and feedback can be done automatically. Another problem may arise when the technical and organizational control loops of the manufacturing system are not well tuned with one another. Figure 11.4 shows how these two control loops interact with the manufacturing process, also discussed in Storr (1989). For example, when a manufacturing alternative has to be considered, which is a common occurrence in any manufacturing operation, different types of control parameter may have to be provided for different types of machining setups.

Both of these previous examples show that even in a small area of a manufacturing system, in this case the production floor, much thought has to be given to an integrated quality assurance effort.

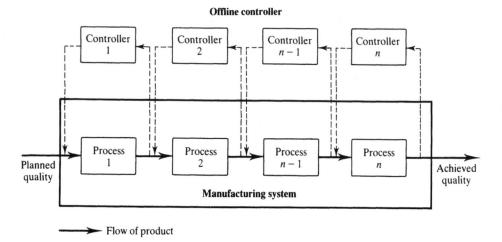

Figure 11.3
A manually controlled
manufacturing system
will react slowly to
production problems.

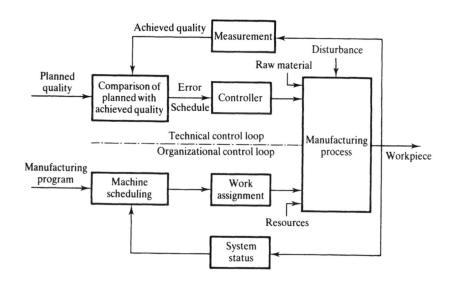

Figure 11.4
Technical and
organizational control
loops for quality
control.

The final inspection of the product in the manufacturing system is the performance test, to verify that it actually renders the service it was designed for. The test procedures and the test equipment used in this operation are often similar to those applied by engineering. Here again, a close cooperation is required between engineering and quality assurance.

11.2.4 The cost of securing quality

Like any other manufacturing operation quality assurance will be competing for scarce company resources. To make things worse, any quality assurance work will be appearing as an expense, and it is extremely difficult to show on a balance sheet the benefits obtained from the quality efforts. As with most automation endeavors, in a manufacturing system the improvements of one operation usually transmit to others. Such intangible benefits are of course extremely difficult to assess. Figure 11.5 shows the result of an industry survey where 94 firms which use computer-operated test systems reported the benefits they obtained from them. The most ostensible advantages are improved productivity of the test, more objective test results, better product quality and improved throughput through the test system. These responses actually verify that the installment of a computer-operated quality aid must be approached from a system point of view.

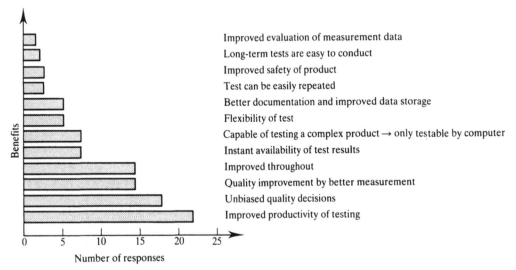

Figure 11.5
Frequency distribution of benefits of a computer-controlled test system.

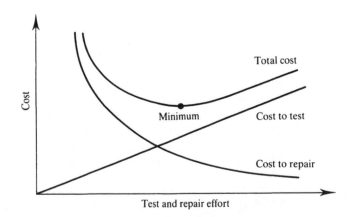

Figure 11.6
When adding the cost of testing to the cost of repairing, a minimum is obtained.

The highest assurance for good quality is to engineer the product well and 100% test every component and every product. In addition, the test limits can be set very tight to intercept any problem before it leaves the factory. If this policy is assumed, it is certain that the cost of quality will be prohibitive. It is also very important that any defect of a product must be detected and repaired in the factory. Problems occurring with the customer will be extremely expensive to repair compared to those found inhouse during product testing. Figure 11.6 shows how the cost of testing increases with increasing test efforts. In a second curve the cost to repair is shown. This cost is high for little testing and low for thorough testing. If the two costs are added a total cost curve will be obtained, which has a minimum. Such curves are not easy to establish since cost figures from field repair usually come back to the factory after a considerable delay, often a year or more.

The cost of quality operations comes mainly from four sources:

(1) preventative quality control

(s) testing

(3) inhouse repair

(4) field repair

The introduction of the computer to quality control has significantly helped to improve measures for preventative quality control in engineering and planning. The computer was also responsible for highly automating conventional test systems. Often, this automation was extremely expensive and contributed heavily to high quality assurance costs. On the other hand, the computer has helped to significantly systemize quality control and to integrate this function into the manufacturing process. Unfortunately, the benefits of this integration are extremely difficult to assess, if possible at all. However, it is important that the cost of quality assurance must be seen from a system point of view, since it is not just a local activity.

11.3 Tasks of Quality Assurance

With an integrated quality assurance system there are various tasks to be performed to build a high-quality product and to ensure its field performance. Quality planning starts with the conception of the product, whereby every part and every manufacturing process has to be studied and reviewed to verify its contribution to a successful product. Since many factors which have an influence on the quality cannot be anticipated before the product has been placed in service, the process of quality planning is an iterative one and will be a continuous endeavor over the service life of the product. Quality assurance is concerned with various aspects, including

• quality planning

- quality testing of the product
- controlling the manufacturing process
- controlling the quality control operations

In an integrated quality control system, an attempt must be made to tie these operations together into one system. This will need a master quality assurance database and designing of a distributed computer network, to access this control source of information. In the following section we will be discussing the various aspects of quality control.

11.3.1 Quality planning

This is probably the most difficult task of quality assurance. We have already discussed some aspects of this activity in Section 11.2.2.

If the new product is a variant of one that has been built already, the data for quality planning can be taken from performance records of a product variant which has a proven quality record. Data from engineering tests and the field performance is retrieved from the master quality assurance database and carefully evaluated. Then a plan is made to use known engineering principles, materials, manufacturing processes and quality control test procedures for the new product. In case design changes are made to the product, or new manufacturing processes, the possible influence of these changes on the quality has to be considered. Engineering and field tests have to be performed with units made as engineering samples or manufactured in pilot production runs, to ensure the durability of the product. The results of these tests will be maintained in the master database for future use.

If the product is completely new, and little engineering and manufacturing skills are available to design and produce it, then the task of quality assurance will be quite difficult. An experienced designer is needed to design the product and to calculate its performance characteristics. The product has to be subjected to short and long-term engineering test to determine its performance and that of its components. When satisfactory results have been obtained, the necessary production methods are devised and their influence on the quality of product must be anticipated. The next step is the manufacturing of the product in a pilot production run, to try out the new manufacturing method, and to produce sample products which can be subjected to field testing. The results of the field test may make product changes necessary and new engineering tests and pilot production runs may have to be done again. Eventually, the process of designing, testing, and manufacturing converges and an acceptable product is created. For this product a production test system has to be built to verify its performance in the factory before it is sent to the customer. Usually, the procedures and test systems are taken from engineering, and applied in a modified form on the production line.

11.3.2 Testing the product

Test procedures and test equipment must be available for the product and its components. If it is standard parts, for example, screws, the testing is usually done somewhere else, and their adherence to specification will only be done on a sample basis. Depending on the part or component, testing may consist of verifying physical parameters, performance data, a chemical composition or others. It is important that all tests are correctly specified and that the test results can be compared with the engineering specifications. Test procedure and the test equipment must be available for every test. In piece part manufacturing, the following tests are usually performed:

- receiving inspection
- part testing
- assembly testing
- product acceptance testing

The special features of these various tests will be discussed in the following sections.

Receiving inspection

The most important engineering parameters are specified for all raw materials, parts, and components ordered from the vendor, and it is expected that the goods received are in accordance with the specification. There are various test principles applied which may be necessary due to the complexity of the part, reliability of the vendor, or the function of the part. Most parts are inspected on a sample basis; but important parts may require 100% testing. The majority of tests are to check the dimensions of a part. However, there is a myriad of other tests which have to be applied, using mechanical, electrical, thermal, color, metallurgical, chemical, and other measuring principles. With modern flowline manufacturing systems, the receiving inspection test may actually be done by the vendor. His test reports will be honored and the parts enter the production stream in the delivery state.

Part testing

The quality of a part to be made depends on the capability of the machine tool, the condition of the tools, the available measuring system and many other factors. When a workpiece is set up in the machine tool it must be made certain that the desired quality of the parts can be made with the equipment and set up. In most machining operations the making of the product is not supervised online. The quality check is made after the part has been finished. This type of operation can result in a high scrap rate, for example, if a part is under-dimensioned it is normally thrown away.

Many manufacturing processes can be supervised online with a sensor system which is directly connected to a computer; this is a desirable feature of part manufacturing. However, it presupposes that measurements can be made online, which is not possible for all manufacturing processes. Online measurements eliminate quite a lot of part handling and speed up the process. In most cases the computer is fast enough to evaluate the test data immediately and to introduce corrective actions, if necessary. If, however, the production rate of the machine tool is high and the algorithms for evaluating measurement data and calculating new set points is quite lengthy, it is possible that the computer will not follow the production. In such a case, special parallel computer architectures must be installed in the measuring and control system.

This computer can also be used to measure the condition of the tools online, to check for sharpness and tool breakage, to observe quality drifts, and to monitor the operation of the machine tool. This can be done in a continuous or periodic mode. If a malfunction is observed it is immediately reported and the problem pinpointed.

Assembly testing

The progress of an assembly is mainly tested visually by the assembler. He has a supervision system (his eyes), a memory with unsurpassed capacity (his brain), and a knowledge base he has built up from experience. Vision is also a desirable feature for automated assembly systems. There are numerous applications publicized where machine vision systems supervise assembly operations. However, in most cases they only do simple observances and are often limited to 2-D problems. In addition, the camera usually has to be orthogonal to the workpiece. Better results can be obtained with special measuring fixtures for assembled products. In this case, an error can only be detected after the assembly has occurred.

The acceptance test

Before a product leaves the factory it is subjected to an acceptance test, usually at the end of the assembly line. These types of tests, often, were the first computer applications in a factory. Typically, for automobiles and appliances the tests comprise a check of all functions, performance and product appearance. The test system may contain computer-controlled actuators to initiate the functions and to operate the product through an entire load spectrum. The test system can be reprogrammed quickly if another model variant is tested. The test results are usually calculated immediately and reported to the inspector.

11.3.3 Controlling the manufacturing process

The capabilities of the manufacturing process have a direct influence on the quality of the product and the scrap rate. The factors which influence the manufacturing process are the operator, material, method and machine. For this discussion we assume that the operator is well trained and that he knows how to run the machines which are in his sphere of responsibility.

The material is specified by engineering; if, however, it is not properly selected, it may have an adverse effect on the product quality. For this reason the selection of the material is a system problem and must be coordinated between engineering and manufacturing. We have already discussed the required cooperation between these two activities in Section 11.2.3.

Experience is needed for the selection of the manufacturing methods and machines. The selection process is done by the process planner and factory scheduler; it may be supported by results of process capability studies, expert systems and literature. We have learned in Section 2.3 that the way of controlling production equipment and of arranging it to a configured system may greatly affect the quality of a product. The best results are usually obtained with a fixed production set up. Reconfigurating a system for a new schedule often results in a production alternative with marginal performance.

In a well-operated production system an attempt must be made to monitor delicate and complex production equipment with the help of the computer and to relate equipment problems to manufacturing problems. This information can be used in the future for process capability studies, the results of which are vital to the process planner and the factory scheduler. If enough data is available on process capabilities it can be used to build up an expert system for selecting manufacturing processes and sequences.

The monitoring data accumulated from ongoing manufacturing operations should also be used for observing the aging or wear out of machines and processes. If the scrap rate becomes too high a repair or replacement of the equipment may be necessary.

11.3.4 Controlling the quality control operations

A quality assurance system must be embedded in the entire manufacturing system. It consists of numerous components which are linked together by a computer network. When the system is being set up and operated, it must be controlled to check its capability of rendering the defined objectives and assessing the desired quality. This is a difficult task, because there is no real standard against which the quality control system can be measured. In addition, quality problems are often extremely difficult to trace and may have their origin in any part of the manufacturing process. However, with a carefully designed and built up system, thoroughly tested components, and responsible operating personnel, assurance can be given that the system works properly.

By looking at the components of an integrated quality assurance system we can roughly distinguish between:

- hardware
- software
- interfaces

The hardware consists of the measuring instruments and the overlying computer systems. For the measuring instrument, established calibration methods have to be followed. However, the computer is not so easy to test. Its operation depends on the computer, its peripherals and the software. The computer and peripherals can be checked periodically with the help of test programs. In critical situations it may even be necessary to provide redundant computing equipment to protect against failures.

Often, the software is a more difficult problem to assess. Frequently, it occurs that software is functioning well over many months or over years and suddenly a problem arises due to an unforeseen event, which could not be anticipated during the design of the system. Usually, this type of problem can cause havoc with the entire system and is extremely difficult to locate; but it is a very typical software problem and may not be eliminated completely. To hedge against these occurrences, well proven software engineering methods must be used to generate and maintain a good software system.

The interfaces between the hardware, software and the various modules for quality assurance are another object of concern. For the operation of a quality control system and for servicing it, as well designed man–machine interface must be available. The operator should be able to communicate with the system in a language he is familiar with. He should not be burdened with the task of learning specific computer languages or with jargon. This would only introduce many operating errors, and jeopardize the integrity of the system. Another man–machine interface is needed for the maintenance of the software and its updating. Of course this service will be done by a computer expert who is knowledgeable about the intricacies of computer hardware and software, and the communication system.

11.4 Performing the Quality Control Operations

When a product is developed and engineered, the tests, test methods, and the test equipment must be selected. This involves the definition of the parameters to be controlled, the selection of the test procedure and test equipment, the performance of the test, and the evaluation and reporting of the test results. We will be discussing these activities under the following headlines:

- quality control planning
- testing
- evaluation of test results
- quality reporting

11.4.1 Quality control planning

The planning of the quality control operation is usually product specific. Occasionally, a manufacturer will be able to purchase or adapt a test system which had been developed for a similar product. A new product may have extended features for which new tests must be provided. The quality control planning is similar to process planning and should be done in conjunction with it. Presently, most manufacturers draft these plans by hand because it is very difficult to develop an automated planning system which has enough knowledge to cover the entire task of testing. However, planning by computer will become an accepted method with the development of good expert systems for quality control planning. It is certain that there will be no universal generic planning system to handle the testing of all products. The future planning systems for quality control will be domain-specific for families of products or product variants.

The ideal quality control planning system would be directly interfaced to the design database to draft a test plan from the product definition. Figure 11.7 shows how such a test plan can be automatically generated for a simple workpiece. The system will be inspecting the three-dimensional presentation of the workpiece and be drafting via the test part the test definition and the test program for the coordinate measuring machine. The data and programs for the planning process are located in the master database.

One of the basic difficulties of the planner is in deciding how to test, determine the features to be tested, specify the extent of testing and select the test aids. All

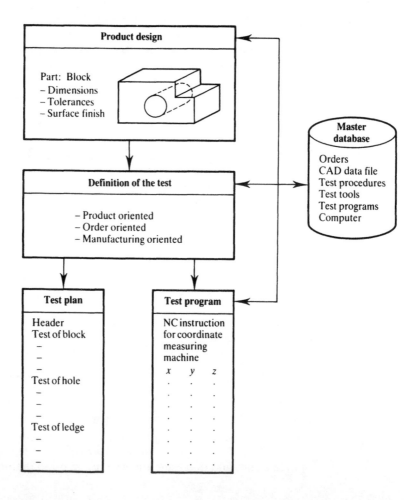

Figure 11.7

Automatic generation of a test plan.

of these parameters are influenced by the design and functions of the product, the rate and type of production, the available test systems, and many others. An important factor to be considered is to minimize the cost of testing. Typical questions to be asked for preparing the test plan are:

- What is the objective of the test?
- How should the testing be done?
- What is the test method, sample or 100% testing?
- What is the duration of the test, short or long term test?
- What are the test parameters and test limits?
- What are the test tools?
- Where in the manufacturing process is the test to be installed?
- Who does the testing?
- Who gets the test results?
- How long have the test results to be stored?

There are two approaches to drafting a test plan: it is either generated completely new, or assembled from variant test plans of similar products.

The first method works in a generative fashion. There must be functional test primitives available for each test step. The planner describes the test with the help of an implicit test description language. The planning system must have enough intelligence to interpret the task-oriented instructions and to infer from them the test plan. This method, of course, presupposes that an expert system is available with knowledge of the product domain and the test procedure for which the testing is done (see Section 8.8 on process planning).

With the variant method, sample test plans of product variants are stored in the computer. The planner retrieves these plans and fills in the parameters needed for describing the test object. The variant method is easy to use. However, it is usually very difficult to maintain the variants and to add new product features, if required. This method also has problems when the test used is dependent on the piece rate and the size of the product.

The programming of the test equipment is also part of test planning. Various programming systems are available.

If the test is done on a universal test system the program will probably be written in Basic, Fortran or a special purpose language. For many applications, the language ATLAS (Abbreviated Test Language for All Systems) is used. Coordinate measuring machines can be programmed with NC-type languages. This programming can easily be done by personnel responsible for part programming of machine tools.

11.4.2 Testing

The test plan and test program are the input to testing. When a computer operates the test, the special capabilities of the computer offer various advantages of performing the test and evaluating the test results. One of the biggest advantages of the computer is that it tests consistently in a repetitive mode and produces unbiased test results.

Usually, all testing is done in a similar fashion. The steps necessary to performing the test are:

- set up of the object to be tested,
- positioning of the sensors,
- initiation of the test,
- control of the test run,
- acquisition of the test data,
- conversion of test data to engineering units,
- statistical evaluation of test results,
- calculation of test limits and quality index,
- output of go/no go signal to test station,
- storing of test results,
- halting the test,
- release of the test unit and sensors,
- output of test report.

The degree to which these functions should be automated depends on factors such as number of products to be tested, complexity of test, size of test object, types of measurements performed and many others. When the computer is part of the test system the test equipment may be configured by three different arrangements:

(1) Manual test system; the test is set up and sequenced by the operator, and the computer acquires and evaluates the test data.

(2) Semi-automatic test system; the test is set up by the operator and the computer performs the sequencing and testing.

(3) Automatic test system; the set up of the test object and testing is done completely under computer control.

The most reliable tests are obtained with a fully automated test system, where man is taken out of the test loop. However, this type of operation can in general only be afforded for mass production.

There are various problems to be solved if the test is completely automated. First, the unit has to be recognized by the test system. This can be done with a light barrier, bar code reader, machine vision, mechanical feeler or other devices (see Section 10.3). Second, the test object may have to be placed in a preferred test position to attach test accessories (for example, a hose for cooling water) and sensors. Third, the test object may be moving on a conveyor line, whereby the power supply leads and test leads have to be pulled along during testing. Fourth, it may be impossible to attach sensors to a moving test object. In this case non-contact sensors may have to be used, which are less accurate than contact sensors. There are numerous other problems associated with automatic testing. Several of them will be discussed later.

11.4.3 Evaluation of the test results

A well-conceived test system should take advantage of the high calculation speed of the computer to evaluate, list, plot and store the test results. Often, with conventional test methods, the evaluation of measurement data and the performance of statistical analysis of the test results are very time consuming, and for this reason are not properly done or short cuts occur. This is frequently the reason why 100% testing is not done, even with complex test objects.

Commercially available all-purpose test systems provide test evaluation software, which can be called up by special call commands in the programming language. Below we list several evaluation routines which are useful for test applications.

- plausibility analysis of test data,
- filtering of data,
- regression analysis,
- statistical distribution analysis,
- calculation of lower and upper test limits,
- automatic creation of test tolerances,
- calculation of quality indices,
- transformation of electrical or physical test data to engineering units,
- calculation of non-measurable parameters,
- linearization of test data and curve fitting,
- interpolation and extrapolation of test data,
- drift analysis and adjustments,
- determination of quality and process trends,
- automatic reduction of test data to standard test conditions.

The evaluation of the test data is done in several stages, using several of the above mentioned routines. It is helpful to build up a library of commonly used programs and routines, and to give the operator of the test equipment easy access to this software by providing him with a well designed man–machine interface to the test computer.

11.4.4 Quality reporting

This is the last stage of the test operation. Here the test data is arranged in proper order and entered into test reports. As a rule, only that data which is pertinent to the user of the report should be contained in it. The often followed practice of summarizing all test data in a common report leads to the problem that the data is not properly communicated and that important test results may be overlooked or wrongly interpreted.

It is also essential to recognize that the quality reports are used by various levels

Figure 11.8

Online observance of various manufacturing problems on a graphic display.

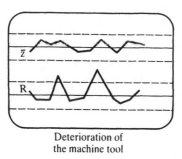

Deterioration of
the machine tool

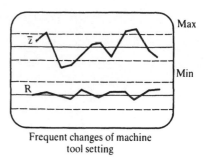

Frequent changes of machine
tool setting

of management. At the lowest level a very detailed knowledge of the defects is needed, whereas on the highest level only failure statistics may be of interest.

Typical test reports contain the following type of information:

- number and types of units which are defective
- types of defect
- manufacturing units where defects occurred
- process which caused the defects
- cost of defects
- cost of rework
- quality trends

In many cases a graphic output will highlight the quality report, containing a statistical distribution of defects, a quality trend over time or other information. Figure 11.8 shows a graphical output where the process trend is shown in two different presentations. The picture on the left side indicates a deterioration of the manufacturing process, and the picture on the right side a problem which is caused by an operator who adjusts the machine too often.

With the availability of graphics terminals the operator can be given tools whereby he is able to produce his own reports or graphics in an interactive way. Thus, he is able to investigate problem areas in detail.

11.5 Components of a Computer-Controlled Test System

11.5.1 Introduction

A computer-controlled test system consists of many components which must be connected to a functional unit. There are numerous sensors and measuring principles available, and since their number is so great, they cannot, and should not, be described in a book on CIM. If properly done, they would fill an entire series of books. In many cases, sensors and measuring methods have to be selected on an individual basis for every test application. If available sensor systems are being used, the measuring principles and methods are described by factory, industry or other standards. The predominant measurements in manufacturing are concerned with dimensional parameters of a workpiece. Coordinate measuring machines of various kinds are available, most of them computer controlled. We will be discussing important features of these machines in this section.

Often, for new or complex workpieces, special sensors, measuring fixtures and test procedures have to be developed, which can be an expensive and time-consuming task. In flowline manufacturing operations it is advantageous to perform the test on the moving production system. However, with this method, it becomes difficult to attach sensors, test and power supply lines to the moving test object.

11.5.2 The test set up

Types of test systems

Test systems may be operated in various modes, which depend on the degree of automation, the cost of the installation and the type of parameters to be measured. Figure 11.9 shows three basic architectures of computer-controlled test systems.

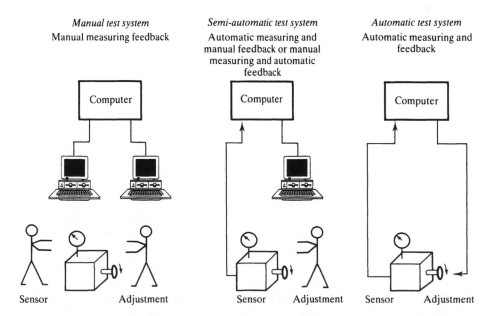

Figure 11.9
The three basic
computer test methods.

They are:

(1) **Manual test system.** With this type of arrangement, testing is done manually and the parameters are entered by hand into the computer. The feedback information is read from a terminal and adjustments to the process are done by hand. This set up is mainly used for low-volume production runs and for process parameters which are difficult or impossible to measure automatically.

(2) **Semiautomatic test system.** With this type of operation, either the measurements are done automatically and the feedback by hand, or the other way around. Usually, equipment is set up in this manner for medium volume production runs, or for complex test objects, or in cases where full automation is difficult, or uneconomical.

(3) **Automatic test system.** This is the most expensive approach to testing. It is used for high-volume production runs or where man has to be taken out of the test loop to improve the reliability of the measurements.

In practice, all three systems are found in factories. The reason why one system is selected over another may not only be governed by economy, but also by management philosophy. A completely automated system can very seldom be designed because many test parameters cannot be acquired automatically. Often, even with highly automated systems several parameters are entered manually.

Connecting the test object with the computer

The sensors and controllers of a test system are connected to the computer via a chain of signal processing components, as in Figure 11.10. Let us assume that we have a pneumatic pressure sensor for measuring a product parameter, and a motor driven valve for adjusting the process with a feedback signal. The pressure is transformed by a signal transformer into a voltage signal and amplified. Before the signal is placed on the bus, it is converted into current to reduce noise and sent to the computer. Here it is converted by an A/D converter into a digital signal and processed. The feedback information is converted into an analog signal and sent to the controller in a similar fashion. If many measuring and control components are connected to the computer, an additional component, the so-called multiplexer, is used selectively to connect and disconnect the peripherals to or from the A/D and D/A converters. Thus it is possible to save electronic components and signal lines.

Contact and non-contact measurements

The best and most consistent measurements are obtained with contact sensors. Here the sensor is brought into direct contact with the test object, and a reading is made after a steady state condition has been reached between the sensor and object. For high-volume production where measurements are made on moving test objects, non-contact sensors have many advantages. There are numerous remote sensing principles and sensor systems available. However, they have the

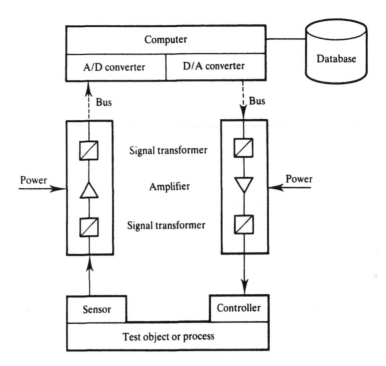

Figure 11.10
Components of a computerized test system.

inherent disadvantage that the test results are often much less accurate than those obtained with contact sensors.

Laser sensors are usually applied for length and distance measurements. Whereas length measurements can be highly accurate, distance measurements frequently do not render the required results. Length measurements can also be made by semiconductor cameras using one- or two-dimensional photodiode arrays. The resolution of these devices depends on the accuracy by which the arrays have been manufactured and on the packing density.

For remote temperature measurements infrared sensors are used. They render good service for many applications, however, they are not suitable for making measurements where the temperature of a small area is of interest.

There are many quality control parameters which cannot be measured by currently available instruments. Typical defects are paint blemishes like discoloration and scratches. Here the human eye is unsurpassed and can detect the smallest problems. The visually recognized parameters are entered via terminals into the computer. The terminals must be able to guide the tester in the operation of the test system and allow him to make corrections in case of a mistake. Recently, various voice communication systems have been available which can be used in connection with manual and visual quality control operations. The operator has to teach the voice recognition system the entire vocabulary it needs for defining all defects. During service, the system will try to match the spoken defects with templates it has formed from the training vocabulary. A major problem with these devices is their teacher dependency; if used in shift operations they must have a complete set of vocabulary learned from each operator. Also physical and mental stresses may change the tone of the operator's voice, resulting in a lower hit rate.

Online correction of the test

Measurements in the factory are done under changing temperature and humidity conditions. The temperature will affect length measurements and performance tests with typical test objects such as engines, refrigeration units, hydraulic drives and so on. The computer can easily correct these problems by converting the test results to standard test conditions obtained in the varying environments. For this purpose, the prevailing ambient temperature has to be monitored, and correction curves must be available; they are normally obtained from laboratory tests. However, these curves can also be accumulated automatically by evaluating test data obtained in the factory over a longer period.

Another problem, which must not disturb online testing, is instrument drift; mostly, it is caused by temperature changes in the environment or in the instrument itself. Such drifts can also be corrected automatically with the computer. If the drift is linear, periodic measurements of a reference voltage from a standard voltage cell can be made and the instrument is automatically set by a software program to a floating zero potential. In other words, the test data are corrected with the offset value measured. With nonlinear drift the correction is not that easy, here the form of the nonlinearity must be known, which may depend on various variables.

Testing units with transient behavior

Frequently, test objects have typical pull down characteristics and reach a steady state condition only after a longer period. Refrigeration units and combustion engines behave in that manner. There are several ways these units can be efficiently tested.

First, only a short pull down test is performed and the steady state condition is predicted with the help of an expert system from these short measurements. This depends on the expert system having reference test data in its knowledge base on the short-term behavior of any defect which may occur. In addition, it must be assumed that all defects will show up during the short-term test.

Second, the units are tested on the moving conveyor system. Figure 11.11 shows the problem with an overhead conveyor system. The test units are loaded on the hanger and connected to the power supply. The power is brought to an electric outlet on the hanger via an open power line running in parallel with the conveyor chain. A sliding contact in the hanger base closes the circuit. Since the signal cables cannot be pulled along with the units, a similar arrangement could also be made to convey the signal cables to a control computer. However, the sliding contacts introduce noise into the measuring system, making this solution unfeasible.

Third, the test computer is hung from the conveyor chain, as shown in Figure 11.11. Its power is supplied via sliding contacts. The noise from contact bouncing

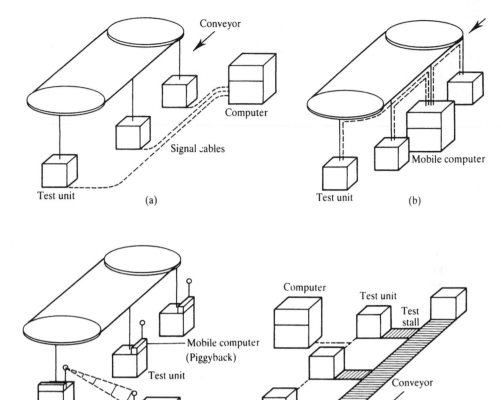

Figure 11.11
Different test
configurations for
moving production
lines: (a) stationary
computer, test unit
pulls signal cables,
(b) mobile computer,
(c) mobile test
computer, stationary
computer,
(d) stationary test stalls.

and arcing is filtered out with capacitors on the power inlet of the computer. With this arrangement the signal lines can be directly connected to the computer.

Fourth, a test computer with a battery power supply is placed in a piggyback fashion on the test unit. It accumulates data from the unit tested and sends these to a control computer by telecommunication at the end of the measurements.

Fifth, the units are pulled from the conveyor system into a test bay where they are tested in a non-moving condition. This is the best solution from the test point of view but it is also the most expensive one. It needs much floor space and duplicated test equipment.

11.5.3 Coordinate measuring machines

Coordinate measuring machines are used widely for measuring dimensions of a workpiece. Figure 11.12 shows several types of measuring machines, see also Kampa and Weckermann (1984). The simplest machine uses a cantilever construction,

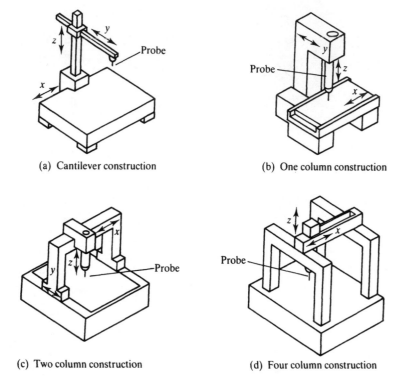

(a) Cantilever construction

(b) One column construction

Figure 11.12

Various construction principles of coordinate measuring machines.

(c) Two column construction

(d) Four column construction

where the probe extends from a boom which can be adjusted in the x, y and z direction, as shown in Figure 11.12(a). The base of the machine is a precision-ground granite table. The workpiece can be slid along the surface of the table for measurement. An improved version of this machine is shown in Figure 11.12(b). Here the position of the table and the boom can be controlled by a program. Figures 11.12(c) and 11.12(d) show two types of two-column universal measuring machines. Modern machines of this type can be automatically operated under computer control using various control principles:

- point to point control,
- linear control along the major axes,
- omnidirectional linear control,
- path control.

The kind of control to be selected depends mainly on the complexity of the workpiece.

The predominantly used measuring probes are of mechanical design, Figure 11.13; although laser or other principles can also be used. The probe is brought in contact with the test piece either by manually operating the x, y, z positioning mechanism or by directing it under program control. A typical measuring of a workpiece is shown in Figure 11.14. Here the probe is brought into contact with the workpiece and the displacement of the measuring probe with respect to a

Figure 11.13
Typical mechanical
probes used with
coordinate measuring
machines.

Figure 11.14
Measuring the
dimensions of a gear
(courtesy of Carl
Zeiss, Oberkochen,
Germany).

reference point is recorded. The reference point can be set by the operator. It is usually a key point of the workpiece. Figure 11.15 shows the principle of the position control of a coordinate measuring machine. Here the workpiece is set up on a measuring table and is brought under hand or program control to the measuring probe. The probe slides along the workpiece and the displacement of the probe tip is recorded. For complex workpieces, a three-dimensional probe head with several probes may be used (see Figure 11.16). The selection of the probes can be done under program control.

The programming support is an important consideration when a coordinate measuring machine is selected for quality control. There are various ways of

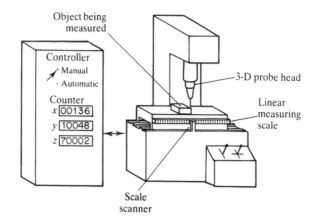

Figure 11.15
Principle of position
control by manual
machine operation
(courtesy of Carl
Zeiss, Oberkochen,
Germany).

Figure 11.16
Three-dimensional
probe head (courtesy
of Carl Zeiss,
Oberkochen,
Germany).

teaching the shape of the workpiece to be measured, including:

- teaching via a master workpiece,
- online programming,
- offline programming.

Teaching via a highly accurate master is the easiest way of acquainting the measuring system with the dimensions of the workpiece. With this method, a probe is led along the contour of the workpiece and the specified measuring points are recorded and stored. For production testing, the finished workpieces are measured and their dimensions are compared with those of the master. This method, however, ties up the measuring machine during teaching.

The second method is online programming. Here the workpiece dimensions are entered into the machine via a keyboard or a terminal. This operation can be quite time consuming for a complex workpiece, and it also ties up the machine.

The third method uses the test program which is generated during the design or test planning activities. For this purpose an offline programming language and its supporting software must be available. The programming is done independently of the measuring machine. A good programming system provides a simulation system, by which the execution of the program can be observed on a graphical terminal.

The work cycle for testing a workpiece goes through the preparation, programming and measuring stages. Figure 11.17 shows for each stage the required tools, activities and information, see Weckermann (1984).

11.5.4 Self-learning test systems

One of the basic problems of online testing is the definition of the upper and lower test limits. In most cases engineering sets the test limits. Often, these test limits cannot readily be used because they were obtained in an ideal test environment with selected engineering samples. In addition, with long or expensive tests not many sample runs can be performed to obtain statistically meaningful tolerances. For this reason, the upper and lower test limits are often obtained right on the production line by taking measurements from a predefined number of good units and by calculating the normal distribution. The test limits are then set to $\pm 3\sigma$ (standard deviation). This method, however, is inaccurate because it uses data taken during the start of a production run and does not consider the learning effect of the improvement of the product over time. The test limits selected are often too wide. With the computer it is possible to constantly learn and update test limits. The various ways this can be done are discussed below.

In the first method, a predetermined set of good units (say, 100 units) is run through the test system and upper and lower test limits are determined. Now the testing of the production run begins and as soon as 100 new good units have been tested their parameters are added to the previous data. New test limits are determined with the aggregate set of data, and the test continues. Updating is done whenever another set of measurements is available from an additional 100 good units, as in Figure 11.18(a). In practice, this learning strategy shows that test data has a tendency of opening up with time.

In the second method, the first set of tolerances is obtained as with the previous test, and it is used until another 100 good units have been tested, as in Figure 11.18(b). With the measurement data of the new units, new tolerances are calculated and used. The old tolerances are discarded, and updating is done in the same cyclic mode. This method has a more stable behavior.

	Tools	Activities	Required information
Preparation	Paper, pencil and computer	• Definition of the test • Drafting the test plan • Defining the clamping and fixturing units • Selection of measuring machine • Selection of test probes • Definition of test models and probing strategy • Selection of workpiece coordinate system • Description of test output	• Workpiece drawing • Test plan • Data of measuring machines • Data of test probes • Data of clamping devices • CNC control program • Evaluation software
Programming	Programming system CNC coordinate measuring machine	**Input of instruction and data** • Design data and tolerances • Measuring trajectories, intermediate points and forces • Definition of workpiece coordinate system • Subroutine call for testing form elements • Comparison of test results with set points • Report generation, plotting, and so on	• Test sequencing plan • Test plan • Workpiece design data • Features of measuring machine • Clamping of workpiece • Features of CNC software • Features of evaluation software
Testing	CNC coordinate measuring machine	• Clamping of workpiece • Attachment of measuring probe • Determination of workpiece position • Start of test • Execution of test • End of test • Workpiece removed	• Control data

Figure 11.17

Work cycle for testing
a workpiece with a
coordinate measuring
machine.

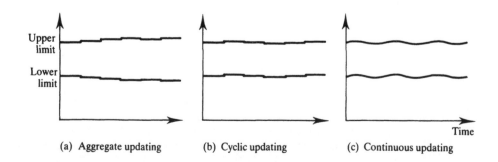

Figure 11.18

Learning methods for
upper and lower test
limits.

Upper limit   Lower limit   Time

(a) Aggregate updating   (b) Cyclic updating   (c) Continuous updating

In the third method, the initiation of the test is done as with the previous methods. However, whenever a new measurement from a good unit is obtained, the data of the first unit measured is discarded. The cycle continues throughout testing. This method showed the best results in practice.

When pursuing learning procedures for test limits caution must be exercised with the stability of the results. We have seen in the first example that the test limits have a tendency to open up with the amount of data used for the aggregated test limits. There are various influences which may cause instability, for example, a non-Gaussian distribution of measurement data and the improved skills of the assembler over time.

11.6 Hierarchical Computer System for Quality Assurance

In the first section of this chapter we discussed the integrated quality assurance system, as shown in Figure 11.2. If such a system is built up in an organization it will be implemented with the help of a hierarchical computer architecture, as shown in Figure 1.30.

An additional requirement for the system is the availability of a quality assurance database, in which all data used for the quality functions is located and from which can be retrieved information on any authorized manufacturing activity. Basically, we distinguish three computer tiers by which the various quality assurance operations are planned and controlled. The tiers can be described as:

- quality assurance supervising tier
- quality assurance planning tier
- quality assurance control tier

We will discuss the principal activities of each level in the following sections.

11.6.1 Quality assurance supervising tier

This is the general planning and coordinating tier where all quality efforts of engineering, customer servicing, manufacturing and testing are tied together and supervised. In a manufacturing system, this activity is done on a large business computer, since a large amount of data has to be processed from all quality operations. The most important tasks on this tier are:

- global quality assurance planning,
- coordination of all quality operations,
- maintenance of the quality assurance database,
- administration and updating of all quality plans and procedures,

- issue of quality plans and quality procedures for inspecting the manufacturing orders,
- processing and documentation of all quality data from lower levels,
- statistical monitoring of the quality of each product and component,
- performance of cost accounting for all quality control activities,
- performance of process capability studies for the manufacturing process.

11.6.2 Quality assurance planning tier

When an order has been placed and is scheduled for production, the quality plans and procedures are received from the upper tier and the test plans will be prepared. The test plans are identified with the order and sent to the lower level for execution. This will include the determination of the test modes (for example, sample or 100% testing) and if necessary, the set up of alternative plans.

In most manufacturing operations, a product goes through a sequence of quality tests; thereby, all measuring data has to be gathered and identified with the product from which they are taken. Since the quality produced in an early operation will affect the quality obtained in a later operation, frequent comparison of measurements and the synchronization of data merging and exchange has to be done. These tasks must be performed on this level and can be quite complex for long production or assembly lines. It will also be the responsibility of the computer to supervise the synchronization, since in many situations it is extremely difficult to follow a product through a plant. For example, if synchronization of data from various tests is done for a product transported through a plant by a chain type conveyor, the hangers may swing; thus, measurements with light barriers or other identification devices may become impossible.

Another important function of this tier is the evaluation and reduction of the test data from the lower level, the calculation of quality trends and the display of the process performance history on a graphical terminal used for the supervision of the process.

The main assignments to this tier can be summarized as follows:

- drafting and distribution of the test plans for the orders,
- determination of the number of tests to be performed,
- synchronization and supervision of the test systems,
- collection, evaluation and compression of test data,
- prediction of quality trends for each product,
- supervision of the performance of the manufacturing process and the prediction of trends,
- functional testing of test equipment on the lower level,
- automatic learning and issue of test parameters,

- automatic loading of test plans and programs for the production runs at lower levels,
- reporting and display of process difficulties,
- location of bottlenecks in the manufacturing system.

The computers used at this level are control or microcomputers with realtime capabilities to interact quickly with the process when difficulties arise.

11.6.3 Quality assurance control tier

The computing equipment on this tier performs the quality control operations. The assignment involves test monitoring, data collection, and evaluation. With highly automated manufacturing operations the direct control of the process may also be done if a quality problem is observed. Since automatic data acquisition may not be possible for all types of defects, data is also entered by hand on this level with portable terminals or other devices.

Tests performed on this level require fast computers with realtime capabilities; in particular with high piece rates where the processes are also controlled. Usually, dedicated microcomputers are employed for these tasks.

A summary of the important functions on this level is as follows:

- activation and deactivation of the test runs,
- control of the measuring process,
- collection and evaluation of test data,
- comparison with test limits,
- plausibility test,
- filtering of test data,
- averaging of data,
- trend analysis,
- drift correction,
- linearization,
- normalization,
- calculation of nonmeasurable parameters,
- periodic testing of the measuring equipment,
- data compression and issue of test reports,
- sending test data to upper levels.

11.6.4 Interfaces of a quality assurance system

When a hierarchical quality assurance system is designed and installed the implementer must be sure to use all available standards for hardware and software communication. Many standards will be identical to those used for other manufacturing operations, however, there are also several standards designed specifically for quality control.

Standards for communication

On the higher level of the hierarchical system the MAP concept may be used for interfaces and communication (see Chapter 6). The lower levels can be interconnected by a field bus or a special instrument communication system, such as the popular IEEE 488 bus. Many instrument manufacturers provide their measuring devices with an interface for this bus. Thus, it is possible to set up measuring isles where instruments of various types and manufacturers are combined in one system. Figure 11.19 shows a typical instrument set up for testing an engine using the IEEE 488 bus. Such a system is supervised by a computer, and it is possible to have communication between the computer and the instruments or just between instruments. Communication is possible over a distance of 20 meters at a bit parallel/byte serial rate of 1 Mbyte/s.

If communication over a longer distance is required, several of these isles can be interconnected by another bus system suitable for such a data transfer.

Software interfaces

There are various software interfaces available; several of them facilitate the communication of the geometrical data of the product with the measuring machine

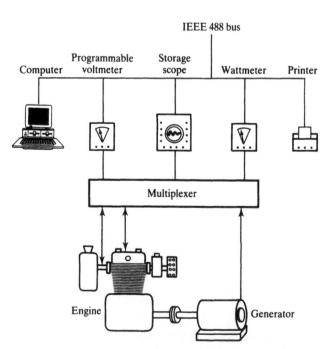

Figure 11.19
IEEE 488 based measuring bus system used for an engine test.

(see Chapter 7). For example, an IGES interface is available to coordinate measuring machines of numerous well-known manufacturers.

For programming of measuring machines, the neutral data file specification (NDFS) was developed within the framework of the CAM-I activities. The NDFS vocabulary is similar to that of APT for NC-machines. It contains elements to define measurement tasks and to describe geometric objects like form elements, tolerances, coordinate systems, and so on. An interface akin to CLDATA was also defined, it is named numerical controlled measuring and evaluation system (NCMES).

11.7 Programming of Test Systems

The usefulness of a computer-controlled test system depends greatly on the available software. A powerful language must be provided and supporting software to configure the test system, to operate it, and to evaluate the test results. It is a desirable feature of the software system to be able to extract the test methods, sequences and operations from the drawing; and also to configure the entire test system. Typical data and programs needed for this task are shown in Figure 11.20. The reader can readily see that it is a formidable task to expect a programming system to set up the entire test on its own and to provide the test programs. If this is done automatically, it would be necessary to extract, from an experienced quality assurance engineer, all the knowledge he has on testing and to set it up in an expert system. With the AI and software tools presently available this is impossible. For this reason, only the most important phases of testing will be automated.

11.7.1 The programming languages

Programming of test systems is usually done by the quality control engineer or the operator of the equipment. In more advanced CIM operations the program is provided by engineering. In general, these people are neither computer scientists nor expert programmers. For this reason, the language must be as simple as possible. Typical criteria are:

- powerful high-level programming language,
- an instruction set which is suitable for defining test applications,
- ease of learning,
- programs must be self-documenting,
- programs must be efficient and reliable,
- portability is required for many applications.

The selection of a language is also governed by factors such as type of test, type of test equipment, availability of control computers, standard languages used in

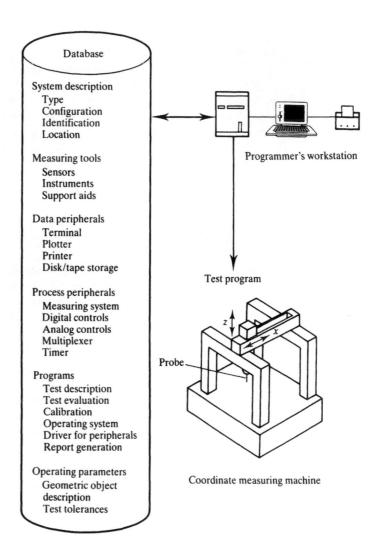

Figure 11.20

Typical data needed for writing a test program for a coordinate measuring machine.

a plant, and many others. Figure 11.21 contains a list of typical programming languages suitable for test applications. We will be looking at them in more detail.

11.7.2 Low-level programming

The low-level languages are mostly used in special-purpose test systems which are sold in large quantities. The program written in such a language controls and operates the test equipment. The user has no access to the operational level of the equipment. To be able to communicate with the test, he is provided with an application-oriented user interface, by which he can describe the test and evaluate the measurement data.

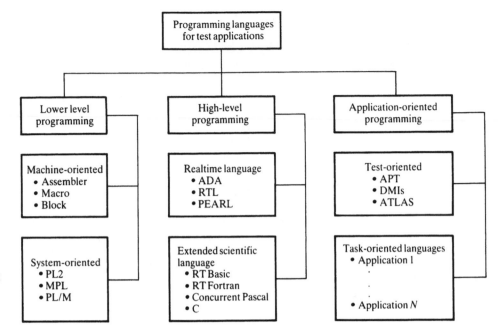

Figure 11.21
Various programming languages used for measuring applications.

The evaluation programs are usually not very powerful. The low-level languages are also well suited for test applications where a considerable amount of digital data has to be processed.

11.7.3 High-level programming

For high-level programming, realtime languages and extended scientific languages are commonly used. The realtime languages can handle the control of the measuring system very well; they are also very powerful for the statistical evaluation of test data and for the issue of test reports. However, good skills are needed to write efficient programs. The extended scientific languages have the advantages that they are known by many engineers. To be useful for realtime control, the extensions consist of special logic instructions for bit and byte manipulation, I/O instructions for digital and analog data, and instructions for task communication. There are test systems on the market which provide well-designed language interfaces, for example, RT Basic. The strength of these systems is that the test can be easily defined and there are numerous statistical and report generation routines available to evaluate and report the test results.

The basic disadvantage of these high-level languages is that they are not universal enough for handling various special-purpose applications, for example, geometric data processing of coordinate measuring machines. In many cases they are also too difficult for the operator to use; usually, Fortran and Basic are exceptions.

11.7.4 Application-oriented programming

The application-oriented languages are either test- or task-oriented. The most important test-oriented languages for coordinate measuring machines are the derivatives of APT, and for more general purpose testing ATLAS. The task-oriented languages are individually designed for special-purpose applications and are an important interface to well-conceived test systems. We will be discussing these three types of languages in more detail.

APT type languages

These languages allow the description of three-dimensional workpieces and also compensate for the diameter of the probe tip, when measurements are taken. When a workpiece is described by APT to be processed on an NC or DNC machine tool, the program for the coordinate measuring machine for testing the workpiece will be very similar. However, the APT language has to be expanded to consider the special requirements of testing. Some of the functions are:

- definition of a reference coordinate system for the workpiece set-up;
- definition of an arbitrary coordinate system in space;
- definition of the measuring angle for the probe;
- selection of various probes;
- selection of measuring increments for index tables;
- definition of measuring points along surface contours such as lines, circles and ellipses;
- reconstruction of geometric form elements from measurements, such as planes, spheres, toruses and so on;
- averaging of measurements;
- support of the calibration of probe tips with the aid of a calibration sphere, as shown in Figure 11.22;
- numerical output of test data in a format like NCMES;
- output of test reports containing measuring data, deviations, extreme values, and so on;
- graphical presentation of test results;
- automatic design of the test cycles, for example, the significant number of measurements to be performed.

Many of the features mentioned are actually provided for commercially available coordinate measuring machines; they may be contained in the programming language or bought as a separate service program.

The reconstruction of the geometric form elements from measurements uses a

Figure 11.22
Calibrating a
measuring probe
(courtesy of Carl
Zeiss, Oberkochen,
Germany).

service program, which can be divided into several routines, see George *et al.* (1984)
and Weckermann (1984):

- Calculation of form elements (straight lines, circles, planes and so on) from the
  measuring data and determination of their location, as shown in Figure 11.23.
  For every form element a minimum number of measuring points must be given
  to identify it;

- Determination of the location of the form elements relative to each other;
  routines are provided to calculate the distance, angle, hole circle, row of holes,
  and points of intersection;

- Calculation of the deviation from form and location. This can be done on the
  basic of a Gaussian distribution;

- Calculation of alignment and transformation to locate the coordinate system
  of the form element with regard to the coordinate system of the workpiece or
  measuring machine;

- Special programs, for example, for measuring gears, splines, and so on.

Form element	Minimum number of measurements	Described form element	Approximated outline of form element
Point	1		
Straight line	2		
Plane	3		
Circle	3		
Sphere	4		
Cylinder	5		
Cone	6		
Torus	7		

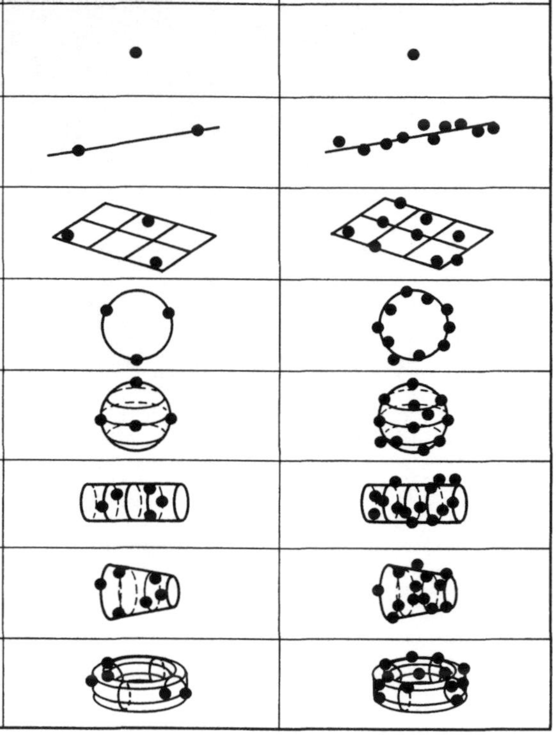

Figure 11.23
Reconstruction of form elements from measurements.

The ATLAS programming language for general purpose applications

ATLAS (Abbreviated Test Language for All Systems) was initially designed for complex test applications in the avionics and space industries. It actually is of such a general design that it can be used for many applications, thus it became an IEEE standard. ATLAS is system-independent. A program designed for one test system should also be operable on another for which an ATLAS compiler is available. Since ATLAS is a very powerful language and not all features are used for all applications, many manufacturers developed subsets of this programming system. This has led to a situation where the subsets are often manufacturer-specific. Example ATLAS instructions are given in Table 1, see also Rembold *et al.* (1985).

Task-oriented languages

For most test applications on the plant floor the power of a general-purpose language is not needed, and there is no requirement for the portability of programs.

Table 11.1 Example instructions from the ATLAS programming language.

BEGIN   Starts a test

TERMINATE   Ends a test

PREAMBLE SECTION   First part of a program; specifies all EQUATE operations and subroutines used

SUBROUTINE   Defines a subroutine; a maximum of 25 are allowed; they are nonrecursive

END   End of a subroutine

EQUATE   Equates the computer I/O channel number with the terminal number of the test station

Procedure section   Defines a series of statements to perform individual test steps

PERFORM   Calls a subroutine or another ATLAS program

CALCULATE   Performs the arithmetic operations:
+   Add
−   Subtract
*   Multiply
/   Divide
**   Exponentiate

COMPARE   Compares results with specified limits

VERIFY   Rates the test results:
GT   Greater than
LT   Lower than
EQ   Equal to
FAIL HIGH/LOW
UNIT FAIL HIGH/LOW
UNIT PASS

GO TO BRANCHING   Unconditional and conditional branching

REPEAT   Defines number of times to repeat a test

MONITOR   Reads and displays test data

WAIT FOR   Delays a test until operator resumes it

PRINT   Output of test data on printer

DISPLAY   Output of test data on CRT terminal

RECORD   Identifies and saves measurement data

READ   Reads analog data and retains them in MEASUREMENT register

FINISH   End of test; places equipment in hold state

CONNECT/DISCONNECT   Connects or disconnects test instruments

SETUP   Program test instruments

APPLY   Compares functional test pattern with reference pattern

MASK   Is used to ignore indeterminate outputs

However, it is important that a test can be programmed and operated by personnel who have no specific computer knowledge. In such a situation, a task-oriented test language renders the best service. Often, such a language only needs about 20 instructions, which can be learned in a couple of hours.

A typical set of instructions may have the following capabilities:

- to initiate and terminate a test,
- to set counters for cyclic testing,
- to provide a facility to input test limits,
- to initiate a measurement and to store the test parameter in memory,
- to filter test data,
- to perform a plausibility check with data,
- to perform a comparison with test parameters,
- to average test results and calculate means,
- to convert the physical or electrical test parameters to engineering units,
- to linearize test data,
- to output feedback control commands to the test system,
- to output test results on a predefined form,
- to graphically display the data on a terminal.

These languages find a high acceptance by operating personnel. Their disadvantage is that they are very rigid and almost impossible to adapt to a new application. They also carry with them a large overhead of system software. Several manufacturers have designed configuration systems for such languages, whereby a given test requirement is specified and the system assembles from a library of test modules the appropriate application instruction set. These languages are equipment-specific and cannot be used with test systems of other manufacturers.

11.8 Problems

(1) Explain how an integrated quality assurance concept can be structured and show the tasks to be performed in various control loops. Why are the control loops necessary and how do they interface with other CIM activities?

(2) Explain the three-phase control loops involved in quality assurance. Why is the customer so important in the global quality control system?

(3) What are the advantages of computer-controlled test systems?

(4) Which tests are usually performed in a production system for discrete products?

(5) Devise a test planning system for rotational parts. What are typical questions to be asked when preparing the test plan? How can test planning be computerized?

(6) How can online testing be supported by the use of process control charts? Design an algorithm to monitor the testing of products on a manufacturing line. How can out of control trends be anticipated?

(7) Which steps must be taken to test a product? Start with the design of the product and end with the customer accepting the test. Where is the computer most useful?

(8) What are the advantages and disadvantages of contact and non-contact measurements? How can the computer help increase the accuracy of non-contact measurements?

(9) How can testing of units with transient behavior be done efficiently? Can an expert system be used effectively for transient tests? How would you design a transient test for testing heating units?

(10) What are the various design principles for coordinate measuring machines and what are the various ways of teaching a machine the shape of a workpiece?

(11) What are the control tiers of a quality control system and which tasks are performed on each tier?

(12) What are typical programming languages for product testing and what are their features?

(13) Why is it often more efficient to design an application-oriented test language rather than using a multipurpose test language.

(14) What are the components of a computer-controlled test system? Explain it with a system using automatic feedback control.

(15) Why is it often so difficult to test products on moving conveyor systems? Design a system for an online test for an internal combustion engine and explain its hardware and software components.

(16) Can a test system be set up to learn its own test limits? Will such a system be stable?

(17) Explain the functions and structure of the IEEE 488 communication bus used for interconnecting test instruments. What are its advantages and disadvantages?

References

Chapter 1   The Manufacturing System

Ausschuß für wirtschaftliche Fertigung e.V. (AWF) (1985). *Integrated Data Processing Applications in Production, CIM – Computer Integrated Manufacturing, Terminology, Definitions, Functions*. AWF Eschborn, Germany

CIM-integrierte rechnerunterstützte Fertigung. *CADCAM Labor* (1987). Kernforschungszentrum Karlsruhe GmbH, Karlsruhe, Germany

DW (1987). Normung von Schnittstellen für die rechnerintegrierte Produktion (CIM). *DIN Fachbericht* (15), Beuth Verlag GmbH, Berlin, Germany

Hirschberg G. (1989). Produktionsautomatisierung und Flexibilität aus der sicht des Informatikers. *VDI Bericht*, **723** VDI Verlag, Düsseldorf, Germany

Hoppen W.G. (1976). Automatic scheduling of manufacturing resources. *Society of Manufacturing Engineers Technical Publication*, EN76-970

Horn K. (1987). Flexible Fertigungssysteme für Flachbaugruppen. *Technische Rundschau*, (36)

Lutz P. (1987). Leitsysteme für die rechnerintegrierte Auftragsabwicklung. *Dissertation*, Universität München, Germany

Milberg J. and Bürstner H. (1986) State of the art and development trends of CIM-concepts. *Rechnerintegrierte Konstruktion und Produktion*, VDI-Bericht (611)

Neunheuser B. (1984). Materialfluß in Fertigung und Montage. *VDI Bericht* (520), VDI Verlag, Düsseldorf, Germany

Nuber C., Köhler C. and Schultz-Wild R. (1989). CIM in der Bundesrepublik. *Technische Rundschau* (8)

Putz U. and Luck D. (1986). Flexible Montage elektronischer Bauteile. *ZWF* (81)

Rembold U., Blume C. and Dillman R. (1985). *Computer Integrated Manufacturing Technology and Systems*. Marcel Dekker, Inc., NY

VDI-Bericht (1971) VDI-Verlag, Düsseldorf, Germany

Wiendahl H.-P. (1986). Development of strategies to shorten the in-shop time of workpieces with the aid of a model. *VDI Bericht, Methods of In-Shop Time Improvements in Single and Small Lotsize Production*, VDI Verlag, Düsseldorf, Germany

Wolf L. (1988). Divergierende Perspektiven bei CIM. *VDI Nachrichten*, **42**, (21) Okt

Chapter 2   CIM Models and Concepts

Albus J.S., Barbara A.J. and Nagel R.N. (1981). Theory and practice of hierarchical control. In *Proc. IEEE Computer Soc. Int. Conf.* Compcon Falls, National Bureau of Standards, Washington DC

CIM-OSA reference architecture specification (1989a). *ESPRIT Project No. 688*, ESPRIT Consortium, AMICE, Brussels

CIM-OSA Open system architecture for CIM. (1989b). *ESPRIT Project No. 688*, **1**, ESPRIT Consortium, AMICE, Springer-Verlag, Berlin, Germany

Dadam D., Dillmann R., Lockemann P.C. and Südkamp N. (1989). Production integration with the aid of databases – requirements, solutions, trends. *VDI Bericht*, 723, VDI-Verlag, Düsseldorf, Germany (in German)

Flatau U. (1988). *CIM Architectural Framework*. Mainard, MA Digital Equipment Corporation

Grabowski H., Anderl R., Pätzold B. and Rude S. (1988). The development of advanced modeling techniques–meeting the challenges of CAD/CAM integration. In *Proc. 4th CIM Europe Conf.* Madrid, Spain, May

IBM (1973). *Communication Oriented Production Information and Control System*. (1973). IBM Corporation, White Plains, NY

Lockemann P. (1988). The database as communication medium in computer integrated manufacturing. *Technische Rundschau*, **45** (in German)

MAP (1985). *Network Users and Manufacturers*. Warren, Michigan: General Motors Technical Center

Milberg J. (1985). *Robots on Wheels*, Würzburg, Germany: Fabrik 2000 Vogel Verlag (in German)

Rembold U. (1986). The automated factory of the Nineties: planning, scheduling, control. In *Proc 7th European Conf. Electronics, EUROCON 86*, Paris, France

Su S.Y.W. (1985). Modeling integrated manufacturing data using SAM. *GI Conf. Database Syst. for Office, Technique and Science* Karlsruhe, March 1985, published in *Informatik-Fachberichte*, 94, Springer-Verlag, Berlin, Germany

Weston R.H., Gascoigne J.D., Sumpter C.M. and Hodgson A. (1989a). The need for generic framework for system integration. In *Advanced Information Technologies for Industrial Material Flow Systems* (Nof S.Y. and Moody C.L. eds) NATO ASI Series, Springer-Verlag, Berlin, Germany

Weston R.H., Harrison R., Booth A.H. and Moore P.R. (1989b). *A New Approach to Machine Control* Department of Manufacturing Engineering, Loughborough University of Technology, Loughborough, Leicester, UK

Further reading

Baumgarter H., Knischewski K. and Wieding H. (1989). *Basic CIM Considerations*. Siemens Aktiengesellschaft, Produktionsautomatisierung und Automatisierungssysteme, Produktionsleittechnik, Nürnberg-Moorenbrun, Germany (in German)

IBM (1985). Modeling of manufacturing systems. *Journal of Research and Development*, **29**, No. 4

Computer Integrated Manufacturing, The CIM Enterprise. (1989). IBM Corporation, White Plains, NY

MacConnaill P. (1988). ESPRIT CIM, achievements and goals. In *Proc. 4th CIM Europe Conf.* Madrid, Spain, May

Metaxas D. and Sellis T. (1991). A frame-based model for large manufacturing database. *Journal of Intelligent Manufacturing*, **2**, No. 1

Next generation database systems. (1991). *Communications*, **34**, No. 10

Waller S. (1989). Computer architecture and communication nets in production. *Technische Rundschau*, **8**, (in German)

Wu D. (1989). Artifical intelligence and scheduling application. *Artifical Intelligence, Manufacturing Theory in Practice* (S.T. Kumars *et al.*, eds) Institute of Industrial Engineers

Chapter 3   Analysis Tools for Manufacturing Systems

Doumeingts G. (1989). GRAI approach to designing and controlling advanced manufacturing systems in CIM environment. *Advanced Information Technologies for Industrial Material Flow System* (Nof S.Y. and Moody C.L. eds) NATO ASI Series, Berlin-Heidelberg: Springer-Verlag

Eversheim W. and Neitzel R. (1989). The interfacing of expert systems with design. In *Proc. INFINA '89 Conf.* Karlsruhe, Germany (in German)

Fox M.S. (1983). *Constraint-Directed Search: A Case Study of Job-Shop Scheduling.* PhD thesis, Ann Arbor: University Microfilm International

Frommherz B. (1989). A concept of a robot action programming system. *PhD Dissertation*, University Karlsruhe, Germany (in German)

Grimm W. (1987). Diagnosis system for detecting faults in the control of manufacturing systems. *PhD Dissertation*, University of Stuttgart, Germany (in German)

Hörmann A. (1989). A Petri net based control architecture for a multi-robot system. In *Proc. 4th IEEE Int. Symp. on Intelligent Control*, Albany, NY

Majumdar L.P. and Rembold U. (1989). 3-D model based robot vision by matching scene description with the object model from a CAD modeler. *Robotics and Autonomous Systems* **5**

Meyer W. and Isenberg R. (1979). Aspects of knowledge-based factory supervision systems ESPRIT 932. *ESPRIT CIM Europe* Athens, Greece

Ochs M. (1988). Design of a programming system for knowledge-based planning and configuration. *PhD Thesis*, University of Karlsruhe, Germany

Ochs M., Ganghoff P. and Wenle H. (1989). Syllogist – an expert system shell for planning and configuring technical systems. In *Proc. INFINA '89 Conf.* Karlsruhe, Germany (in German)

Raczkowsky J. (1990). A blackboard system for the diagnosis of assembly tasks in robotics. In *Proc. Int. Conf. Artifical Intelligence and Expert Syst. in Manufacturing*, London

Rembold U., Blume C. and Dillmann R. (1985). *Computer Integrated Manufacturing Technology and Systems.* New York: Marcel Dekker

Zhon M., Di Cesare F. and Desrochers A.A. (1989). A top-down approach to systematic Petri net models for manufacturing systems. In *Proc. Int. Conf. on Robotics and Automation*, Scottsdale, AZ

Further reading

Ahmed R. (1989). *Concepts for the management of computer aided design databases*. PhD Dissertation, University of Florida, FL

Ahmed R. and Navathe S.B. (1991). Version management of composite objects in CAD databases. In *Proceedings of the ACM-SIGMOD Conference*, **29** (2) 218–27

Albus J., Barbara A.J. and Nagel R.N. (1981). Theory and practice of hierarchical control. In *Proc. IEEE Computer Soc. Int. Conf.*, Compcon Falls, National Bureau of Standards, Washington, DC

Banerjee J., Kim W., Kim H. and Korth H.F. (1987). Semantics and implementation of schema evolution in object-oriented databases. In *Proceedings of the ACM-SIGMOD Conference*, **3**, 311–22

Ben-Arieh D. (1988). A knowledge-based simulation and control system. In *Artificial intelligence: Implications for CIM*, (Kusiak A. ed), Springer-Verlag, 257–68

Ben-Arieh D. (1991). Concurrent modeling and simulation of multi robot systems. *Robotics and Computer Integrated Manufacturing* **8** (4) 67–73

Ben-Arieh D. (1992). Concurrent modeling language (CML) for discrete process modeling, simulation and control. *Journal of Intelligent Manufacturing*, **3** 31–41

Bernhard R. *et al.* (1992). *Integration of Robots into CIM*. Chapman & Hall, London, New York

Biliris A. (1989) A data model for engineering design objects. In *Proceedings of the ACM IEEE 2nd International Conference on Data and Knowledge Systems for Manufacturing and Engineering*, Gaithersburg, Maryland, 49–58

Billington J., Wheeler G.R. and Wilbur-Hum M.C. (1988) Protean: a high level Petri net tool for the specification and verification of communication protocols. *IEEE Transactions on Software Engineering*, **14** 301–16

Black J.T. (1988). The design of manufacturing cells (step one to integrated manufacturing systems). In *Proceedings of Manufacturing International '88*, Atlanta, GA, **3** 143

Caselli S., Papaconstantinou C. and Doty K.L. (1990). Using semantic data models in knowledge-based manufacturing workcell design. In *Proceedings of the 5th IEEE International Symposium on Intelligent Control*, Philadelphia.

Caselli S., Papaconstantinou C., Doty K.L. and Navathe S. (1992). A structure-function-control paradigm for knowledge-based modeling and design of manufacturing workcells. *Journal of Intelligent Manufacturing*, **3** 11–30

Chaturvedi A.R. and Hutchinson G.K. (1990). *A model for simulating FMS congestion in an FMS*, Working Paper No. 975, Purdue University, Indiana

Chaturvedi A.R., Hutchinson G.K. and Nazareth D.L. (1992). A synergistic approach to manufacturing systems control using machine learning and simulation. *Journal of Intelligent Manufacturing*, **3** 43–57

Chaturvedi A.R., Hutchinson G.K. and Nazareth D.L. (1992). Supporting complex real-time decision making through machine learning, *Decision Support Systems*, forthcoming.

Desrochers C.M. (1990). Modeling and Control Using Petri Nets. In *Modeling and Control of Automated Manufacturing Systems*, IEEE Computer Society Press, Washington, DC, 239–51

Duffie N.A. and Piper R.S. (1987). Non-hierarchical control of a flexible manufacturing cell. *Robotics and Computer Integrated Manufacturing*, **3** (2) 175–180

Ehrismann R. and Reissner J. (1987). Intelligent manufacture of bent, punched, and laser cut sheet metal parts. *Technische Rundschau* **27** (in German)

Elmasri R. and Navathe S.B. (1989). *Fundamentals of Database Systems*. Benjamin Cummings Publishing Company Inc., Redwood City, CA, 641–44

Gaines B.R. (1988). Applying expert systems. *Expert Systems, Strategies and Solutions in Manufacturing Design and Planning* (Kusiak A. ed), SME

Hübner Th. (1992). Real-time candidate generation for diagnosis. *Second International Conference on Automation, Robotics and Computer Vision*; Singapore, Proceedings, **3**, Inv-5.4, 1-5.

Jackson P. (1990). *Introduction to Expert Systems*, second edition, Addison-Wesley, Massachusetts.

Jenson K., Rozenberg G. (eds.) (1991). *High-level Petri Nets: Theory and Application.* Springer-Verlag

Joshi S., Vissa N.N. and Chang T.-C. (1988). Expert process planning with solid modeling interface. *International Journal of Production Research*, **26** 863–85

Kasturia E., DiCesare F. and Desrochers A.A. (1988). Real-time control of multilevel manufacturing systems using colored Petri nets. In *Proceedings of the IEEE International Conference on Robotics and Automation*, Philadelphia, 1114–19

Katz R.L. (1990). Business/Enterprise Modeling, *IBM Systems Journal*, **29** (4)

Kumara S.T., Kashyap R.L. and Soyster A.L. (eds) (1988). *Artificial Intelligence: Manufacturing Theory and Practice*, IIE Press, USA

Kusiak A. (1986). Application of operations research models and techniques in flexible manufacturing systems. *European Journal of Operations Research*, **24** 336–45

Levas A. and Jayaraman R. (1989). WADE: an object-oriented environment for modeling and simulation of workcell application. *IEEE Transactions on Robotics and Automation*, **5** 324–336

Misra J. (1986). Distributed discrete event simulation. *Computing Surveys*, **18** 39–65

Murray K.J. and Sallie V.S. (1988). Knowledge based simulation and specification. *Simulation*, **50** 112–19

Nevins J.L. and Whitney D.E. (1989). *Concurrent Design of Products and Processes.* McGraw-Hill Publishing Company, New York

Papaconstantinou C., Fernicola P.F., Doty K.L. and Navathe S.B. (1989). Knowledge-based manufacturing workcell design and modeling tool. In *PROCIEM '89*, Orlando, FL, 103–5

Park S.C., Raman N. and Shar M.J. (1989). Heuristic learning for pattern directed scheduling in a flexible manufacturing system. In *Proceedings of the Third ORSA/TIMS Conference on Flexible Manufacturing Systems*, Elsevier, Amsterdam, 369–76

Patton R., Frank P., Clark R. (1989). *Fault Diagnosis in Dynamic Systems, Theory and Applications*, Prentice Hall International, Series in Systems and Control Engineering, ed. M.J. Grimble

Peterson J.L. (1989) *Petri Net Theory and the Modeling of Systems*, Prentice Hall, Englewood Cliffs, NJ.

Plünnecke H., Reisig W. (1991). Bibliography of Petri Nets. In Rozenberg G. *Advances in Petri Nets*, Springer Verlag, Lecture Notes in Computer Science, (524) 317–572

Ruby D. and Kibler D. (1989). Learning to plan in complex domains. In *Proceedings of the Sixth International Workshop on Machine Learning*, Ithaca, New York, 180–2

Shapiro E. (1989). The family of concurrent logic programming languages. *Computing Surveys*, **21** 412–510

Spooner D.L., Hardwick M. and Liu K.L. (1988). Integrating the CIM environment using object-oriented data management technology. In *Proceedings of the International Conference on C.I.M.*, Troy, NY, 144–52

Stylianou A.C., Madey G.R. and Smith R.D. (1992). Selection criteria for expert system shells. *Communications*, **35** (10)

Teng S. and Black J.T. (1990). Cellular manufacturing systems modeling: the Petri net approach. *Journal of Manufacturing Systems*, **9** 45–54.

Walters J. and Nielsen N. (1988). *Crafting Knowledge-based Systems, Expert Systems made Realistic*, John Wiley and Sons, New York

Willson R.G. and Krogh B.H. (1990). Petri net tools for the specification and analysis of discrete controllers. *IEEE Transactions on Software Engineering*, **16** 39–50

Chapter 4   Flexible Manufacturing and Assembly Equipment

Further reading

Baril R. (1987). *Modern Machining Technology*. Delmar Publishers Inc.

Bernhardt R., Dillmann R., Hörmann K. and Tierney K. (1992). *Integration of Robots into CIM*. London: Chapman & Hall

Bradley D.A., Dawson D., Burd N.C. and Loader, A.J. (1991). *Mechatronics Electronics in Products and Processes*. London: Chapman & Hall

CECIMO. (1985). *Working Party on Standardization: Terminology for Automated Manufacturing and Machining Systems*. CECIMO, Brussels

CIRP Terminal Report. (1990). Nomenclature and Definitions for Manufacturing Systems. *Annals of the CIRP* **2** (32)

Fricke F., Görnke M., Hammer H., Schuster J. and Vieweger, B. (1988). *Flexible Fertigungssysteme in der Praxis*. Firmenschrift Werner und Kolb

Genschow H. (1990). Gesichtspunkte bei der Auswahl eines Bearbeitungszentrums. *VDI-Z* **132** (9) 50–57

Harthy J. (1984). *FMS at Work*. New York: North Holland Publishing Company

Heisel U. (1990). Planung und Auslesung flexibler Fertigungssysteme. In *Leittechnik für verkettete Fertigungssysteme* Dusseldorf: VDI Bildungswerk

Pritschow G. (1985). Die flexible Fertigungszelle. Chance und Herausforderung auch für den Mittelständischen *Betrieb. wt-Z. ind. Fertig.* **75** (11) 663–668

Rembold U. (1990). *Robot Technology and Applications*. New York: Marcel Dekker

Rembold U., Blume C. and Dillmann R. (1985). *Computer-Integrated Manufacturing Technology and Systems*. New York: Marcel Dekker

Serope Kalpakjian (1989). *Manufacturing Engineering and Technology*. Addison-Wesley Publishing Company

Serope Kalpakjian (1985). *Manufacturing Process for Engineering Materials*. Addison-Wesley Publishing Company

Stute G. (1981). *Regelung an Werkzeugmaschinen*. Munich, Vienna: Hanser

Tuffentsamer K., Storr A., Lange K., Pritschow G. and Warnecke H.-J. (1988). *Flexibles Fertigungssystem – Beiträge zur Entwicklung des Produktionsprinzips*. Weinheim: VCH Verlagsgesellschaft

Viehweger B. (1990). Wirtschaftlichkeitsbetrachtung für flexible Fertigungs-systeme. In *Leittechnik für verkettete Fertigungssysteme* Düsseldorf: VDI Bildungswerk

Chapter 5   Control Structures for CAM Systems

Further reading

Bauer A., Bowden R., Browne J., Duggan J. and Lyons G. (1991). *Shop-Floor Control Systems from Design to Implementation.* London: Chapman & Hall

Bedworth D.D., Henderson M.R. and Wolfe P.M. (1991). *Computer-Integrated Design and Manufacturing.* New York: McGraw-Hill

Beier H.H. (1989). Fertigungsleitstand und Fertigungsleittechnik. *ZwF* **84** (3) 133–137

Bellman B. (1991). Betriebsmittelorganisation, Voraussetzung für die wirtschaftliche Fertigung. (1991). In *Wettbewerbsfaktor Zeit in Produktionsunternehmen* (*Proc. Munich Colloq.*) Berlin, Springer-Verlag, 259–274

Granow R. (1989). Rationalisierungspotential Betriebsmittelmanagement. *ZwF*, **84** (6) 316–320

Groha A. and Schönecker W. (1988). Universelles Zellenrechnerkonzept zur nutzungsverbesserung flexibler Fertigungssysteme. *Wt Werkstattstechnik*, **78** 313–318

Lyonnet P. (1991). *Tools of Total Quality An Introduction to Statistical Process Control.* London: Chapman & Hall

Mayer J. (1988). *Werkzeugorganisation für flexible Fertigungszellen und -systeme* Berlin, Heidelberg, New York, Tokyo: Springer-Verlag

Milberg J. and Groha A. (1986). Der Zellengedanke als Strukturierungsprinzip im Informations- und Materialfluss flexibler Fertigungssysteme. *ZwF*, **81** (12) 682–687

Pritschow G., Frager O., Schumacher H. and Wieland E. (1989). Programmierung von roboterbestückten Producktionsanlagen. *Robotersysteme*, **5** 47–56

Rembold U., Blume C. and Dillmann R. (1985). *Computer-Integrated Manufacturing Technology and Systems.* New York: Marcel Dekker

Scheller J. and Sommer E. (1989). Hierarchisches Steuerungskonzept für flexible Montagezellen. *Automatisierungstechnische Praxis* atp, **31** (4) 166–173

Skivington J.J. (1990). *Computerizing Production Management Systems: A Practical Guide for Managers*, London: Chapman & Hall

Storr A. (1983). Programmieren von NC-Maschinen. *wt-Z. ind. Fertig.* **73** (1) 29–39

Storr A. (1990). Rechnerunterstützte Betriebsmittelorganisation und ihre Verknüpfung mit der Leittechnik. In *Leittechnik für verkettete Fertigungssysteme.* Dusseldorf: Bildungswerk

Storr A. and Brantner K. (1991). Das adaptierbare Leitsteuerungssystem ALSYS. In *Leit- und Steuerungstechnik in flexiblen Produktionsanlagen.* Munich, Vienna: Carl Hanser Verlag, 32–52

Storr A. and Donn R. (1983). Möglichkeiten zur Unterstützung des Programmierens von CNC-werkzeugschleifbearbeitungen. *wt-Z. ind. Fertig,* **73**, (8) 641–644

Storr A. and Hake U. (1991). Rechnerunterstützte Fertigungshilfsmittelbereitstellung. In *Leit- und Steuerungstechnik in flexiblen Productionsanlagen* Munich, Vienna: Carl Hanser Verlag, 77–91

Storr A., Mayer J. and Walker M. (1986). Werkzeugorganisation mit Schnittstelle zu Fertigungsleitsystemen. *wt-Z. ind. Fertig.* **76** (5) 287–291

Storr A., Zirbs J. and Hofmeister W. (1987). CAD/NC-Kopplung-Ziele, Probleme und Lösung. *Technische Rundschau*, **79** (39) 138–143

Venkutesk V.C. and McGeouchl J.A. (1991). *Computer-Aided Production Engineering.* Amsterdam: Elsevier

Walker M. (1988). Beitrag zur Schnittstellenfestlegung zwischen PPS- und Leitsystem. *wt Werkstatttechnik*, **78** 445–449

Walker M. and Staudenmayer A. (1990). Fertigungsleitsystemfunktionsbausteine und Softwarestrukturen. In *Leittechnik für verkettete Fertigungssysteme* Düsseldorf: VDI Bildungswerk

Walter W. and Hofmeister W. (1987). Universeller CAD/NC-kopplungsbaustein für NC-Programmiersystem. *wt-Z. ind. Fertig*, **77** (3) 129–133

Weck M. and Pauls A. (1991). Betriebsmittelverwaltung, neue Wege zur Koordination des Betriebsmittelflusses. *wt Werkstatttechnik*, **81** (1) 41–44

Young J.B. (1991). *Modern Inventory Operations Methods for Accuracy and Productivity* New York: Van Nostrand Reinhold

Chapter 6   Communication Nets and Protocol Standards

Further reading

Barth G. and Welsch C. (1988). Objektorientierte Programmierung (Object-oriented programming), *Informationstechnik*, it 6/88, R. Oldenbourg Verlag

Bever M. (1989). OSI application layer, Entwicklungsstand und Tendenzen, Tutorium Kommunikation in verteilten Systemen, Stuttgart

ESPRIT Project 2617 CNNMA. (1989). *Experimental Pilot at the University of Stuttgart, Detailed Functionality Specification*

Halsall H. (1992). *Data Communications, Computer Networks and Open Systems*. Addison-Wesley Publishing Company.

Hollingum J. (1986). *The MAP Report, Manufacturing Automation Protocol*. IFS Publications London: Springer-Verlag

Hordeski M. (1988). *Computer Integrated Manufacturing*. TAB Professional and Reference Books

Hutchison D. (1988). *Local Area Network Architectures*. Reading: Addison-Wesley.

ISO 7498. (1984). *Information Processing Systems Open Systems Interconnection Basic Reference Model*

ISO 9506-1 *Information Processing Systems Open Systems Interconnection Manufacturing Message Specification (Part 1) Service Definition*

ISO 9506-2 *Information Processing Systems, Open Systems Interconnection Manufacturing Message Specification (Part 2) Protocol Specification*

ISO DIS 95-6-3 *Information Processing Systems Open Systems Interconnection Manufacturing Message Specification (Part 3) Robot Specific Message System*

ISO DP 9506-4 *Information Processing Systems Open Systems Interconnection Manufacturing Message Specification (Part 4) Numerical Control Message Specification*

Keil R. (1985). *Map Specification*. Manufacturing Engineering and Development, General Motors Technical Center.

Kugler W. (1988). Fabrikkommunikation. In: *CIM-TT Querschnittsthema 6, Netze und Kommunikationstechnik*, Karlsruhe: Kernforschungszentrum

Further reading

Kühn P., Pritschow G. *et al.* (in prep). Series: CIM Fachmann, *Kommunikationstechnik für den Rechnerunterstützten Fabrikbetrieb* Berlin, Heidelberg, New York: Springer Verlag

Lederhofer, A. and Mottram D. (1989). *ESPRIT Project 2617* CNMA. *CNMA Implementation Guide 4.0*

MAP/TOP Users Group of SME – Training Subcommittee (1986). *MAP/TOP Training Guidelines*. The Computer and Automated Systems Association of SME.

Pritschow G. (1989). Automatisierungstechnik, eine ganzheitliche steuerungstechnische Aufgabe, *Produktionstechnisches Kolloquium* Vorträge, Berlin, Germany

Valenzano A., Demartini C. and Ciminiera L. (1992). *MAP and TOP Communications*. Reading: Addison-Wesley.

Withnell S. and Van Puymbroeck W. *et al.* (1990). *Proc. Open Communications Congress* Stuttgart: Springer-Verlag

Chapter 7   CAD: Its Role in Manufacturing

Ando K., Takeshige A. and Yoshikawa H. (1988). An approach to computer integrated production management. *Int. J. Product Research* **26** (3) 333–350

Boothroyd G. and Dewhurst P. (1983). *Design for assembly: A designer's handbook*. Technical report, Dept. of Mech. Eng., Univ. of Massachusetts at Amherst

Boothroyd G. (1975). *Fundamentals of Metal Machining and Machine Tools*. New York: McGraw-Hill

Boyes J.W. (1975). Interference detection among solids and surfaces. *Comms. ACM* **22** (1)

Chen J.P. (1991). *Feature Reasoning for Automatic Process Planning in Manufacturing*. Ph.D. dissertation, Department of Industrial Engineering and Operations Research, University of Massachusetts at Amherst

Dixon J.R. (1988). Designing with features: building manufacturing knowledge into more intelligent CAD systems. In: *Proc. ASME Manufacturing International*, Atlanta, GA

Fabrycky W.J. and Blanchard B.S. (1991). Life-cycle cost and economic analysis. In *Int. Series in Industrial and Systems Engineering*, Englewood Cliffs, NJ: Prentice-Hall

Frankfurt Verband der Automobilindustrie (1983). *VDA-Flächenschnittstelle (VDAFS), Version 1.0*. Verband der Automobilindustrie, Frankfurt, Germany

Grabowski H. and Glatz, R. (1986). Interfaces for the exchange of data to describe a product. *VDI-Zeitschrift*, **128** (10)

Groover M.P. and Zimmers E.W. (1984). *CAD/CAM: Computer-Aided Design Computer Aided Manufacturing*. Englewood Cliffs, NJ: Prentice-Hall

IGES. (1988). Initial graphics exchange specification (IGES) Version 4.0. Technical report, National Institute of Standards and Commerce

Johnson T.E. (1963). *Sketchpad iii: A computer program for drawing in 3-dimensions*. Technical report, MIT Electron System Lab.

Kusiak A. (1990). *Intelligent Manufacturing Systems*. Englewood Cliffs, NJ: Prentice-Hall

Milberg J. and Peiker S. (1987). Geometry and technology oriented interconnection of programming and NC systems. *Wt-Werkstattstechnik*, **77** (10)

Mund A. *et al.* (1987). *VDA Flächenschnittstelle (VDAFS), Version 2.0*. Berband der Automobilindustrie, Frankfurt, Germany

Nnaji B.O. (1988). A framework for CAD-based geometric reasoning for robot assembly planning. *The Int. J. Production Research Special Issue*, **26** (5) 735–764

Nnaji B.O. (1990). CAD-driven machine programming. In: *Proc. JAPAN-USA Symp. Automation and Robotics*, Kyoto, Japan

Nnaji B.O. (1992). *Theory of Automatic Robot Programming and Assembly.* London: Chapman & Hall

Nnaji B.O. and Chen J.P. (1991). *A Generic Feature Classification Scheme for Machining Parts and Sheet Metal Components, Report 250.* Technical report, Automation and robotics Laboratory, University of Massachusetts at Amherst

Nnaji B.O. and Kang T. (1990). Interpretation of CAD models through IGES interface. *Artificial Intelligence for Design, Analysis and Manufacturing,* **4** (1) 15–45

Nnaji B.O. and Liu H. (1989). Product assembly modeler. In *Proc. 4th Int. Conf. CAD/CAM* Delhi, India

Nnaji B.O. and Liu H. (1990). Feature reasoning for automatic robotic assembly. *Int. J. Production Research,* **28** (3) 517–540

Nevins J.L., Whitney D.E. *et al.* (1989). *Concurrent Design of Products and Processes: A Strategy for the Next Generation in Manufacturing* NY: McGraw-Hill Publishing Company

PDES The content, plan and schedule for the first version of the product data exchange specification (PDES). (1985). Technical report, National Institute of Standards and Technology

Prinz F.B., Pinilla J.M. and Finger S. (1989). Shape feature description and recognition using augmented topology graph grammar. Reprints of *NSF Engineering Design Research Conf.* University of Massachusetts at Amherst, pp. 285–300

Rembold U., Blume C. and Dillmann R. (1985). *Computer Integrated Manufacturing Technology and Systems.* New York: Marcel Dekker

Requicha, A.A.G. (1980). Representations for rigid solids: theory, method, and system. *ACM Computing Surveys.* **12** (4)

Requicha A.A.G. and Voelker H.B. (1982). Solid modeling: A historical summary and contemporary assessment. *IEEE CG&A*

Requicha A.A.G. and Voelker H.B. (1983). Solid modeling: current status and research directions. *IEEE CG&A*

Salvendy G. (1982). Value Engineering. In *Handbook of Industrial Engineers.* New York: John Wiley

Schilli B., Grabowski H., Anderl R. and Schmitt M. (1989). Step – development of an interface for the exchange of manufacturing data. *VDI-Zeitschrift,* **131** (9)

Schlechtendahl E.G. (1988). *Specification of a VAD*I Neutral File for CAD Geometry.* Berlin, Heidelberg, New York: Springer-Verlag

Schlechtendahl E.G. (1989). *CAD Data Interface for Solid Models.* Berlin, Heidelberg, New York: Springer-Verlag

Schlechtendahl E.G. (1991). STEP/EXPRESS/STEP datafile. *Informatik spetitrum,* **14**(2)

Shah J.J. (1988). Feature transformations between application specific feature spaces, *Computer-Aided Engineering,* **5** (6) 247–255

Shilperoot B.A. (1975). Classification, coding and automated process planning. *CAM-I Proceedings*

Subbarao P. and Jacobs C. (1978). Application of nonlinear goal programming to machine variable optimization. *6th NAMRC Proceedings*

Swift K.G. (1987). *Knowledge-Based Design for Manufacture.* Englewood Cliffs, NJ: Prentice-Hall

Taguchi G. and Wu Y. (1986). *Off-line quality control.* Romulus, MI: American Supplier Institute

Whitney D.E., DeFazio T.L., Gustavson R.E. *et al.* (1989). Tools for strategic product design. In: *Reprints of the NSF Engineering Design Research Conference,* Amherst, MA,

Weiss I.A. (1984). *Product definition data interface: the solution.* Technical report, ICAM Project Report

Chapter 8 Process Planning and Manufacturing Scheduling

Berra P.B. (1968). Investigation of automated planning and optimization of metal working process processes. *Master's thesis,* Purdue University

Challa K. and Berra P. (1976). Automated planning and optimization of machining procedures – a systems approach. *Computers and Industrial Engineering,* (1) 35–46

Chang T.C. (1982). TIPPS – a totally integrated process planning system. *Master's thesis,* Virginia Polytechnic Institute and State University

Chang T.C. and Wysk R.A. (1985). *An Introduction to Automated Process Planning Systems.* Englewood Cliffs, NJ: Prentice-Hall

Conway R.W., Maxwell W.L. and Miller L.W. (1967). *Theory of Scheduling.* Reading, MA: Addison-Wesley

Davis R.P., Hayes G.M. Jr and Wysk R.A. (1981). A dynamic programming approach to machine requirements planning. *AIIE Trans,* **13** (2)

Day J.E. and Hottenstein M.P. (1970). Review of sequencing research. *Naval Research Logistics Quarterly,* (17) 11–39

Ehrismann R. and Reissner J. (1987). Intelligent manufacture of bent, punched, and laser cut sheet metal parts. *Technische Rundschau,* (27)

Eversheim W. and Fuchs H. (1980). Integrated generation of drawings, process-plans and NC-tapes. In *Advanced Manufacturing Technology,* IFIP/IFAC New York

Fiacco A.V. and McCormick G.P. (1968). *Nonlinear Programming.* Chichester: Wiley and Sons

Goyal S.K. (1975). Scheduling a single machine system: A multiproduct multi-item case. *Operational Research Quarterly,* **26** 619–627

Groover M.P. (1987). *Automation, Production Systems, and Computer-Integrated Manufacturing.* Englewood Cliffs, NJ: Prentice-Hall

Groover M.P. and Zimmers E.W. (1984). *CAD/CAM: Computer Aided Design Computer Aided Manufacturing.* Englewood Cliffs, NJ: Prentice-Hall

Gupta J.N.D. (1971). M-state scheduling problem–a critical appraisal. *Int. J. Prod. Res.,* **9** 267–281

Hati S. and Rao S. (1976). Determination of the optimum machining conditions – deterministic and probabilistic approaches. *J. Engineering for Industry,* **98**

Iwata K. *et al.* (1977). Optimization of cutting conditions for multipass operations considering probabilistic nature in machining processes. *J. Engineering for Industry,* **99**

Jackson J.R. (1956). An extension of Johnson's results on job-lot scheduling. *Naval Research Logistics Quarterly,* **3** (3)

Johnson L.A. and Montgomery D.C. (1974). *Operations Research in Productions Planning, Scheduling, and Inventory Control.* New York: John Wiley and Sons

Kanumury, M., Shah J. and Chang T.C. (1989). *An automatic process planning system for OTC–an integrated CAD and RAM system.* Technical report, School of Engineering, Purdue University

Kusiak A. (1988). *Artificial Intelligence: Implications for CIM*. IFS, Springer-Verlag

Kusiak A. (1990). *Intelligent Manufacturing Systems*. Englewood Cliffs, NJ: Prentice-Hall

Kilbridge M.D. and Wester L. (1961). A heuristic method of assembly line balancing. *J. Industrial Engineering*, **12** (4)

Nnaji B.O. and Davis R.P. (1988). Multi-stage multi-product lot size sequencing of operations. *Applied Mathematical Modeling*, **12** 593–600

Nnaji B.O. and Liu H. (1990). Feature reasoning for automatic robotic assembly. *Int. J. Prod. Res.*, **28** (3) 517–540

Chapter 9   Robotics

Ambler A.P. and Popplestone R.J. (1975). Inferring the positions of bodies from specified spatial relationships. *Artificial Intelligence*, **6** (2) 157–174

Brady M.J., Hollerbach J.M., Johnson T.L., Lozano-Perez T. and Mason M.T. (1983). *Robot Motion – Planning and Control*. MIT Press

Frommherz B. (1989). *A Concept for a Robot Action Planning System*. PhD thesis Institute for Real-time Computer Systems and Robotics, University of Karlsruhe, Germany

Frommherz B. and Werling G. (1988). Specifying configurations of 3-D objects by a graphical definition of spatial relationships. *Artificial Intelligence in Engineering: Robotics and Processes*, **2** 77–98

Frommherz B. and Werling G. (1989). A graphical implicit programming system for robot action planning. In *Proc. Int. Conf. Intelligent Autonomous Syst. IAS2*, **2** 196, Amsterdam, The Netherlands

Fu K.S., Gonzales R.C. and Lee C.S.G. (1987). *Robotics: Control, Sensing, Vision and Intelligence*. New York: McGraw-Hill

Grossman D.D. and Taylor R.H. (1978). Interactive generation of object models with a manipulator. *IEEE Trans. Syst., Man, Cybernatics SMC*, **8** (9) 667–679

Groover M.P., Weiss M., Nagel R.N. and Odrey N. (1986). *Industrial Robotics: Technology, Programming, and Applications*. New York: McGraw-Hill

Hoermann K. (1989). *Programming of the Robot Cell*. Master's thesis, University of Karlsruhe, Germany

Hoermann K. and Rembold U. (1988). A robot action planner for automatic parts assembly. *IEEE Int. Workshop on Intelligent Robots and Syst.*

Homem de Mello L.S. and Sanderson A.C. (1986). And/or graph representation of assembly plans. In *AAAI-86* American Association for Artificial Intelligence

IGES. (1988). *Initial graphics exchange specification (IGES) version 4.0*. Technical report, National Institute of Standards and Commerce

Libardi E.C. and Dixon J.R. (1986). Designing with features: Design and analysis of extrusions as an example. In *Spring National Design Engineering Conference and Show* Chicago, IL. March

Lieberman L. and Wesley M. (1977). AUTOPASS: An automatic programming system for computer controlled mechanical assembly. *IBM J. Res. and Dev.*

Lozano-Perez (1983). Robot programming. *Proc. IEEE*, **71** 821–841

Lozano-Perez T. and Winston P.H. (1977). Lama: A language for automatic mechanical assembly. In *Proc. 5th Int. Joint Conf. Artificial Intelligence*, Cambridge, MA,

Nnaji B.O. (1988). A framework for CAD-based geometric reasoning for robot assembly planning. *The Int. J. Production Research, Special Issue*, **26** (5) 735–764

Nnaji B.O. (1992). *Theory of Automatic Robot Programming and Assembly*. London: Chapman & Hall

Nnaji B.O. and Chu J. (1990). RALPH static planner: CAD based robotic assembly task planning for CSG-based objects. *Int. J. Intelligence Syst.*, **5** (2) 153–181

Nnaji B.O. and Davis R.P. (1985). Summary of robotic components and their characteristics. *Material Flow*, **2** 245–261

Nnaji B.O. and Liu H. (1990). Feature reasoning for automatic robotic assembly. *Int. J. Production Research*, **28** (3) 517–540

Nnaji B.O. and Liu H.C. (1993). *Product Modeling for Sheet Metal, Assembly and Machining*. London: Chapman & Hall

Paul R.P. (1981). *Robot Manipulators*. MIT Press

Popplestone R.J., Ambler A.P. and Bellos I.M. (1978). RAPT: A language for describing assemblies. *The Industrial Robot*

Rist A. (1990). *Geometry-based Robotic Simulator for Ralph*. Master's thesis, University of Massachusetts, Amherst, MA

Snyder W.E. (1985). *Industrial Robots: Computer Interfacing and Control*. Englewood Cliffs, NJ: Prentice-Hall

Unimation Inc. (1984). *Programming manual user's guide to VALII*, document 398t1. Technical report, Westinghouse Co., Danbury, CT

Voelker H.B., Requicha A., Hartquist E. *et al.* (1978). The padl-1.0/2 system for defining and displaying solid objects. *ACM Computer Graphics*, **12** (3)

Whitney D.E., DeFazio T.L., Gustavson R.E. *et al.* (1989). Tools for strategic product design. In *Reprints of the NSF Engineering Design Research Conference*, Amherst, MA

Yeh S. (1990). *Translation of IGES data from CAD into a Structure for Object Reasoning*. Master's thesis, University of Massachusetts, Amherst, MA

Chapter 10   Material Handling

Börnecke G. (1989). Die Rolle der Kommunikationstechnik in Materialflußsystemen, *Förder-technik* 2/89

Demag Carrier System DCS (1989). Mannesmann Demag Fördertechnik, 6050 Offenbach 16 (Catalog Be/0489/3T)

Eisenmann (1989). *Robot Vehicles*. Fördertechnik KG, 7038 Holzgerlingen (catalog)

Jünemann R. (1989). *Material Flow and Logistics*. Berlin: Springer-Verlag (in German)

Rembold, U., Blume C. and Dillmann R. (1985). *Computer Integrated Manufacturing Technology and Systems*. New York, Marcel Dekker, Inc.

Rembold U. and Hörmann A. (1991). Development of an Advanced Robot for Autonomous Assembly. *IEEE Int. Conf. on Robotics and Automation*, Sacramento, California

Rupper, P. (1987). *Enterprise Logistics* Verlag Industrielle Organization, Zürich. (in German)

Siemens (1). *Computer-Aided Manufacturing in the Automotive Industry*. Prozeßrechner und-systemtechnik mit Siemens Systemen 300, Siemens AG, 8520 Erlangen (Order No. E-355/1007)

Siemens (2). *Systems of Manufacturing Control for the Appliance Industry*. Datentechnik im Produktionsbereich mit Siemens Systemen 300, Siemens AG, 8520 Erlangen (Order No. 135031PA118—3)

Stübing H. (1988). *Production and Logistics Can Only be Changed Together*, Audi AG, Germany

Tompkins J.A. and Cramer M.C. (1987). Just-in-time: the real story. *CIM Technology*

Further reading

Bauer A., Bowden R., Browne J., Duggan J., Lyons G. (1991) *Shop Floor Control Systems*. London: Chapman & Hall

Eade R. (1989). AGVs make their move, *Manufacturing Engineering* **9**

Eade R. (1989). Material handling first step toward automation. *Manufacturing Engineering* **9**

Hollier R.H. (ed) (1987). *Automated Guided Vehicle Systems*, IFS Publications, Springer-Verlag

Kupec T. (1989). Integration of autonomous, mobile robots in flexible manufacturing Systems. In *Proceedings of 21st International Symposium on Automotive Technology and Automation*, 1087–1102

Lubben R.T. (1988). *Just-in-Time Manufacturing*. McGraw-Hill Book Company

Meinberg (1989) Effective design of control software for highly flexible AGV systems. In *Proceedings of 21st International Symposium on Automotive Technology and Automation*, 1119–1126.

Meystel A. (1991). *Autonomous Mobile Robots*. World Scientific Series in Automation, 1, World Scientific

Rembold U. and Hörmann A. (1992). Autonomous mobile robots. In *Proceedings of 9th Congresso Brasleiro de Autmática*, 14-18. Vitória, Brasil

Skivington J.J. (1990). *Computerizing Production Management Systems*. London: Chapman & Hall

Young J.B. (1991). *Modern Inventory Operations*. New York: Van Nostrand Reinhold

Zhang J. and Bohner P. (1992). A fuzzy control approach for executing subgoal guided motion of a mobile robot in a partly-known environment. *IEEE International Conference on Robotics and Automation*, Atlanta, GA

Chapter 11   Quality Assurance

George B., Goch G., Schwerz N. and Weckermann W. (1984). Data processing for the coordinate measuring technique. In *Handbuch für Industrie und Wissenschaft, Fertigungstechnik* (eds Warnecke, H.-H. and Dutschke, J.) Berlin: Springer-Verlag (in German)

Kampa H. and Weckermann A. (1984). Design of coordinate measuring machines. In *Handbuch für Industrie und Wissenschaft, Fertigungstechnik*, (eds Warnecke H.-J. and Dutschke J). Berlin: Springer-Verlag

Rembold U. *et al.* (1989). *CAM Handbook*. Berlin: Springer-Verlag

Rembold U., Blume C. and Dillmann R. (1985). *Computer-Integrated Manufacturing Technology and Systems* New York: Marcel Dekker

Stehle J. *An Investigation of Computer Automated Test Systems*. Master's Thesis, University of Karlsruhe, Germany

Weckermann A. (1984). Programming of computer controlled coordinate measuring machines. *Technische Messen*, **51** (6) (in German)

Weckermann, A. and Nordhorst H.-J. (1984). Requirements of the measurement technology for computer integrated production. *VDI Zeitschrift*, **129** (1) (in German)

Index

A

access
  procedure 268
  rights 96
accessories, organization of 216
accounting 32
  payroll 372
ACSE (Association Control Service
    Element) 267
action planner 556
active
  guide wire 547
  wheels 563
aids
  graphic simulation 127
  simulation 126
analysis
  data preparation and 107
  economical 107
  network 120
  requirements 107–8, 113
  return on investment 115
AND/OR graph 169, 506
application layer standards 268
application-oriented subsets 325
APT 73, 294, 604, 607
artificial intelligence (AI) 44, 135,
    172, 393
assembly 351
  automatic 361
  classification methodology 350
  equipment 241
  features 352
  line 444
  line balancing 429
  planning 363
  process 502
  process planning 442
  sequence 165, 363
  stations 207
ATLAS 586, 607, 609
attendance reporting 29
attributes
  assembly operation 504
  operation 582
AUTAP 398
automated warehouses 533

automatic
  dispatching 546
  handling devices 526
automatically guided vehicles (AGV)
    546
automation
  degree of 106
  flexible 16
  potential 17
autonomous
  material movement 549
  system 549
  vehicles 546
AUTOPASS 498, 500
axes
  linear 182
  rotating 182

B

backtracking 167
bar code
  levels 547
  reader 540
Basic 586, 606
batch
  shop 372
  size 421, 531
beacons 546
beam refresh 309
bending sequences 390
Bezier curves 303
bill of materials 66, 157, 320–1,
    363, 408
  indented 322
  information 323
  product structures 321
  single level 322
  summarized 321
  tree 505, 508
blackboard architecture 173
boundary
  information 318
  representation (B-rep) 316, 318,
    320
branch and bound technique 76,
    437
B-representation 333

B-spline 303
buffer 209
buses 47
  structures 259
  ring 259
  star 259
  tree 259
systems
  IEEE 490, 603
  open 86
  ring 86
  star 86

C

CAD 234, 238, 288–9, 295, 323,
    325, 392, 460, 500, 506
  /CAD exchange 328
  /CAM exchange 333
  * ESPRIT CAD interface 71
  /NC
    coupling 254
    integration 253
CAM 234, 323, 604
cameras
  semiconductor 592
  2D 160, 557
candidate rules 138
capital-intensive facilities 198
CAPP 288–9, 398
CAQ 234
carrier sense multiple access (CSMA)
    262
cells
  boring-and-milling 188, 191
  controller levels 214, 218
  decomposition 314
  turning 188, 191
centers
  of gravity 459
  machining 181–2, 194, 290
  metal-bending 185
  nibbling 185
  punching 185
  turning 181–2, 194, 290
centralized database 66
CIM 234
  Amherst-Karlsruhe model 63

629

linear control 595
machines 590, 594, 596
path control 595
point-to-point control 595
transformations 464
COPICS 49
corporate level 84
correction memory 250
costs
computer hardware 106
control 207
material-flow 207
objective function 424
of quality operations 579
of securing quality 577
software 106
testing 579
CPM 441
critical path methods (CPM) 429,
438
backward pass 440
computation 440
forward pass 440
CRC
commitment 269
concurrency end 269
recovery 269
CSG 320, 364, 501
CSMA/CD 268
C-space 510
obstacle 510
cube space skeleton (CSS) 512
customer service 571
cutting tool 335

D
d'Alembert approach 482
data
accessory 113
acquisition system 538
acquisitioning 29, 602
collection 602
engineering analysis 300
evaluation 602
exchange 39
fault 223
format 64
integrity 95
loss of 95
machine-independent 251
maintaining 113
maintenance 236
management 236
for manufacturing 359
master 229–30
models
hierarchical 92
network 92
relational 92
planning 113
preparation and analysis 113

product design 113
production 323
of process 359
quality 223
scheduling 229
state 230, 233
temporary 230
database 228
for CAD/CAM 97
CAD 76, 100, 160, 310
collection of independent 99
communication 94
language 94
consistency 98
design of 99
distribution 102
independence 98
information 101
interfaced 101
machine 80
master manufacturing 80
organization for CAD 311
quality 580, 600
transparency of 113
dataflow 223
decisions
centers 109–10
flow of 108
frames 110, 112
processes 113
defect entry 592
degrees of mobility 562
delivery dates 531, 533
Denavit-Hartenberg representation
477
dependent demand 409
design 23–4, 37, 66
automated drafting 301
bottom up 298
computer-aided (CAD) 10
evaluation 301
export systems for 152
features 342
of fixtures 153
geometric modeling 299
hierarchy 298
life cycle costs in 36
for manufacturability 356
modified 153
new 153
parameters 366
phases 296
previous 359
problem definition 299
process 295, 353
product 353, 359
quality methods in 364
shape 359
top down 296
variant 153
designer 363

destination address 543
diagram 136
direct
numerical control (DNC) 179
view storage (DVST) 310
dispatching 536, 549
algorithm 437
sequencing 547
distribution computer network 66
division of labour 2
DNC 217, 221, 226, 243
docking 550–1
manoeuvre 551
domain independent tools 150
drafting 339
drawing 295
engineering 295
machine 295
drilling 182
dynamic
control 129
planning 129

E
economic
feasibility 117
justification 531
order quantity (EOQ) 410–11
end-effector 484
engineer
design 356–7
manufacturing 356–7
engineering 7, 23–4, 37, 66, 380, 573
concurrent 355
development 36
knowledge 358
principles 580
tests 575, 580
value 365
enterprise
engineering 58
operation 58
equipment 168
flexible
assembly 176
manufacturing 176
material-flow 202
redundant 533
error
recognition 535. 556
recovery routines 535
ESPRIT Project 335
Ethernet 283
Euler
angles 469, 510
Formula 342, 346
EXAPT 250, 399
expert
of the domain 145
knowledge 562
systems 44, 142–3, 170

expert *continued*
  systems *continued*
    building of 144, 148
      computer architecture 150
      conception phase 147
    control structure of 150
    developing language 150
    diagnosis 161
    fundamentals 138
    implementation language 150
    implementation phase 148
    life cycle of 145
    for manufacturing 151, 157
    operating system 150
    realtime 173
    shell 150
    structure phase 147
    support environment 149
    test phase 148
  explainer 138
explanation facility 150
explicit free space 512

F
fabrication schedules 372
factories
  floor layout 135, 525
  model 4–5
  planning 172
  programmable 2, 47
failure search 78
features
  classification 343, 345, 348–9
  descriptors 459
  design by 386
  design with 384
  extraction 386
  grouping 347
  indirect 459
  matrices 363
  object 459
  patterns 347
  recognition 347, 460
  symmetry 349
feedback 76, 576
feedforward 576
FEM elements 334
FEMGEM 160
FFIM 71
field
  bus 86, 283
    application areas 284
    requirements 285
  performance 580
file transfer access and management
  (FTAM) 269
Findpath problems 510
Findspace 509
finished goods inventory 79

finite element analysis (FEA) 311,
  339
fixed period quarterly (FPQ) 410
flexible
  assembly 222
  manufacturing
    equipment 181
    requirements 16
  transfer lines 191
floating zero potential 593
flow
  production 534
  type operation 526
flowline
  inventory 17
  operation 531
  principle 2
  production 17
force sensors 457
Fortran 121, 225, 586, 606
frames 140, 386, 502
  relative 463
free slack 441
freedom
  degree of 475, 478
  of motion 502
FTAM 290
function
  -oriented structure 214
  view 59
Function Entities (FE) 61
functional
  classification 233
  entity 63

G
GEMOS 506
generate and test method 144, 158
generic
  classification 345
  data classes 381
  machine controller 82
  manufacturing model 47–8
  model 89
  planning system 585
genus 346
geographical control 53
geometric elements 301, 342, 353
  complex 339
  conics 303
  curves and surfaces 303
  forms 389
  primitive 324
geometry, 3D 331
GKS 71
global planner 557
GPSS 125
gradient search method 397
GRAI
  grid 111, 151
  method 109

Graphic Kernel System (GKS) 325
graphical scheduling methods 431
graphics
  aids 307, 324
  database 311
  display terminals 308
  elements 327
  modeling 306
  non-programming methods 43
  operations 327
  raster 307–8
  representation scheme 318
  software 307
  solid modeling 309
  test reports 589
  vector 307–9
  wireframe 311
graphs 388
  precedence 506
grippers 191, 457, 460–2, 510–11
  actuators 461
  designs 462
  drivers 461
gripping
  face 513
  plane 513
gross
  positioning 553
  requirement 408–9
group technology 376, 398
  code 376
guided vehicles 192–3

H
handling
  device 555
  strategy 527
  system 362
hardware control structures 217
hard-wired control 179
heuristic methods 437
hierarchical
  computer 52
    network 78
  control 66, 213, 550
  decomposition 166, 170
  planning 49, 213
  quality control system 592
  senior system 52
  structures 33
  systems 33, 56
hierarchy
  database 35
  management 34
  organizational 34
historical performance data 78
hoists 559
hole-in-peg 517
homogeneous
  transformations 472–4
  transforms 469

host language   325
hypothesis   161
  and test method   143
  space   147

I
ICAM project   331
identified
  alphanumeric   539
  code   539
  mass flow   539
  of parts   16, 541, 567
  programmable devices   541-2
  special coding   541
  station   544
  of tools   16
idle time   444
IGES   336, 388, 501, 604
image
  binary   459
  characteristics   459
  processing   459
independent demands   409
in-depth investigation   108
industrial robots   251
in-factory time   533
inference
  engine   138, 141
  mechanisms   150
  rules   148
information
  flow   39-40, 530
  view   61
infrared
  communication   547
  light   541
  responder   544
  sensor   456
inheritance   139
Initial Graphics Exchange
    Specification (IGES)   71-2, 328
  elements of   330
in-process buffers   518
instructions
  geometric   177
  technological   177
instrument draft   593
integrated quality assurance   606
integration
  business function   57
  CAD/CAM   301
  data structure   57
  infrastructure   57, 59, 62
  supplier   527
interactive
  computer graphic   296
  graphical editor   145
  graphics suppport   258
interfaces   47, 49, 56, 59, 84, 228,
    328
  analog 4-20 mA   283

CAD*I   335
CAD/CAM   323-4
  man/machine   172
  open-ended   324
  product definition (PDDI)   331
  software   603
  types   325
interference protection   96
interpolation routines   556
interpolator   180
inventory   567
  ABC   404
  acquision   406
  control   403
  cost   411-12
  dependent demand   409
  disposition   406
  independent demand   407
  in-process   419, 531, 534-5
  management   30-1, 405
  manufacturing   405-6
  master file   567
  perpetual   534-5
  planning   406
  policies   405
  stock-keeping   406
  types   405
  zero   530
inversion   305
  2D   305
IRDATA   75, 167, 368
ISO/OSI
  data link layer   265
  physical layer   265
  reference model   23-4, 286
ISO-workgroup   336

J
job
  assignment   29
  execution   216
  management   215
  preparation   217
  priority   435
joints   473
  angle   477
  axis   476
  prismatic   475, 478
  revolute   475
  rotary   477
  trajectory   488
just-in-time (JUT)   576
  operations   564
  philosophy   53
  planning   532

K
KANBAN principle   84, 576
kinematics   339, 359
  chains   479
    dynamics of   481

income   484
  states   511
knowledge
  acquisition   145, 556
    tools   149
  base   151
  -based
    configuration of a
        manufacturing system   167
    selector   76
  declarative   138
  engineer   145, 147, 150, 172
  expert   556
  factual   138
  functional   168
  heuristic   138
  inferred   138
  meta-   138-9
  presentation   149-50
  procedure   168
  quality   168
  repertoire   145
  representative   138

L
Lagrange-Euler formulation   481-2
Lagrangian mechanics   481
LAMA   498
LAN   84
languages
  automation integration   81
  C-   225
  explicit   44
  EXPRESS   337, 340
  formal   43
  HDSC (High-level Data
      Specification Language)   336
  implicit   44
  knowledge representation   140
  operation   8
  programming   225
  robot programming   491
  specification   142
  system configuration   68
  task
    -level   496
    -oriented   44
  uniformity   95
  VAL   493
lasers   595
latest
  finish time   441
  start time   441
layout   198
lead time   404, 408, 533
learning   172
  active   549
  by analogue   174
  automatic   601
  capabilities   174
  from example   174

learning *continued*
 from instruction 174
 from observation 174
 passive 549
 by rote 174
 of test
  aggregate updating 598
  continuous updating 598
  cyclic updating 598
  limits 576
light barrier 587
light-pen 295
limit switches 457
line balancing 442
 objective 443
linear axis controller 180
links 475
LISP 148, 172
loading 550, 555
local area network (LAN) 84, 88
'logic blocking' 546
logic decoupling 325
logistics 206, 210, 523, 527
 manufacturing 525
 material handling system 524
 materials 525
 purchasing 525
 sales 525
long-range forecasting 23
look-up tables 393
loops
 position control 180
 speed control 180
lot sizing techniques 410, 415
low level coding 408

M
machine tools 523, 534
machines
 assignment 421
 configuration 114
 control 29, 33, 79
 controller 167
  configuration 83
 cost model for 395
 five-axis milling 288
 loading 514
 monitoring 582
 numerically controlled (NC) 12,
  73, 279
 programs 115
 requirements 426
 sequencing 397
 shop operation 530
 vision 582, 587
machining
 cells 181
 centers 181
 multi-axis 376
 optimization 393, 397
 single-pass 395

table 557
mailboxes 66, 225–6
maintenance 31–2, 135
management
 business process 62
 information 62
  system 52, 89
 tool 219
manipulator 523
 kinematics 475
manufacturing
 accessories 240
 alternatives 119, 121, 168, 576
 assembly 37
 automatic 530
 cell 12
 computer-aided (CAM) 10
 computer-integrated (CIM) 3, 10
 control 7–8, 26, 33, 36, 78, 289, 321
  hierarchy 35
  structure 40
 controlling the process 580
 cost 115
 data acquisition 78
 engineer 144
 engineering 380
 flexible 105, 214, 219, 224, 521,
  523, 536
 flowline operations 590
 high-volume 429
 knowledge 358
 message specification (MMS)
  270
 methods selection 583
 monitoring 26, 28, 78
 operations
  identifying 538
  typical 11
 processes 580
  control of 220, 582
  manufacturing 226
 resources 403, 529
 scheduling 36
 sequence 168, 419
 strategies 117
  CAD/CAM 43
  CAD 43
  CIM 43
  PP&C 43
 system functions 19–20
 systems 12
 time 114
manufacturing automatic process
 (MAP) 66–7, 78, 87, 281, 603
 applications 281
 broadband network 82
 interface 83
 mini 282
MAP/TOP 267, 280
 North American users 282
market research 21

marketing 26, 37, 380
mask techniques 250
mass production 106
master files 90
 manufacturing database 79
 production schedule 403
Material Requirements Planning
 (MRP) 403
materials
 allocation 29
 coding 537
 concepts 522
 configurable handling 535
 control 380
 dispatching of carriers 538
 distribution 28, 521
 flexible handling 522
 flow 49, 529, 533, 547
  control 231, 534, 538
  equipment 535
  operation 535
  structures 208
  system 535, 542
 handling 512, 530–1, 534, 536,
  557
 hangers 543
 inventory systems 403
 movement 536
 release 372
 requirements planning 406, 410,
  531
 specification 583
 supply route 525
 synchronization 534
 tracking of 536, 538
 transfer 514
 transportation 556, 564
  vehicles 546, 560
 vehicles
  automatically guided 501
  autonomous 562
  chain pulled 561
  driver operated 562
matrixes
 homogeneous transfer 478
 inverse 472
 multiplying 486
 orientation 465
 rotation 465, 468
 translation 473, 478–9
maximum output 530
measurements 219
 automatic 576
 contact 591
 non-contact 591
 offline 582
 online 582
 temperature 592
measuring
 fixtures 590
 methods 590

probes 596
mechanical feeler 587
methods
  air spray 516
  electrostatic spray 516
menu system 363
milling 182
MMS 286, 290
  client 272
    access 275
  companion standards to 276
  core 277
  domain 274
    management 275
  environment 275
  event 274
    management 275
  journal 274
    management 276
  management 275
  object class 274, 278
  operator
    class 274, 278
    communication 275
    station 274
  program invocation 274, 275
  protocol 271
  provider 272
  semaphor 274
    management 275
  services 275
    classes 275
  variable 274
    access 275
mobile stacker facility 567
modeling
  CAD 308, 335
  of data objects 97
  methods 119
  of recursive operations 98
  world 500
models
  B-rep 336
  CSG 336
  enterprise 54
  FEM 333, 336
  formal mathematical 136
  form-feature 337
  generic 58
  geometry 339
  graphical 119, 120
  heuristic 136
  hierarchical 54
  informal symbolic 136
  information technology 56
  inventory 120
  manufacturing system 66
  materials 337
  mathematical 119–20
  multipass 397
  physical 121

pictorial 136
presentation 337
process planning 71
query 120, 437
reference 151
sequencing 423
shape representation 337
single pass 393
surface 336–7
topology 339
wireframe 333, 336
Modula 121
modular
  hardware 53, 59
  software 53, 59, 224
modules
  behavioral 147
  docking 551
  emulation 125
  functional 229
  knowledge requisition 137
  learn 76
  loading 552
  logic control 57
  manufacturing knowledge 138
  modeling 125
  perception 167
  process planning 71
  product 68
  programming 125
  standard 56
  user 224
monitoring 29, 79
motion
  parameters 556
  planner 165
move
  card 541
  tickets 537
moving test object 587
multiple
  products 436
  resources 436
multiuser operation 96

N
navigation 550
  hardware 547
  level 552
  principles
    compass 550
    dead reckoning 550
    vehicle-based external sensors
      550
NC 217, 221, 290
  control 245
  controller 2
  data organization 216
  machines 200
  processor 251
  program classifiication 248

programming 234
  constants 248
programs 202, 206
requirements 179
tape, standard code 246
technology 206
NCMES 604
NDFS 604
net requirements 409
network standards 87
Newton–Euler formulation 481
n/m job shop 436
non-Gaussian distribution 600
numeric
  control (NC) 177–9
  continuous path 178
  point-to-point 178
  straight-line 178
$n/2$
  flow shop 432–3
  job shop 434
$n/3$ job shop 435

O
object
  -attribute-value-presentation 139
  geometry 318
  interpretation 318
online
  calculations 572
  optimization 530
open system architecture 43
Open Systems Interconnection
  (OSI) 259
operation
  parameters 556
  sequencing 372
operational sheet 374
operator communication 584
operators sequencing 157, 403, 417,
  420
optical
  proximity detectors 458
  quality control 592
    automatic 592
    visual 592
  range sensor 458
  triangulation 460
optimal planning of production 415
optimization criteria 422
order
  control 8
  picking 523
  -reduced information 90
  release 8
organization
  criteria 114
  strategy 119
organizational view 61
OSI 87

OSI/ISO
  application layer 267
  network layer 266
  presentation layer 267
  session layer 266
  transport layer 266
overhead conveyors
  fixed hangers 559
  monorail 559
  power and free 559

P
pallet
  changing 182
  delivery of 567
  handler 557
  handling 221
palletizing operations 514
PAM 502
  parallel processes 533
  part
    feeders 557
    interchangeability 2
part
  family concept 526
  life cycle 19
  orientation 362
parts
  cubical 392
  cylindrical 379
  feeding of 361
  identifying 539
  nonrotational 383
  orientation of 361
  presentation of 361
  prismatic 188, 395, 398
  recognition 551
  rotational 188, 191, 383, 398
  semi-finished 567
  sheet metal 398
Pascal 121
passive
  guide line 546
  rollers 563
pattern matching 347
PDES (Product Data Exchange
    Specification) 71, 332, 362
PEARL 225
performance 112
  file 29
  monitoring 601
  time 443
peg-in-hole 515
peripherals 536
Petri nets 127, 230, 505
  conflict 129
  mutual exclusion 129
  parallel 129
  sequence 129
  timing 129
photodiode arrays 592

physical
  layout 115
  parameters 581
  plant 108
'pick-and-place' operations 509–10
picking facility 567
pictorials 119
picture
  evaluation 551
  file 327
  pilot level 555
planner 552
planning 210, 505
  action 553
  AI 506
  assembly 105, 114, 225, 506
  automatic manufacturing 418
  capacity 157, 400, 403
  capital equipment 22
  centralized 33
  computer-aided (CAP) 10, 104
  control function 198
  cycle 106
  detailed 215, 231
  dynamic 522
  facility 22, 422
  fine motion 511
  flow systems 418
  grasp 166, 508, 512
  gross motion 508, 510, 553
  hierarchical 32
  in-depth 104
  intermediate range 400
  inventory 400
  knowledge-based 549
  level 554
  long-range 36–7, 89
  of machine components 202
  manufacturing resource 400
  market 398
  motion 505, 508–9
  operational 79, 198
  operations 33, 105
  order release 157, 400, 403
  organizational 8, 36
  part-linking 508
  path 553
  production 372, 419
    control 214, 217
  project 198
  resources 36, 400
  route 553
  sequence 206
  short range 89
  short term 400
  stages 200
  strategic 8, 33, 36, 198
  subtasks 203
  technical 8
  test 105, 114, 598
  time 104

trajectory 481, 490–1
tools 105
top down 109
plant control level 84
PLC 217–18, 221, 290
points
  interpolation 487
  knot 487
polyhedral envelope 513
portable terminals 537, 544
position
  conditions 488
  constraints 489
  control 490
post-processor 250, 330
PP&C 100
precedence
  assembly 508
  constraint 507
  graph 506
  relations 506
  requirements 552
  restrictions 422
predicate calculus 142
pre-processor 330
primitive instancing scheme 314
principles
  batch flowline 14
  flowline 576
  flying master 78
  laser 550
  pull production 533
  raster scan 310
  radiotelemetry 550
priority queuing models 437
prismatic parts 182
probability distribution 121
probes
  recovery 595
  test 595
problems
  forward kinematic 481
  inverse kinematic 481
  traffic 552
procedures
  access 261
  policing 261
  time-division multiplexing 261
Process Fortran 225
process
  capability studies 583
  parameters 383
  plan 115, 138, 380
  planner 363
  planning 7, 24–5, 36–7, 66,
    105, 114, 214, 230, 323, 371,
    583
    automated 372, 385
    of bent-sheet metal parts 390
    expert systems for 155
    generative 155–6, 373

variant 155–6, 373
  selection 387
  sequencing 392
  trends 78
processing time 115
processor
  parallel 47
  sensor data 556
  vector array 117
product
  assembly
    model 367
    modeler 362
  design 585
  development 135
    cycle 332
  features 575
  flow 49
  functional analysis 359
  life cycle 336, 337, 366, 573
  mix 531
  model 45, 341
    administrative description 341
    functional description 341
    process information 341
    shape description 341
  new 580, 584
  performance 575
  structure 339, 361, 407–8
  variant 580, 584
production
  capacity 115
  control 36–7, 517, 519
  cycle 332
  data entry (PDE) 214
  equipment controlling 583
  facility 114
  lines 181
  mass 530
  pilot run 580
  planning 527, 529
    and control (PP&C) 10
  programme 202
  pull 531
  push 531
  small batch size 186
productivity of tests 578
programming
  automatic 243
  cell
    assembly 255
    machine 255
  computer-aided 243
  dynamic 76, 120, 396–7
  explicit 164
    language 235
  generative process 72
  goal 397
  implicit 164, 173, 255
  integer 437
  interactive 173

linear and integer 76, 120
  machine 173, 325
  mathematical 396
  manufacturing 73
  NC equipment 242
  offline 598
  online 598
  process 325
  robots 251
    direct 252
    explicit 252
    implicit 252
    lead through 252
    manual 252
    teach-in 252
    textual 252
  sensor-assisted 255
  shop-floor 74, 248
  task
    -level 492, 510
    -oriented 72
  of test equipment 586
  test machines
    via a master workpiece 597
    offline 597
    online 597
  uniform 255, 259
  variant process 73
  workstation-oriented 250
programs
  EXAPT 249
  NC 247
  part 244, 247
  processing 244
project evaluation and review
    techniques (PERT) 429, 441
Prolog 148, 172
protocols 47, 49
  application 339
  facilities 121
proximity sensors 457
pull production 534
PUMA robot 477, 509
punched holes 541
  patterns 537
purchasing 28, 380, 576

Q
quality
  application-oriented programming
      607
  assurance 36–7, 571–2, 575, 577,
      583, 600
  ATLAS programming language
      609
  of computer hardware 584
  control 29, 31, 37, 323, 573, 579
    APT type languages 607
    export systems for 159
    high-level programming 606
    operations 583

planning 572, 584
  processing of multisensor data
      160
  programming languages for 604
  task-oriented languages 609
  test planning 159
  test procedures 580
  visual data 160
data evaluation and interpretation
    159
evaluation 589
feedback 573
master database 585
planning 579–80
plans 601
reporting 584, 588
standards 576
testing 580, 584
trends 601
query language 65

R
radio waves 121
random number generator 121
range finders 550
RAPT 498, 500
raster scan CRT 307
reading station 542
real-time
  behavior 84
  requirements 324
reasoning
  backward 142
  forward 142
  mechanism 141
receiving
  inspection 29, 566
  station 564
recursions
  backward 482
  forward 482
reflection 305
  2D 305
relational databases 65
relationships
  'adjacency' 348
  spatial 348, 502–4, 506, 512
  symbolic spatial 499
re-order point 401
repairs
  field 579
  inhouse 579
representation scheme 320
requester 63
requirements qualification 572
resources
  production 521
  view 61
responder 63
retooling 186
ring buffer 543

robots   73, 186, 194, 197, 209, 449,
    534, 557
    applications   514
    assembly
        adhesive joining   519
        application   517
        fastening screws   518
        press fits   518
        retainers   518
        welding processes   518
    Cartesian geometry   449
    components   454
    control   279
        classification   454
        system   455
    coordinate systems   463
    device   555
    drive system   455
    end-effector   460
    geometric classification   449
    industrial   449
    jointed or articulated geometry
        453
    kinematics   455, 475
    planning   330
        explicit   491–2
        graphics-based techniques   497
        language-based techniques   492
        lead-through   499
        synthesis   505
        task-level programming   498
        task-specification   502
        textual   499
    position and orientation   463
    processing operations   515
    SCARA   453
    simulation   495
    spherical geometry   452
    world coordinates   474
rotational parts   181, 378
rotations   304, 307, 470
    2D   304
    3D   307
route
    optimizer   535
    planner   556
RT Basic   606
rules   139
    dispatching   437

S
sales   571
scaling
    2D   306
    3D   307
schedule simulator   76
scheduling   76, 114, 135, 226, 333,
    371
    activities   66
    detailed   215

internal   231
job shop   429
jobs   431
manufacturing   36, 75
master production   321, 400
problems
    n/1   432
    n/2   432
    n/3   435
    2/*m*   435
production   37
    order   26
project   429, 438
shop   437
systems   429
sculpted surfaces   303
search
    space   142
    strategies   142
security methods   268
selection
    of blanks or stock   389
    cutting
        parameters   393
        path   397
    fixture   373
    fixturing   390
    machine   372
        tool   390
    material   372
    operation   372
    semantic
        nets   139
        networks   388
    tool   378, 392
sense of touch   454
sensors   536, 556, 587, 590
    contact   587, 591
    hypothesis   556
    non-contact   587, 591
sequence
    optimal   426
    search procedure   427
    selection   426
sequencing   418–19, 443
services
    communication   62
    data
        processing   57
        storage   57
    and display   216
    front-end   62
    networking   57
    system-wide exchange   62
servicing, customer order   22
SET (Standard d'Exchange et de
    Transfert)   71, 333, 336
set up
    cost   531
    times   533
shared memory   225

shearing   306
    2D   306
shop-floor control   84
shortest processing time (SPT)   437
signal processing   591
simulation   110, 114, 119, 135, 204,
    437, 497
    analytical   204
    database   125
    of discrete events   121
    dynamics   204
    emulation   44
    graphic   118
    methods   121
    numeric   118, 204
    systems   44, 119, 204
    tools   119, 125
Sketchpad   293
SLAM   125
Smalltalk   172
software
    -oriented structures   224
    -wired central processing   179
solid
    objects   313
    representation schemes   313
space
    Cartesian   484
    free configuration   511
    joint   484
    representation   511
spray coating   516
stacker crane   567
standardization   16
standards
    companion   271, 280
    components   358
    data link layer   267
    ISO   270–1
    MMS   270
    physical layer   267
statistical test   121
STEP   71, 328, 336
storage
    cycle   567
    facility   188, 518
    high-rise   197
    places   564
structural entities   330
subject-oriented data presentation
    97
symmetry
    reflective   349
    rotational   349
synchronization   64, 546
    of test data   601
systems
    assembly   200
    CAD
        control   290
        driven programming   501

cell control 292
classification 201
configuration 107, 114, 116
control 229, 562
conveyor 543, 567
data acquisition 226
database 238
design 365
  and documentation 107, 118
drive 562
engineering 366
expert 68, 135, 153, 551, 556
flexible
  assembly 194, 206, 208
  manufacturing (FMS) 181,
    186, 191, 194, 197, 200–2,
    210, 212
  production 530
gantry handling 191
integrated
  logistics 529
  quality control 580
knowledge-based 112, 135, 153,
  157
logic control 67
machine vision 529
model-based diagnostic 163
monitoring 544
multisensor 551
operating 224
overhead material transportation
  558
pallet 197
physical control 67
realization 107, 118
robot control 279–80
self-learning test 598
sensor 550, 562, 581
simulation 522
supervisory control 214, 290
testing 118
tool supply 192
tracking 544
transport 192
vision 456, 460, 544
VMS operating 231

T
tactile sensors 456–7
Taguchi method 365
task-level reasoning 318
technical
  calculations 325
  evaluation 114
  and office protocol (TOP) 281
technological 177
telephone network (PBX) 260
test
  acceptance 582
  definition 585
  devices 231

drifts 593
equipment 73, 581, 587
evaluation 588
information 589
laboratory 593
modes 601
monitoring 602
on-line correction 593
plan 585–6
  generative 586
  variant 586
procedures 575–6, 581, 587, 590
program 585
pull down 593
receiving inspection 581
results 588
semi-automatic 587, 590–1
set up 50–1
simulation 598
steady state 593
synchronization 601
systems
  automatic 587, 590–1
  correction 593
  hardware 589
  interfacing quality 602
  manual 587, 590–1
  programming of 604
  semi-automatic 587, 590–1
temperatures 593
transient behavior 593
work cycle 599
testing 29, 586
  assembly 581–2
  automatic 16
  computer-aided (CAT) 10
  on conveyors 593
  field 580
  functional 601
  part 581
  performance 572
  product 581
    acceptance 581
    variants 572
  tool 582
  variant 582
time-control mechanism 121
time
  phasing 28, 408
  set-up 210
  throughput 210
Token Ring 260
tokens 127
tolerance 331, 359
tools 231, 237
  change 188, 193
  data
    geometric 234
    organization 234
    statistical 234
    technological 234

database 236, 238
  model 238
  storage 193
  supply 207
TOP 67, 78, 88
topology
  adjacent 331
  bus 260
  object 316
  ring 260
  star 260
  tree 260
total slack time 441
touch forces 457
tracking
  analogue 544
  beacons 546
  camera 547
  carriers 548
  direct 542
  indirect 542
  infrared 546
  laser 547
  of objects 542
  optical 543
  radio 547
traffic flow 547
trajectory 484, 515
  polynomial 485
transactions 567
transfer line 14–16, 372, 530
transition 127, 130
translation 304, 306, 468, 470
  3D 306
transmission medium 267
transport
  devices 188, 221
  facilities 194
transportation 218
  chain 527
  order 539
  schedule 551
  vehicle 523, 557
travel route 548
travelling sales problems 397
triangulation 546
  proximity sensors 458
trollies 559
turn tables 557
turrets 182
TV cameras 550

U
ultrasonic ranging 458
ultrasound 541
undocking 551
uniform code 256
unit machine surface 392
unloading 555
user of tutor interface 137

V
variable sequence problems   423
VDAFS (Verband der
    Automobilindustrie
    Flächenschnittstelle)   289, 334,
    336
  CAD   71
vehicle
  controller   546
  navigation   546
    global planning and
      supervision   551
verification activity   7
virtual
  fabrication facility   522
  manufacturing device   273
  modeling device (VMD)   271
vision   459, 504
  machine   459
  stereo   460

VMD   277, 280
voice communication   592

W
Wagner–Whitin algorithm   410, 414
waiting queue   551
warehousing   564
  automatic   564
  computer   564, 567
  manual   564
  semi-automatic   564
welding
  continuous arc   515
  spot   515
wide area network (WAN)   84
window technique   65
wireless transmission   537
work order sorting   372
workpiece

automatic handling   206
carrier   206, 543, 544
clamping   221
classification   200
families   185
handling   186
prismatic   191
selection   198
supply   207
3D   160
workstations   202
number of   443
world
  model   498, 549, 551, 555–6
    docking   556
    dynamic   152, 556
    navigation   556
    part handling   556
    static   152, 556
  perception   167